ELC International Meeting on Inference, Computation, and Spin Glasses

(ICSG 2013)

Journal of Physics: Conference Series Volume 473

Sapporo, Japan
28-30 July 2013

Editor:

Yoshiyuki Kabashima

ISBN: 978-1-63266-701-4
ISSN: 1742-6588

Printed from e-media with permission by:

Curran Associates, Inc.
57 Morehouse Lane
Red Hook, NY 12571

CURRAN ASSOCIATES INC. proceedings.com

Some format issues inherent in the e-media version may also appear in this print version.

Printed by Curran Associates, Inc. (2014)

For permission requests, please contact the Institute of Physics at the address below.

Institute of Physics
Dirac House, Temple Back
Bristol BS1 6BE UK

Phone: 44 1 17 929 7481
Fax: 44 1 17 920 0979

techtracking@iop.org

Additional copies of this publication are available from:

Curran Associates, Inc.
57 Morehouse Lane
Red Hook, NY 12571 USA
Phone: 845-758-0400
Fax: 845-758-2634
Email: curran@proceedings.com
Web: www.proceedings.com

TABLE OF CONTENTS

ELC International Meeting on Inference, Computation, and Spin Glasses (ICSG2013)

Yoshiyuki Kabashima, Koji Hukushima, Jun-ichi Inoue, Toshiyuki Tanaka and Osamu Watanabe

PREFACE

The close relationship between probability-based inference and statistical mechanics of disordered systems has been noted for some time. This relationship has provided researchers with a theoretical foundation in various fields of information processing for analytical performance evaluation and construction of efficient algorithms based on message-passing or Monte Carlo sampling schemes.

The ELC International Meeting on 'Inference, Computation, and Spin Glasses (ICSG2013)', was held in Sapporo 28–30 July 2013. The meeting was organized as a satellite meeting of STATPHYS25 in order to offer a forum where concerned researchers can assemble and exchange information on the latest results and newly established methodologies, and discuss future directions of the interdisciplinary studies between statistical mechanics and information sciences. Financial support from Grant-in-Aid for Scientific Research on Innovative Areas, MEXT, Japan 'Exploring the Limits of Computation (ELC)' is gratefully acknowledged.

We are pleased to publish 23 papers contributed by invited speakers of ICSG2013 in this volume of *Journal of Physics: Conference Series*. We hope that this volume will promote further development of this highly vigorous interdisciplinary field between statistical mechanics and information/computer science.

Editors and ICSG2013 Organizing Committee:
Koji Hukushima
Jun-ichi Inoue (Local Chair of ICSG2013)
Yoshiyuki Kabashima (Editor-in-Chief)
Toshiyuki Tanaka
Osamu Watanabe (General Chair of ICSG2013)

Peer review statement

All papers published in this volume of *Journal of Physics: Conference Series* have been peer reviewed through processes administered by the proceedings Editors. Reviews were conducted by expert referees to the professional and scientific standards expected of a proceedings journal published by IOP Publishing.

ELC International Meeting on Inference, Computation, and Spin Glasses (ICSG2013) IOP Publishing
Journal of Physics: Conference Series **473** (2013) 012001 doi:10.1088/1742-6596/473/1/012001

Statistical-mechanics analysis of Gaussian labeled-unlabeled classification problems

Toshiyuki Tanaka

Graduate School of Informatics, Kyoto University,
36-1 Yoshida Hon-machi, Sakyo-ku, Kyoto-shi, Kyoto 606-8501, Japan

E-mail: tt@i.kyoto-u.ac.jp

Abstract. The labeled-unlabeled classification problem in semi-supervised learning is studied via statistical-mechanics approach. We analytically investigate performance of a learner with an equal-weight mixture of two symmetrically-located Gaussians, performing posterior mean estimation of the parameter vector on the basis of a dataset consisting of labeled and unlabeled data generated from the same probability model as that assumed by the learner. Under the assumption of replica symmetry, we have analytically obtained a set of saddle-point equations, which allows us to numerically evaluate performance of the learner. On the basis of the analytical result we have observed interesting phenomena, in particular the coexistence of good and bad solutions, which may happen when the number of unlabeled data is relatively large compared with that of labeled data.

1. Introduction

In the basic framework of classification problems, one is given a dataset of labeled data $\mathcal{D}_l = \{(\boldsymbol{x}^\mu, y^\mu) : \mu = 1, \ldots, L\}$, where $\boldsymbol{x}^\mu$ and y^μ are the feature vector and the class label of datum μ, respectively, and infers the rule of the classification underlying the dataset $\mathcal{D}_l$. This framework is categorized as supervised learning in learning theory, because one can regard that a supervisor has provided the class labels y^μ on the basis of the features $\boldsymbol{x}^\mu$. Although classification problems have mainly been studied as supervised learning, it might not be reasonable to expect in real-world classification problems that a dataset containing a sufficient number of labeled data is available. In some applications only human experts can serve as the supervisor providing the class labels in the dataset. In some other cases it is very expensive and/or time-consuming to obtain the class labels. Hence, a typical situation in this respect is that only a limited number of labeled data are available, whereas a relatively larger number of unlabeled data are easily obtained. In the era of "big data," one should face with such a situation more than before, in a wide variety of applications. Classification problems in such situations are generically called labeled-unlabeled classification problems [1].

One may consider two extreme strategies for a labeled-unlabeled classification problem. One extreme is to just ignore unlabeled data and to make use of available labeled data only in learning. This strategy is of course not optimal in typical problem settings, since unlabeled data may provide some additional information about the underlying classification rule. The other extreme is first to perform learning on labeled data, to estimate labels for unlabeled data on the basis of the result of the learning, and then to perform learning using the labeled data as well as the unlabeled data accompanied with the estimated labels. The process may be iterated until convergence is achieved. This strategy can be problematic if only a few number of labeled data are available so that the result of the initial learning, on the basis of which one estimate labels of unlabeled data, is unreliable.

Several methods have been proposed to deal with labeled-unlabeled classification problems. Readers are referred to surveys [1, 2] and a book [3] for reviews of these methods. They have been tested empirically, often with significant performance gains compared with the extreme strategy of ignoring unlabeled data, and in some cases performing almost as good as the fully-supervised cases. To the author's knowledge, however, there are only a few analytical studies in the literature on the theoretical upper limit of how well one can utilize unlabeled data in labeled-unlabeled classification problems.

In this paper we report some results of our preliminary analysis on the labeled-unlabeled problem. We focus on the problem with two classes, and assume that the classes are represented with multivariate Gaussian distributions. In contrast to the standard asymptotic theory in statistics, where one fixes the dimension of the feature space and takes the limit of large numbers of data [4, 5], we study in this paper the case where the dimension of the feature space tends to infinity proportionally to the numbers of labeled and unlabeled data, which is a typical problem setting in the framework of the statistical-mechanics approach to information processing [6, 7].

2. Formulation

We consider the two-class labeled-unlabeled classification problem defined as follows. We assume that feature vectors of positively- and negatively-labeled data are generated from N-dimensional Gaussian distributions centered at $N^{-1/2}\boldsymbol{w}_0$ and $-N^{-1/2}\boldsymbol{w}_0$, respectively, where $\boldsymbol{w}_0 \in \mathbb{R}^N$. The covariance matrices of these Gaussian distributions are assumed to be equal to $\lambda_0^{-1}I$, where I denotes the identity matrix. The dataset of labeled data is given by

$$\mathcal{D}_l = \{(\boldsymbol{x}^\mu, y^\mu) \in \mathbb{R}^N \times \{-1,1\} : \mu = 1,\ldots,L\}, \tag{1}$$

where $\boldsymbol{x}^\mu \in \mathbb{R}^N$ and $y^\mu \in \{-1,1\}$ denote the feature vector and the class label of datum μ, respectively. The feature vector $\boldsymbol{x}^\mu$ is assumed to have been generated according to the conditional distribution $p(\boldsymbol{x}^\mu|y^\mu) = \mathcal{N}(y^\mu N^{-1/2}\boldsymbol{w}_0, \lambda_0^{-1}I)$. The dataset of unlabeled data is

$$\mathcal{D}_u = \{\boldsymbol{x}^\mu \in \mathbb{R}^N : \mu = 1,\ldots,U\}, \tag{2}$$

where $\boldsymbol{x}^\mu$ is assumed to have been generated according to the distribution $p(\boldsymbol{x}^\mu) = \sum_{y=\pm1}(1/2)\mathcal{N}(yN^{-1/2}\boldsymbol{w}_0, \lambda_0^{-1}I)$. All the labeled and unlabeled data are assumed independent.

The learner in our labeled-unlabeled classification problem assumes an equal-weight mixture of symmetrically-located Gaussian distributions:

$$\begin{aligned} p(\boldsymbol{x}, y|\boldsymbol{w}) &= p(\boldsymbol{x}|y,\boldsymbol{w})p(y), \\ p(\boldsymbol{x}|y,\boldsymbol{w}) &= \left(\frac{\lambda}{2\pi}\right)^{N/2} e^{-\lambda\|\boldsymbol{x}-yN^{-1/2}\boldsymbol{w}\|^2/2}, \\ p(y) &= \frac{1}{2}, \quad y \in \{-1,1\}, \end{aligned} \tag{3}$$

where $\boldsymbol{w} \in \mathbb{R}^N$ is the parameter vector of the model assumed by the learner. The learner is supposed to estimate $\boldsymbol{w}$ on the basis of the dataset of labeled and unlabeled data $\mathcal{D} = \mathcal{D}_l \cup \mathcal{D}_u$. This is the formulation of the labeled-unlabeled classification problem we discuss in this paper.

3. Replica analysis

The likelihood function of the parameter $\boldsymbol{w}$ given the dataset of labeled and unlabeled data $\mathcal{D}$ is

$$p(\mathcal{D} \mid \boldsymbol{w}) = \left(\frac{\lambda}{2\pi}\right)^{N(L+U)/2} \prod_{\mu \in \mathcal{D}_l} e^{-\lambda \|\boldsymbol{x}^\mu - y^\mu N^{-1/2}\boldsymbol{w}\|^2/2} \\ \times \prod_{\mu \in \mathcal{D}_u} \left[\frac{1}{2}\left(e^{-\lambda\|\boldsymbol{x}^\mu - N^{-1/2}\boldsymbol{w}\|^2/2} + e^{-\lambda\|\boldsymbol{x}^\mu + N^{-1/2}\boldsymbol{w}\|^2/2}\right)\right]. \quad (4)$$

We also assume that the learner has as the prior distribution of the parameter vector $\boldsymbol{w}$ the Gaussian distribution $\mathcal{N}(\boldsymbol{0}, \kappa^{-1}I)$, where $\kappa^{-1} = \|\boldsymbol{w}_0\|^2/N$. The posterior distribution of $\boldsymbol{w}$ given the dataset $\mathcal{D}$ is thus given via the Bayes formula as

$$p(\boldsymbol{w} \mid \mathcal{D}) = \frac{p(\boldsymbol{w})p(\mathcal{D} \mid \boldsymbol{w})}{\int p(\boldsymbol{w}')p(\mathcal{D} \mid \boldsymbol{w}')\, d\boldsymbol{w}'}. \quad (5)$$

Quantities of our interest are those represented as posterior means

$$\mathbb{E}_{\boldsymbol{w}|\mathcal{D}}[f(\boldsymbol{w})] = \int f(\boldsymbol{w})\, p(\boldsymbol{w} \mid \mathcal{D})\, d\boldsymbol{w}. \quad (6)$$

Such quantities include the squared error of the posterior mean estimate of $\boldsymbol{w}_0$. Those posterior means also depend on the dataset $\mathcal{D}$ used by the learner. Since the dataset $\mathcal{D}$ is assumed to be generated randomly in our problem setting, the posterior means are also random quantities. We are thus interested in their averages over randomness of the dataset $\mathcal{D}$, that is,

$$\mathbb{E}_{\mathcal{D}}[\mathbb{E}_{\boldsymbol{w}|\mathcal{D}}[f(\boldsymbol{w})]] = \iint f(\boldsymbol{w})\, p(\boldsymbol{w} \mid \mathcal{D}) p(\mathcal{D} \mid \boldsymbol{w}_0)\, d\boldsymbol{w}\, d\mathcal{D} \\ = \int \frac{\int f(\boldsymbol{w})\, p(\boldsymbol{w}) p(\mathcal{D} \mid \boldsymbol{w})\, d\boldsymbol{w}}{\int p(\boldsymbol{w}')p(\mathcal{D} \mid \boldsymbol{w}')\, d\boldsymbol{w}'} p(\mathcal{D} \mid \boldsymbol{w}_0)\, d\mathcal{D}. \quad (7)$$

Analytically evaluating the averages with respect to the randomness of the dataset $\mathcal{D}$ poses the major challenge. We adopt the statistical-mechanics approach and apply the replica method, in which we evaluate

$$\Xi_n = \mathbb{E}_{\mathcal{D}}\left[\left(\int p(\boldsymbol{w})p(\mathcal{D} \mid \boldsymbol{w})\, d\boldsymbol{w}\right)^n\right]. \quad (8)$$

The dataset $\mathcal{D}$ serves as the quenched randomness in our problem. The standard prescription of the replica method is that one evaluates the free energy[1] averaged over the randomness of the dataset $\mathcal{D}$ as

$$\mathcal{F} = \lim_{N\to\infty} \mathbb{E}_{\mathcal{D}}\left[\frac{1}{N}\log \int p(\boldsymbol{w})p(\mathcal{D} \mid \boldsymbol{w})\, d\boldsymbol{w}\right] = \lim_{N\to\infty} \frac{1}{N} \lim_{n\to 0} \frac{\partial \log \Xi_n}{\partial n} \\ = \lim_{n\to 0} \frac{\partial}{\partial n} \lim_{N\to\infty} \frac{1}{N} \log \Xi_n, \quad (9)$$

where in the last equality we have interchanged the order of the operations on n and the thermodynamic limit $N \to \infty$, assuming that this interchange does not affect the end result.

[1] Note that in the definition above the overall sign is reversed from the conventional physics definition of "free energy". This does not affect the following analysis.

We then evaluate $\lim_{N\to\infty}(1/N)\log\Xi_n$ via the saddle-point method. In evaluating this quantity we temporarily assume that n is a natural number and rewrite Ξ_n as

$$\begin{aligned}\Xi_n &= \int\cdots\int\prod_{a=1}^{n}[p(\boldsymbol{w}_a)p(\mathcal{D}\mid\boldsymbol{w}_a)]p(\mathcal{D}\mid\boldsymbol{w}_0)\,d\mathcal{D}\prod_{a=1}^{n}d\boldsymbol{w}_a \\ &= \int\cdots\int\left(\int\prod_{a=0}^{n}p(\mathcal{D}\mid\boldsymbol{w}_a)\,d\mathcal{D}\right)\prod_{a=0}^{n}p(\boldsymbol{w}_a)\,d\boldsymbol{w}_a, \end{aligned} \tag{10}$$

where we have formally introduced $p(\boldsymbol{w}_0)$ to make the resulting formula symmetric in the replica index a. One can use as $p(\boldsymbol{w}_0)$ the "true prior" if one is interested in quantities which are further averaged over randomness of $\boldsymbol{w}_0$.

Let $W=(\boldsymbol{w}_0,\boldsymbol{w}_1,\ldots,\boldsymbol{w}_n)$. A key observation is that the quantity

$$\int\prod_{a=0}^{n}p(\mathcal{D}\mid\boldsymbol{w}_a)\,d\mathcal{D}, \tag{11}$$

appearing in the integrand of (10), depends on W only through its normalized Gram matrix $Q=N^{-1}W^TW$. Omitting details of derivations, we have

$$\lim_{N\to\infty}\frac{1}{N}\log\Xi_n=\sup_Q[\mathcal{G}(Q)-\mathcal{I}(Q)] \tag{12}$$

as the result of applying the saddle-point method, where

$$\mathcal{G}(Q)=-\frac{\alpha+\alpha'}{2}\operatorname{tr}Q\Lambda+\alpha\frac{\mathbf{1}^T\Lambda Q\Lambda\mathbf{1}}{2(\lambda_0+n\lambda)}+\alpha'\log\left(\frac{1}{2^{n+1}}\sum_{\boldsymbol{s}\in\{-1,1\}^{n+1}}e^{\boldsymbol{s}^T\Lambda Q\Lambda\boldsymbol{s}/2(\lambda_0+n\lambda)}\right), \tag{13}$$

$$\mathcal{I}(Q)=\sup_{\tilde{Q}}\left(\sum_{a\le b}\tilde{Q}_{ab}Q_{ab}+\sum_{a=1}^{n}\tilde{Q}_{0a}Q_{0a}-\log\mathcal{M}(\tilde{Q})\right), \tag{14}$$

$$\mathcal{M}(\tilde{Q})=\mathbb{E}\left(e^{\sum_{a\le b}\tilde{Q}_{ab}w_aw_b+\sum_{a=1}^{n}\tilde{Q}_{0a}w_0w_a}\right), \tag{15}$$

and where we let $\alpha=L/N$, $\alpha'=U/N$, and $\Lambda=\operatorname{diag}(\lambda_0,\lambda,\ldots,\lambda)$. The expectation in the last equation is taken over $(w_1,\ldots,w_n)\sim\mathcal{N}(\mathbf{0},\lambda^{-1}I)$. We have also introduced $\tilde{Q}$ as the $(n+1)\times(n+1)$ symmetric order-parameter matrix conjugate to the Gram matrix Q.

The function $\mathcal{G}(Q)$ represents the exponent of the integrand (11) in terms of the Gram matrix Q. Large-deviations theory [8] tells us that the rate function $\mathcal{I}(Q)$ is obtained via the Legendre transform of the cumulant generating function $\log\mathcal{M}(\tilde{Q})$ for the multivariate Gaussian random variable $(w_1,\ldots,w_n)\sim\mathcal{N}(\mathbf{0},\lambda^{-1}I)$.

The saddle-point conditions are

$$\tilde{Q}_{aa}=-(\alpha+\alpha')\frac{\lambda(\lambda_0+(n-1)\lambda)}{2(\lambda_0+n\lambda)}, \tag{16}$$

$$\tilde{Q}_{0a}=\frac{\lambda_0\lambda}{\lambda_0+n\lambda}\left(\alpha+\alpha'\frac{\sum_{\boldsymbol{s}\in\{1,-1\}^{n+1}}s_0s_ae^{\boldsymbol{s}^T\Lambda Q\Lambda\boldsymbol{s}/2(\lambda_0'+n\lambda)}}{\sum_{\boldsymbol{s}\in\{1,-1\}^{n+1}}e^{\boldsymbol{s}^T\Lambda Q\Lambda\boldsymbol{s}/2(\lambda_0'+n\lambda)}}\right), \tag{17}$$

$$\tilde{Q}_{ab}=\frac{\lambda^2}{\lambda_0+n\lambda}\left(\alpha+\alpha'\frac{\sum_{\boldsymbol{s}\in\{1,-1\}^{n+1}}s_as_be^{\boldsymbol{s}^T\Lambda Q\Lambda\boldsymbol{s}/2(\lambda_0'+n\lambda)}}{\sum_{\boldsymbol{s}\in\{1,-1\}^{n+1}}e^{\boldsymbol{s}^T\Lambda Q\Lambda\boldsymbol{s}/2(\lambda_0'+n\lambda)}}\right), \tag{18}$$

$$Q_{ab}=\langle w_aw_b\rangle, \tag{19}$$

where $\langle\cdots\rangle$ is defined as

$$\langle\cdots\rangle := \frac{\mathbb{E}\left[(\cdots)e^{\sum_{a\le b}\tilde{Q}_{ab}w_aw_b+\sum_{a=1}^n\tilde{Q}_{0a}w_0w_a}\right]}{\mathbb{E}\left[e^{\sum_{a\le b}\tilde{Q}_{ab}w_aw_b+\sum_{a=1}^n\tilde{Q}_{0a}w_0w_a}\right]}. \tag{20}$$

In order to proceed further we make the assumption of replica symmetry (RS), where we assume that the order parameters are invariant under permutations of replica indices $a = 1,\ldots,n$. We introduce RS order parameters by letting

$$\begin{aligned} Q_{aa} &= p, & Q_{ab} &= q, & Q_{0a} &= m, \\ \tilde{Q}_{aa} &= \tilde{p}, & \tilde{Q}_{ab} &= \tilde{q}, & \tilde{Q}_{0a} &= \tilde{m}. \end{aligned} \tag{21}$$

Under the RS assumption one has

$$\begin{aligned} \mathcal{F} = &-\frac{(\alpha+\alpha')\lambda}{2}\left(\frac{\lambda_0-\lambda}{\lambda_0}p+\frac{\lambda}{\lambda_0}q\right)+\alpha\lambda m+\alpha'\sqrt{\frac{\lambda_0}{2\pi q}}\int e^{-\frac{\lambda_0}{2q}(z-m)^2}\log\cosh\lambda z\,dz \\ &- m\tilde{m}-p\tilde{p}+\frac{1}{2}q\tilde{q}+\frac{1}{2}\frac{\tilde{m}^2+\kappa\tilde{q}}{\kappa(\kappa-2\tilde{p}+\tilde{q})}+\frac{1}{2}\log\left(\frac{\kappa}{\kappa-2\tilde{p}+\tilde{q}}\right). \end{aligned} \tag{22}$$

The saddle-point equations for the RS order parameters are

$$\partial_m\mathcal{F} = \alpha\lambda+\alpha'\lambda\sqrt{\frac{\lambda_0}{2\pi q}}\int e^{-\frac{\lambda_0}{2q}(z-m)^2}\tanh\lambda z\,dz-\tilde{m}=0, \tag{23}$$

$$\partial_p\mathcal{F} = -\frac{(\alpha+\alpha')\lambda(\lambda_0-\lambda)}{2\lambda_0}-\tilde{p}=0, \tag{24}$$

$$\partial_q\mathcal{F} = -\frac{(\alpha+\alpha')\lambda^2}{2\lambda_0}+\frac{\alpha'\lambda^2}{2\lambda_0}\sqrt{\frac{\lambda_0}{2\pi q}}\int e^{-\frac{\lambda_0}{2q}(z-m)^2}(1-\tanh^2\lambda z)\,dz+\frac{\tilde{q}}{2}=0, \tag{25}$$

$$\partial_{\tilde{m}}\mathcal{F} = -m+\frac{\tilde{m}}{\kappa(\kappa-2\tilde{p}+\tilde{q})}=0, \tag{26}$$

$$\partial_{\tilde{p}}\mathcal{F} = -p+\frac{\tilde{m}^2+\kappa\tilde{q}}{\kappa(\kappa-2\tilde{p}+\tilde{q})^2}+\frac{1}{\kappa-2\tilde{p}+\tilde{q}}=0, \tag{27}$$

$$\partial_{\tilde{q}}\mathcal{F} = \frac{q}{2}-\frac{1}{2}\frac{\tilde{m}^2+\kappa\tilde{q}}{\kappa(\kappa-2\tilde{p}+\tilde{q})^2}=0. \tag{28}$$

In the following of this paper we focus on the case where the learner assumes the true noise variance, that is, where $\lambda_0=\lambda$ holds. One has $\tilde{p}=0$ under this condition. Rescaling the RS order parameters as $\lambda m\to m$, $\lambda q\to q$, $\tilde{m}/\lambda\to\tilde{m}$, and $\tilde{q}/\lambda\to\tilde{q}$, in order to simplify the argument, the saddle-point equations take the following simple form:

$$\tilde{m} = \alpha+\alpha'\sqrt{\frac{1}{2\pi q}}\int e^{-(z-m)^2/2q}\tanh z\,dz, \tag{29}$$

$$\tilde{q} = \alpha+\alpha'\sqrt{\frac{1}{2\pi q}}\int e^{-(z-m)^2/2q}\tanh^2 z\,dz, \tag{30}$$

$$m = \frac{a^2\tilde{m}}{1+a\tilde{q}},\quad q = \frac{a^3\tilde{m}^2+a^2\tilde{q}}{(1+a\tilde{q})^2}, \tag{31}$$

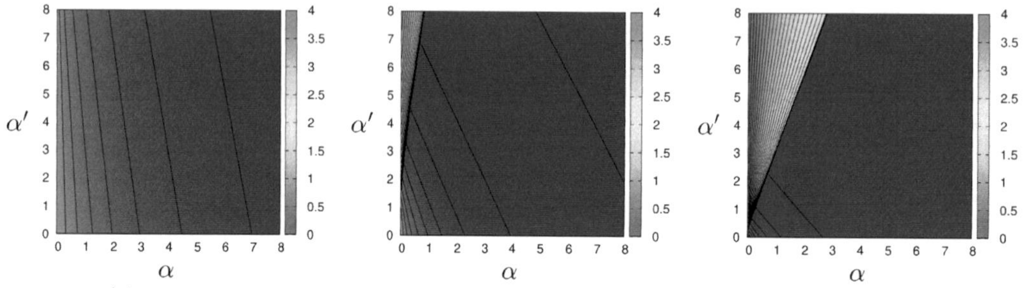

(a) $a = 0.33$ (b) $a = 1$ (c) $a = 3.33$

Figure 1. Mean-squared errors of posterior-mean estimates.

where we have let $a = \lambda/\kappa$, which represents the signal-to-noise ratio of the problem. The normalized mean-squared error of the posterior mean estimate $\boldsymbol{w}$ against $\boldsymbol{w}_0$ is given in terms of a saddle-point solution as

$$\mathcal{E} = \frac{q - 2m + a}{a}. \tag{32}$$

Specifying the signal-to-noise ratio a, as well as the normalized numbers of labeled and unlabeled data α and α', one can numerically solve the above saddle-point equations to evaluate the theoretical performance of the labeled-unlabeled classification problem in terms of the normalized mean-squared error. This is the main result of this paper.

4. Numerical evaluations

We have numerically evaluated the normalized mean-squared error of the posterior mean estimate $\boldsymbol{w}$ using the analytical result presented in the previous section. The result is summarized in figure 1. One can observe in these figures that $\boldsymbol{w}_0$ can be estimated accurately when α' is small. As α' increases, mean-squared error becomes even smaller, implying that unlabeled data are utilized in learning. Let us define utility r of unlabeled data as a function of (α, α') as

$$r = \frac{\partial \mathcal{E}}{\partial \alpha'} \Big/ \frac{\partial \mathcal{E}}{\partial \alpha}, \tag{33}$$

that is, the utility of unlabeled data represents how effective an unlabeled datum is in reducing the mean-squared error in comparison with a labeled data. The utility of unlabeled data in the small-α' regime depends on the signal-to-noise ratio a, in such a way that the utility approaches 0 and 1 as a becomes smaller and larger, respectively. One can show that when $a\alpha \gg 1$ and $\alpha' \ll 1$ the utility of unlabeled data is given by

$$r = f(a) = \int \sqrt{\frac{a}{2\pi}} e^{-a(z-1)^2/2} \tanh az \, dz. \tag{34}$$

The shape of the function $f(a)$ is depicted in figure 2. It shows that the utility is close to 1 when the signal-to-noise ratio a is 10 or larger, and that it drops below 0.1 when a is less than 0.1.

One can also observe in figure 1 that coexistence of good and bad solutions occurs when α is small and α' is large enough. Indeed, in the α-α' plane, there is a "spinodal" line $\alpha' = \alpha'_c(\alpha)$, below which only a solution with good performance exists, and above which the coexistence of good and bad solutions takes place. It should be noted that in figure 1 only the bad solution is

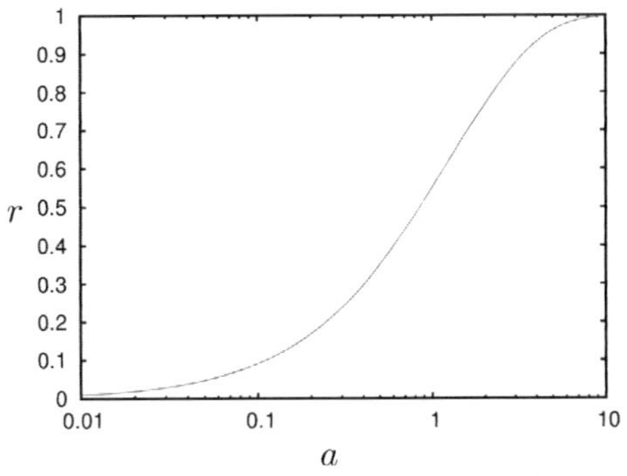

Figure 2. Utility r of unlabeled data as a function of the signal-to-noise ratio a.

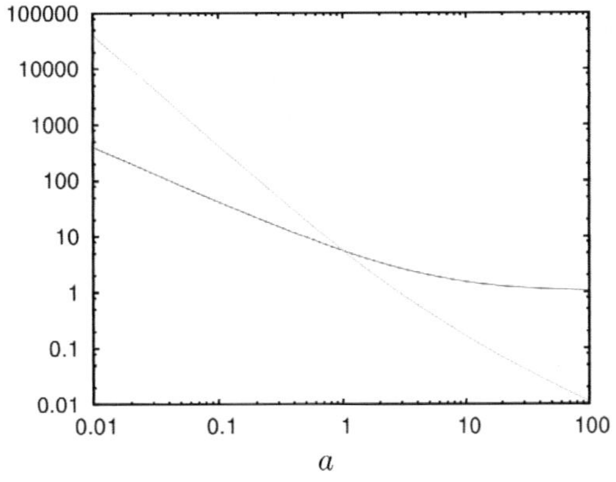

Figure 3. Slope and intercept of asymptotic of spinodal line when α and α' are large enough, versus the signal-to-noise ratio a. Red and green curves represent the slope and the intercept, respectively.

shown where the coexistence occurs. With perturbation analysis, one can show that the spinodal line has the following asymptotic form for small enough α:

$$\alpha'_c(\alpha) = \frac{1}{a^2}\left(1 + \sqrt{8a(a+1)\alpha}\right) \tag{35}$$

This asymptotic formula for the spinodal line tells us that, even when $L = o(N)$, that is, when the number of labeled data is vanishingly small compared with the dimension of the feature space, the coexistence of multiple solutions does not occur when $\alpha' < 1/a^2$. It also shows that a small increase in α from 0 causes a significant increase in the critical α' value, above which the coexistence of multiple solutions occurs. Indeed, the increase is superlinear, being proportional to $\alpha^{1/2}$.

When α and α' are large enough, on the other hand, one can show, again via perturbation analysis, that the spinodal line is asymptotically linear. In figure 3 we show the slope and the intercept of the asymptotic spinodal line as a function of a. Details of the perturbation analysis will be presented elsewhere.

5. Summary and outlook

We have reported some results of our preliminary statistical-mechanics-based analysis on the labeled-unlabeled classification problem. In our analysis we have dealt with the simplest setting, in which a learner with the equal-weight mixture of two symmetrically-located Gaussians is assumed. Even in this simplest setting we have observed interesting phenomena, in particular the coexistence of good and bad solutions, which may happen when the number of unlabeled data is relatively large compared with that of labeled data. We have obtained explicit expressions describing asymptotic behaviors of the spinodal line that marks the boundary between the region with the coexistence and the region without it.

On the basis of the basic results presented so far, one can pose several problems to be explored. First of all, the unequal-weight case, in which a learner is to estimate the class weights as well as the class centers, should be explored to make the setting a bit more realistic. Another interesting problem would be to study the case where unlabeled data are generated from a probability model which is similar to but not the same as that for labeled data. The utility of unlabeled data would decrease as the probability model of unlabeled data becomes less similar to that of labeled data. It would also be important to consider the multi-class problem as well. From the viewpoint of statistical-mechanics analysis, it is necessary to study stability of the RS assumption in order to see if the RS assumption is valid. Finally, algorithms to efficiently solve the labeled-unlabeled classification problem should be studied as well, in view of practical applications.

Acknowledgments

The author would like to thank Professor Seiji Miyoshi with Kansai University, Osaka, Japan, for the discussion they had in March, 2013, which has inspired me to do the study.

References

[1] Seeger M 2002 Learning with labeled and unlabeled data Tech. Rep. EPFL-REPORT-161327 École Polytechnique Fédérale de Lausanne URL `http://infoscience.epfl.ch/record/161327/files/review.pdf`

[2] Zhu X 2008 Semi-supervised learning literature survey Technical Report 1530 Department of Computer Sciences, University of Wisconsin Madison

[3] Chapelle O, Schölkopf B and Zien A (eds) 2006 *Semi-Supervised Learning* (MIT Press)

[4] O'Neill T J 1978 *Journal of the American Statistical Association* **73** 821–6

[5] Ganesalingam S and McLachlan G J 1978 *Biometrika* **65** 658–62

[6] Nishimori H 2001 *Statistical Physics of Spin Glasses and Information Processing: An Introduction* (Oxford University Press)

[7] Mézard M and Montanari A 2009 *Information, Physics, and Computation* (Oxford University Press)

[8] Dembo A and Zeitouni O 1998 *Large Deviations Techniques and Applications* 2nd ed (Springer)

Free products of large random matrices - a short review of recent developments

Zdzislaw Burda

Marian Smoluchowski Institute of Physics and Mark Kac Center for Complex Systems Research, Jagiellonian University, Reymonta 4, PL30-059 Cracow, Poland

E-mail: `zdzislaw.burda@uj.edu.pl`

Abstract. We review methods to calculate eigenvalue distributions of products of large random matrices. We discuss a generalization of the law of free multiplication to non-Hermitian matrices and give a couple of examples illustrating how to use these methods in practice. In particular we calculate eigenvalue densities of products of Gaussian Hermitian and non-Hermitian matrices including combinations of GUE and Ginibre matrices.

1. Introduction

Products of random matrices arise in many areas of research and engineering [1]. In studies of dynamical systems one uses products of random matrices to derive limiting laws for Lyapunov exponents [2]; in information theory and wireless telecommunication – to calculate channel capacities for serial MIMO (multiple-input multiple-output) transmission [3]. With the help of products of random matrices one investigates unitary evolution [4], matrix diffusion [5] or phase transitions in quantum Yang-Mills theories [6]. One encounters products of random matrices in studies of quantum entanglement [7], financial engineering [8], and many other fields of research [9, 10].

The problem of random matrix multiplication has attracted attention since the sixties [11, 12]. Originally the studies were mostly concentrated on deriving limiting laws for products of infinitely many finite random matrices [10] with some exceptions where also limiting laws for infinite matrices were formulated [2]. Recently the focus has shifted to products of infinitely large matrices due to the discovery of a correspondence between infinitely large invariant random matrices and free random variables [13, 14] which have been studied in the framework of free probability [15] – a new branch of probability theory that has been intensively developed since the eighties [16, 17]. Many results derived in this theory have been applied to large random matrices in the limit of infinite matrix size $N \to \infty$.

In this paper we review some of these results, and give practical recipes to calculate eigenvalue densities of products of large random matrices. Instead of giving formal proofs we gradually build up intuition passing from matrix addition through multiplication of Hermitian random matrices to multiplication of non-Hermitian matrices. Free addition of large random matrices is a counterpart of addition of independent random variables in classical probability. We start with addition because it is much simpler than multiplication but it already has essential ingredients needed to understand a special role of invariant random matrices played in the formulation of limiting laws for eigenvalue distributions in the large N limit. The key feature of invariant random matrices is that their eigenvectors are not correlated with the eigenvalues, and that they are uniformly distributed. The last statement means that the unitary matrix constructed from the normalized eigenvectors is uniformly distributed on the unitary group. In effect, randomness of such matrices is entirely shaped by the probability distribution of the eigenvalues. The sum of independent invariant random matrices is an invariant random matrix. In the limit $N \to \infty$ the eigenvalue density of the sum depends only on eigenvalue densities of the matrices added. This law is very similar to the law of addition in classical probability where the probability distribution of the sum of independent random variables can be calculated from the probability distributions of these variables. Similarities to classical probability go far beyond that. One can write a dictionary between classical and free probability. In this dictionary, random variable corresponds to invariant random matrix for $N \to \infty$, probability density function to eigenvalue density, characteristic function to Green's function, cumulant generating function (logarithm of the characteristic function) to R transform, Mellin transform of the probability density function to S transform, independence to freeness, etc. Using this correspondence one can find a full analogy between free addition and free multiplication of large invariant matrices and addition and multiplication of independent random variables.

It is worth emphasizing that the laws of free addition and free multiplication can be independently derived using field theoretical methods, including perturbation theory of invariant matrix models [18, 19], without referring to free probability. Perturbation theory is a way of graphical enumeration of planar Feynman diagrams that arise in the large N limit [20]. A great advantage of the field theoretical approach is that it also works in the case of non-Hermitian matrices and it allows to derive a non-Hermitian version of the multiplication law [21]. To the best of our knowledge this result is beyond the reach of free probability so far. We shortly recall this result here.

The paper is organized as follows. In section 2 we discuss eigenvalues of a sum of random matrices. In particular we define a slightly modified version of matrix addition – such that it is independent of the eigenvectors of the matrices added. This brings us to the class of invariant random matrices that are discussed in section 3. In section 4 we recall the definition of the Green's function, the moment generating function and the R transform. These functions are used in the derivation of limiting laws for invariant matrices in the large N limit. We employ these functions in the formulation of the law of free addition. In section 5 we discuss the law of free multiplication for invariant Hermitian matrices. In section 6 we discuss consequences of this law for multiplication of infinitely large isotropic non-Hermitian matrices. Finally in section 7 we show how to calculate eigenvalue densities for products of large independent non-Hermitian matrices. We illustrate calculations with examples. We conclude the paper with a short summary in section 8.

2. Eigenvalues of the sum of random Hermitian matrices

Let a and b be $N \times N$ Hermitian matrices. We are interested in the eigenvalues of the sum $a + b$. To calculate them it is not sufficient to know the eigenvalues of a and b. One would also need to know their eigenvectors. Actually it would be enough to know relative positions of the eigenvectors of a and b: for example the decomposition of b-eigenvectors in the eigenbasis of a. One can, however, define a non-standard addition of matrices that depends only on the eigenvalues of individual matrices in the sum and not on their eigenvectors. This new addition is defined in a probabilistic way as a sum

$$a \boxplus b = u^\dagger a u + v^\dagger b v \ , \tag{1}$$

where u, v are independent Haar unitary matrices and $u^\dagger$ denotes the Hermitian conjugate of u. Let us recall that Haar unitary matrix is a random matrix distributed according to the uniform measure on the unitary group $U(N)$. The matrix $a \boxplus b$ is not a concrete matrix but it is a random matrix or equivalently an ensemble of matrices with a certain probability measure. The eigenvalues of $a \boxplus b$ are random variables. One is interested in the joint probability distribution of these random variables. The marginal distribution of the joint probability distribution is referred to as the eigenvalue distribution. One should note that the transformations $a \to u^\dagger a u$ and $b \to v^\dagger b v$ preserve the eigenvalues of the individual terms, but uniformly randomize their eigenvectors. One can therefore expect that the eigenvalue distribution of $a \boxplus b$ depends only on eigenvalues of a and b. Moreover, since the sum $u^\dagger a u + v^\dagger b v$ can be written as $u^\dagger(a + w^\dagger b w)u$, where $w = vu^{-1}$ is also a Haar unitary matrix, the eigenvalue distribution of the matrix $a + w^\dagger b w$ averaged over w is the same as the eigenvalue density of $a \boxplus b$. In other words it is sufficient to randomize the eigenvectors of one matrix against the other.

One can consider a more general version of the problem in which a and b are independent random matrices instead of being deterministic ones. In this case the question is whether one can determine the eigenvalue distribution of $a \boxplus b$ from the eigenvalue distributions of individual matrices a and b. The answer is affirmative for $N \to \infty$. In this limit the eigenvalue density $\rho_{a \boxplus b}(\lambda)$ of the sum $a \boxplus b$ can indeed be determined solely from the eigenvalue densities $\rho_a(\lambda)$ and $\rho_b(\lambda)$. For finite N the answer is more complicated since it depends also on eigenvalue correlations. The correlations can be neglected only in the limit $N \to \infty$. The addition (1) is called free addition when $N \to \infty$ and a and b are independent. In the next section we shall see that the addition (1) is equivalent to the standard matrix addition $a + b$ for invariant random matrices.

3. Invariant Hermitian matrices and free matrices

Random matrices are formally defined by matrix ensembles equipped with probability measures. Let us denote the measure of an $N \times N$ Hermitian random matrix x by $\mu(x)$. Random matrix x is called invariant if the measure is invariant under the transformation $x \to U^\dagger x U$, where U is an arbitrary unitary matrix: $\mu(x) = \mu(U^\dagger x U)$. Let us give two standard examples.

One can construct an invariant random matrix from a diagonal random matrix $\Lambda = \mathrm{diag}(\lambda_1, \ldots, \lambda_N)$ and Haar unitary random matrix u

$$a = u^\dagger \Lambda u \ . \tag{2}$$

The invariance of the measure for a: $\mu(a) = \mu(U^\dagger a U)$ is a direct consequence of the invariance of the Haar measure $\mu_{\mathrm{Haar}}(u) = \mu_{\mathrm{Haar}}(uU)$ under the right multiplication by a unitary matrix U. One can choose the diagonal elements λ_i's of Λ to be independent, identically distributed real random variables. In such a case the joint eigenvalue density factorizes and $\rho_N(\lambda_1, \ldots, \lambda_N) = p(\lambda_1) \ldots p(\lambda_N)$ where $p(\lambda)$ is the probability density function for the random variables representing diagonal elements of Λ. The factorization of the joint eigenvalue density reflects the independence of the eigenvalues. The eigenvalue density is $\rho(\lambda) = p(\lambda)$.

In physical applications one frequently encounters invariant measures of the form [22, 23]

$$d\mu(a) \propto e^{-N \mathrm{Tr} V(a)} da \ , \tag{3}$$

where $V(a)$ is a polynomial or a power series in a and $da = \prod_{1 \le i \le N} da_{ii} \prod_{1 \le i < j \le N} d\mathrm{Re} a_{ij} d\mathrm{Im} a_{ij}$, denotes flat measure for $N \times N$ Hermitian matrices. The proportionality sign in (3) means that the right hand side is displayed up to a constant. This implicit constant ensures the probabilistic normalization $\int d\mu(a) = 1$.

The joint probability eigenvalue distribution is given by a well-known Dyson formula [22, 24]

$$\rho_N(\lambda_1, \ldots, \lambda_N) \propto e^{-N \sum_i V(\lambda_i)} \prod_{i<j} (\lambda_j - \lambda_i)^2 \ , \tag{4}$$

which can be interpreted as a partition function that describes a gas of N particles in one dimension. We again skipped a normalization constant that ensures $\int_{R^N} \rho_N(\lambda_1, \ldots, \lambda_N) d\lambda_1 \ldots d\lambda_N = 1$. In this picture the i-th eigenvalue λ_i corresponds to the position of the i-th particle in one-dimensional space. Particles are in the potential $V(\lambda)$. They repel each other with a logarithmic potential depending of the distance $|\lambda_i - \lambda_j|$. The repulsion makes the eigenvalues interact. As a result they are correlated. The eigenvalue density can be calculated by integrating all but one eigenvalue from the joint eigenvalue distribution $\rho(\lambda) = \int \rho(\lambda, \lambda_2, \ldots, \lambda_N) d\lambda_2 \ldots d\lambda_N$. In the limit $N \to \infty$ one can relate the eigenvalue density to the potential by an integral equation [18]

$$V'(x) = 2 \text{ P.V.} \int d\lambda \frac{\rho(\lambda)}{x - \lambda} \tag{5}$$

obtained by applying the saddle point approximation to (4). P.V. stands for principal value. One can use this equation to calculate the eigenvalue density for the given potential or to find the potential corresponding to a given eigenvalue density. In particular one can choose the potential to reproduce the same eigenvalue density as in the first ensemble (2) where the eigenvalue density is just equal to the probability density function for the diagonal elements of Λ. The two random matrix ensembles (2) and (3) have completely different statistics of eigenvalues even if they have identical eigenvalue densities.

Let us come back to the general discussion of invariant random matrices. The standard addition $a + b$ of invariant random matrices is equivalent to free addition $a \boxplus b$ (1). This follows from the invariance of the measure, which implies that the matrix a is identically distributed as $u^\dagger a u$. The same holds for b and $v^\dagger b v$. For a and b being invariant random matrices, the two sums $a + b$ and $a \boxplus b$ have identical probability measures. They represent the same random matrix ensemble. Therefore from here on we shall not distinguish between $a + b$ and $a \boxplus b$ while discussing invariant random matrices. The sum of invariant random matrices is also an invariant random matrix. An important property of the sum $a + b$ of invariant matrices is that the eigenvalue density depends in the limit $N \to \infty$ solely on the eigenvalue densities of a and b, as follows from the discussion in section 2. Independent invariant random matrices for $N \to \infty$ are called free. There is one-to-one correspondence between them and the free random variables known from free probability [15].

To summarize this section, the standard addition of invariant random matrices is equivalent to free addition (1). In the limit $N \to \infty$ independent invariant random matrices are called free and the eigenvalue density $\rho_{a+b}(\lambda)$ of the free sum can be calculated from eigenvalues $\rho_a(\lambda)$ and $\rho_b(\lambda)$ of a and b. The resulting density is independent of the detailed statistics of eigenvalues. For example, it does not depend on whether eigenvalues are correlated or not or whether they are generated using the ensemble of type (2), (3) or any other. The only thing that matters is the eigenvalue densities of the individual matrices in the sum. In the next section we briefly summarize the algorithm to calculate $\rho_{a+b}(\lambda)$.

4. Addition of free matrices

For $N \to \infty$ the eigenvalue density of an invariant random matrix a can be described by an eigenvalue density $\rho_a(\lambda)$ having support I on the real axis. Typically I is a finite interval. In calculations it is convenient to use the Stieltjes transform of the density

$$G_a(z) = \int_I \frac{\rho_a(\lambda) d\lambda}{z - \lambda} \,, \tag{6}$$

that is a complex function defined on the complex plane outside I. This function is often called Green's function. The density can be reconstructed from the Green's function $G_a(z)$ by calculating the imaginary part of the Green's function close to the real axis

$$\rho_a(\lambda) = -\frac{1}{\pi} \lim_{\epsilon \to 0^+} \mathrm{Im} G_a(\lambda + i\epsilon) . \tag{7}$$

The power expansion $G_a(z)$ at infinity

$$G_a(z) = \sum_{n=0}^{\infty} \frac{m_{a,n}}{z^{n+1}} \tag{8}$$

generates moments of the eigenvalue distribution

$$m_{a,n} = \int_I d\lambda \rho_a(\lambda) \lambda^n . \tag{9}$$

These moments are equivalent to the trace moments

$$m_{a,n} = \left\langle \frac{1}{N} \mathrm{Tr} a^n \right\rangle = \int d\mu(a) \frac{1}{N} \mathrm{Tr} a^n, \tag{10}$$

calculated with respect to the probability measure $\mu(a)$ for the entire matrix a. The zeroth moment is fixed $m_{a,0} = 1$ by the normalization of the measure. One often defines another moment generating function as a power series at zero

$$\phi_a(z) = \sum_{n=1}^{\infty} m_{a,n} z^n . \tag{11}$$

It is related to the Green's function

$$\phi_a(z) = \frac{1}{z} G_a\left(\frac{1}{z}\right) - 1 . \tag{12}$$

The eigenvalue density $\rho_a(\lambda)$, the moment generating function $\phi_a(z)$, and the Green's function $G_a(z)$ contain roughly speaking the same information about the eigenvalue distribution of the invariant matrix a and one can reproduce these functions from each other. The idea is to calculate these functions for $a + b$ when the corresponding functions for a and b are given. Let us first calculate the moments of $a + b$

$$m_{a+b,n} = \left\langle \frac{1}{N} \mathrm{Tr}(a+b)^n \right\rangle = \int \int d\mu_a(a) d\mu_b(b) \frac{1}{N} \mathrm{Tr}(a+b)^n . \tag{13}$$

The measure factorizes, since a and b are independent by assumption. For $n = 1, 2, \ldots$ the last equation gives

$$\begin{aligned} m_{a+b,1} &= m_{a,1} + m_{b,1} , \\ m_{a+b,2} &= m_{a,2} + m_{b,2} + 2m_{a,1}m_{b,1} , \\ &\ldots . \end{aligned} \tag{14}$$

In the second equation we replaced $\langle \frac{1}{N} \mathrm{Tr} ab \rangle$ by $m_{a,1}m_{b,1}$. The equality $\langle \frac{1}{N} \mathrm{Tr} ab \rangle = m_{a,1}m_{b,1}$ follows from the equation $\int \int d\mu_a(a) d\mu_b(b) \frac{1}{N} \mathrm{Tr}(a - \mathbb{1} m_{a,1})(b - \mathbb{1} m_{b,1}) = 0$, which holds in the limit $N \to \infty$ as a result of invariance and independence of a and b. The relations between moments (14) can be written in a compact way using free cumulants which are certain (specific) combinations of moments [15]: $\kappa_{a,1} = m_{a,1}$, $\kappa_{a,2} = m_{a,2} - m_{a,1}^2$, $\kappa_{a,3} = m_{a,3} - 3m_{a,2}m_{a,1} + 2m_{a,1}^3$, etc. Equations (14) correspond to

$$\kappa_{a+b,n} = \kappa_{a,n} + \kappa_{b,n} \ , \tag{15}$$

for $n = 1, 2, \ldots$. The concept of free cumulants is analogous to the standard cumulants in classical probability. Let us recall that cumulants for a real random variable are generated as coefficients of the power series expansion of the logarithm of the characteristic function. The logarithm of the characteristic function for a sum of independent random variables is equal to the sum of logarithms of the characteristic functions of these random variables. Therefore cumulants for the sum of independent variables are additive, exactly as free cumulants (15). Free cumulants are, however, generated in a different way. The generating function for free cumulants is called R transform [15, 16]. Free cumulants are equal to coefficients of the power series expansion

$$R(z) = \sum_{n=1}^{\infty} \kappa_n z^{n-1} \ . \tag{16}$$

Note that there is a shift between the power of z and the index of the corresponding cumulant. The R transform is related to the Green's function which generates moments of the eigenvalue distribution. The relation reads

$$G(z) = \frac{1}{z - R(G(z))} \ . \tag{17}$$

We give it without a proof [15]. One can find a diagrammatic interpretation of this equation [21] using field theoretical methods for planar graphs enumeration [18, 19]. Using (17) one can find $G(z)$ if $R(z)$ is known. One can also invert (17)

$$G\left(R(z) + \frac{1}{z}\right) = z \ . \tag{18}$$

This form is useful when one wants to determine the R transform for the given Green's function. The R transform for the sum of free matrices $a + b$ is a sum of R transforms for a and b [15, 16]

$$R_{a+b}(z) = R_a(z) + R_b(z) \ , \tag{19}$$

exactly as the logarithm of the characteristic function for the sum of independent real random variables in classical probability. The additivity of R transform reflects the invariance and independence of the measures for a and b in (13). For $N \to \infty$ these two properties correspond to freeness.

Explicit relations between free cumulants κ_n and moments m_n can be found by inserting the power series $R(z) = \kappa_1 + \kappa_2 z + \ldots$ and $G(z) = 1/z + m_1/z^2 + \ldots$ into (18). The first three free cumulants are identical as the standard ones $\kappa_1 = m_1$, $\kappa_2 = m_2 - m_1^2$, $\kappa_3 = m_3 - 3m_2 m_1 + 2m_1^3$. From the fourth one on they are different. The fourth free cumulant is $\kappa_4 = m_4 - 4m_1 m_3 - 2m_2^2 + 10m_1^2 m_2 - 5m_1^4$ while the standard one is $\kappa_4 = m_4 - 4m_1 m_3 - 3m_2^2 + 12m_1^2 m_2 - 6m_1^4$.

The R transform was first introduced in free probability by Voiculescu [16]. Later it was observed that free cumulants are related to a combinatorial problem of non-crossing lines on a plane [14, 25]. For this reason free cumulants are sometimes called non-crossing cumulants. For completeness we mention that (18) can be deduced [21, 26] using field theoretical methods of planar graph enumeration [18]. The Green's function and the R transform correspond to sums over certain classes of diagrams, and equations like (17) - to Dyson-Schwinger equations [18]. The combinatorics for planar Feynman diagrams for matrix models is equivalent to the combinatorial approach to freeness [14, 25].

We close this section by giving a standard example. Recall that in classical probability the normal random variable with the probability density function $p(x) = \frac{1}{\sqrt{2\pi}} e^{-x^2/2}$ has all cumulants equal zero except $\kappa_2 = 1$. What is the corresponding random matrix eigenvalue density in free probability that has all free cumulants equal zero except the second one $\kappa_2 = 1$? We want to find the eigenvalue density corresponding to the R transform equal $R(z) = z$ (16). Using (17) we first find an equation for the Green's function $G(z) = 1/(z - G(z))$. It can be solved for $G(z)$: $G(z) = \frac{1}{2}(z - \sqrt{z^2 - 4})$. We select the branch of the square root to match the asymptotic behavior $G(z) \sim 1/z$ for large z. The eigenvalue density is (7) $\rho(\lambda) = \frac{1}{2\pi}\sqrt{4 - \lambda^2}$ for $\lambda \in [-2, 2]$ and zero otherwise. It is the Wigner semicircle law. Odd moments of this distribution are equal zero while even moments $m_{2k} = \int_{-2}^{2} d\lambda \rho(\lambda) \lambda^{2k} = \frac{1}{k+1}\binom{2k}{k}$ are equal to the Catalan numbers. One can use (5) to find that the corresponding potential is quadratic $V(x) = x^2/2$, so the invariant measure (3) is Gaussian $d\mu(a) \propto e^{-N\mathrm{Tr}a^2/2}$. Let us slightly modify the R transform and consider $R(z) = \alpha + \sigma z$. Without repeating the calculations one can easily argue on general grounds that the corresponding eigenvalue density is shifted as compared to the previous case by the first cumulant $\kappa_1 = \alpha$ and rescaled by the second one $\kappa_2 = \sigma$ leading to $\rho(\lambda) = \frac{1}{2\pi\sigma}\sqrt{4 - (\lambda - \alpha)^2/\sigma^2}$ and $V(x) = (x - \alpha)^2/2\sigma^2$.

Let us now calculate the eigenvalue density of the sum of two independent large ($N \to \infty$) invariant matrices a and b that have eigenvalue densities given by the Wigner semicircle law with different shift and scale parameters. Denote these parameters by α_a, σ_a for a and α_b, σ_b for b, respectively. The R transforms are $R_a(z) = \alpha_a + \sigma_a z$ and $R_b(z) = \alpha_b + \sigma_b z$. To find the eigenvalue density of the sum $c = a + b$ we first calculate R transform that is equal to $R_c(z) = \alpha_c + \sigma_c z$ with $\alpha_c = \alpha_a + \alpha_b$ and $\sigma_c = \sigma_a + \sigma_b$. The R transform is exactly of the same form as for a and b. Therefore the eigenvalue density for c is also given by the Wigner semicircle law. The situation is analogous to that in classical probability where the sum of two random Gaussian variables is known to be a Gaussian random variable. One says that the semicircle law is stable with respect to free addition in the same way as the Gaussian law is stable with respect to addition in classical probability. Actually there exist a bijection between stable laws in classical and free probability that allows one to classify all stable laws in free probability [27]. Just as a curiosity we mention that the only distribution that is stable with respect to addition (random variables) and free addition (random matrices) is the Cauchy distribution.

5. Products of large invariant matrices

As we have seen the law of addition for independent large invariant matrices can be concisely expressed in terms of the R transform (19). Before discussing the law of multiplication it is convenient to recall the corresponding law in classical probability. Let x and y be independent non-negative real random variables. What is the distribution of the random variable xy? A way to derive this distribution is via the Mellin transform of the probability density function. The Mellin transform $M_{xy}(s)$ for the product xy is the product of Mellin transforms $M_{xy}(s) = M_x(s)M_y(s)$ for x and y. Clearly this multiplication law is analogous to the addition law discussed in the previous section (19). It remains to find the corresponding law for products of large invariant matrices. Let a and b be independent invariant matrices. For the moment we also assume a and b to be positive semi-definite. Later we will weaken this restriction. The first trivial observation is that even if a and b are Hermitian, the product ab is generically non-Hermitian, so its eigenvalues are complex. Therefore, it is convenient to introduce a slightly modified version of the multiplication [28]

$$a \cdot b = \sqrt{a}\, b\, \sqrt{a} \,. \tag{20}$$

We use dot in the notation to distinguish $a \cdot b$ from the standard matrix multiplication ab. Square root of a positive semi-definite matrix is defined as $\sqrt{a} = U^\dagger \mathrm{diag}(\sqrt{\lambda_1}, \ldots, \sqrt{\lambda_N})U$

where U is a unitary matrix that diagonalizes a: $a = U^\dagger \mathrm{diag}(\lambda_1, \ldots, \lambda_N)U$. The product $a \cdot b$ of Hermitian matrices is Hermitian. The product $a \cdot b$ of invariant random matrices is an invariant random matrix. Moreover, the probability measure $\mu_{a \cdot b}$ for the product $a \cdot b$ is identical as for $b \cdot a$: $\mu_{a \cdot b} = \mu_{b \cdot a}$ so this product is in a sense commutative. One should also note that the trace moments for the product $a \cdot b$ (20) are identical as for the standard product: $\frac{1}{N}\mathrm{Tr}(a \cdot b)^k = \frac{1}{N}\mathrm{Tr}(ab)^k$.

The law of free multiplication was formulated by Voiculescu in free probability [15, 17]. Later it was shown that it applies also to multiplication (20) of invariant random matrices in the limit $N \to \infty$ [28]. The multiplication law is expressed in terms of S transform

$$S_{a \cdot b}(z) = S_a(z) S_b(z) \tag{21}$$

where $S_a(z)$ is a complex function. The S transform $S_a(z)$ can be constructed from the moment generating function $\phi_a(z)$ (11) or actually the inverse function of ϕ_a that is usually denoted by χ_a: $\chi_a(\phi_a(z)) = \phi_a(\chi_a(z)) = z$. The relation reads [17]

$$S_a(z) = \frac{z+1}{z} \chi_a(z) \ . \tag{22}$$

The S transform can be viewed as a series in z

$$S_a(z) = \frac{1}{m_{a,1}} + \frac{m_{a,1}^2 - m_{a,2}}{m_{a,1}^3} z + \frac{2m_{a,2}^2 - m_{a,2} m_{a,1}^2 - m_{a,1} m_{a,3}}{m_{a,1}^5} z^2 + \ldots \tag{23}$$

that is obtained by inserting the inverse series in place of $\phi_a(z)$ in equation (22). The coefficients of this series are related to the moments $m_{a,k}$. The coefficients are singular for $m_{a,1} = 0$, in which case the S transform is ill-defined.

The S transform is related to the R transform as follows [15, 21]

$$S(z) = \frac{1}{R(zS(z))} \ . \tag{24}$$

If one knows the R transform then the last equation can be used to calculate the S transform. One can invert the last equation [21]

$$R(z) = \frac{1}{S(zR(z))} \ , \tag{25}$$

which allows one to calculate the R transform if the S transform is known. Using these relations one can show that the multiplication law (21) can be alternatively written in the R transform form [21]:

$$\begin{aligned} R_{a \cdot b}(z) &= R_a(w) R_b(v) \ , \\ v &= z R_a(w) \ , \\ w &= z R_b(v) \ . \end{aligned} \tag{26}$$

This set of equations looks more complicated than (21) but in some situations it is more advantageous. For example, it can be used when the first moments vanish, that is when the S transforms are ill-defined. Moreover, this form can be generalized to the case of non-Hermitian matrices, as we discuss in section 7. The two forms of the multiplication law are in a sense complementary. The original S transform form of (21) has a very simple and appealing structure that reflects important properties of free products, for example that the multiplication $a \cdot b$ (20) of invariant random matrices is commutative and associative: $S_{(a\cdot b)\cdot c}(z) = S_{c\cdot(a\cdot b)}(z)$. This follows from the fact that multiplication on the right hand side of (21) is just multiplication of complex numbers which is commutative and associative.

Let us give an example to illustrate how these formal relations work in practice. Consider the product $a \cdot b$ of two independent identical Wishart matrices [29]. The Wishart matrix can be thought of as a square of a Gaussian matrix from the GUE ensemble [22]. The matrices a and b have the eigenvalue densities $\rho_a(\lambda) = \rho_b(\lambda) = \rho(\lambda)$

$$\rho(\lambda) = \frac{1}{2\pi}\sqrt{\frac{4-\lambda}{\lambda}} \tag{27}$$

for $\lambda \in [0,4]$ and $\rho(\lambda) = 0$ for λ outside this interval. The Green's function (6) corresponding to this density is

$$G(z) = \frac{1}{2} - \frac{1}{2}\sqrt{\frac{z-4}{z}} \tag{28}$$

and the moment-generating function (12)

$$\phi(z) = \frac{1}{2z}\left(1-\sqrt{1-4z}\right) - 1\ . \tag{29}$$

The inverse function of ϕ fulfills the equation $z = \frac{1}{2\chi(z)}\left(1-\sqrt{1-4\chi(z)}\right) - 1$ which can be solved for $\chi(z)$

$$\chi(z) = \frac{z}{(1+z)^2}\ . \tag{30}$$

Thus the S transform is (22)

$$S(z) = \frac{1}{1+z}\ . \tag{31}$$

We now want to find the S transform for the product $S_{a\cdot b}(z)$ from $S_a(z) = S_b(z) = S(z)$. Using the multiplication law (21) we obtain

$$S_{a\cdot b}(z) = \frac{1}{(1+z)^2}\ . \tag{32}$$

We now proceed to find the eigenvalue density $\rho_{a\cdot b}(\lambda)$ that corresponds to the S transform given above. First we write the equation for the inverse function of the moment generating function $\chi_{a\cdot b}(z) = z/(1+z)^3$ using (22). We invert it for $\phi_{a\cdot b}(z)$:

$$z = \frac{\phi_{a\cdot b}(z)}{(1+\phi_{a\cdot b}(z))^3}\ . \tag{33}$$

It is a cubic equation for $\phi_{a\cdot b}(z)$. Replacing $\phi_{a\cdot b}(z)$ with the Green's function (12) we have

$$z^2 G^3_{a\cdot b}(z) - zG_{a\cdot b}(z) + 1 = 0\ . \tag{34}$$

This equation has three solutions given by the Cardano formulas. We select the one which asymptotically behaves as $1/z$ for large z. Using (7) we find the density

$$\rho_{a\cdot b}(\lambda) = \frac{2^{1/3}3^{1/2}}{12\pi}\frac{2^{1/3}\left(27+\sqrt{27(27-4\lambda)}\right)^{2/3} - 6\lambda^{1/3}}{\lambda^{2/3}\left(27+\sqrt{27(27-4\lambda)}\right)^{1/3}} \tag{35}$$

for $\lambda \in \left[0, \frac{27}{4}\right]$ and zero otherwise.

We close this section with a few general comments. The multiplication law (21) or (26) can be derived using solely perturbation expansion for invariant matrix models in the limit $N \to \infty$ without referring to free probability. We refer the interested reader to [21]. The second comment concerns the multiplication law for matrices which are not positive semi-definite. In the discussion above we restricted the class of matrices to positive semi-definite ones because we wanted to use $\sqrt{a}$ in the definition of the product $a \cdot b$ (20). For (invariant) Hermitian matrices the product $a \cdot b$ is (invariant) Hermitian. One can also consider the multiplication law for the standard product ab of Hermitian matrices. In this case a and b do not have to be positive semi-definite. The matrix ab is non-Hermitian and thus has complex eigenvalues. Its trace moments $\frac{1}{N}\mathrm{Tr}(ab)^k$ are, however, real. Since the S transform depends only on the moments, all calculations are identical as discussed before for $a \cdot b$ except that the final result $\rho_{ab}(\lambda)$ cannot be interpreted as an eigenvalue density function, but merely as a function to calculate the trace moments of the matrix ab:

$$m_{ab,n} = \left\langle \frac{1}{N}\mathrm{Tr}(ab)^n \right\rangle = \int\int d\mu(a)d\mu(b)\frac{1}{N}\mathrm{Tr}(ab)^n = \int_R d\lambda \rho_{ab}(\lambda)\lambda^n \; . \tag{36}$$

The eigenvalue density has in this case generically a two-dimensional support on the complex plane. For example, the product of two independent Hermitian matrices from the GUE ensemble is a non-Hermitian matrix. Each GUE matrix has real eigenvalues with the eigenvalue distribution given by the Wigner semicircle law. The R transform $R_a(z) = R_b(z) = z$ (we consider standardized GUE with the mean zero and unit variance). Using the R transform form of the multiplication law (26) one finds that $R_{ab}(z) = 0$ and thus $G_{ab}(z) = 1/z$. This means that all the moments $m_{ab,n} = 0$. It does not mean, however, that eigenvalue density of the product is zero. To calculate the eigenvalue density in this case one has to use methods that allow one to treat non-Hermitian matrices. We discuss them in the subsequent sections. Here we only mention that the eigenvalue density for the product is

$$\rho_{ab}(z) = \begin{cases} \frac{1}{2\pi}\frac{1}{|z|} \,, & |z| \le 1 \\ 0 \,, & |z| > 1 \end{cases} \tag{37}$$

where z is a complex variable. This result was first derived in [30] with the help of field theoretical methods. Note that the result is invariant under rotations around the origin of the complex plane.

6. Product of isotropic random matrices

The extension from Hermitian to non-Hermitian random matrices is similar to the extension from real to complex random variables in classical probability. In this section we employ this similarity to define isotropic random matrices [31] being probably the simplest extension of invariant random matrices to non-Hermitian ones.

Consider a complex random variable with a probability measure that is invariant under rotation around the origin of the complex plane $\mu(z) = \mu(ze^{i\alpha})$. The probability density is a function of the distance from the origin $d\mu(z) = d^2z\rho\left(|z|\right)$. Such a random variable $z = re^{i\phi}$ can be constructed as a product of a real non-negative random variable r and a uniformly distributed random phase $\phi \in [0, 2\pi)$. When one knows the probability density function for the radial variable r, one can easily derive the probability density for z on the complex plane.

Isotropic random matrices are constructed in a way that imitates this construction. A random matrix A is called isotropic if the probability measure $\mu(A)$ is invariant under the left and right multiplication by unitary matrices U, V: $\mu(A) = \mu(U^\dagger A V)$. In Math and Engineering community such random matrices are called Bi-Unitarily Invariant (BUI) and in general they can be rectangular. Here we concentrate on square matrices and use the name isotropic in reference to them.

The simplest way of constructing isotropic random matrices is to build them from real semi-positive diagonal random matrices a and Haar unitary matrices u, v independent of a (free of a for $N \to \infty$)

$$A = v^{\dagger} a u \; . \tag{38}$$

Clearly, the diagonal matrix a corresponds to the radial variable r of $z = re^{i\phi}$ and the Haar unitary matrices v and u to the uniform phase ϕ. There are two of them to ensure the invariance under multiplication by a unitary matrix on both sides. Alternatively one could consider a one-sided invariance, in which case the last equation would take the form $A = au$. Most of the results given below hold in either case.

Another way to construct such matrices is to define them by a probability measure of the form [32]

$$d\mu(A) = e^{-N\mathrm{Tr}V(A^{\dagger}A)} dA \tag{39}$$

where $dA = \prod_{ij} d\mathrm{Im}A_{ij} d\mathrm{Re}A_{ij}$ is flat measure for $N \times N$ complex matrices and $V(x)$ is a polynomial or a power series in x. There are many other ways of defining isotropic matrices. An interesting property of isotropic random matrices is that independently of the way they are defined, in the limit $N \to \infty$ there is a one-to-one correspondence between the eigenvalue density of the matrix A and the eigenvalue density of the Hermitian matrix $A^{\dagger}A$ associated with A [33]. By construction the matrix $A^{\dagger}A$ is an invariant positive semi-definite random matrix. We denote the square root of this matrix by $a = \sqrt{A^{\dagger}A}$. Note that the eigenvalues of a correspond to singular values of A. We can use the whole arsenal of methods discussed in the previous section to calculate the eigenvalue distribution of such matrices or their products.

The correspondence between the eigenvalue distribution of a large isotropic matrix and the distribution of its singular values was found by Feinberg and Zee [32] who applied field theoretical perturbative methods to solve the matrix model (39) and independently by Haagerup and Larsen [33] who used methods of free probability. Below we quote the result in the form given by Haagerup and Larsen [33]. In the terminology of free probability isotropic matrices in the limit $N \to \infty$ correspond to variables that are called R-diagonal [34]. Here we prefer to call them large isotropic matrices. The invariance of the measure implies that the eigenvalue density of an isotropic matrix is invariant under rotation around the origin of the complex plane. For an isotropic matrix A it is convenient to define radial cumulative density function as the probability that a random eigenvalue of the isotropic matrix lies on the complex plane within the distance x from the origin: $F_A(x) = \mathrm{Prob}(|\lambda_A| \leq x)$. The eigenvalue density is related to the radial cumulative function as

$$\rho_A(z) = \frac{1}{2\pi|z|} F'_A(|z|) + p_0 \delta^{(2)}(z) \; , \tag{40}$$

where $p_0 = F_A(0)$ is a possible point mass located at the origin. It corresponds to the probability of finding zero eigenvalues in the eigenvalue spectrum of A. For $p_0 > 0$ the number of zero eigenvalues is a finite fraction of all eigenvalues.

Knowing $F_A(x)$ one can calculate the density $\rho_A(z)$ and vice versa. With the help of this function one can write an equation which relates the eigenvalue distributions of A and $a = \sqrt{A^{\dagger}A}$ [33]

$$S_{a^2}\left(F_A(x) - 1\right) = \frac{1}{x^2} \; , \tag{41}$$

for $x > 0$. This equation does not fix the value of the radial cumulative function for $x = 0$. It is fixed by an additional equation

$$p_0 = F_A(0) = F_a(0) \; , \tag{42}$$

where $F_a(x) = \int_0^x \rho_a(\lambda)d\lambda$ is the cumulative eigenvalue density function for a. This equation merely means that the number of zero eigenvalues of A is equal to the number of zero eigenvalues of a. The value $F_a(0)$ is generically zero except the situation when there is a point measure (Dirac delta) at zero in the eigenvalue density $\rho_a(\lambda)$, as for instance in the distribution

$$\rho_a(\lambda) = (1-r)\delta(0) + \frac{1}{2\pi\lambda}\sqrt{(\lambda_+ - \lambda)(\lambda - \lambda_-)}\ , \tag{43}$$

with $\lambda_\pm = (1 \pm \sqrt{r})^2$, $0 < r < 1$ and the support being $\{0\} \cup [\lambda_-, \lambda_+]$. This distribution naturally arises in the context of random matrices and it is called anti-Wishart distribution in physics literature and free Poissonian in math and engineering literature. The first term on the right hand side means that the fraction of zero eigenvalues of the corresponding random matrix a is $(1-r)$. The corresponding isotropic random matrix $A = au$ 'inherits' all zero eigenvalues from a and thus the corresponding fraction of zero eigenvalues of A is also $1-r$.

Equations (41) and (42) provide an implicit relation between $\rho_A(z)$ and $\rho_a(\lambda)$ that allows one to calculate one from another.

A few comments are in order. Equation (41) was first published in physical literature in a slightly different (less transparent) albeit equivalent form [32]. In this paper it was also observed that for $N \to \infty$ the support of the eigenvalue density of an isotropic random matrix has the shape of a single ring on the complex plane and that the radii of the ring can be derived from the eigenvalue density of a. The external radius is equal to $R_e = (\int_0^\infty \rho_a(\lambda)\lambda^2 d\lambda)^{1/2}$ and the internal one to $R_i = (\int_0^\infty \rho_a(\lambda)\lambda^{-2}d\lambda)^{-1/2}$. The ring reduces to a disk when $R_i = 0$. In the paper [33] the discussion was formalized and all statements were proven rigorously using concepts of free probability. In addition to that the S transform form of (41) and the discussion of the point measure for zero eigenvalues $p_0 \neq 0$ (42) were given. From the discussion [33] it follows that in general there are three different types of solutions for F_A. The ring solution that corresponds to $F_A(x)$ growing monotonically from 0 to 1 for $x \in [R_i, R_e]$ and for $R_i > 0$, the disk solution that corresponds to $F_A(x)$ growing monotonically from 0 to 1 for $x \in [0, R_e]$ and the disk solution with a finite fraction of zero eigenvalues $p_0 > 0$ that corresponds to $F_A(x)$ growing monotonically from p_0 to 1 for $x \in [0, R_e]$.

Let us illustrate how the Haagerup-Larsen theorem (41) works in practice by giving an example. Consider a random matrix A from the Ginibre ensemble [36, 37]. It is an $N \times N$ random matrix whose elements are independent identically distributed, normally distributed, centered complex random variables. The probability measure for A is

$$d\mu(A) \propto e^{-N\mathrm{Tr}A^\dagger A}dA \tag{44}$$

where dA is flat measure for $N \times N$ complex matrices. Obviously A is an isotropic random matrix. The limiting eigenvalue density for this matrix for $N \to \infty$ is given by the uniform distribution on the unit disk [36, 37]

$$\rho_A(z) = \frac{1}{\pi}\ , \ |z| \leq 1\ . \tag{45}$$

It is zero outside the disk. The cumulative density $F_A(x) = \int_0^x dr 2\pi r \rho_A(r)$ is

$$F_A(x) = x^2\ ,\ x \leq 1\ . \tag{46}$$

Inserting it into (41) we obtain an equation for the S transform

$$S_{a^2}(z) = \frac{1}{1+z}\ . \tag{47}$$

Next we find (22) $\chi_{a^2}(z) = \frac{z}{(z+1)^2}$, the moment generating function (the inverse function for χ_{a^2}) and the Green's function (12)

$$G_{a^2}(z) = \frac{1}{2} - \frac{1}{2}\sqrt{\frac{z-4}{z}} \ . \tag{48}$$

Finally we calculate the eigenvalue density of a^2 (7)

$$\rho_{a^2}(\lambda) = \frac{1}{2\pi}\sqrt{\frac{4-\lambda}{\lambda}} \ , \ \lambda \in [0,4] \ , \tag{49}$$

and correspondingly of a

$$\rho_a(\lambda) = \frac{1}{\pi}\sqrt{4-\lambda^2} \ , \ \lambda \in [0,2] \ . \tag{50}$$

This is the density of singular values of the Ginibre matrix. We see that it is given by the quarter-circle law.

Let us now discuss the product of n independent isotropic matrices $A_1, A_2, \ldots, A_n$

$$P = A_1 A_2 \ldots A_n \tag{51}$$

where A_i's are independent isotropic matrices [30, 35]. The goal is to calculate the eigenvalue density and the singular value density in the limit $N \to \infty$ having the knowledge of eigenvalue densities for the matrices in the product. Since the product itself is an isotropic matrix we can use (41) to write

$$S_{p^2}\left(F_P(x) - 1\right) = \frac{1}{x^2} \ . \tag{52}$$

The S transform for $p^2 = PP^\dagger$ can be calculated using the multiplication law (21)

$$S_{p^2}(z) = S_{a_1^2 \cdot a_2^2 \cdot \ldots \cdot a_n^2}(z) = \prod_{i=1}^{n} S_{a_i^2}(z) \ . \tag{53}$$

Since the right-hand side of this equation is a product of complex numbers, it does not depend on the order of multiplication. If instead of P we considered a product of matrices permuted with a permutation π

$$P_\pi = A_{\pi(1)} A_{\pi(2)} \ldots A_{\pi(n)} \ , \tag{54}$$

we would obtain the S transform $S_{p_\pi^2}(z)$ identical to $S_{p^2}(z)$ because multiplication on the right hand side of (53) is commutative. Therefore $F_P(x) = F_{P_\pi}(x)$ are equal for any permutation π. In other words, the product of isotropic matrices is spectrally commutative in the limit $N \to \infty$.

Isotropic random matrices possess a striking property that has no counterpart in classical probability. Consider the product $P = A_1 \ldots A_n$ of independent identically distributed isotropic random matrices. This can be realized by picking up n independent matrices from the same ensemble and multiplying them. It turns out that in the limit $N \to \infty$ such a product has exactly the same eigenvalue density as the n-th power $Q = A^n$ of one matrix picked up at random from this ensemble [31]. Actually it was already observed in [33] that singular values of Q and P are identically distributed. This observation can be trivially extended to eigenvalues if one additionally assumes that $p_0 = 0$ (42). This is a sort of spectral self-averaging of multiplication in the limit $N \to \infty$ because a single matrix from the ensemble is sufficiently representative in the n-fold product to give the same result as n independent random realizations. The proof is short so we recall it here in the form given in [31] where it was tacitly assumed that $p_0 = 0$. All constituents needed for this proof can be found already in [33].

We assume that $p_0 = F_A(0) = F_a(0) = 0$ (42). The S transform for the product of independent identically distributed matrices is the n-th power of the S transform for a single one $S_{p^2}(z) = S^n_{a^2}(x)$ as follows from (53) in which all $S_{a_i^2}(z)$ are replaced by $S_{a^2}(z)$. Inserting $S^n_{a^2}(x)$ in place of $S_{p^2}(z)$ in (52) and taking the n-th root of both sides we get

$$S_{a^2}\left(F_P(x) - 1\right) = \frac{1}{x^{2/n}} \ . \tag{55}$$

Comparing the last equation to (41) we see that $F_P(x) = F_A(x^{1/n})$. We can now argue that the radial cumulative density $F_Q(x)$ for the power Q is given by the same expression. Indeed, using directly the definition of radial cumulative density we have

$$F_Q(x) = \text{Prob}(|\lambda_Q| \leq x) = \text{Prob}(|\lambda_A|^n \leq x) = \text{Prob}(|\lambda_A| \leq x^{1/n}) = F_A(x^{1/n}) \tag{56}$$

for $x > 0$, and for $F_Q(x) = F_P(x) = 0$ for $x = 0$ and hence $F_P(x) = F_Q(x) = F_A\left(x^{1/n}\right)$. This completes the proof. If $F_A(0) = p_0 > 0$ then the proof breaks down at $x = 0$. Recall that p_0 is the probability of drawing at random a zero-eigenvalue from the spectrum of random matrix A. The corresponding probability for the product of n independent A matrices is $F_P(0) = 1-(1-p_0)^n$, while for the n-th power $F_Q(0) = 1 - p_0$, so we see that $F_P(0) \neq F_Q(0)$ for $p_0 > 0$.

In particular, one can use the product-power equivalence to calculate the eigenvalue density of the product $P = A_1 \dots A_n$ of n independent Ginibre matrices in the limit $N \to \infty$. We know that $F_A(x) = x^2$ for $x \in [0,1]$ for one Ginibre matrix (46) thus for the product of n

$$F_P(x) = x^{\frac{2}{n}} \ , \ x \in [0,1] \ . \tag{57}$$

The eigenvalue density for the product is (40)

$$\rho_P(z) = \frac{1}{n\pi}|z|^{\frac{2(1-n)}{n}} \ , \ |z| \leq 1 \tag{58}$$

and zero outside the unit circle. This calculation reproduces in a very simple way the result that was originally obtained by different methods [30]. This result can be generalized also to products of rectangular Gaussian matrices [38] or products [40, 41] of Wigner matrices [39] with non-Gaussian entries.

One can also find the distribution of singular values for the product of Ginibre matrices [7, 35]. Inserting radial cumulative density function (57) to the Haagerup-Larsen formula (41) one can find the Green's function for $p^2 = P^\dagger P$ [35]

$$z^n G^{n+1}_{p^2}(z) = zG_{p^2}(z) - 1 \tag{59}$$

and calculate the density function $\rho_{p^2}(\lambda)$ by using (7). For $n = 1$ it gives the Wishart distribution (27); for $n = 2$ the distribution given in (35). One can find an explicit expression for any n [42]. The last step is to change variables $\lambda \to \lambda^2$ to go from $\rho_{p^2}(\lambda)$ to $\rho_p(\lambda)$.

To summarize, large isotropic random matrices are exceptional in many respects. There exists a one-to-one correspondence between the densities of eigenvalues and singular values [33]. Multiplication of independent isotropic matrices is spectrally commutative and self-averaging in the large N limit [31, 35]. One can also show that if the eigenvalue density for an isotropic matrix has a power-law singularity at zero $\rho_A(z) \sim |z|^{-s}$ with $0 < s < 2$, then the density of singular values behaves at zero as $\rho_p(\lambda) \sim \lambda^{-\frac{s}{4-s}}$ [35].

7. General multiplication law

So far we have discussed invariant and isotropic random matrices. We are now going to discuss the more general case of products of non-Hermitian matrices. The multiplication law for non-Hermitian matrices was derived in reference [21] by field theoretical methods.

To build some intuition let us discuss Gaussian non-Hermitian random matrices. One can construct such matrices from independent identically distributed Hermitian Gaussian matrices from the GUE ensemble. The probability measure for GUE matrices is $d\mu(a) \sim e^{-N\mathrm{Tr}a^2/2}da$, where da is flat measure for Hermitian matrices. Let us take two independent such matrices a and b. The following combination

$$X = \frac{1}{\sqrt{2}}(a + ib) \tag{60}$$

is a complex random matrix; a gives rise to Hermitian sector of X and ib to the anti-Hermitian one. The resulting matrix X is a Ginibre random matrix [36, 37] with the uniform eigenvalue distribution on the unit disk (45). One can also define elliptic random matrices [37, 43]

$$X = \cos(\alpha)a + i\sin(\alpha)b \tag{61}$$

being linear combinations of Hermitian and anti-Hermitian random matrices with different coefficients controlled by a mixing parameter α. For $\alpha = 0$ the random matrix X reduces to the Hermitian matrix a, for $\alpha = \pi/4$ to the Ginibre matrix and for $\alpha = \pi/2$ to the anti-Hermitian matrix ib. The eigenvalue density of the matrix X is constant on the support being an ellipse whose oblateness depends on α. The probability measure for matrix X can be calculated from the measure $d\mu(a,b) = d\mu(a)d\mu(b) \propto e^{-N\mathrm{Tr}(a^2+b^2)/2}dadb$ that factorizes since a and b are independent. Changing variables to $X = \cos(\alpha)a + i\sin(\alpha)b$ and $X^\dagger = \cos(\alpha)a - i\sin(\alpha)b$ one finds the probability measure for X [44, 45]

$$d\mu(X) \propto \exp\left(-N\frac{1}{1-\tau^2}\mathrm{Tr}\left(XX^\dagger - \tau(XX + X^\dagger X^\dagger)\right)\right)dX\ , \tag{62}$$

where $\tau = \cos(2\alpha)$. For this measure one can calculate two-point correlation functions $\langle X_{ab}X_{cd}\rangle = \int d\mu(X)X_{ab}X_{cd}$. There are four such functions :

$$\begin{aligned} \left\langle X^\dagger_{ab}X^\dagger_{cd}\right\rangle &= \tau\frac{1}{N}\delta_{ad}\delta_{bc}\ , \quad \left\langle X^\dagger_{ab}X_{cd}\right\rangle = \frac{1}{N}\delta_{ad}\delta_{bc}\ , \\ \left\langle X_{ab}X^\dagger_{cd}\right\rangle &= \frac{1}{N}\delta_{ad}\delta_{bc}\ , \quad \langle X_{ab}X_{cd}\rangle = \tau\frac{1}{N}\delta_{ad}\delta_{bc}\ , \end{aligned} \tag{63}$$

that can be compactly written in the matrix form

$$\begin{pmatrix} \left\langle X^\dagger_{ab}X^\dagger_{cd}\right\rangle & \left\langle X^\dagger_{ab}X_{cd}\right\rangle \\ \left\langle X_{ab}X^\dagger_{cd}\right\rangle & \langle X_{ab}X_{cd}\rangle \end{pmatrix} = \begin{pmatrix} \tau & 1 \\ 1 & \tau \end{pmatrix} \otimes \frac{1}{N}\delta_{ad}\delta_{bc}\ . \tag{64}$$

The matrix on the right hand side of the last equation defines the correlation structure of the model [21]. It plays an important role, as we shall see below. The parameter τ takes values between -1 and 1. For Ginibre matrix it is 0, and for GUE it is 1.

Without going into details we quote the law of non-Hermitian matrix multiplication and refer the interested reader to [21]. We present it in the R transform form (26). The main difference as compared to the Hermitian case (26) is that the R transform is not a complex function but a quaternionic one. We denote quaternions by calligraphic letters. The R transform maps quaternions $\mathcal{G}$ onto quaternions $\mathcal{G} \to \mathcal{R}(\mathcal{G})$. The multiplication law reads

$$\begin{aligned} \mathcal{R}_{AB}(\mathcal{G}_{AB}) &= [\mathcal{R}_A(\mathcal{G}_B)]^L\,[\mathcal{R}_B(\mathcal{G}_A)]^R \ , \\ [\mathcal{G}_A]^R &= \mathcal{G}_{AB}\,[\mathcal{R}_A(\mathcal{G}_B)]^L \ , \\ [\mathcal{G}_B]^L &= [\mathcal{R}_A(\mathcal{G}_B)]^R\,\mathcal{G}_{AB} \ . \end{aligned} \tag{65}$$

Here $\mathcal{G}_A = \mathcal{G}_A(z)$, $\mathcal{G}_B = \mathcal{G}_B(z)$ and $\mathcal{G}_{AB} = \mathcal{G}_{AB}(z)$ correspond to generalized quaternionic Green's functions for A, B and AB. They have the structure:

$$\mathcal{G}(z) = \begin{pmatrix} g(z) & ib(z) \\ i\bar{b}(z) & \bar{g}(z) \end{pmatrix} , \tag{66}$$

where $b(z)$ and $g(z)$ are complex functions. We suppressed the indices of the corresponding matrix (A, B or AB) in the last equation. The density is related to the upper left element $g(z)$ of the generalized quaternionic Green's function

$$\rho(z) = \frac{1}{\pi}\partial_{\bar{z}} g(z) \ . \tag{67}$$

The superscripts L and R refer to the left and right adjoint actions of the following unitary matrix

$$U = \begin{pmatrix} \left(\frac{z}{\bar{z}}\right)^{\frac{1}{8}} & 0 \\ 0 & \left(\frac{\bar{z}}{z}\right)^{\frac{1}{8}} \end{pmatrix} \tag{68}$$

on matrices in the square brackets

$$[X]^L = UXU^\dagger \ , \quad [X]^R = U^\dagger XU \ . \tag{69}$$

The diagonal elements of U are constructed from z being the argument of the Green's functions (66). The generalized R transform is related to the Green's function by an equation analogous to (17)

$$\mathcal{G}(z) = (\mathcal{Z} - \mathcal{R}(\mathcal{G}(z)))^{-1} \tag{70}$$

where

$$\mathcal{Z} = \begin{pmatrix} z & 0 \\ 0 & \bar{z} \end{pmatrix} . \tag{71}$$

This law looks complicated but in practice it reduces to a set of algebraic equations for $g(z)$ from which one can calculate the eigenvalue density by taking the derivative with respect to $\bar{z}$ (67). Before we give an example let us mention an electrostatic analogy [45]. In this analogy eigenvalues of a random matrix represent positions of charges in two-dimensions in the given potential, the function $g(z)$ is interpreted as the electric field $\vec{g} = (g_x, g_y)$ with components $g_x = \mathrm{Re}g(z)$ and $g_y = -\mathrm{Im}g(z)$ and (67) as the Gauss law in two dimensions [45]. It takes a more familiar form when written in the components of $z = x + iy$:

$$\rho(z) = \frac{1}{2\pi}(\partial_x g_x + \partial_y g_y) = \frac{1}{2\pi}\vec{\nabla}\cdot\vec{g} \ . \tag{72}$$

We now illustrate how to use the multiplication law (65) to calculate the eigenvalue density of the product. We consider as an example the multiplication of Gaussian matrices constructed from the centered elliptic matrix X (62) by shifting and rescaling

$$A = q\mathbb{1} + \sigma X \ . \tag{73}$$

The scale parameter σ is a real number and the shift parameter q is a complex number. The quaternionic R transform for A is

$$\mathcal{R}_A\left(\left(\begin{array}{cc} g & ib \\ i\bar{b} & \bar{g} \end{array}\right)\right)=\left(\begin{array}{cc} q & 0 \\ 0 & \bar{q} \end{array}\right)+\sigma\left(\begin{array}{cc} \tau g & ib \\ i\bar{b} & \tau\bar{g} \end{array}\right) . \tag{74}$$

There are only two terms on the right-hand side because the matrix is Gaussian and thus all higher order cumulants vanish (16). The second term on the right-hand side is directly related to the matrix on the right hand side of (64) that defines the two-point correlation structure of the model [21, 30].

The simplest example is $q = 0$, $\sigma = 1$ and $\tau = 0$ (73). It corresponds to the Ginibre matrix. Let us calculate the eigenvalue density for the product of two such matrices. The first equation in the multiplication law (65) gives

$$\mathcal{R}_{AB}(\mathcal{G}_{AB})=\left(\begin{array}{cc} 0 & icb_B \\ i\bar{c}\bar{b}_B & 0 \end{array}\right)\left(\begin{array}{cc} 0 & i\bar{c}b_A \\ ic\bar{b}_A & 0 \end{array}\right)=\left(\begin{array}{cc} -c^2 b_B\bar{b}_A & 0 \\ 0 & -\bar{c}^2\bar{b}_B b_A \end{array}\right) . \tag{75}$$

We used a shorthand notation $c = (z/\bar{z})^{1/4}$. Because A and B are identically distributed, we search a symmetric solution $b_A(z) = b_B(z)$ and $g_A(z) = g_B(z)$.

$$\mathcal{R}_{AB}(\mathcal{G}_{AB})=\left(\begin{array}{cc} -c^2|b_A|^2 & 0 \\ 0 & -\bar{c}^2|b_A|^2 \end{array}\right) . \tag{76}$$

Now we can use the two remaining equations (65). Because of the symmetry it is enough to use only one of them, for example the first one. Replacing $\mathcal{G}_{AB}^{-1}$ by $\mathcal{Z}-\mathcal{R}_{AB}(\mathcal{G}_{AB})$ (70) we can write this equation as

$$(\mathcal{Z}-\mathcal{R}_{AB}(\mathcal{G}_{AB}))\,[\mathcal{G}_A]^R=[\mathcal{R}_A(\mathcal{G}_B)]^L . \tag{77}$$

We can now eliminate $\mathcal{R}_{AB}$ in the last equation using (76). Writing also $[\mathcal{G}_A]^R$ and $[\mathcal{R}_A(\mathcal{G}_B)]^L$ in the explicit matrix form we have

$$\left(\begin{array}{cc} z+c^2|b_A|^2 & 0 \\ 0 & \bar{z}+\bar{c}^2|b_A|^2 \end{array}\right)\left(\begin{array}{cc} g_A & i\bar{c}b_A \\ ic\bar{b}_A & \bar{g}_A \end{array}\right)=\left(\begin{array}{cc} 0 & icb_A \\ i\bar{c}\bar{b}_A & 0 \end{array}\right) , \tag{78}$$

where $c = (z/\bar{z})^{1/4}$, as before. It is a set of equations for $g_A = g_A(z)$ and $b_A = b_A(z)$. It has a trivial solution $g_A(z) = b_A(z) = 0$ and a non-trivial one $g_A(z) = 0$ and $|b_A(z)|^2 = 1 - |z|$. The two solutions coincide at the unit circle $|z| = 1$. The trivial solution holds for $|z| > 1$ and the nontrivial one for $|z| \leq 1$. Note that for both of them $g_A(z) = 0$. Inserting these solutions to (76) we obtain $\mathcal{R}_{AB}$ and then using (70) we derive the final result for the generalized Green's function of the product

$$\mathcal{G}_{AB}(z)=\left(\begin{array}{cc} \sqrt{\frac{\bar{z}}{z}} & 0 \\ 0 & \sqrt{\frac{z}{\bar{z}}} \end{array}\right) \tag{79}$$

inside the unit circle $|z| \leq 1$, and

$$\mathcal{G}_{AB}(z)=\left(\begin{array}{cc} \frac{1}{z} & 0 \\ 0 & \frac{1}{\bar{z}} \end{array}\right) \tag{80}$$

outside. Applying the Gauss law (67) to the upper left element of $\mathcal{G}_{AB}(z)$ (denoted by $g_{AB}(z)$) we find the density

$$\rho_{AB}(z)=\frac{1}{\pi}\partial_{\bar{z}}g_{AB}(z)=\frac{1}{2\pi|z|} \tag{81}$$

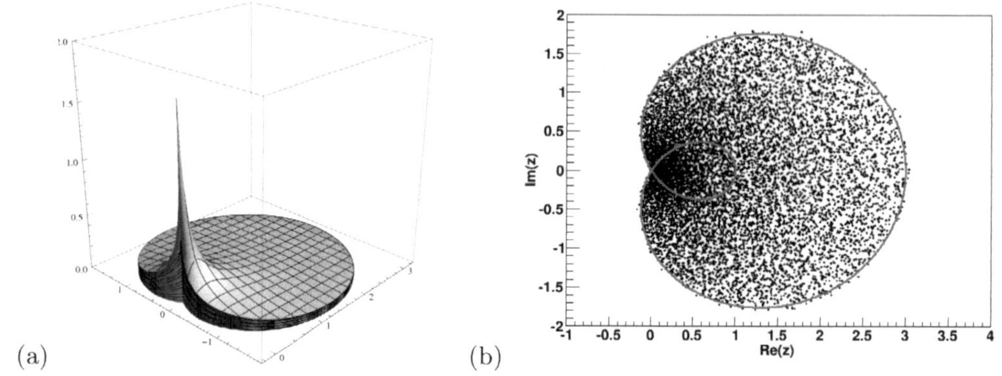

Figure 1. (a) Eigenvalue density for the product $(\mathbb{1} + X_1)(\mathbb{1} + X_2)$ where X_1 and X_2 are independent standardized Ginibre matrices, each having the eigenvalue density uniformly distributed on the unit circle. The result is derived analytically [21]. The horizontal axes correspond to the real and imaginary axes of the complex plane and the vertical one to the value of the eigenvalue density $\rho(z)$. The eigenvalue density has a support surrounded by a contour given by the Pascal's limaçon. (b) Solid line represents the Pascal's limaçon. Dots correspond to eigenvalues collected for ten matrices of size 1000×1000 generated by Monte-Carlo simulations. A few eigenvalues lie outside the limiting contour but this effect is expected by finite size analysis. The number of outliers decreases when the matrix size increases.

for $|z| \leq 1$, and $\rho_{AB}(z) = 0$ otherwise. The calculations were done for $\tau_A = \tau_B = 0$. One, however, (quite surprisingly) obtains the same result for the product of matrices with arbitrary τ_A and τ_B [30]. The reason for this is the following. Looking at the R transform (74) we see that τ appears only in the combination τg (or, writing it separately for A and B, in combinations $\tau_A g_A$ and $\tau_B g_B$). The dependence on τ's disappears for $g_A(z) = g_B(z) = 0$. Since in the solution discussed above we had $g_A(z) = g_B(z) = 0$, we conclude that this solution is independent of τ_A and τ_B and thus it holds for any τ_A and τ_B. In particular, the product AB of two Hermitian GUE matrices ($\tau_A = \tau_B = 1$) or any elliptic random matrices has exactly the same eigenvalue density (81), which is spherically symmetric. In a similar way one can find the eigenvalue density for the product of matrices with non-zero constants q (73) but the calculations are tedious. In figure 1 (a) we show an analytic result for the eigenvalue density for $(\mathbb{1} + X_1)(\mathbb{1} + X_2)$ where X_1 and X_2 are independent Ginibre matrices [21]. The eigenvalue support is surrounded by a contour $r = 1 + 2\cos\phi$ that corresponds to the the Pascal's limaçon. In figure 1 (b) we compare this theoretical prediction to Monte Carlo simulations of large, finite matrices. The agreement is very good. As another example, in figure 2 we show a comparison between Monte Carlo results and the analytic prediction for the contour of the support for the eigenvalue density of the product $(\mathbb{1} + H)(\mathbb{1} + X)$, where H is the GUE Hermitian matrix and X is the Ginibre matrix. Again the comparison shows a good agreement. The details of the calculations and other examples can be found in [46]. We summarize this section shortly by emphasizing that there exists an algorithm to calculate eigenvalue densities for products of non-Hermitian matrices. This algorithm works for large random matrices ($N \to \infty$) with probability measures of the type $d\mu(X) \propto e^{-N\mathrm{Tr}V(X,X^\dagger)}dX$ with a self-adjoint potential $\left(V(X,X^\dagger)\right)^\dagger = V(X,X^\dagger)$ which is bounded from below [21].

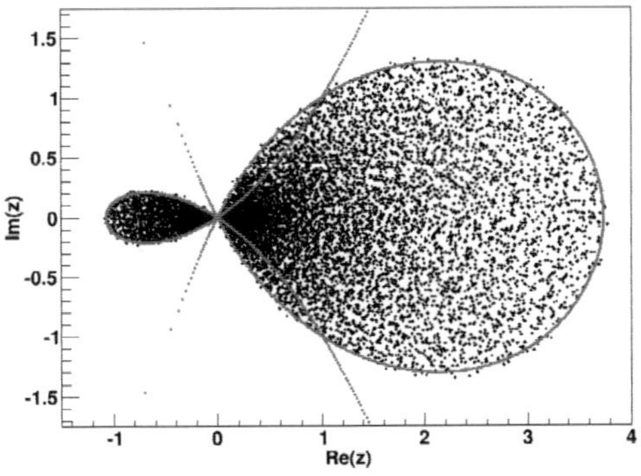

Figure 2. Solid line represents the contour of the eigenvalue density for the product $(\mathbb{1}+X)(\mathbb{1}+H)$ where X is a standardized Ginibre matrix and H is an independent standardized Hermitian GUE matrix. The curves forming the contour are calculated analytically [46] using the multiplication law (65). The extrapolation of these curves outside the support is shown in red dots. Black dots correspond to eigenvalues collected for ten matrices of size 1000×1000 generated by Monte-Carlo simulations.

8. Conclusions

Using free probability one can find an explicit formula for the moment generating function for the product of invariant random Hermitian matrices in the limit $N \to \infty$. This formula allows to calculate trace moments $\langle \frac{1}{N}\mathrm{Tr}(ab)^k \rangle$ for independent (free) random matrices a, b and to determine the eigenvalue density of the matrix $\sqrt{a}b\sqrt{a}$ (if a is positive semi-definite) from eigenvalue densities of a and b. We discussed how to extend the formalism to non-Hermitian matrices to calculate the eigenvalue density of the product ab, which is generically a non-Hermitian matrix. The extension has been worked out using field theoretical methods of planar graphs enumeration. The multiplication law (65) is conveniently written in the R transform form (26). The R transform is a quaternionic function. One can use this law to find eigenvalue densities for products of large non-Hermitian random matrices.

Isotropic matrices are a special class of non-Hermitian random matrices. These matrices have two exceptional features as far as matrix multiplication is concerned. Isotropic random matrices are spectrally commutative and self-averaging in the limit $N \to \infty$. The first notion means that the eigenvalue distribution of the product of such matrices does not depend on the order of matrix multiplication. For example, the matrix $abcd$ has the same eigenvalue density as $bdac$. The second one means that the product of n independent isotropic identically distributed matrices has the same eigenvalue distribution as the n-th power of a single matrix.

We finish by mentioning that recently a progress has been also made in the understanding of microscopic properties of eigenvalues statistics for products of random matrices. For example, an explicit form of the joint probability distribution for eigenvalues and singular values of the product of Ginibre matrices has been explicitly found [47, 48]. Using these results one can determine not only the eigenvalue density but also joint eigenvalue distribution and eigenvalue correlations for products of large random matrices. One can also infer finite size corrections to the limiting laws discussed in this paper.

Acknowledgments

The author would like to thank Romuald Janik, Andrzej Jarosz, Giacomo Livan, Maciej A. Nowak, Gabor Papp, Artur Swiech and Ismail Zahed for many interesting discussions and a fruitful collaboration on the subject and Jeremi Ochab for critical reading of the manuscript. The author also wants to thank the organizers of the conference "Inference, Computation, and Spin Glasses" held in Sapporo, July 28th–30th 2013, for a kind invitation to give a talk on this conference. The research presented in the paper was financed by Grant No. DEC-2011/02/A/ST1/00119 of National Centre of Science in Poland. The participation in the conference was supported by the grant: Grant-in-Aid for Scientific Research on Innovative Areas, MEXT, Japan.

References

[1] Crisanti A, Paladin G and Vulpiani A 1993 *Products of Random Matrices in Statistical Physics*, (Berlin, Heidelberg: Springer-Verlag)
[2] Newman C M 1986 *Commun. Math. Phys.* **103** 121–6
[3] Mueller R R 2002 *IEEE Trans. Inf. Theor.* **48** 2086–91
[4] Janik R A and Wieczorek W 2004 *J. Phys. A: Math. Gen.* **37** 6521–9
[5] Gudowska-Nowak E, Janik R, Jurkiewicz J, Nowak M A and Wieczorek W 2010 *Chem. Phys.* **375** 380–5
[6] Lohmayer R, Neuberger H and Wettig T 2009 *JHEP* **0905** 107
[7] Collins B, Nechita I and Zyczkowski K 2010 *J. Phys. A: Math. Theor.* **43** 275303
[8] Bouchaud J-P, Laloux L, Miceli M A and Potters M 2007 *EPJ* B **55** 201–7
[9] Bougerol P and Lacroix J 1985 *Products of Random Matrices with Applications to Schrödinger Operators* (Boston: Birkjäuser)
[10] ed Cohen J E, Kesten H and Newman C M 1986 *Random Matrices and Their Applications* (Contemporary Mathematics vol 50) (Providence, RI: American Mathematical Society)
[11] Furstenberg H and Kesten H 1960 *Ann. Math. Statist.* **31** 457–69
[12] Oseledec V I 1968 *Trans. Moscow Math. Soc.* **19** 197–231
[13] Voiculescu D V 1991 *Invent. Math.* **104** 201–20
[14] Speicher R 1994 *Math. Ann.* **298** 611–28
[15] Voiculescu D V, Dykema K J and Nica A 1992 *Free Random Variables* (CRM Monograph Series vol 1) (Providence, RI: American Mathematical Society)
[16] Voiculescu D V 1986 *J. Funct. Anal.* **66** 323–46
[17] Voiculescu D V 1987 *J. Operator Theory* **18** 223–35
[18] Brezin E, Itzykson C, Parisi G and Zuber J-B 1978 *Commun. Math. Phys.* **59** 35–51
[19] Bessis D, Itzykson C and Zuber J-B 1980 *Adv. Appl. Math.* **1** 109–57
[20] 't Hooft G 1974 *Nucl. Phys.* B **72** 461–73
[21] Burda Z, Janik R A and Nowak M A 2011 *Phys. Rev.* E **84** 061125
[22] Mehta M L 2004 *Random Matrices* (Amsterdam: Elsevier)
[23] ed Akemann G, Baik J and Di Francesco P 2011 *The Oxford Handbook of Random Matrix Theory* (Oxford: Oxford University Press)
[24] Dyson F J 1962 *J. Math. Phys.* **3** 140–56
[25] Nica A and Speicher R 2006 *Lectures on the Combinatorics of Free Probability* (London Mathematical Society Lecture Note Series vol 335) (Cambridge: Cambridge University Press)
[26] Zee A 1996 *Nucl. Phys.* B **474** 726–44
[27] Bercovici H and Pata V 1995 *Math. Res. Lett.* **2** 791–5
[28] Rao N R and Speicher R 2007 *Elect. Comm. in Probab.* **12** 248–58
[29] Wishart J 1928 *Biometrika* **20A**(1-2) 32–52
[30] Burda Z, Janik R A and Waclaw B 2010 *Phys. Rev.* E **81** 041132
[31] Burda Z, Nowak M A and Swiech A 2012 *Phys. Rev.* E **86** 061137
[32] Feinberg J and Zee A 1997 *Nucl. Phys.* B **501** 643–69
[33] Haagerup U and Larsen F 2000 *J. Funct. Anal.* **176** 331–67
[34] Nica A and Speicher R 1997 *Free Probability Theory* (Fields Institute Communications vol 12) ed Voiculescu D V pp 149–88
[35] Burda Z, Livan G and Swiech A 2013 *Phys. Rev.* E **88** 022107
[36] Ginibre J 1965 *J. Math. Phys.* **6** 440–9
[37] Khoruzhenko B A and Sommers H-J 2009 *Non-Hermitian Random Matrix Ensembles* Chapter 18 in reference [23] (Preprint arXiv:0911.5645 [math.ph])

[38] Burda Z, Jarosz A, Livan G, Nowak M A and Swiech A 2010 *Phys. Rev.* E **82** 061114
[39] Wigner E P 1955 *Ann. of Math.* **62** 548–64
[40] Alexeev N, Goetze F and Tikhomirov A 2010 *Lith. Math. J.* **50** 121–32
[41] O'Rourke S and Soshnikov A 2011 *Electron. J. Probab.* **16** 2219–45
[42] Penson K A and Zyczkowski K 2011 *Phys. Rev.* E **83** 061118
[43] Girko V L 1985 *Theor. Prob. Appl.* **29** 694–706
[44] Fyodorov Y V, Khoruzhenko B A and Sommers H-J 1997 *Phys. Lett.* A **226** 46–52
[45] Sommers H-J, Crisanti A, Somplinsky H and Stein Y 1988 *Phys. Rev. Lett.* **60** 1895–8
[46] Swiech A 2013 *Applications of generalized free addition and multiplication laws to large non-Hermitian random matrices* (MSc thesis, Jagiellonian University, Cracow)
[47] Akemann G and Burda Z 2012 *J. Phys. A: Math. Theor.* **45** 465201
[48] Akemann G, Kieburg M and Wei L 2013 *J. Phys. A: Math. Theor.* **46** 275205

Reconstruction algorithm in compressed sensing based on maximum a posteriori estimation

Koujin Takeda[1] and Yoshiyuki Kabashima[2]

Department of Intelligent Systems Engineering, Ibaraki University[1]
Department of Computational Intelligence and Systems Science, Tokyo Institute of Technology[2]

E-mail: ktakeda@mx.ibaraki.ac.jp

Abstract. We propose a systematic method for constructing a sparse data reconstruction algorithm in compressed sensing at a relatively low computational cost for general observation matrix. It is known that the cost of ℓ_1-norm minimization using a standard linear programming algorithm is $O(N^3)$. We show that this cost can be reduced to $O(N^2)$ by applying the approach of posterior maximization. Furthermore, in principle, the algorithm from our approach is expected to achieve the widest successful reconstruction region, which is evaluated from theoretical argument. We also discuss the relation between the belief propagation-based reconstruction algorithm introduced in preceding works and our approach.

1. Introduction

Nowadays, use of the compressed sensing (CS) approach [1, 2, 3] is rapidly spreading to various fields in information technology, where the sparsity of the original data plays a crucial role [4].

In this article, we present our study on a very basic problem of CS. Let us consider a linear observation process expressed as

$$\boldsymbol{y} = \boldsymbol{F}\boldsymbol{x}, \tag{1}$$

where $\boldsymbol{y} \in \mathbb{R}^M$ is the observed data and $\boldsymbol{x} \in \mathbb{R}^N$ is the original sparse data with many zero entries. The (averaged) number of nonzero entries is denoted by K. The $M \times N$ matrix $\boldsymbol{F}$ describes the process of observation, and the limit of large M, N with compression rate $\alpha := M/N < 1$ is taken into account. (Throughout this paper, bold symbols denote a vector/matrix.)

We investigate the basic ℓ_1-norm minimization

$$\min_{\boldsymbol{x}} \|\boldsymbol{x}\|_1 \ \text{ subject to } \ \boldsymbol{y} = \boldsymbol{F}\boldsymbol{x}. \tag{2}$$

This problem is written as linear programming and can be solved at a computational cost of $O(N^3)$ using a standard algorithm, such as the interior point method. However, many alternative algorithms have been proposed for reducing the computational cost [5, 6]. Among these, Approximate Message Passing (AMP) [7], which is a thresholding algorithm that is based on the message passing approach [8], is noteworthy. In this algorithm, each entry of the matrix $\boldsymbol{F}$ is assumed to be random with identical Gaussian distribution. Then, it is shown that the original data can be reconstructed at a computational cost of $O(N^2)$. Furthermore, this algorithm has theoretical significance: for the AMP, the successful reconstruction threshold in terms of compression rate α and data sparsity $\rho := K/N$ is analytically shown to be the same as for ℓ_1-norm minimization [9] using a state evolution technique [7, 10, 11, 12]. However, it is not intuitively evident from their argument why the threshold of the AMP agrees with that of ℓ_1-norm minimization.

In this article, we propose an approach for constructing a sparse data reconstruction algorithm by using maximum *a posteriori* probability (MAP) for general matrix $\boldsymbol{F}$ with as small an M as possible. Using this approach, we obtain an algorithm whose computational cost is $O(N^2)$. (For a special case, when the matrix $\boldsymbol{F}$ is sparse and has only $O(1)$ nonzero entries in each column/row, the computational cost is reduced to $O(N)$.) In addition, our approach explains from another perspective why the reconstruction thresholds of the AMP and ℓ_1-norm minimization are analytically equivalent.

2. Reconstruction algorithm using the MAP

To construct our MAP algorithm, first, we prepare the posterior probability. We need to infer the original data $\boldsymbol{x}$ from the observed data $\boldsymbol{y}$ and the matrix $\boldsymbol{F}$; therefore, we define the posterior probability $P(\boldsymbol{x}|\boldsymbol{y},\boldsymbol{F})$ for $\boldsymbol{x}$ in accordance with the ℓ_1-norm minimization problem as (2)

$$P(\boldsymbol{x}|\boldsymbol{y},\boldsymbol{F}) := \lim_{\beta\to\infty}\frac{1}{Z(\beta)}\exp\left(-\beta\sum_i |x_i|\right)\prod_\mu \delta\left(y_\mu - \sum_i F_{\mu i}x_i\right), \tag{3}$$

where $Z(\beta)$ is the normalization,

$$\begin{aligned} Z(\beta) &:= \exp(-\beta C), \\ C &:= \min_{\boldsymbol{x}}\|\boldsymbol{x}\|_1 \text{ subject to } \boldsymbol{y}=\boldsymbol{F}\boldsymbol{x}. \end{aligned} \tag{4}$$

(Roman subscripts run from 1 to N, and Greek from 1 to M). This probability is zero unless the constraint $\boldsymbol{y} = \boldsymbol{F}\boldsymbol{x}$ is satisfied. In addition, it is unity only if ℓ_1-norm of $\boldsymbol{x}$ is minimum; otherwise zero. Then, it is found that the MAP solution is equivalent to the ℓ_1-norm minimum. Note that we do not need to take the limit of $\beta\to\infty$ for a MAP solution; however, we take it for clarifying the relation between the MAP algorithm and the reconstruction threshold analysis using a statistical mechanical replica method [13] that gives the exact expression of the threshold [9], where this posterior probability is defined as the Boltzmann weight, and zero temperature limit $\beta\to\infty$ is taken for technical reasons related to the analysis.

An obstacle to dealing with posterior probability is the singular delta function; we regularize this by the quadratic term

$$P(\boldsymbol{x}|\boldsymbol{y},\boldsymbol{F}) = \lim_{\beta\to\infty}\frac{1}{Z(\beta)}\exp\left(-\beta\frac{\sum_\mu(y_\mu-\sum_i F_{\mu i}x_i)^2}{2} - \beta k\sum_i |x_i|\right). \tag{5}$$

This posterior reproduces the original ℓ_1 minimization problem by taking the limit $\beta\to\infty$. (The readers can find a similar form of the probability density in [10] through the discussion of the AMP algorithm.) In this formulation, we need to introduce a constant $k(>0)$ that represents the "relative significance" of constraint and minimization. For ℓ_1-norm minimum solution, we must consider the limit $k\to 0$ because we are attempting to find the minimum solution *under* the constraint at present. This framework is essentially the same as that used in the lasso [14]; however, the crucial point here is that we need to take the limit $k\to 0$ for the MAP solution appropriately.

For the MAP solution, we differentiate the term in the exponential with respect to x_i, and write the stationary condition

$$\sum_\mu F_{\mu i}y_\mu - \sum_{j(\neq i)}\sum_\mu F_{\mu i}F_{\mu j}x_j - \sum_\mu F_{\mu i}^2 x_i - k\Theta(x_i) = 0 \tag{6}$$

(where $\Theta(x)$ is the Heaviside function), which can be rewritten as

$$x_i = \frac{1}{\sum_\mu F_{\mu i}^2} \eta\left(\sum_\mu F_{\mu i}\left(z_\mu + F_{\mu i}x_i\right); k\right),$$
$$z_\mu := y_\mu - \sum_i F_{\mu i}x_i. \tag{7}$$

Here, for convenience, we define the function for thresholding

$$\eta(x;k) := \begin{cases} x-k & k<x \\ 0 & -k \le x \le k \\ x+k & x<-k \end{cases}, \tag{8}$$

and also introduce the variable z_μ, which represents the residual error of the constraint $\boldsymbol{y} = \boldsymbol{F}\boldsymbol{x}$. Basically, we can obtain the ℓ_1-norm minimum solution by solving this stationary condition. We construct an iterative algorithm by adding the iteration step superscript (t):

$$x_i^{(t)} = \frac{1}{\sum_\mu F_{\mu i}^2} \eta\left(\sum_\mu F_{\mu i}\left(z_\mu^{(t)} + F_{\mu i}x_i^{(t-1)}\right); k\right),$$
$$z_\mu^{(t)} = y_\mu - \sum_i F_{\mu i}x_i^{(t-1)}. \tag{9}$$

The MAP solution is obtained by finding the fixed point of $x_i^{(t)}$. Remember that the original ℓ_1-norm minimization is a convex optimization, and accordingly there is no fixed point of local minimum.

We now give some remarks about this algorithm (9).

- Computational cost
  In algorithmic equations (9), only a single summation appears, and the numbers of the equations for $x_i^{(t)}$ and $z_\mu^{(t)}$ are N, M, respectively. Then, the computational cost is $O(N^2)$ when the compression rate M/N is $O(1)$. In addition, if the matrix $\boldsymbol{F}$ is sparse (having only $O(1)$ nonzero entries in each column/row, which is discussed in [15, 16, 17]), the computational cost is reduced to $O(N)$. Note that the number of iteration steps for convergence is assumed to be much smaller than the orders of M and N. However, in general, it does not hold near the region of the reconstruction threshold, where the speed of $k \to 0$ limit must be slow for convergence and the number of iteration steps for $k \to 0$ becomes dominant in comparison with the orders of M and N.
- Applicability to general $\boldsymbol{F}$
  We do not make any assumption for $\boldsymbol{F}$. Therefore, in principle, this algorithm can be used for general matrix $\boldsymbol{F}$, whereas in the original AMP [7] or in the statistical mechanical analysis [13] i.i.d. random entries are assumed.
- $k \to 0$ limit
  We must take the $k \to 0$ limit to the final iteration step. Empirically, k should be decreased exponentially with the step as $k^{(t)} \propto \exp(-t/\text{const.})$. The decay constant in the exponential is significant because too small a constant leads to a wrong solution. This constant is found to be the threshold parameter in the AMP algorithm equation as will be elucidated later, which is significant for convergence as stated in [7]. In the original work of the AMP [7] the authors discussed the appropriate limit of $k \to 0$ by choosing k as the mean squared error, and, in conjunction with the state evolution technique, they arrive at the

reconstruction threshold for convergence, which is represented by the relation between α and ρ with auxiliary variable z,

$$\rho = \alpha \max_{z\geq 0}\left(\frac{1-\frac{2}{\alpha}\left\{(1+z^2)H(z) - z\frac{e^{-\frac{z^2}{2}}}{\sqrt{2\pi}}\right\}}{1+z^2-2\left\{(1+z^2)H(z) - z\frac{e^{-\frac{z^2}{2}}}{\sqrt{2\pi}}\right\}}\right), \tag{10}$$

where $H(z) := \int_z^{+\infty} dx e^{-x^2/2}/\sqrt{2\pi}$. This expression is *analytically* equivalent to that of ℓ_1-norm minimization evaluated by combinatorial geometry [9, 18]. Replica method can also give the threshold equations [13] as

$$2(1-\rho)\left(H(z) - \frac{1}{z}\frac{e^{-\frac{z^2}{2}}}{\sqrt{2\pi}}\right) + \rho = 0,$$
$$\alpha = 2(1-\rho)H(z) + \rho, \tag{11}$$

which is also analytically equivalent. (Readers can check the equivalence between (10) and (11) after some algebra.)
As elucidated above, our initial step for the algorithm is the probability distribution in (5), which appears as Boltzmann weight in replica analysis for computation of the threshold. In the AMP, the convergence condition (11) is discussed using the thresholding algorithm *without* Onsager term (namely, from the naïve algorithm (9)), which also originates from the distribution (5). This relation explains why we arrive at the same threshold equation.

- Partial update
As is widely recognized, the naïve thresholding algorithm (9) still presents a problem. We cannot always obtain a correct ℓ_1-norm minimum solution using (9), even in successful reconstruction region from (10, 11) and even when the limit $k \to 0$ is suitably designed. Such a failure occurs near the reconstruction threshold of ℓ_1-norm minimization. (Readers will find the details below.) The AMP successfully solves this problem by the introduction of Onsager reaction term, which is reduced from perturbative analysis in the original work [7]. We attempt to address this problem from another point of view, which is applicable for general matrix $\boldsymbol{F}$.
The strategy is as follows. In the original algorithm, we fully update the variable $x_i^{(t)}$ in each step. Now, we introduce a partition ratio in the algorithm, and modify the algorithm to include the partial update rule as

$$x_i^{(t)} = \frac{\gamma^{(t)}}{1+\gamma^{(t)}}x_i^{(t)} + \frac{1}{1+\gamma^{(t)}}\frac{1}{\sum_\mu F_{\mu i}^2}\,\eta\left(\sum_\mu F_{\mu i}\left(z_\mu^{(t)} + F_{\mu i}x_i^{(t-1)}\right); k\right),$$
$$z_\mu^{(t)} = y_\mu - \sum_i F_{\mu i}x_i^{(t-1)}, \tag{12}$$

where the partition ratio $\gamma^{(t)}$ is dependent on the step in general. Obviously, the fixed point of the algorithm is the same as that of the original; nevertheless, the convergence to a correct solution is improved by choosing $\gamma^{(t)}$ properly.
Here $\gamma^{(t)}$ can be chosen arbitrarily. We can discuss how to design $\gamma^{(t)}$ to achieve better convergence by the stability analysis around the fixed point of the algorithm. However, at present we do not have the best prescription of how to choose $\gamma^{(t)}$. We omit the details of the discussion here due to limited space.

3. Discussion on the MAP algorithm

Let us move on to the properties of the proposed algorithm. Here, we use the matrix $\boldsymbol{F}$ with i.i.d random entries drawn from Gaussian distribution with zero mean and variance $1/M$, which is the same as in the AMP [7] and statistical mechanical analysis [13]. For large M, N, we have $\lim_{M\to\infty}\sum_\mu F_{\mu i}^2 = 1$ and the algorithm equation (9) is simplified to

$$\begin{aligned} x_i^{(t)} &= \eta\left(\sum_\mu F_{\mu i} z_\mu^{(t)} + x_i^{(t-1)}; k\right), \\ z_\mu^{(t)} &= y_\mu - \sum_i F_{\mu i} x_i^{(t-1)}, \end{aligned} \tag{13}$$

which can also be found in the introductory part of [7], where it is mentioned that the performance of this naïve thresholding algorithm is worse than that of ℓ_1-norm minimization. We introduce a partition ratio in the first line of (13) (here γ is not dependent on the step t) to improve it:

$$x_i^{(t)} = \frac{\gamma}{1+\gamma} x_i^{(t-1)} + \frac{1}{1+\gamma}\eta\left(\sum_\mu F_{\mu i} z_\mu^{(t)} + x_i^{(t-1)}; k\right). \tag{14}$$

We rewrite this equation as

$$x_i^{(t)} = \frac{\gamma}{1+\gamma} x_i^{(t-1)} + \eta\left(\frac{1}{1+\gamma}\left\{\sum_\mu F_{\mu i} z_\mu^{(t)} + x_i^{(t-1)}\right\}; k\right), \tag{15}$$

using the property of η. (We rescale the parameter k, which is irrelevant to the fixed point because k should be taken to zero, finally.) Instead of (15), we now introduce slightly modified update rule

$$x_i^{(t)} = \eta\left(\frac{1}{1+\gamma}\sum_\mu F_{\mu i} z_\mu^{(t)} + x_i^{(t-1)}; k\right) \tag{16}$$

by moving the first term on r.h.s. into the argument of η. This gives the same fixed point as (15), which can be verified by eliminating the step superscript (t) from both update rules (15, 16) and solving them with respect to x. With the rescaled residual error $\hat{z}_\mu^{(t)} = z_\mu^{(t)}/(1+\gamma)$, we obtain

$$\begin{aligned} x_i^{(t)} &= \eta\left(\sum_\mu F_{\mu i} \hat{z}_\mu^{(t)} + x_i^{(t-1)}; k\right), \\ \hat{z}_\mu^{(t)} &= \frac{1}{1+\gamma}\left\{y_\mu - \sum_i F_{\mu i} x_i^{(t-1)}\right\}. \end{aligned} \tag{17}$$

This expression of the algorithm indicates that the partition ratio is associated with the scaling of the residual error z_μ.

Next, we consider the step-dependent $\gamma^{(t)}$. In particular, we choose

$$\gamma^{(t)} = \frac{1}{M}\sum_i \eta'\left(\sum_\mu F_{\mu i} z_\mu^{(t)} + x_i^{(t-1)}; k\right), \tag{18}$$

where $\eta'(x;k) := \partial_x \eta(x;k)$. This $\gamma^{(t)}$ is chosen in order to achieve faster convergence according to the second order stability analysis around the solution, however the details are omitted. The point is that $\gamma^{(t)}$ is expressed by the function for thresholding, namely equation (8), which also appears in the algorithm. Note that $\gamma^{(t)}$ approaches the value $K/M = \rho/\alpha$ for $t \to \infty$ when the reconstruction is successful. For this step-dependent $\gamma^{(t)}$, we can also reach another expression of the algorithm in a similar manner as constant γ, by using that $\gamma^{(t)}$ is represented by the function (8),

$$\begin{aligned} x_i^{(t)} &= \eta\left(\sum_\mu F_{\mu i}\widehat{z}_\mu^{(t)} + x_i^{(t-1)}; k\right), \\ \widehat{z}_\mu^{(t)} &= \frac{1}{1+\frac{1}{M}\sum_j \eta'\left(\sum_\mu F_{\mu j} z_\mu^{(t)} + x_j^{(t-1)}; k\right)} z_\mu^{(t)}, \\ z_\mu^{(t)} &= y_\mu - \sum_i F_{\mu i} x_i^{(t-1)}. \end{aligned} \tag{19}$$

This expression indicates that the scaling of the residual error z_μ varies as a function of the step t.

4. Relation to the AMP

Based on the observation above, we choose the step-dependent partition ratio $\gamma^{(t)}$ with negative sign

$$\gamma^{(t)} = -\frac{1}{M}\sum_i \eta'\left(\sum_\mu F_{\mu i} z_\mu^{(t)} + x_i^{(t-1)}; k\right), \tag{20}$$

which leads to partial update rule by external division. In this case we have almost the same equations as in (19); the only difference is the negative sign in the denominator in the second equation. Here, we also introduce the partition ratio $\widehat{\gamma}^{(t)}$ for the update rule of the rescaled residual error $\widehat{z}_\mu^{(t)}$,

$$\widehat{z}_\mu^{(t)} = \widehat{\gamma}^{(t-1)}\widehat{z}_\mu^{(t-1)} + \frac{1-\widehat{\gamma}^{(t-1)}}{1-\frac{1}{M}\sum_j \eta'\left(\sum_\mu F_{\mu j} z_\mu^{(t)} + x_j^{(t-1)}; k\right)} z_\mu^{(t)}. \tag{21}$$

By choosing $\widehat{\gamma}^{(t)} = -\gamma^{(t)} = \sum_j \eta'(\sum_\mu F_{\mu j} z_\mu^{(t)} + x_j^{(t-1)})/M$, the algorithm is changed to

$$\begin{aligned} x_i^{(t)} &= \eta\left(\sum_\mu F_{\mu i}\widehat{z}_\mu^{(t)} + x_i^{(x-1)}; k\right), \\ \widehat{z}_\mu^{(t)} &= z_\mu^{(t)} + \frac{1}{M}\widehat{z}_\mu^{(t-1)}\sum_j \eta'\left(\sum_\mu F_{\mu j} z_\mu^{(t)} + x_j^{(t-1)}; k\right), \\ z_\mu &= y_\mu - \sum_i F_{\mu i} x_i. \end{aligned} \tag{22}$$

We arrive at this expression just by introducing the partition ratios and rewriting the update rule. When we replace $z_\mu^{(t)}$ with $\widehat{z}_\mu^{(t-1)}$ in the second equation above (approximation by $z \to \widehat{z}$, which does not change the fixed point because z and $\widehat{z}$ finally approach zero), and shift the step number of $x_i^{(t-1)}$ in η' to $t-2$ (which does not change the fixed point either), we finally obtain

$$\begin{aligned} x_i^{(t)} &= \eta\left(\sum_\mu F_{\mu i}\widehat{z}_\mu^{(t)} + x_i^{(t-1)}; k\right), \\ \widehat{z}_\mu^{(t)} &= y_\mu - \sum_i F_{\mu i} x_i^{(t-1)} + \frac{1}{M}\widehat{z}_\mu^{(t-1)} \sum_j \eta'\left(\sum_\mu F_{\mu j}\widehat{z}_\mu^{(t-1)} + x_j^{(t-2)}; k\right), \end{aligned} \tag{23}$$

which is nothing but the update rule of the variables in the AMP [7], where the parameter k controls the thresholding. Thus, we can find the relation between the AMP and the partial update MAP algorithm. In the AMP, the last term in the second equation in (23) is derived from the discussion of message passing, and is interpreted as the Onsager reaction term in statistical mechanics [7, 12]. Using the logic here, we can provide a viewpoint that this term is introduced by way of a partition ratio in order to achieve better convergence of the algorithm.

5. Numerical experiment

5.1. Reconstruction threshold

First, we evaluated the reconstruction threshold of our algorithm. In the experiment, the original data dimension was fixed as $N = 10^3$. The numbers of observations M and (averaged) nonzero data components K were varied. Each entry of $\boldsymbol{F}$ was drawn randomly and independently from Gaussian distribution with zero mean and variance $1/M$ (in Figures 1-3). Each nonzero element of $\boldsymbol{x}^0$ was similarly drawn from Gaussian distribution with zero mean and unit variance. The initial value of $x_i^{(t)}$ was set to be zero. The parameter k approached zero in each step exponentially by multiplying the factor 0.999 in each update, that is, a very slow approach to the limit of $k \to 0$ was used to guarantee convergence. We took the value $x_i^{(t)}$ as the reconstruction result after 5×10^3 steps in Figures 1 and 2, and 1×10^4 in Figures 3 and 4, respectively. For evaluating the reconstruction threshold, we conducted the experiment 50 times and computed the success rate for fixed M, K. In each trial, the reconstruction was judged to be a success if the mean squared error per data was less than 10^{-3}. Then, we computed the success rate for every pair of M, K by changing their values. (After the experiment for one pair of M, K, we increased/decreased M or K by 25.)

The results are shown in Figures 1-4. The success rate obtained in the experiment is expressed by gray-scale in each figure. In Figure 1, the result by the algorithm without partition ratio is shown, which exhibits narrower success region than that of ℓ_1-norm minimization. (This threshold was also evaluated in [7].) The result with the partition ratio $\gamma^{(t)} = 1$ is shown in Figure 2, where a reconstruction threshold close to ℓ_1-norm threshold curve is observed. Then, for improvement we used the step-dependent partition ratio in (18). With this modification, we can obtain much closer threshold to ℓ_1-norm curve as in Figure 3. In lower compression rate (=large α) region, the performance seems slightly worse than that of ℓ_1-norm minimization. However, we expect that larger number of iteration steps, appropriate choice of $k \to 0$ limit, and suitable partition ratio $\gamma^{(t)}$ will improve the performance. Next, for checking the applicability to general matrix $\boldsymbol{F}$, we study the case where the random matrix $\boldsymbol{F}$ is generated by a different rule: we first generate a dense random matrix $\boldsymbol{F}$ using the same rule as before, and then randomly eliminate 90% of the entries, which are set to be zero. In this experiment we use the algorithm (12) with the partition ratio $\gamma^{(t)} = 1$. The result in Figure 4 exhibits a success region similar to that of ℓ_1-norm, which can be understood from reconstruction threshold universality for wide class of random $\boldsymbol{F}$, as indicated in [19, 7] and theoretically discussed in [13, 20, 21].

From these results, we confirm that the MAP algorithm is essentially equivalent to ℓ_1-norm minimization, and through a suitable design of the algorithm, we can expect a performance that

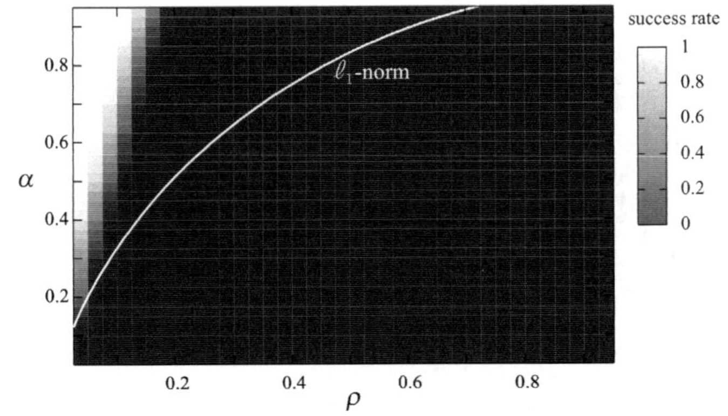

Figure 1. Profile of the success rate using the algorithm without partition ratio (9). Success rate is displayed by gray-scale. As seen, the success region is not as wide as with ℓ_1-norm minimization.

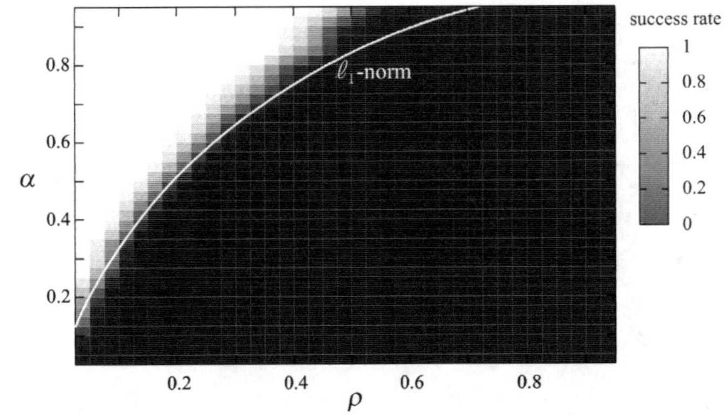

Figure 2. Profile of the success rate using the algorithm (12) with the partition ratio $\gamma^{(t)} = 1$. The area of the success region is almost the same as that of ℓ_1-norm minimization.

is almost the same as that of ℓ_1-norm. We stress that the algorithm proposed here will also be successful under the general matrix $\boldsymbol{F}$, as observed in Figure 4.

5.2. Convergence of the algorithm

Next, we examined the speed of convergence to the solution. We compared two algorithms: the MAP algorithm with step-dependent partition ratio (18), and the AMP (23). We set the parameters $N = 2000, M = 1000$ (compression ratio $\alpha = 0.5$) and $K = 200$ (fraction of nonzero entries $\rho = 0.1$). The rules for generating the matrix $\boldsymbol{F}$ and the original data $\boldsymbol{x}^0$ were the same as in Figures 1-3. For the parameter k, the exponential decay limit to zero was taken by multiplying 0.95 in each step for both algorithms. (Note that the original AMP uses the mean squared error for the update of k as elucidated before, which shows much better convergence. Here we adopt slow exponential update rule for both algorithms in the experiment for simpler experimental condition.) We conducted the reconstruction experiment 100 times and observed the behavior of the mean square error per data. The results are shown in Figure 5. It can be seen that the AMP exhibits a better performance than the MAP algorithm. From the relation between the MAP and the AMP algorithms, as discussed in the previous section, this difference

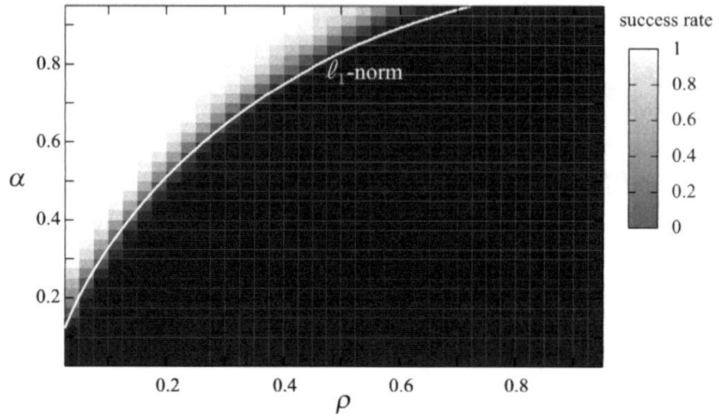

Figure 3. Profile of the success rate using the algorithm with step-dependent partition ratio $\gamma^{(t)}$ in (18). We can obtain a reconstruction threshold that is much closer to that of ℓ_1-norm minimization.

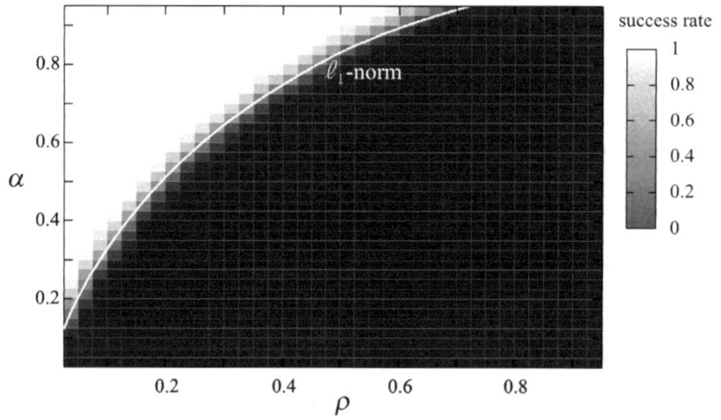

Figure 4. Profile of the success rate for a random matrix with a 10% nonzero matrix element. We use the algorithm with the constant partition ratio $\gamma^{(t)} = 1$. We can obtain a success region similar to that of ℓ_1-norm minimization also in this case.

in convergence speed might be understood from the introduction of the partition ratio to the residual error z_μ to improve the convergence in the AMP (as in (21)), whereas in the MAP algorithm in (12) this point is not taken into consideration. Therefore, it will be necessary to consider faster convergence of the residual error z_μ to improve the MAP algorithm.

6. Conclusion and perspective

In this article, we presented a methodology for constructing an algorithm using the MAP approach, discussed its relation with a known thresholding algorithm, and evaluated its performance through numerical experiments. We verified that, by designing the algorithm appropriately, almost the same reconstruction threshold as that of ℓ_1-norm minimization can be achieved. In the case of the i.i.d. random matrix $\boldsymbol{F}$, we clearly presented a viewpoint on the reason why the AMP has the same analytical expression of the reconstruction threshold as ℓ_1-norm minimization. The significant point is that we do not make any assumption for the matrix $\boldsymbol{F}$, and accordingly our algorithm's construction, whose computational cost is relatively low, is applicable to a general matrix.

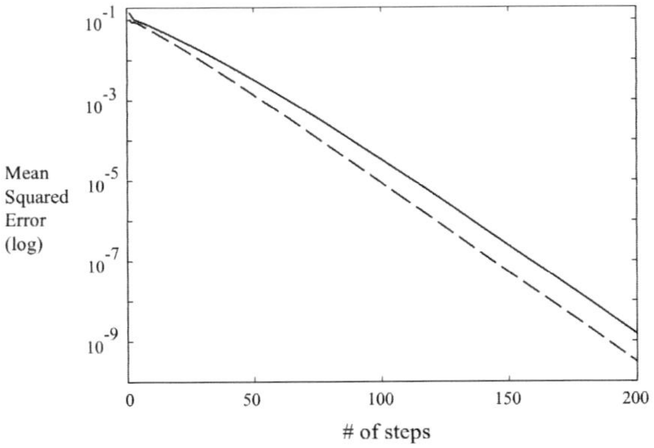

Figure 5. Convergence speed of the algorithm. We compare the convergence speed of the algorithms, the MAP with a step-dependent partition ratio (solid) and the AMP (broken). The behavior of the mean squared error per single data point is depicted. The AMP exhibits a better performance than the MAP in terms of convergence speed.

We also emphasized that it is still possible that faster convergence to the correct solution can be achieved by a more suitable design of the partition ratio. In the case of the random i.i.d. Gaussian matrix, the optimality of the algorithm is discussed in [12]. For a general matrix such as sparse or structured matrix, for which feasible algorithm is proposed and discussed in [15, 16, 17], future work will address a systematic method for designing an algorithm that will achieve faster convergence using the discussion presented in this paper.

Acknowledgments

This work is supported by KAKENHI Nos. 24700007 (KT), 22300003 and 25120013 (YK). YK also acknowledges the ELC project (Grant-in-Aid for Scientific Research on Innovative Areas MEXT Japan) for encouraging the research presented in this paper.

References

[1] Donoho D L 2006 *IEEE Trans. Inform. Theory* **52** 1289
[2] Candès E J and Tao T 2005 *IEEE Trans. Inform. Theory* **51** 4203
[3] Candès E J, Romberg J and Tao T 2006 *IEEE Trans. Inform. Theory* **52** 489
[4] Candès E J and Wakin M B 2008 *IEEE Signal Process. Mag.* **25** 21
[5] Elad M 2010 *Sparse and redundant representations* (New York: Springer)
[6] Tropp J A and Wright S J 2010 *Proc. of IEEE* **98** 948
[7] Donoho D L, Maleki A and Montanari A 2009 *Proc. Natl. Acad. Sci.* **106** 18914
[8] Pearl J 1988 *Probabilistic reasoning in intelligent systems: networks of plausible inference* (San Francisco: Morgan Kaufmann)
[9] Donoho D L and Tanner J 2005 *Proc. Natl. Acad. Soc.* **102** 9452
[10] Donoho D L, Maleki A and Montanari A 2010 *Proc. of Information Theory Workshop, Cairo* (*Preprint* arXiv:0911.4219)
[11] Donoho D L, Maleki A and Montanari A 2010 *Proc. of Information Theory Workshop, Cairo* (*Preprint* arXiv:0911.4222)
[12] Bayati M and Montanari A 2011 *IEEE Trans. Inform. Theory* **57** 764
[13] Kabashima Y, Wadayama T and Tanaka T 2009 *J. Stat. Mech.* L09003
[14] Tibshirani R 1996 *J. Royal Statist. Soc.* B **58** 267
[15] Kabashima Y and Wadayama T 2011 *Preprint* arXiv:1102.3220
[16] Krzakala F, Mézard M, Sausset F, Sun Y and Zdeborová L 2012 *Phys. Rev.* X **2** 021005
[17] Krzakala F, Mézard M, Sausset F, Sun Y and Zdeborová L 2012 *J. Stat Mech.* P08009
[18] Donoho D L and Tanner J 2009 *J. Amer. Math. Soc.* **22** 1
[19] Donoho D L and Tanner J 2009 *Phil. Trans. R. Soc.* A **367** 4273

[20] Bayati M, Lelarge M and Montanari A 2012 *Proc. of IEEE International Symposium on Information Theory* 1643 (*Preprint* arXiv:1207.7321)

[21] Takeda K and Kabashima Y 2012 *Proc. of 46th Annual Conference on Information Sciences and Systems* (*Preprint* arXiv:1203.6246)

Simplifying generalized belief propagation on redundant region graphs

Chuang Wang and Hai-Jun Zhou

State Key Laboratory of Theoretical Physics, Institute of Theoretical Physics, Chinese Academy of Sciences, Zhong-Guan-Cun East Road 55, Beijing 100190, China

E-mail: chuangphy@gmail.com, zhouhj@itp.ac.cn

Abstract. The cluster variation method has been developed into a general theoretical framework for treating short-range correlations in many-body systems after it was first proposed by Kikuchi in 1951. On the numerical side, a message-passing approach called generalized belief propagation (GBP) was proposed by Yedidia, Freeman and Weiss about a decade ago as a way of computing the minimal value of the cluster variational free energy and the marginal distributions of clusters of variables. However the GBP equations are often redundant, and it is quite a non-trivial task to make the GBP iteration converges to a fixed point. These drawbacks hinder the application of the GBP approach to finite-dimensional frustrated and disordered systems.

In this work we report an alternative and simple derivation of the GBP equations starting from the partition function. Based on this derivation we propose a natural and systematic way of removing the redundance of the GBP equations. We apply the simplified generalized belief propagation (SGBP) equations to the two-dimensional and the three-dimensional ferromagnetic Ising model and Edwards-Anderson spin glass model. The numerical results confirm that the SGBP message-passing approach is able to achieve satisfactory performance on these model systems. We also suggest that a subset of the SGBP equations can be neglected in the numerical iteration process without affecting the final results.

1. Introduction

Short loops are abundant in finite-dimensional ferromagnetic spin models and spin glass models. They cause complicated and strong local correlations in the systems. The accumulation and propagation of these local correlations then lead to long-range correlations and the emergence of various collective behaviors. The cluster variation method (CVM) is a general theoretical framework for treating local correlations in many-body statistical systems. The original idea of the cluster variation method was conceived by Professor Ryoichi Kikuchi (1919-2003) in 1951 [1]. Since then CVM has been applied to many different types of systems and has been further developed and generalized [2, 3, 4, 5, 6]. The basic idea of CVM is to decompose the entropy of the whole system into the residual entropy contributions of various clusters of spin variables. After this decomposition, the true entropy of the system is approximated by the sum of residual entropy contributions from a properly chosen subset of spin clusters, see [2] or section 2.2 of [7] for a brief introduction.

The celebrated Bethe-Peierls tree approximation [8, 9, 10, 11] is an important limiting case of the cluster variation method. This tree approximation plays a central role in the mean-field theory of spin glasses [12, 13], and it is also underlying the widely used belief propagation (BP) algorithm in information science [14]. For various spin glass problems defined on finite-connectivity random graphs, due to the absence of short loops, the Bethe-Peierls approximation or its extended version (which considers the possibility of ergodicity breaking) can give asymptotically exact results in the thermodynamic limit (see [13] for a comprehensive review). But the Bethe-Peierls approximation is inadequate in treating strong local correlations and therefore it performs poorly on finite-dimensional spin glass systems.

Considering more local correlations beyond the level of Bethe-Peierls approximation is conceptually easy within the CVM framework, but for strongly disordered and frustrated spin glass systems this task is practically quite challenging. There are two major issues: (a) How to construct a suitable variational marginal probability distribution for each chosen cluster of spin variables? (b) How to efficiently minimize the total Kikuchi cluster variational free energy, a complicated function of a large set of parameters? Yedidia and co-authors proposed in [15] a particular way of constructing cluster marginal probability distribution functions, and then they suggested a generalized belief propagation (GBP) message-passing approach to minimize the resulting Kikuchi variational free energy. This GBP message-passing approach outperforms the BP message-passing approach considerably in terms of numerical precision, but its widespread applications on finite-dimensional spin glass systems are still hindered by two important drawbacks: first, the GBP equations are often redundant; and second, it is quite a non-trivial task to make the GBP iteration converges to a fixed point.

In this work we understand the cluster marginal probability distribution functions of [15] from the viewpoint of the equilibrium partition function, and give an alternative and simple derivation of the GBP equations. Based on this derivation we propose a natural and systematic way of removing the redundance of the GBP equations. The resulting simplified generalized belief propagation (SGBP) equations are much more convenient for numerical implementation compared with the original GBP equations. We also point out that, for a given system, a subset of the SGBP equations can be safely ignored in the numerical iteration process. As redundance is minimized, the iteration process based on SGBP is much more easier to converge to a fixed point. We demonstrate the good performance of the SGBP equations by applying these equations to the two-dimensional (2D) and the three-dimensional (3D) ferromagnetic Ising model and Edwards-Anderson spin glass model.

The next section defines the general model system. The GBP equations are derived in section 3, and are then simplified in section 4. Some numerical results obtained by the SGBP equations are discussed in section 5. We conclude this work in section 6.

2. The general model system

2.1. Energy

Consider a system with N vertices and M two- or many-body interactions among the vertices. The indices of the vertices are denoted as $i, j, k, \ldots \in [1, N]$ and those of the interaction terms as $a, b, c, \ldots \in [1, M]$. Each vertex i has a state x_i which, for notational simplicity, is assumed to be a discrete scalar variable. A microscopic configuration of the system is denoted as $\underline{x}$, with $\underline{x} \equiv \{x_1, \ldots, x_i, \ldots, x_N\}$. The energy function has the following general form

$$H(\underline{x}) = \sum_{i=1}^{N} E_i(x_i) + \sum_{a=1}^{M} E_a(\underline{x}_{\partial a}) \, . \tag{1}$$

In the above expression, $E_i(x_i)$ and $E_a(\underline{x}_{\partial a})$ are, respectively, the self-energy of vertex i and the energy of interaction a; ∂a denotes the set of vertices involved in interaction a, and $\underline{x}_{\partial a} \equiv \{x_i | i \in \partial a\}$ is a microscopic sub-configuration for this particular set of vertices.

To give a concrete example, let us mention the Edwards-Anderson (EA) spin glass system on a finite-dimensional regular lattice [16]. The vertices are then the lattice sites, each of them

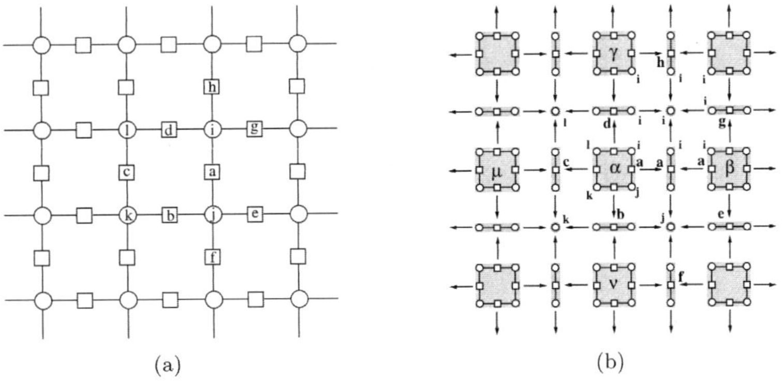

Figure 1. (a) Part of the factor graph for the Edwards-Anderson model on a square lattice with periodic boundary conditions. Each small circle (such as i) denotes a vertex (lattice site); each small square (such as a) denotes an interaction. An edge (i, a) between a circle i and a square a means that vertex i participates in interaction a. (b) Part of a redundant region graph for the factor graph of (a). Each square region (highlighted in cyan, such as α and β) contains four vertices and four interactions; each rod region (highlighted in green, such as a and b) contains two vertices and one interaction; and each vertex region (highlighted in red, such as i and j) contains only one vertex. A directed edge points from a parent region to a child region. Each vertex appears in 9 regions (see the example of vertex i); and each interaction appears in 3 regions (see the example of interaction a between vertex i and j). The counting number for a square region, a rod region and a vertex region is $c = 1$, $c = -1$ and $c = 1$, respectively.

having a binary spin state ($x_i \in \{-1, +1\}$). There is a spin coupling interaction between each pair of nearest neighboring lattice sites. The energy of a given spin configuration is

$$H(\underline{x}) = -\sum_{i=1}^{N} h_i^0 x_i - \sum_{(i,j)} J_{ij} x_i x_j \ , \tag{2}$$

where h_i^0 is the external field on vertex (lattice site) i, and (i, j) denotes a pair of nearest-neighboring vertices i and j, with J_{ij} being the coupling constant between them.

A conventional graphical representation for the general model system (1) is a bipartite graph of N small circles (representing the vertices) and M small squares (representing the interactions). Such a bipartite graph is referred to as a factor graph in the literature [17]. Each edge in the factor graph is between a small circle and a small square. If and only if a vertex i is involved in an interaction a, then there will be an edge connecting the corresponding small circle and small square in the factor graph. Figure 1(a) shows part of the factor graph for the EA model (2) on a periodic square lattice.

2.2. Partition function and free energy

The partition function of system (1) has the sum-product form

$$Z(\beta) \equiv \sum_{\underline{x}} e^{-\beta H(\underline{x})} = \sum_{\underline{x}} \prod_{i=1}^{N} \psi_i(x_i) \prod_{a=1}^{M} \psi_a(\underline{x}_{\partial a}) \ . \tag{3}$$

The inverse temperature $\beta \equiv \frac{1}{k_B T}$, with T being the temperature (we shall set Boltzmann's constant $k_B = 1$ in the following discussions). The functions $\psi_i(x_i)$ and $\psi_a(\underline{x}_{\partial a})$ are, respectively, the Boltzmann factor for the self-energy E_i and the interaction energy E_a,

$$\psi_i(x_i) \equiv e^{-\beta E_i(x_i)} , \qquad \psi_a(\underline{x}_{\partial a}) \equiv e^{-\beta E_a(\underline{x}_{\partial a})} . \tag{4}$$

The equilibrium free energy $F(\beta)$ is related with the partition function $Z(\beta)$ as

$$F(\beta) \equiv -\frac{1}{\beta} \ln Z(\beta) . \tag{5}$$

The free energy and the partition function have fundamental importance in equilibrium statistical mechanics. But exactly computing the partition function (and the free energy) is an impossible task generically. Many numerical schemes have been developed over the years to obtain good approximate values for $Z(\beta)$ [4]. A well-known theoretical framework for performing approximation is Kikuchi's cluster variation method [1, 2, 3, 5]. However, the variational problem of minimizing the Kikuchi cluster free energy is rather difficult to solve, especially for spin glass systems with quenched random parameters. Yedidia, Freeman, and Weiss [15] demonstrated that the minimal points of the Kikuchi variational free energy correspond to fixed points of a set of self-consistent generalized belief propagation (GBP) equations. Minimisation of the Kikuchi cluster free energy was therefore turned into the problem of constructing a fixed point for the GBP equations, which can be achieved through an iterative message-passing process on a region graph (see next subsection). Unfortunately, the GBP iterative process is still computationally demanding, especially when the underlying region graph are redundant.

2.3. Region graph representation

The basic motivation of the region graph concept is to facilitate treatments of local correlations through distributing vertices into different overlapping groups, with each group containing a subset of vertices that are believed to be strongly correlated [1, 2, 15].

A region graph R is a graph composed of regions and directed edges between pairs of regions [15]. The regions are generally denoted by Greek symbols. Each region γ contains a subset of the vertices and a subset of the self-energies and interaction energies. If there is a directed edge from a region μ to another region ν, denoted as $\mu \rightarrow \nu$, then ν must be contained in μ (namely all the vertices and all the energy contained in ν must also be contained in μ). When the edge $\mu \rightarrow \nu$ is present, we say that μ is a parent of ν and ν a child of μ. If there is a directed path from a region α to another region γ, we say that α is an ancestor of γ and γ a descendant of α, and denote this ancestor–descendant relationship by $\alpha > \gamma$ and $\gamma < \alpha$. The notation $\alpha \geq \gamma$ ($\alpha \leq \gamma$) is understood as either α and γ are identical to each other or α is an ancestor (descendant) of γ.

Figure 1(b) shows part of a region graph for the EA model of figure 1(a). There are three types of regions. Each square region contains four vertices and four interactions, and it is parent of four rod regions; each rod region contains two vertices and one interaction, and it is parent of two vertex regions; each vertex region contains a single vertex. In this particular example, for notational simplicity, each rod region is denoted by the index of the single interaction in it, and each vertex region is denoted by the index of the single vertex in it.

Each region γ is assigned a counting number c_γ [2], which is constructed recursively by

$$c_\gamma = 1 - \sum_{\{\alpha\,:\,\alpha > \gamma\}} c_\alpha . \tag{6}$$

If a region α has no directed edges pointing to it, its counting number is $c_\alpha = 1$. For a region μ with ancestors, it is obvious from the construction (6) that $c_\mu + \sum_{\gamma > \mu} c_\gamma \equiv 1$.

The vertices and energy terms of system (1) are clustered into various regions of R. This clustering, however, is not exclusive. A vertex and an energy term may be assigned to more

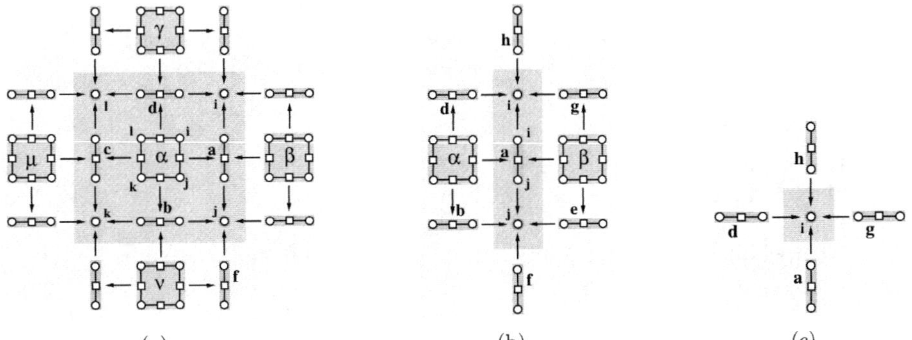

Figure 2. The interior and boundary for the square region α (a), the rod region a (b), and the vertex region i (c) of figure 1(b). The interior set I_γ of any region γ is marked by a pink-colored domain, while the boundary set B_γ of region γ is marked by a yellow-colored domain.

than one region. To ensure that each energy term contributes only one Boltzmann factor to the partition function (3), the region graph R is required to satisfy the following two constraints [15]: (1) For any vertex i, the induced subgraph R_i (i.e., the region subgraph formed by all the regions containing i and all the directed edges between pairs of these regions) is connected, and the sum of counting numbers within R_i is unity:

$$\sum_{\gamma \in R_i} c_\gamma = 1 \; ; \tag{7}$$

and (2) for any interaction a, the induced subgraph R_a (i.e., the region subgraph formed by all the regions containing a and all the directed edges between pairs of these regions) is connected, and the sum of counting numbers within R_a is unity:

$$\sum_{\gamma \in R_a} c_\gamma = 1 \; . \tag{8}$$

Because of (7) and (8), the partition function can be expressed as

$$Z(\beta) = \sum_{\underline{x}} \prod_{\alpha \in R} \left[\prod_{i \in \alpha} \psi_i(x_i) \prod_{a \in \alpha} \psi_a(\underline{x}_{\partial a}) \right]^{c_\alpha} \; . \tag{9}$$

In our earlier work [18, 7], the expression (9) was the starting point for performing partition function expansion on the region graph R.

For each region α of a given region graph R, let us denote by $I_\alpha \equiv \{\gamma : \gamma \leq \alpha\}$ the set formed by region α and all its descendants, by $A_\alpha \equiv \{\gamma : \gamma > \alpha\}$ the set formed by all the regions ancestral to region α, and by B_α the set that contains all the regions not belonging to set I_α but parental to at least one region of set I_α [7]. We refer to the set I_α as the interior of region α and set B_α as the boundary of region α. As some examples, we show in figure 2 the interiors and boundaries of the square region α, the rod region a and the vertex region i of the region graph of figure 1(b).

A region graph R is referred to as *non-redundant* if the region subgraph R_i induced by any vertex i is a tree (containing no loops), otherwise R is referred to as *redundant* [18, 7]. In a non-redundant region graph there is only one directed path from a region α to a descendant region γ, while in a redundant region graph there may exist multiple directed paths from an ancestral region α to a descendant region γ. The region graph shown in figure 1(b) is redundant, since there are two directed paths ($\alpha \to a \to i$ and $\alpha \to d \to i$) from the square region α to the vertex region i.

3. Generalized belief propagation equations

We give an alternative derivation of the generalized belief propagation (GBP) equations [15] in this section. Let us introduce on each directed edge $\mu \to \nu$ of the region graph R an arbitrary probability distribution function $m_{\mu\to\nu}(\underline{x}_\nu)$, with the only constraints that this function is positive and is properly normalized, $\sum_{\underline{x}_\nu} m_{\mu\to\nu}(\underline{x}_\nu) \equiv 1$. The function $m_{\mu\to\nu}(\underline{x}_\nu)$ is a probability measure on the microscopic states $\underline{x}_\nu \equiv \{x_i | i \in \nu\}$ of region ν. This measure is applied on ν by the parent region μ. The probability $m_{\mu\to\nu}(\underline{x}_\nu)$ can be regarded as a message from the parent region μ to the child region ν concerning the microscopic sub-configuration $\underline{x}_\nu$.

We observe that, for each directed edge $\mu \to \nu$,

$$\sum_{\{\alpha \,:\, \mu\in B_\alpha, \nu\in I_\alpha\}} c_\alpha = \sum_{\alpha\geq\nu} c_\alpha - \sum_{\alpha\geq\mu} c_\alpha = 1 - 1 = 0 \,. \tag{10}$$

This identity guarantees that the partition function (9) can be expressed as

$$Z(\beta) = \sum_{\underline{x}} \prod_{\alpha\in R} \Big[\prod_{i\in\alpha} \psi_i(x_i) \prod_{a\in\alpha} \psi_a(\underline{x}_{\partial a}) \prod_{\{\mu\to\nu \,:\, \mu\in B_\alpha, \nu\in I_\alpha\}} m_{\mu\to\nu}(\underline{x}_\nu) \Big]^{c_\alpha} \,. \tag{11}$$

For each region α let us define a Boltzmann factor z_α as

$$z_\alpha = \sum_{\underline{x}_\alpha} \prod_{i\in\alpha} \psi_i(x_i) \prod_{a\in\alpha} \psi_a(\underline{x}_{\partial a}) \prod_{\{\mu\to\nu \,:\, \mu\in B_\alpha, \nu\in I_\alpha\}} m_{\mu\to\nu}(\underline{x}_\nu) \,. \tag{12}$$

Then the partition function can be re-written as

$$Z(\beta) = Z_0 \times \sum_{\underline{x}} \prod_{\alpha\in R} \omega_\alpha(\underline{x}_\alpha)^{c_\alpha} \,, \tag{13}$$

where $Z_0 \equiv \prod_{\alpha\in R} z_\alpha^{c_\alpha}$, and the weight ω_α is defined as

$$\omega_\alpha(\underline{x}_\alpha) \equiv \frac{1}{z_\alpha} \prod_{i\in\alpha} \psi_i(x_i) \prod_{a\in\alpha} \psi_a(\underline{x}_{\partial a}) \prod_{\{\mu\to\nu \,:\, \mu\in B_\alpha, \nu\in I_\alpha\}} m_{\mu\to\nu}(\underline{x}_\nu) \,. \tag{14}$$

By exploiting the expression (13) we can write the free energy $F(\beta)$ as $F(\beta) = F_0 + \Delta F$, with

$$F_0 \equiv -\frac{1}{\beta} \ln Z_0 = -\frac{1}{\beta} \sum_{\alpha\in R} c_\alpha \ln \Big[\sum_{\underline{x}_\alpha} \prod_{i\in\alpha} \psi_i(x_i) \prod_{a\in\alpha} \psi_a(\underline{x}_{\partial a}) \prod_{\{\mu\to\nu \,:\, \mu\in B_\alpha, \nu\in I_\alpha\}} m_{\mu\to\nu}(\underline{x}_\nu) \Big] \,, \tag{15}$$

and correction contribution $\Delta F = -\frac{1}{\beta} \ln[\sum_{\underline{x}} \prod_{\alpha\in R} \omega_\alpha(\underline{x}_\alpha)^{c_\alpha}]$.

Let us assume that ΔF can be safely neglected in comparison with F_0. Then the free energy can be approximated as $F(\beta) \approx F_0$. A rigorous justification of this approximation is absent. In the case of non-redundant region graphs, it was shown in [18, 7] that ΔF is the sum of correction contributions from looped region sub-graphs. Another interesting scheme of computing ΔF was recently proposed by Welling and co-authors [19].

The expression (15) for F_0 still contains all the auxiliary probability distribution functions. Naturally we require the free energy F_0 to be stationary with respect to the chosen set of probability functions $\{m_{\mu\to\nu}\}$. In other words, the first derivative of F_0 with respect to any $m_{\mu\to\nu}$ should be zero,

$$\frac{\delta F_0}{\delta m_{\mu\to\nu}} = 0 , \qquad \forall(\mu \to \nu) \in G . \tag{16}$$

This condition is satisfied if we require the auxiliary functions to be chosen in such a way that, for each directed edge $\mu \to \nu$ of the region graph G,

$$\sum_{\underline{x}_\mu \backslash \underline{x}_\nu} \omega_\mu(\underline{x}_\mu) = \omega_\nu(\underline{x}_\nu) . \tag{17}$$

Equation (17) ensures the consistency of the two marginal probability distributions $\omega_\mu(\underline{x}_\mu)$ and $\omega_\nu(\underline{x}_\nu)$ of each parent–child pair (μ, ν). Equations (14) and (17) lead to the following generalized belief-propagation equation on each directed edge $\mu \to \nu$ of the region graph R:

$$\prod_{\{\alpha\to\gamma\,:\,\alpha\in B_\nu\cap I_\mu,\gamma\in I_\nu\}} m_{\alpha\to\gamma}(\underline{x}_\gamma) = C \sum_{\underline{x}_\mu\backslash\underline{x}_\nu} \prod_{j\in\mu\backslash\nu} \psi_j(x_j) \prod_{b\in\mu\backslash\nu} \psi_b(\underline{x}_{\partial b}) \prod_{\{\eta\to\tau\,:\,\eta\in B_\mu,\tau\in I_\mu\backslash I_\nu\}} m_{\eta\to\tau}(\underline{x}_\tau) , \tag{18}$$

where C is an adjustable constant to ensure that $m_{\mu\to\nu}(\underline{x}_\nu)$ is properly normalized. The set of GBP equations (18) was first derived in [15] through a different approach.

If the region graph R is non-redundant, then for each directed edge $\mu \to \nu$, $B_\nu \cap I_\mu = \{\mu\}$ and $\{\alpha \to \gamma : \alpha \in B_\nu \cap I_\mu, \gamma \in I_\nu\} = \{\mu \to \nu\}$. Then equation (18) is simplified as

$$m_{\mu\to\nu}(\underline{x}_\nu) = C \sum_{\underline{x}_\mu\backslash\underline{x}_\nu} \prod_{j\in\mu\backslash\nu} \psi_j(x_j) \prod_{b\in\mu\backslash\nu} \psi_b(\underline{x}_{\partial b}) \prod_{\{\eta\to\tau\,:\,\eta\in B_\mu,\tau\in I_\mu\backslash I_\nu\}} m_{\eta\to\tau}(\underline{x}_\tau) . \tag{19}$$

It can be shown that equation (19) is equivalent to the region graph belief propagation (rgBP) equation of [18, 7] obtained through partition function expansion.

If the region graph R is redundant, then there are multiple directed paths between some pairs of regions. As a consequence, the consistency conditions (17) on some of the directed edges must be redundant. For example, because there are two directed paths $\alpha \to a \to i$ and $\alpha \to d \to i$ from the square region α to the vertex region i in figure 2(b), then if the consistency conditions on the edges $\alpha \to a$, $\alpha \to d$ and $a \to i$ are all satisfied, the consistency condition on the edge $d \to i$ is satisfied automatically. For a redundant region graph, some of the consistency conditions (17) can therefore be dropped without affecting the final theoretical results. Consequently, some of the GBP equations (18) do not need to be considered.

4. Simplified GBP on a redundant region graph

In this paper we are interested in the case of the region graph R being redundant. For example, figure 1(b) shows a redundant region graph for the 2D Edwards-Anderson model on a square lattice with period boundary conditions. In this region graph, the region subgraph R_i induced by each vertex i is not a tree. It contains nine regions and twelve directed edges, and loops exist in R_i at the level of regions.

According to the definition (12), each boundary edge $(\mu \to \nu)$ of region α contributes a product term $m_{\mu\to\nu}(\underline{x}_\nu)$ to the Boltzmann factor z_α. As an example, consider the rod region a

of figure 2(b). Its Boltzmann factor z_a is

$$z_a = \sum_{x_i,x_j} \psi_i(x_i)\psi_j(x_j)\psi_a(x_i,x_j)m_{\alpha\to a}(x_i,x_j)m_{\beta\to a}(x_i,x_j)m_{h\to i}(x_i)m_{f\to j}(x_j) \times m_{d\to i}(x_i)m_{g\to i}(x_i)m_{b\to j}(x_j)m_{e\to j}(x_j) . \quad (20)$$

Notice that the square region α and its two child rod regions b and d all send a message to region a or its descendants, and these three messages are assumed to be mutually independent in the probability product form of equation (20). Similarly, the square region β and its two child rod regions e and g contribute a product term of three probability messages to z_a. However, because the rod regions b and d are children of region α, the probability message $m_{\alpha\to a}(x_i,x_j)$ already contains the effects of region b and d to region a. In other words, the probability distributions $m_{d\to i}(x_i)$ and $m_{b\to j}(x_j)$ very likely are strongly related to the probability distribution $m_{\alpha\to a}(x_i,x_j)$. Similarly, we expect the probability distributions $m_{g\to i}(x_i)$ and $m_{e\to j}(x_j)$ to be strongly related to the probability distribution $m_{\beta\to a}(x_i,x_j)$.

All such possibly strong dependence effects among the input probability distributions are completely neglected in (20) and in the more general expression (12). In this sense the redundance in the region graph structure causes redundance in the Boltzmann factor expressions (12), which in turn contributes at least part of the redundance in the set of GBP equations (18) and increases the complexity of numerical computations. The issue of removing the GBP redundance has been discussed by several recent studies [7, 6, 20, 21, 22].

We now propose a new way of removing redundance in the GBP equations. The basic idea is very simple: for a region ν in the boundary set B_α of region α, the input probability messages from ν to the set I_α should be removed as many as possible if an ancestor of ν also belongs to the boundary set B_α. Let us first discuss the ideal situation, namely the following equality is valid on each directed edge $\mu \to \nu$ of the region graph R:

$$\sum_{\{\alpha\,:\,\mu\in B_\alpha,\nu\in I_\alpha,A_\mu\cap B_\alpha=\emptyset\}} c_\alpha = 0 , \quad (21)$$

where A_μ is the ancestor set of region μ as defined in section 2.3. The summation on the left-hand side of the above expression is over all the regions α satisfying the following conditions: (1) the boundary set B_α contains region μ but not any of the ancestors of μ ($A_\mu \cap B_\alpha = \emptyset$); and (2) the interior set I_α contains region ν. As a simple example, let us mention that identity (21) is satisfied on each directed edge of the redundant region graph of figure 1(b). This particular region graph is therefore redundant and 'ideal'.

If (21) holds on each directed edge $\mu \to \nu$, then the partition function expression (11) can be simplified as

$$Z(\beta) = \sum_{\underline{x}} \prod_{\alpha\in R} \Big[\prod_{i\in\alpha} \psi_i(x_i) \prod_{a\in\alpha} \psi_a(\underline{x}_{\partial a}) \prod_{\{\mu\to\nu\,:\,\mu\in B_\alpha,\nu\in I_\alpha,A_\mu\cap B_\alpha=\emptyset\}} m_{\mu\to\nu}(\underline{x}_\nu) \Big]^{c_\alpha} . \quad (22)$$

We can then define a simplified Boltzmann factor for region α as

$$\tilde{z}_\alpha \equiv \sum_{\underline{x}_\alpha} \prod_{i\in\alpha} \psi_i(x_i) \prod_{a\in\alpha} \psi_a(\underline{x}_{\partial a}) \prod_{\{\mu\to\nu\,:\,\mu\in B_\alpha,\nu\in I_\alpha,A_\mu\cap B_\alpha=\emptyset\}} m_{\mu\to\nu}(\underline{x}_\nu) . \quad (23)$$

A simplified Boltzmann weight for region α is defined correspondingly:

$$\tilde{\omega}_\alpha(\underline{x}_\alpha) \equiv \frac{1}{\tilde{z}_\alpha} \prod_{i\in\alpha} \psi_i(x_i) \prod_{a\in\alpha} \psi_a(\underline{x}_{\partial a}) \prod_{\{\mu\to\nu\,:\,\mu\in B_\alpha,\nu\in I_\alpha,A_\mu\cap B_\alpha=\emptyset\}} m_{\mu\to\nu}(\underline{x}_\nu) . \quad (24)$$

For the ideal redundant region graph of figure 1(b), the Boltzmann factor $\tilde{z}_a$ of the rod region a is readily written down as (see figure 2(b))

$$\tilde{z}_a = \sum_{x_i, x_j} \psi_i(x_i)\psi_j(x_j)\psi_a(x_i, x_j) m_{\alpha\to a}(x_i, x_j) m_{\beta\to a}(x_i, x_j) m_{h\to i}(x_i) m_{f\to j}(x_j) \, , \tag{25}$$

which is much simpler than the expression (20).

The partition function is then expressed as

$$Z(\beta) = \tilde{Z}_0 \times \sum_{\underline{x}} \prod_{\alpha} \tilde{\omega}_\alpha(\underline{x}_\alpha)^{c_\alpha} \, , \tag{26}$$

where $\tilde{Z}_0 \equiv \prod_\alpha \tilde{z}_\alpha^{c_\alpha}$. A new approximate free energy $\tilde{F}_0$ is defined as

$$\tilde{F}_0 \equiv -\frac{1}{\beta} \ln \tilde{Z}_0 = -\frac{1}{\beta} \sum_{\alpha \in R} c_\alpha \ln \Big[\sum_{\underline{x}_\alpha} \prod_{i\in\alpha} \psi_i(x_i) \prod_{a\in\alpha} \psi_a(\underline{x}_{\partial a}) \prod_{\{\mu\to\nu \,:\, \mu\in B_\alpha, \nu\in I_\alpha, A_\mu\cap B_\alpha = \emptyset\}} m_{\mu\to\nu}(\underline{x}_\nu) \Big] \, . \tag{27}$$

Requiring $\tilde{F}_0$ to be stationary with respect to the set of auxiliary probability distributions $\{m_{\mu\to\nu}(\underline{x}_\nu)\}$, the following set of simplified consistency conditions are obtained:

$$\sum_{\underline{x}_\mu \backslash \underline{x}_\nu} \tilde{\omega}_\mu(\underline{x}_\mu) = \tilde{\omega}_\nu(\underline{x}_\nu) \, , \qquad \forall (\mu\to\nu) \in R \, . \tag{28}$$

This consistency condition leads to the following simplified generalized belief propagation (SGBP) equation on each directed edge $\mu \to \nu$:

$$\prod_{\{\alpha\to\gamma \,:\, \alpha\in B_\nu\cap I_\mu, \gamma\in I_\nu, A_\alpha\cap B_\nu=\emptyset\}} m_{\alpha\to\gamma}(\underline{x}_\gamma) = C \sum_{\underline{x}_\mu\backslash\underline{x}_\nu} \prod_{j\in\mu\backslash\nu} \psi_j(x_j) \prod_{b\in\mu\backslash\nu} \psi_b(\underline{x}_{\partial b}) \prod_{\{\eta\to\tau \,:\, \eta\in B_\mu, \tau\in I_\mu\backslash I_\nu, A_\alpha\cap B_\mu=\emptyset\}} m_{\eta\to\tau}(\underline{x}_\tau) \, , \tag{29}$$

where C is again an adjustable normalisation constant.

For a given spin glass model (1), it may be a relatively easy task to construct a redundant region graph that has the property (21) on each of its directed edges. The set of SGBP equations (29) can then be iterated on such an 'ideal' redundant region graph R, and the approximate free energy $\tilde{F}_0$ can be computed accordingly.

To be complete, let us also briefly discuss the 'non-ideal' case in which the identity (21) holds on some but not all of the directed edges. A simple solution would be to divide the directed edges into two classes (say C_I and C_N), one (C_I) with the property (21) and the other one (C_N) with only the weaker property (10). For an edge $(\mu \to \nu) \in C_I$, the probability message $m_{\mu\to\nu}(\underline{x}_\nu)$ is received by all the regions α in the set $\{\alpha : \mu \in B_\alpha, \nu \in I_\alpha, A_\mu \cap B_\alpha = \emptyset\}$. While for an edge $(\eta \to \tau) \in C_N$, we require the probability message $m_{\eta\to\tau}(\underline{x}_\tau)$ to be received by all the regions in a properly constructed non-empty subset of the set $\{\gamma : \eta \in B_\gamma, \tau \in I_\gamma\}$ (this constructed subset has the property that the sum of counting numbers of its regions is equal to zero). Following the theoretical approach of this section we can then obtain a modified set of SGBP equations for the non-ideal region graph R.

As there are multiple directed paths between some pairs of regions in the redundant region graph R, the simplified consistency conditions (28) on some of the directed edges must be redundant (see the discussion at the last paragraph of section 3). Therefore some of these simplified consistency conditions and the corresponding SGBP equations do not need to be considered in the actual numerical iteration process.

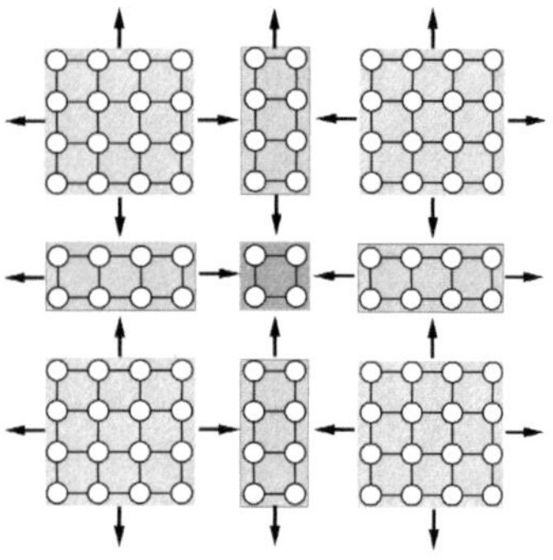

Figure 3. The local structure of another redundant region graph for the 2D model of figure 1(a). Each square region (cyan color) contains 4×4 vertices and is parental to four rod regions, its counting number is $c=1$; each rod region (green color) contains 2×4 vertices and is parental to two plaquette regions, its counting number is $c=-1$; each plaquette region (pink color) contains 2×2 vertices, its counting number is $c=1$. For simplicity we do not show the small squares of figure 1(a) here but use a small edge between two vertices to indicate a coupling interaction.

5. Applications to the Ising and the Edwards-Anderson model

We now apply the SGBP equations to the ferromagnetic Ising model and the Edwards-Anderson spin glass model (2) on a square lattice or a cubic lattice. Periodic boundary condition is assumed on each dimension of the lattice systems. For the Ising model, each coupling constant $J_{ij}\equiv J$, while for the EA model J_{ij} is assigned a value $+J$ or $-J$ independently and uniformly at random and then is fixed in time (we shall set $J=1$ in the following discussions).

For the ideal redundant region graph (later referred to as $R_{2D}^{n=2}$) of figure 1(b), the simplified Boltzmann weights of a square region α, a rod region a and a vertex region i can be easily written down as (see figure 2)

$$\tilde{\omega}_\alpha(x_i,x_j,x_k,x_l) \propto e^{\beta(h_i^0 x_i + h_j^0 x_j + h_k^0 x_k + h_l^0 x_l + J_{ij}x_i x_j + J_{jk}x_j x_k + J_{kl}x_k x_l + J_{li}x_l x_i)} \times m_{\beta\to a}(x_i,x_j) m_{\nu\to b}(x_j,x_k) m_{\mu\to c}(x_k,x_l) m_{\gamma\to d}(x_l,x_i)\,, \quad (30a)$$

$$\tilde{\omega}_a(x_i,x_j) \propto e^{\beta(h_i^0 x_i + h_j^0 x_j + J_{ij}x_i x_j)} m_{\alpha\to a}(x_i,x_j) m_{\beta\to a}(x_i,x_j) m_{h\to i}(x_i) m_{f\to j}(x_j)\,, \quad (30b)$$

$$\tilde{\omega}_i(x_i) \propto e^{\beta h_i^0 x_i} m_{a\to i}(x_i) m_{h\to i}(x_i) m_{d\to i}(x_i) m_{g\to i}(x_i)\,. \quad (30c)$$

The SGBP equations on the directed edges $\alpha\to a$ and $a\to i$ are, respectively

$$m_{\alpha\to a}(x_i,x_j) m_{h\to i}(x_i) m_{f\to j}(x_j) \propto \sum_{x_k,x_l} e^{\beta(h_k^0 x_k + h_l^0 x_l + J_{jk}x_j x_k + J_{kl}x_k x_l + J_{li}x_l x_i)} \times m_{\nu\to b}(x_j,x_k) m_{\mu\to c}(x_k,x_l) m_{\gamma\to d}(x_l,x_i)\,, \quad (31a)$$

$$m_{a\to i}(x_i) m_{d\to i}(x_i) m_{g\to i}(x_i) \propto \sum_{x_j} e^{\beta(h_j^0 x_j + J_{ij}x_i x_j)} m_{\alpha\to a}(x_i,x_j) m_{\beta\to a}(x_i,x_j) m_{f\to j}(x_j)\,. \quad (31b)$$

To include more local correlations in the SGBP equations, we also consider another ideal redundant region graph (referred to as $R_{2D}^{n=4}$, see figure 3) for the 2D system (2). This region graph has the same topology as $R_{2D}^{n=2}$. The only difference is that each vertex region now becomes a plaquette region of 2×2 vertices, and each square region now contains 4×4 vertices. The SGBP equations for this region graph are similar to equations (31a) and (31b).

For the cubic lattice systems we consider a simple 3D extension of the region graph of figure 1(b). This extended region graph (later referred to as $R_{3D}^{n=2}$) has four types of regions: cube regions, surface regions, rod regions, and vertex regions. Each $2\times 2\times 2$ cubic cell of the 3D lattice forms a cube region of the region graph, it is parental to six surface regions, and its counting number is $c=1$; each 2×2 plaquette of the lattice forms a surface region, it is parental to four rod regions and has counting number $c=-1$; each pair of nearest neighboring vertices of the lattice forms a rod region (it is parental to two vertex regions, and its counting number is $c=1$); and each vertex region contains only one single vertex, with counting number $c=-1$.

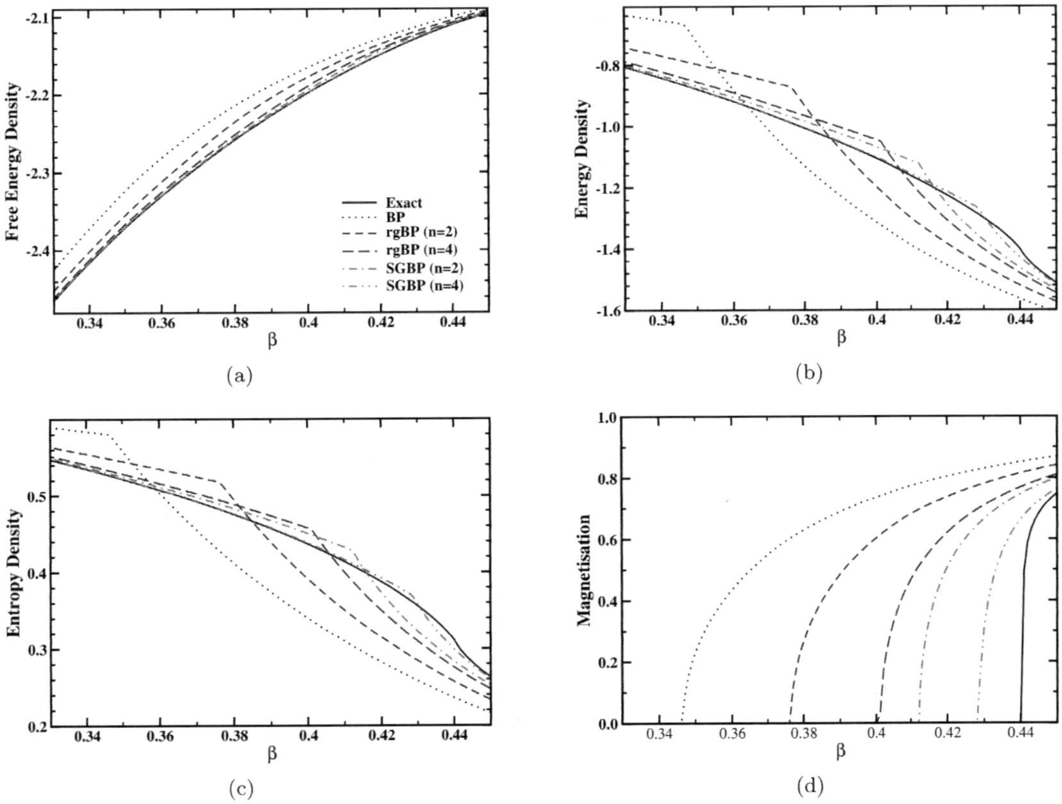

Figure 4. Results on the 2D ferromagnetic Ising model obtained by the SGBP equations, the rgBP equations of [7], and the conventional BP equation: (a) free energy density; (b) mean energy density; (c) entropy density; (d) magnetisation (mean spin value of a single vertex). The solid lines denote the exact results of [23]. The parameter n of the SGBP equations and the rgBP equations is the number of vertices on each side of the maximal square region of the region graph. For the SGBP curves, $n = 2$ and $n = 4$ correspond to the region graph of figure 1(b) and figure 3, respectively. The region graphs at $n = 2$ and $n = 4$ for the rgBP equations are described in [7].

The SGBP equations for such a 3D region graph are much more simplified in comparison with the original GBP equations.

We are mainly interested in the spontaneous emergence of collective behaviors, therefore we set the external field on each vertex to be zero ($h_i^0 = 0$ for all the vertices i).

5.1. The 2D Ising model

First we consider the simplest case, namely the ferromagnetic Ising model on an infinite square lattice with periodic boundary conditions. As the system has no disorder the SGBP equations can be easily solved at all temperatures. In figure 4 we compare the results obtained by the SGBP equations with the exact results of Onsager [23] and with the results obtained through the region graph belief propagation (rgBP) equations (see [7, 18] for details).

For the SGBP equations on a given region graph, we perform linear stability analysis to determine precisely the instability point of the paramagnetic solution, similar to that performed

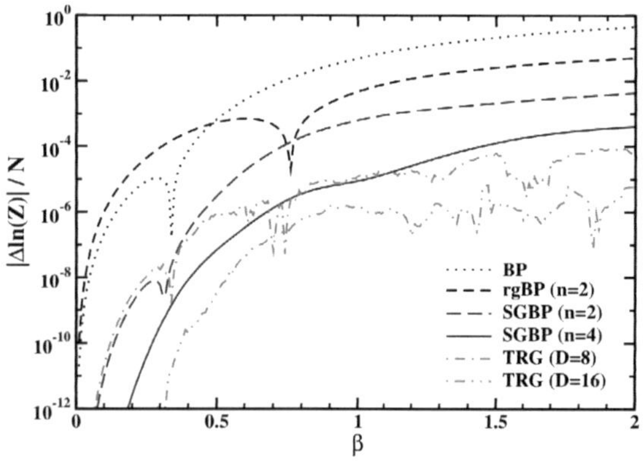

Figure 5. Difference $\Delta \ln Z$ between the exactly calculated value of $\ln Z$ and the value of $\ln Z$ obtained by BP, rgBP with $n = 2$, SGBP with $n = 2$ and $n = 4$, and tensor renormalisation group (TRG) with cutoff parameter $D = 8$ and $D = 16$. All the numerical calculations are performed on a single instance of the 2D EA model with side length $L = 64$ and periodic boundary conditions (the total number of vertices is $N = 64^2$).

in [7]. The predicted transition inverse temperatures by the SGBP equations using the region graph $R_{2D}^{n=2}$ and $R_{2D}^{n=4}$ are, respectively, $\beta_c \approx 0.4126$ and $\beta_c \approx 0.429$. These values are rather close to the exactly known critical inverse temperature value of $\beta_c^{(exact)} \approx 0.4407$ [24], they are much improved over the predicted value of $\beta_c \approx 0.3466$ by the conventional BP method and the predicted values of $\beta_c \approx 0.3765$ and $\beta_c \approx 0.401$ by the rgBP equations on two non-redundant region graphs [7].

The SGBP equations also give better predictions on the free energy density, the mean energy density, the entropy density, and the mean magnetisation of the system, see figure 4. Using the region graph $R_{2D}^{n=4}$, the predicted values of the SGBP equations on the free energy density, energy density, and entropy density become very close to the exactly known results. The prediction power of the SGBP equations will be further improved if we consider more local correlations by enlarging each region of figure 3 while keeping unchanged the topology of this region graph.

5.2. The 2D Edwards-Anderson model

Because of the existence of multiple directed paths between some pairs of regions in the redundant region graph $R_{2D}^{n=2}$ or $R_{2D}^{n=4}$, the SGBP equations on some of the directed edges are redundant and can be dropped from the numerical iteration process (see discussion in the last paragraph of section 4). If instead the SGBP equations on all the directed edges of the region graph are used in the iteration process, the paramagnetic solution of the SGBP equations will be unstable even at $\beta = 0$ ($T = \infty$).

In our numerical implementations for the 2D Edwards-Anderson model we still keep the SGBP equations on all the directed edges but add a damping effect to the SGBP equations to facilitate the convergence. We notice that the paramagnetic solution of the SGBP equations on the region graphs $R_{2D}^{n=2}$ and $R_{2D}^{n=4}$ fail to be stable at low temperatures even for optimally adjusted magnitude of the damping effect. However, the $\pm J$ EA model on 2D square lattice is believed to have no finite-temperature spin glass phase [25]. Therefore only the paramagnetic solution of the SGBP equations are qualitatively correct, even if it is unstable at low temperatures.

Since the free energy of a given instance of the EA model on a periodic square lattice with side length L can be computed exactly in polynomial time [26], we can compare the free energy $\tilde{F}_0$ of the paramagnetic solution of the SGBP equations with the exact value, see figure 5. The SGBP results on the region graph $R_{2D}^{n=4}$ are comparable to the results obtained by the tensor renormalisation group (TRG) method [27] with cutoff parameter $D = 8$ but are worse than the TRG results with cutoff parameter $D = 16$. We expect that the performance of the SGBP equations will be further improved by enlarging the side length n of the square regions. An advantage of the SGBP message passing approach is that it can simultaneously give the

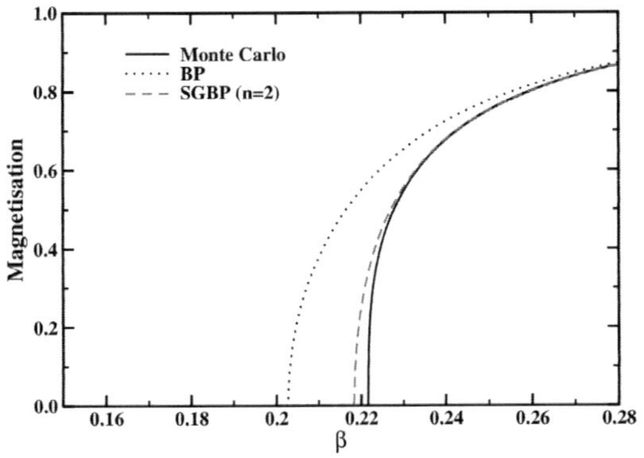

Figure 6. Mean magnetisation of the 3D Ising model as obtained by the SGBP equations (using region graph $R_{3D}^{n=2}$), the BP equations, and Monte Carlo simulations [28].

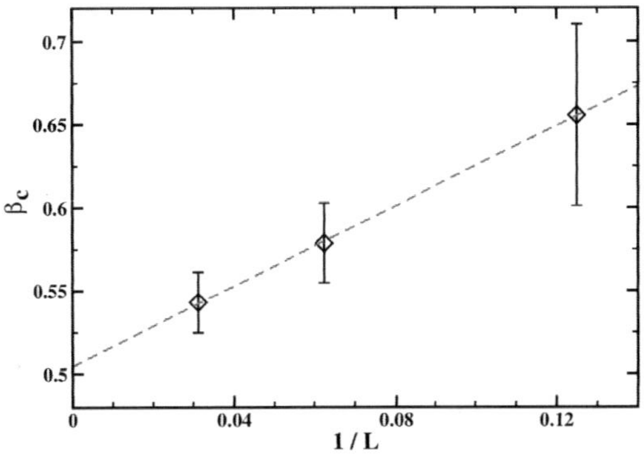

Figure 7. The inverse temperature β_c for the instability of the paramagnetic solution of the SGBP equations on the 3D EA model with lattice side length L. The dashed line is the fitting curve of $\beta_c = 0.505 + 1.20/L$.

marginal probability distributions $\tilde{\omega}_\mu(\underline{x}_\mu)$ for all the regions μ.

5.3. The 3D Ising model and Edwards-Anderson model

The structure of a simple redundant region graph $R_{3D}^{n=2}$ for the 3D Ising model and EA model is described at the beginning of section 5. This region graph is formed by cube regions, surface regions, rod regions and vertex regions. Each cube region contains $n \times n \times n$ vertices with $n = 2$. Linear stability analysis of the SGBP equations predicts that the ferromagnetic transition occurs at the critical inverse temperature $\beta_c \approx 0.2183$, which is very close to the value of $\beta_c \approx 0.2217$ obtained by Monte Carlo simulations [28] and the higher-order tensor renormalisation group (TRG) method [29]. As a comparison, the critical inverse temperature predicted by the BP approximation is $\beta_c \approx 0.2027$.

The magnetisation values predicted by the BP and SGBP equations are compared with the Monte Carlo simulation results of [28] in figure 6. It is evident that the performance of SGBP improves considerably over that of BP.

We also perform linear stability analysis on the paramagnetic solution of the SGBP equations for the 3D EA model (2). The stability threshold β_c of the SGBP paramagnetic solution on the region graph $R_{3D}^{n=2}$ depends on the particular disorder instance. For each cubic lattice of side length L under the periodic boundary condition, we generate 32 independent realisations of the coupling constants $\{J_{ij}\}$ and then obtain the value of β_c for each disorder realisation. The relationship between the mean value of β_c and lattice size L is shown in figure 7. The mean β_c value decreases with lattice side length L and appears to approach the value of $\beta_c \approx 0.505$ at $L \to \infty$. As a comparison, Monte Carlo simulations predicted that the spin glass phase transition occurs at $\beta_c \approx 0.893$ (see [30] and references therein). We believe that if we enlarge the side length n of the maximal cube region of the used region graph, the critical value β_c at $L = \infty$ will increase and be much more closer to the value predicted by Monte Carlo simulations.

6. Conclusion

In summary, we presented an alternative way of deriving a set of generalized belief propagation equations for a general lattice statistical physical system. Based on this derivation we proposed a way of simplifying the GBP equations. We also pointed out that, due to the existence of redundance in the region graph, some of the simplified generalized belief propagation (SGBP) equations can be ignored in the numerical iteration process. We applied the set of SGBP equations to the two-dimensional and three-dimensional Ising model and Edwards-Anderson model. The numerical results confirmed that the SGBP message-passing approach significantly outperforms the conventional BP approach in treating short-range correlations. Hopefully this work will stimulate further developments of the SGBP message-passing approach and the applications of this approach to finite-dimensional disordered systems and finite-dimensional optimisation problems.

The three-dimensional Edwards-Anderson model (2) is in the spin glass phase at low temperatures, with broken ergodicity. To describe the low-temperature spin glass property of the EA model, the effects of ergodicity-breaking need to explicitly considered in the SGBP equations. This issue remains to be solved.

Acknowledgments

This work was supported by the Knowledge Innovation Program of Chinese Academy of Sciences (No. KJCX2-EW-J02), and the National Science Foundation of China (grant No. 11121403 and 11225526).

References

[1] Kikuchi R 1951 *Phys. Rev.* **81** 988–1003
[2] An G 1988 *J. Stat. Phys.* **52** 727–34
[3] Morita T, Suzuki M, Wada K and Kaburagi M (eds) 1994 *Foundations and Applications of Cluster Variation Method and Path Probability Method* (*Prog. Theor. Phys. Suppl.* vol 115) (Kyoto: Yukawa Institute for Theoretical Physics and the Physical Society of Japan)
[4] Opper M and Saad D (eds) 2001 *Advanced Mean Field Methods: Theory and Practice* (Cambridge, MA: MIT Press)
[5] Pelizzola A 2005 *J. Phys. A: Meth. Gen.* **38** R309–39
[6] Rizzo T, Lage-Castellanos A, Mulet R and Ricci-Tersenghi F 2010 *J. Stat. Phys.* **139** 375–416
[7] Zhou H J and Wang C 2012 *J. Stat. Phys.* **148** 513–47
[8] Bethe H A 1935 *Proc. R. Soc. London* A **150** 552–75
[9] Peierls R 1936 *Proc. Camb. Phil. Soc.* **32** 477–81
[10] Peierls R 1936 *Proc. R. Soc. London* A **154** 207–22
[11] Chang T S 1937 *Proc. R. Soc. London* A **161** 546–63
[12] Mézard M, Parisi G and Virasoro M A 1987 *Spin Glass Theory and Beyond* (Singapore: World Scientific)
[13] Mézard M and Montanari A 2009 *Information, Physics, and Computation* (New York: Oxford Univ. Press)
[14] Pearl J 1988 *Probabilistic Reasoning in Intelligent Systems: Networks of Plausible Inference* (San Franciso, CA, USA: Morgan Kaufmann)
[15] Yedidia J S, Freeman W T and Weiss Y 2005 *IEEE Trans. Inf. Theory* **51** 2282–312
[16] Edwards S F and Anderson P W 1975 *J. Phys. F: Met. Phys.* **5** 965–74
[17] Frey B J 1998 *Graphical Models for Machine Learning and Digital Communication* (Cambridge, MA: MIT Press)
[18] Zhou H J, Wang C, Xiao J Q and Bi Z 2011 *J. Stat. Mech.: Theo. Exp.* L12001
[19] Welling M, Gelfand A and Ihler A T 2012 *Proc. of 28th Conf. on Uncertainty in Artificial Intelligence (Catalina Island, CA, USA)* (Corvallis, Oregon: AUAI Press) p 883
[20] Lage-Castellanos A, Mulet R, Ricci-Tersenghi F and Rizzo T 2011 *Phys. Rev.* E **84** 046706
[21] Domínguez E, Lage-Catellanos A, Mulet R, Ricci-Tersenghi F and Rizzo T 2011 *J. Stat. Mech.: Theor. Exp.* P12007
[22] Lage-Castellanos A, Mulet R, Ricci-Tersenghi F and Rizzo T 2013 *J. Phys. A: Math. Theor.* **46** 135001
[23] Onsager L 1944 *Phys. Rev.* **65** 117–49
[24] Kramers H A and Wannier G H 1941 *Phys. Rev.* **60** 252–62
[25] Thomas C K, Huse D A and Middleton A A 2011 *Phys. Rev. Lett.* **107** 047203
[26] Barahona F 1982 *J. Phys. A: Math. Gen.* **15** 3241–53
[27] Levin M and Nave C P 2007 *Phys. Rev. Lett.* **99** 120601
[28] Talapov A L and Blöte H W J 1996 *J. Phys. A: Math. Gen.* **29** 5727–33
[29] Xie Z Y, Chen J, Qin M P, Zhu J W, Yang L P and Xiang T 2012 *Phys. Rev.* B **86** 045139
[30] Katzgraber H G, Körner M and Young A P 2006 *Phys. Rev.* B **73** 224432

Belief propagation with multipoint correlations and its application to inverse problem

Masayuki Ohzeki

Department of Systems Science, Graduate School of Informatics, Kyoto University,

E-mail: mohzeki@i.kyoto-u.ac.jp

Abstract. We give explicit formulas of the Bethe approximation with multipoint correlations for systems with magnetic field. The obtained formulas include the closed form of the magnetization and the correlations between adjacent degrees of freedom. On the basis of our results, we propose a new iterative algorithm of the improved Bethe approximation. We confirm that the proposed technique is available for the random spin systems and indeed gives more accurate locations of the critical points. We discuss the possibility of the application of our method to the Inverse Ising model by use of these equations.

1. Introduction

Inverse Ising problem, which is originally known as Boltzmann machine learning, has received a lot of attention and been in the active research field. The problem is to infer both pairwise interactions and external fields of an Ising model given the vector of magnetizations and the matrix of pairwise correlations. In statistical mechanics, we often deal with the so called forward (direct) process, which consists of the computation of the free energy in order to evaluate several quantities as the magnetization and correlations. On the other hand, the inverse Ising problem is the backward process. This can be viewed as the dual counterpart with respect to the direct problem. The standard approach of the backward process consists of the iterative manipulation to solve the forward process while matching the obtained provisional quantities with the known data. Therefore we demand relatively fast techniques to compute various quantities in the forward process.

One of the qualitatively poor but very fast approach is the mean-field approximations, which approximately describes a physical system in terms of few parameters. A variant of the mean-field approximation, the Bethe approximation, which is originally derived for the ferromagnetic model on regular lattices [1, 2], has been extended, under the name of cavity method, to models with arbitrary couplings and topologies [3]. Although the Bethe approximation is exact only for tree-like topologies without loops, its application to models defined on random graphs has been very successful asymptotically in large number of components [4]. In order to mitigate the computational cost in the forward process, both of the approximations are often employed. The mean-field approximation in the inverse Ising problem [5, 6] is very fast but poor description without correlations between the adjacent components (i.e. Ising spins). By applying the linear response theorem, Kappen and Rodrìguez improve the mean-field approximation to give the two-point correlation [7], and its generalization [8] was performed in consistent with the Thouless-Anderson-Palmer approach [9] by use of Plefka expansion [10]. In the Bethe approximation, the two-point correlation as well as the expectation of the single component, which are matched with the given vectors, can be computed in conjunction with the linear response theorem [11, 12, 13, 14]. The Bethe approximation is usually implemented as some iterative algorithm known as the belief propagation [15], which suffers from lack of convergence in complicated system such as the spin glass model. Therefore it is also important to improve the precision of the mean-field approximation with skillful techniques [16, 17].

In recent years, many researchers have attempted to improve the Bethe approximation with correction of the correlation between adjacent degrees of freedom. One of the approaches is the loop calculus, which is the reduction of the partition function into a summation of contributions from subgraphs that are free of dangling edges [18, 19]. Recently it is generalized to the case with multiple solutions of the Bethe approximation [20]. In a different way, a method with correction of loops to the Bethe approximation is proposed by systematically considering correlations between adjacent degrees of freedom [21, 22]. This technique is simple to implement an iterative technique as in reference [23]. In the present study, we improve the method to the level with multipoint correlation. The multipoint correlation was dealt with in the previous study [23] by some approximative procedure, but the explicit form of the correction coming from them was absent. We give several formulas to be available to compute the various quantities with correction of multipoint correlations. This is an important step to investigate a nontrivial aspect of the discrepancies between the Bethe approximation and exact calculations of the thermodynamic quantities such as the partition function and free energy. As an application of our method to the Inverse Ising model, we also give closed equations to estimate the magnetization and pairwise correlations while taking into account the correction of the multipoint correlations. In addition, we propose an iterative algorithm such as belief propagation to compute the magnetization and correlation on the basis of the obtained formulas.

2. Standard description of Bethe approximation

Before going to the central part in this study, it is convenient for readers to find a short review on the Bethe approximation. Here let us take a very generic instance and explain the motivation to use the Bethe approximation. We deal with the random-field and random-bond Ising model with the following Hamiltonian

$$H = -\sum_{\langle ij \rangle} J_{ij} S_i S_j - \sum_i h_i S_i, \tag{1}$$

where S_i is the Ising spin taking ± 1. The summation denoted by $\langle ij \rangle$ is taken over all pairs of the adjacent spins. The interaction is denoted as J_{ij}, and the magnetic field is represented as h_i. In order to directly calculate the expectation of the local spin and the correlation between adjacent spins, we must obtain the following marginals

$$P(S_i, S_j) = \sum_{\{S_k\}/S_i,S_j} P(S_1, S_2, \cdots, S_N) \tag{2}$$

$$P(S_i) = \sum_{S_j} P(S_i, S_j), \tag{3}$$

where $P(S_1, S_2, \cdots, S_N)$ is the joint probability of the spin configuration, and N is the number of spins. The manipulation of the direct evaluation of the marginals (forward process) is intractable in general. Therefore we often need some approximation to mitigate its difficulty. The Bethe approximation is a very reasonable approximation in a moderate computational cost and usually employed to evaluate the marginals in the step of the forward process of the Inverse Ising problem.

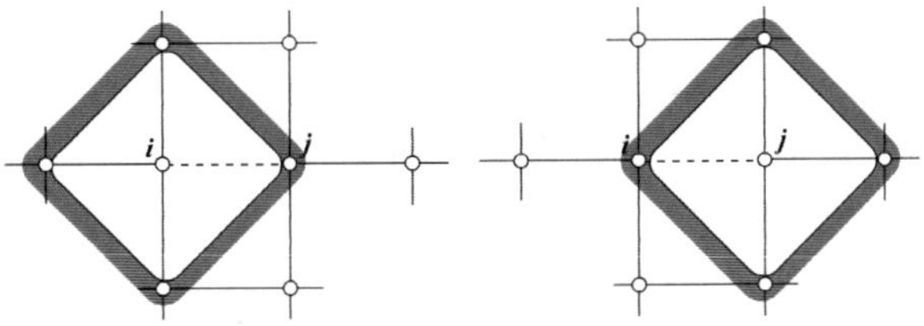

Figure 1. sets of the spins adjacent to a particular site i and j. The left (right) panel shows the region that ∂i (∂j) cover with. We consider the two type of marginals $P_{\partial i}(S_{\partial i})$ and $P_{\partial j}(S_{\partial j})$ in the absence of the interaction between two adjacent spins denoted by i and j.

There is a qualitatively poorer method than the Bethe approximation, the mean-field approximation. In the mean-field approximation, the joint probability is regarded as the product of the local distribution of the single spin as

$$P(S_1, S_2, \cdots, S_N) \approx \prod_{i=1}^{N} P_i(S_i). \tag{4}$$

The Bethe approximation is a simple improvement of the mean-field approximation by considering the distribution of the two adjacent spins as well as that of the single spins as

$$P(S_1, S_2, \cdots, S_N) \approx \prod_{\langle ij \rangle} \frac{P_{ij}(S_i, S_j)}{P_i(S_i)P_j(S_j)} \prod_{i=1}^{N} P_i(S_i). \tag{5}$$

The marginals of the single spin in the denominator exist in order to avoid the multiple count of its effect.

3. Systematic improvement of Bethe approximation

In order to further improve the Bethe approximation, we show a different derivation of the Bethe approximation. Let us introduce a marginal probability $P_i(S_{\partial i})$ for a set of the spins adjacent to a specific site i as depicted in figure 1. We find the exact expressions of the following marginal probabilities for two adjacent spins as

$$P_j^{(i)}(S_i, S_j) = \frac{1}{Z_i} \sum_{S_{\partial i}/j} P_i(S_{\partial i}) \exp\left(\beta H_i S_i + \beta \sum_{l \in \partial i/j} J_{il} S_i S_l \right), \tag{6}$$

and

$$P_i^{(j)}(S_i, S_j) = \frac{1}{Z_j} \sum_{S_{\partial j}/i} P_j(S_{\partial j}) \exp\left(\beta H_j S_j + \beta \sum_{k \in \partial j/i} J_{jk} S_j S_k \right). \tag{7}$$

The difference between two marginals stems from the choice of $P_i(S_{\partial i})$ and $P_j(S_{\partial j})$. We consider the expectation of the local spin by use of the above marginals as

$$A_j^{(i)} = \sum_{S_i, S_j} S_i P_j^{(i)}(S_i, S_j) \tag{8}$$

$$B_j^{(i)} = \sum_{S_i, S_j} S_j P_j^{(i)}(S_i, S_j). \tag{9}$$

We call the former quantity as the exact cavity magnetization and the latter one as the exact cavity field. We can relate two of the expectations by the following equation

$$B_j^{(i)} = A_i^{(j)}. \tag{10}$$

This relation holds exactly since any approximations are not employed yet. Once we obtain the explicit form of two expectations $B_j^{(i)}$ and $A_i^{(j)}$, we can calculate the desired expectations as the magnetizations and correlations.

To connect the above exact cavity magnetization and field to the physical quantities, we consider the partition function. It is convenient to use the partial partition function instead of the marginal $P_i(S_{\partial i})$ as

$$Z_i(S_{\partial i}) = \sum_{\{S_k\}/S_{\partial i},S_i} \exp(\beta \sum_{k\neq i} h_k S_k + \beta \sum_{\langle jk\rangle/\langle \partial i,i\rangle} J_{jk}S_jS_k). \tag{11}$$

This is related to the marginal probability $P_i(S_{\partial i})$ as

$$P_i(S_{\partial i}) = \frac{Z_i(S_{\partial i})}{\sum_{S_{\partial i}} Z_i(S_{\partial i})}. \tag{12}$$

Then the partition function can be written as

$$Z = \sum_{S_{\partial i},S_i} \exp(\beta h_i S_i + \beta \sum_{j\in\partial i} J_{ij}S_iS_j) Z_i(S_{\partial i}). \tag{13}$$

In addition, the joint probability of the nearest-neighboring spins is rewritten as

$$P(S_i,S_j) = \frac{1}{Z}\sum_{S_{\partial i/j}} \exp(\beta h_i S_i + \beta \sum_{j\in\partial i} J_{ij}S_iS_j) Z_i(S_{\partial i}) = \frac{\exp(\beta J_{ij}S_iS_j)P_j^{(i)}(S_i,S_j)}{\sum_{S_i,S_j}\exp(\beta J_{ij}S_iS_j)P_j^{(i)}(S_i,S_j)}. \tag{14}$$

Notice that these relations hold exactly.

In order to connect the obtained quantity as the cavity field in the Bethe approximation to the desired one, we evaluate the magnetization and disconnected correlation by use of the above exact expressions of the marginals. It is straightforward to obtain the following expression of the magnetization if one employs the same procedure as the above manipulation in the previous subsection.

$$m_i = \frac{A_j^{(i)} + t_{ij}B_j^{(i)}}{1 + t_{ij}D_{ij}^{(i)}}, \tag{15}$$

where

$$D_{ij}^{(i)} = \sum_{S_i,S_j} S_iS_jP_j^{(i)}(S_i,S_j). \tag{16}$$

In addition, the disconnected correlation can be evaluated as

$$\tilde{c}_{ij} = \frac{D_{ij}^{(i)} + t_{ij}}{1 + t_{ij}D_{ij}^{(i)}}. \tag{17}$$

Equations (15) and (17) are the exact formulas since any approximations have not yet performed.

In the Inverse Ising problem, the performance of the applied method relies on the precision of the estimation of the magnetization and correlations. Therefore we have to improve the precision beyond that of the Bethe approximation. According to the idea proposed by Montanari and Rizzo [21], let us take the effect of multipoint correlation between adjacent spins to the Bethe approximation.

3.1. Bethe approximation

In order to evaluate $B_j^{(i)}$ and $A_i^{(j)}$, we employ some approximation. One of the simplest way corresponds to the Bethe approximation.

The direct manipulations of three types of the expectations $A_j^{(i)}$, $B_j^{(i)}$, and $D_{ij}^{(i)}$ are

$$A_j^{(i)} = \frac{t_i \sum_{\mathcal{A}:\text{even}} t_{\mathcal{A}}^{(i)} \tilde{C}_{\mathcal{A}} + \sum_{\mathcal{A}:\text{odd}} t_{\mathcal{A}}^{(i)} \tilde{C}_{\mathcal{A}}}{\sum_{\mathcal{A}:\text{even}} t_{\mathcal{A}}^{(i)} \tilde{C}_{\mathcal{A}} + t_i \sum_{\mathcal{A}:\text{odd}} t_{\mathcal{A}}^{(i)} \tilde{C}_{\mathcal{A}}} \tag{18}$$

$$B_j^{(i)} = \frac{\sum_{\mathcal{A}:\text{even}} t_{\mathcal{A}}^{(i)} \tilde{C}_{\mathcal{A}\cup j} + t_i \sum_{\mathcal{A}:\text{odd}} t_{\mathcal{A}}^{(i)} \tilde{C}_{\mathcal{A}\cup j}}{\sum_{\mathcal{A}:\text{even}} t_{\mathcal{A}}^{(i)} \tilde{C}_{\mathcal{A}} + t_i \sum_{\mathcal{A}:\text{odd}} t_{\mathcal{A}}^{(i)} \tilde{C}_{\mathcal{A}}} \tag{19}$$

$$D_{ij}^{(i)} = \frac{\sum_{\mathcal{A}:\text{odd}} t_{\mathcal{A}}^{(i)} \tilde{C}_{\mathcal{A}\cup j} + t_i \sum_{\mathcal{A}:\text{even}} t_{\mathcal{A}}^{(i)} \tilde{C}_{\mathcal{A}\cup j}}{\sum_{\mathcal{A}:\text{even}} t_{\mathcal{A}}^{(i)} \tilde{C}_{\mathcal{A}} + t_i \sum_{\mathcal{A}:\text{odd}} t_{\mathcal{A}}^{(i)} \tilde{C}_{\mathcal{A}}}, \tag{20}$$

where $\mathcal{A} = \partial i/j$. Here we define the (disconnected) correlation function for a subset $\mathcal{A}$ as

$$\tilde{C}_{\mathcal{A}}^{(i)} \equiv \sum_{S_{\partial i}} P_i(S_{\partial i}) \prod_{j\in\mathcal{A}} S_j. \tag{21}$$

Notice that the connected correlation $C_{\mathcal{A}}^{(i)}$ is related to the above disconnected one as

$$\tilde{C}_{\mathcal{A}}^{(i)} = \sum_{[\mathcal{A}_1,\cdots,\mathcal{A}_n]} C_{\mathcal{A}_1}^{(i)} \cdots C_{\mathcal{A}_n}^{(i)}. \tag{22}$$

Here the summation is taken over all possible combinations of the subset $\mathcal{A}$. We use abbreviations as $t_i = \tanh(\beta h_i)$ and $t_{\mathcal{A}}^{(i)} = \prod_{l\in\mathcal{A}} t_{il}$, where $t_{il} = \tanh(\beta J_{il})$. In a strict sense, the above correlation should be called the cavity correlation (i. e. the correlation without the specified spin S_i). In the present study, we distinguish the cavity correlation from the standard correlation for clarity.

In the Bethe approximation, we restrict ourselves to the case that two adjacent spins are correlated. In other words, the adjacent spins to S_i are independent. Thus $P_i(S_{\partial i}) \approx \prod_{j\in\partial i} P_j^{(i)}(S_j)$. In this case, the disconnected correlations $\tilde{C}_{\mathcal{A}}^{(i)}$ can be factorized as $\prod_{j\in\mathcal{A}} M_j^{(i)}$. Here we define the cavity field, which is the expectation of S_j in the absence of the neighboring spin S_i, as

$$M_j^{(i)} = \sum_{S_j} S_j P_j^{(i)}(S_j). \tag{23}$$

Then we find the approximate expressions of the expectations as

$$A_j^{(i)} = \frac{t_i \sum_{\mathcal{A}:\text{even}} \prod_{j\in\mathcal{A}} t_{ij} M_j^{(i)} + \sum_{\mathcal{A}:\text{odd}} \prod_{j\in\mathcal{A}} t_{ij} M_j^{(i)}}{\sum_{\mathcal{A}:\text{even}} \prod_{j\in\mathcal{A}} t_{ij} M_j^{(i)} + t_i \sum_{\mathcal{A}:\text{odd}} \prod_{j\in\mathcal{A}} t_{ij} M_j^{(i)}} \tag{24}$$

$$B_j^{(i)} = M_j^{(i)} \tag{25}$$

$$D_{ij}^{(i)} = M_j^{(i)} M_i^{(j)}. \tag{26}$$

Therefore we find the self-consistent equation of the cavity field in the Bethe approximation from equation (10) as

$$M_j^{(i)} = T_j^{(i)}(h_i). \tag{27}$$

For simplicity, we define the following quantity

$$T^{(i)}_{j_1,j_2,\cdots,j_k}(h_i) = \tanh\left\{\beta h_i + \sum_{l\in\partial i/j_1,j_2,\cdots,j_k} \tanh^{-1}\left(t_{il}M^{(i)}_l\right)\right\}. \tag{28}$$

In the following, we omit the expression of dependence on the magnetic field unless it appears. Consequently we obtain the well-known results for the magnetization and correlation at the level of the Bethe approximation as

$$m_i = T^{(i)}(h_i) \tag{29}$$

$$\tilde{c}_{ij} = \tanh\left(\beta J_{ij} + \tanh^{-1}(M^{(i)}_j M^{(j)}_i)\right). \tag{30}$$

3.2. Improvement of Bethe approximation

Let us consider the improvement of the Bethe approximation by dealing with the correlation between adjacent spins. The disconnected cavity correlation is reduced to the product of the cavity field in the Bethe approximation. We take the effect of the multipoint correlation by considering the disconnected cavity correlation. For instance, the 3-body disconnected cavity correlation can be written as

$$\tilde{C}^{(i)}_{jkl} = M^{(i)}_j M^{(i)}_k M^{(i)}_l + C^{(i)}_{jk} M^{(i)}_l + C^{(i)}_{kl} M^{(i)}_j + C^{(i)}_{lj} M^{(i)}_k + C^{(i)}_{jkl}. \tag{31}$$

These multipoint cavity correlation can be estimated by the linear response theorem as

$$C^{(i)}_{j,j_1,\cdots,j_k} = \frac{1}{\beta^k}\frac{\partial^k M^{(i)}_j}{\partial h_{j_1}\cdots\partial h_{j_k}}. \tag{32}$$

Indeed, the two-point cavity correlation follows

$$C^{(i)}_{j,k} = \left\{1-(M^{(i)}_j)^2\right\}\left(\delta_{j,k} + \sum_{k\in\partial j/i}\frac{t_{ik}C^{(j)}_{ik}}{1-t^2_{ik}(M^{(j)}_i)^2}\right). \tag{33}$$

This is a kind of the self-consistent equation over the two-point cavity correlations. Notice that the self-consistent equation consists only of the cavity field and two-point cavity correlations. The three-point cavity correlations follows a similar equation only with the cavity fields, two-point cavity correlations and themselves. Once we compute k-point cavity correlations, we also estimate the $k+1$-point cavity correlations by successive use of the linear response theorem.

By use of the multipoint cavity correlations, let us consider the correction of the Bethe approximation. We assume that the multipoint cavity correlation should be small in the following calculation. We expand several quantities up to the first order of the multipoint correlation in order to evaluate more precise values of the magnetic field and correlation. The detailed calculation is summarized in Appendix. The resultant self-consistent equation for the cavity field is

$$\begin{aligned} M^{(i)}_j &= T^{(j)}_i(h_j) + \sum_{k=2}\sum_{l_k\in\partial j/i} C^{(j)}_{l_1,\cdots,l_k}\Gamma^{(j)}_{i,l_1,\cdots,l_k}\frac{1-t^2_j}{(1+t_jT^{(j)}_i)^2} \\ &\quad - \sum_{k=1}\sum_{j_k\in\partial i/j} C^{(i)}_{j,j_1,\cdots,j_k}\Omega^{(i)}_{j,j_1,\cdots,j_k}, \end{aligned} \tag{34}$$

where $\Gamma^{(j)}_{i,l_1,\cdots,l_k}$ and $\Omega^{(i)}_{j,j_1,\cdots,j_k}$ are polynomials consisting of the cavity fields. This gives a solution $B^{(i)}_j$ (i.e. $A^{(i)}_j$) to the exact relation (10) up to the first order of the multipoint cavity correlation function. In reference [21], the solution for the case of the homogenous Ising model ($J_{ij} = 1$) without magnetic field up to the first-order of the two-point cavity correlation. The cavity correlation then was exactly calculated by the Fourier transformation since the system holds translational invariance. For the two-dimensional case, the infrared divergence avoids the straightforward manipulation to give a explicit solution.

Instead of the analytical approach, in the present study, we propose an iterative algorithm, such as the belief propagation, to find the solution of the improved Bethe approximation. The algorithmic procedure is as follows. First, we compute the cavity field by the standard belief propagation. Then we regard the self-consistent equation (27) as the update recursion for the cavity field. Second, we recursively estimate the k-point cavity correlations (i. e. susceptibility propagation) through the linear response theorem (32) by use of the obtained cavity fields and the resultant $k-1$-point cavity correlations. For instance, we use equation (33) for the case on the $k = 2$-point cavity correlations. Notice that we keep the cavity fields as the obtained ones in the belief propagation. Finally, by use of equation (34), we calculate the improved cavity field. We then set the cavity correlations as the obtained ones in the susceptibility propagation. The improved cavity fields are available for estimation of the magnetization and correlations as

$$\begin{aligned} m_i = T^{(i)}(h_i) &+ \sum_{k=2}\sum_{l_k\in\partial j/i} C^{(j)}_{l_1,\cdots,l_k}\Gamma^{(j)}_{i,l_1,\cdots,l_k}\frac{\left\{1-t_{ij}^2(M_j^{(i)})^2\right\}(1-t_j^2)}{(1+t_jT_i^{(j)})^2(1+t_{ij}M_j^{(i)}T_i^{(j)}(h_i))^2} \\ &-\sum_{k=1}\sum_{j_k\in\partial i/j} C^{(i)}_{j,j_1,\cdots,j_k}\Omega^{(i)}_{j,j_1,\cdots,j_k}\frac{t_{ij}(1-(T_j^{(i)})^2)(1-t_it_{ij}M_j^{(i)})\tanh^{\delta_k}(\beta h_i-\tanh^{-1}t_{ij}M_j^{(i)})}{(1+t_iT_j^{(i)})(1+t_{ij}M_j^{(i)}T_i^{(j)}(h_i))^2} \end{aligned} \tag{35}$$

$$\begin{aligned} \tilde{c}^{(i)}_{ij} &= \tanh\left(\beta J_{ij}+\tanh^{-1}(M_j^{(i)}T_j^{(i)})\right) \\ &+\sum_{k=2}\sum_{j_k\in\partial i/j} C^{(i)}_{j_1,\cdots,j_k}\Gamma^{(i)}_{j,j_1,\cdots,j_k}\frac{(1-t_{ij}^2)(1-t_i^2)M_j^{(i)}}{(1+t_iT_j^{(i)})^2(1+t_{ij}M_j^{(i)}T_i^{(j)}(h_i))^2} \\ &+\sum_{k=1}\sum_{j_k\in\partial i/j} C^{(i)}_{j,j_1,\cdots,j_k}\frac{(1-t_{ij}^2)(1-t_i^2)(T_j^{(i)})^{1-2\delta_k}\Omega^{(i)}_{j,j_1,\cdots,j_k}}{(1+t_iT_j^{(i)})^2(1+t_{ij}M_j^{(i)}T_i^{(j)}(h_i))^2}, \end{aligned} \tag{36}$$

where $\delta_k = 2$ for even k and $\delta_k = 1$ for odd k. Equivalently, one may directly insert these results into equation (15) and (17).

We test our algorithm for the homogeneous Ising model without magnetic field on two and three-dimensional lattice (i.e square and cubic ones). We first generate $L = 24$ lattice and assign the cavity fields and cavity correlations randomly. Following the our algorithm, we update the cavity fields, and then cavity correlations. We then obtain the results at the level of the Belief propagation. Furthermore we compute the corrected estimations following equation (34). In figure 2, we show an interesting result of the magnetization for the case of the two-dimensional lattice. Our iterative method does not suffer from any divergent effect, which avoids the analysis in reference [21], to estimate the magnetization for the case of the two-dimensional lattice. We have not found dependence of the estimations on the linear size of the system. The obtained magnetization reveals a closer estimation of the critical point to the exact solution than those of the Bethe approximation. We also reproduce the result $\beta_c \approx 0.238$ obtained from equation (35) in an analytical way [21] for the case on the three-dimensional lattice. The absence of the divergent behavior, which was found in the previous study, stems from the fact that our

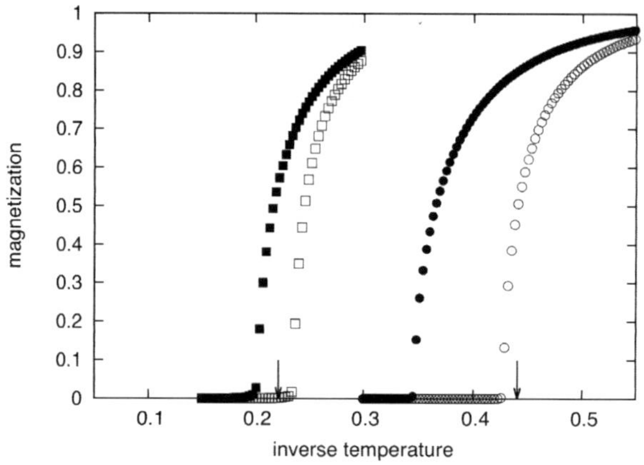

Figure 2. Plots of the magnetization for the case of the homogeneous Ising model without magnetic field on the two (white circles) and three-dimensional lattices (white squares). For comparison, we also describe the results by the ordinary belief propagation (black circles and squares). The left arrow indicates the existing result ($\beta_c = 0.221654(6)$[24]) for the cubic lattice. The other arrow denotes the exact solution ($\beta_c = 0.440687$ [25]) for the square lattice.

method deals with both of the effect from the cavity fields and cavity correlations. The analysis performed in the previous study was under the assumption that the cavity field is very small even in the ferromagnetic phase in the level of the Bethe approximation. The correction coming from the cavity correlation would compete with effect of increase in the cavity field around the critical point given by the Bethe approximation. Therefore the sudden change at the wrong critical point as found in the Bethe approximation vanishes, and we observe the critical behavior at another point, which is the corrected critical point. Our method is available to approximately investigate the spin glass model in the two-dimensional space. For the $\pm J$ Ising model on the square lattice with $P(J_{ij}) = p\delta(J_{ij}-1)+(1-p)\delta(J_{ij}+1)$, we test our proposed algorithm for several values of p and the linear size $L = 12$. We plot the values of the magnetization for each case with p in figure 3. We find the fact that we can describe the more correct behavior of the magnetization for each value of p, since the beginning of the standing behavior of the magnetization is observed at the closer point to the precise location given by the duality analysis [26, 27]. However, if we increase the value of p, we would need more correction to our estimation. In addition, we estimate the magnetization across the special critical point, which known as the multicritical point on the Nishimori line [28, 29]. Along the Nishimori line, the inverse temperature is related to the concentration of the ferromagnetic interaction as $\exp(-2\beta) = (1-p)/p$. The expected location of the multicritical point is at $p_c \approx 0.8908$ [26, 27]. We find improvement of the critical behavior around a closer point than that given by the belief propagation as shown in figure 4

4. Discussion

We show the derivation of the generic formulas for the improved Bethe approximation for the case with magnetic field by dealing with correction from not only two-point correlations but also multipoint ones. Our analysis starts from the exact relationship between cavity fields. Therefore, if we systematically take the effects of the higher-order multipoint correlations, we asymptotically obtain the exact results of the magnetization and correlations. Our formulas are available for the Inverse Ising model since we expect precise and fast calculations in the forward process. In order to implement our formulas to indeed compute the magnetization

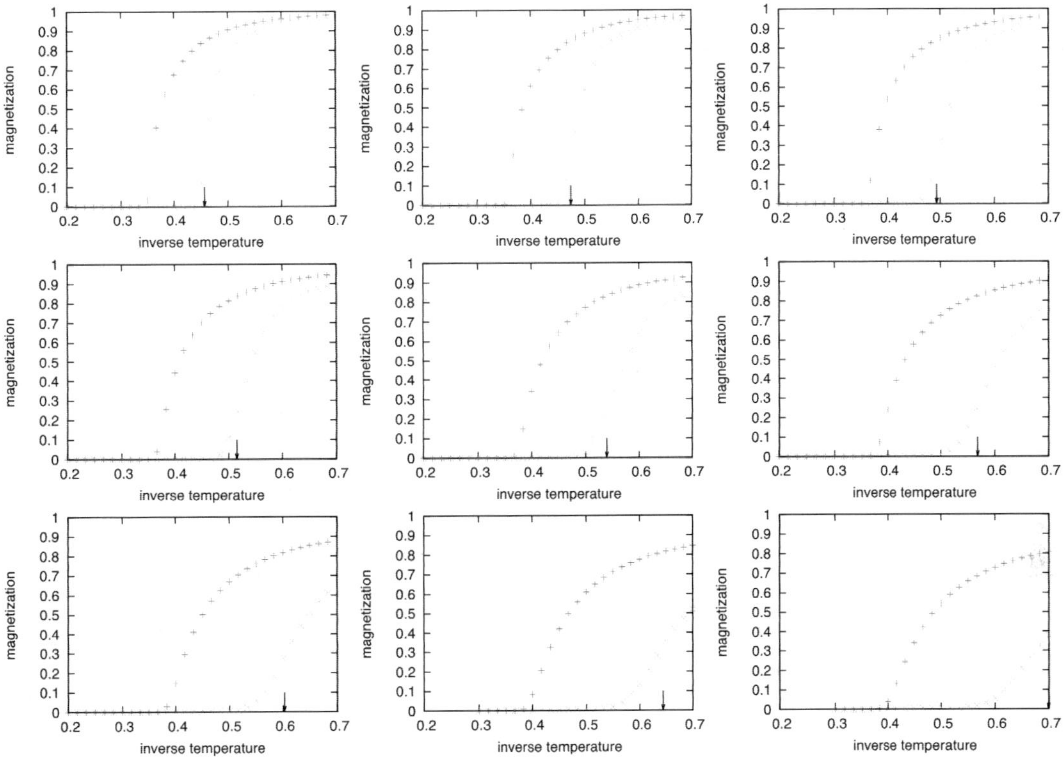

Figure 3. Plots of the magnetization for each case with p (tilted-cross green points). For comparison, we also describe the results by the ordinary belief propagation (cross red points). The arrow denotes the expected location given by a precise analysis [26, 27]). We test our algorithm in a small system $L = 12$ and use 1000 samples for J_{ij}.

and correlations, we propose an iterative algorithm such as the belief propagation. However this is based on the Bethe approximation to compute the multipoint cavity correlations. The lack of the convergent behavior in the iterative manipulation, belief propagation, of the Bethe approximation prevents us from investigating a special class of the problems. Therefore we must develop the method to avoid the poor convergence of the belief propagation to achieve the precise estimations following the current of this study.

In addition, the relationship to the loop calculus should be clarified. In reference [23], it is pointed out that the direction of the present study and that of the loop calculus might be different. However the computation by use of the multipoint cavity correlations are also available for estimations of the partition function with loops and reveals the discrepancies from the Bethe approximation. We hope that the future work shed light on this issue and deepen the understanding of the Bethe approximation and further analysis in statistical mechanics.

Acknowledgments

The author thanks the hospitality during the main part of this study in Universit'a di Roma, "La Sapienza". He also appreciate the fruitful discussions with Tommaso Rizzo, Ulisse Ferrari, Federico Richi-Tersenghi, and Giorgio Parisi. This work was partially supported by MEXT in Japan, Grant-in-Aid for Young Scientists (B) No.24740263.

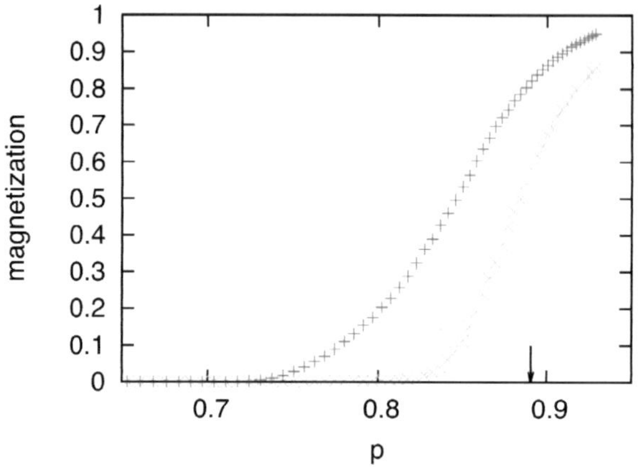

Figure 4. Plots of the magnetization along the Nishimori line (tilted-cross green points). For comparison, we also describe the results by the ordinary belief propagation (cross red points). The arrow denotes the expected location given by a precise analysis [26, 27]). We test our algorithm in a small system $L = 12$ and use 1000 samples for J_{ij}.

Appendix A. Correction from cavity correlations

In order to evaluate the correction of the multipoint correlation, we have to deal with four types of the summation, namely $\sum_{\mathcal{A}:\mathrm{even}} t_{\mathcal{A}}\tilde{C}^{(i)}_{\mathcal{A}}$, $\sum_{\mathcal{A}:\mathrm{odd}} t_{\mathcal{A}}\tilde{C}^{(i)}_{\mathcal{A}}$, $\sum_{\mathcal{A}:\mathrm{even}} t_{\mathcal{A}}\tilde{C}^{(i)}_{\mathcal{A}\cup j}$, $\sum_{\mathcal{A}:\mathrm{odd}} t_{\mathcal{A}}\tilde{C}^{(i)}_{\mathcal{A}\cup j}$. Notice that $\mathcal{A}$ is a subset except for S_j in the absence of the spin S_i. Let us first write down the explicit form of each quantity up to the first order of the multipoint correlation as $C^{(i)}_{ij}$ and $C^{(i)}_{ijk}$.

$$
\begin{aligned}
\sum_{\mathcal{A}:\mathrm{even}} t_{\mathcal{A}}\tilde{C}^{(i)}_{\mathcal{A}} &= 1 + \sum_{j_1,j_2\in\partial i/j} t_{ij_1}t_{ij_2}\left(M^{(i)}_{j_1}M^{(i)}_{j_2} + C^{(i)}_{j_1,j_2}\right) \\
&\quad + \sum_{j_k\in\partial i/j} t_{ij_1}t_{ij_2}t_{ij_3}t_{ij_4}\left(M^{(i)}_{j_1}M^{(i)}_{j_2}M^{(i)}_{j_3}M^{(i)}_{j_4} + C^{(i)}_{j_1,j_2}M^{(i)}_{j_3}M^{(i)}_{j_4} + C^{(i)}_{j_1,j_2,j_3}M^{(i)}_{j_4}\right) \\
&\quad + \cdots \\
&\equiv f_0 + \sum_{j_k\in\partial i/j} C^{(i)}_{j_1,j_2} f_{j_1,j_2} + \sum_{j_k\in\partial i/j} C^{(i)}_{j_1,j_2,j_3} f_{j_1,j_2,j_3} + \cdots, \qquad \text{(A.1)}
\end{aligned}
$$

where f_0 is the remaining term in the Bethe approximation, and its explicit form is $\sum_{\mathcal{A}:\mathrm{even}} t^{(i)}_{\mathcal{A}} \prod_{j\in\mathcal{A}} M^{(i)}_j$. Here $f_{j_1,\cdots,j_k}$ denotes a function only with the cavity field except for the components appearing in the multipoint cavity correlation $C^{(i)}_{j_1,\cdots,j_k}$. Similarly the summation of the odd-number components is

$$
\begin{aligned}
\sum_{\mathcal{A}:\mathrm{odd}} t_{\mathcal{A}}\tilde{C}^{(i)}_{\mathcal{A}} &= \sum_{j_1\in\partial i/j} t_{ij_1}M^{(i)}_{j_1} \\
&\quad + \sum_{j_k\in\partial i/j} t_{ij_1}t_{ij_2}t_{ij_3}\left(M^{(i)}_{j_1}M^{(i)}_{j_2}M^{(i)}_{j_3} + C^{(i)}_{j_1,j_2}M^{(i)}_{j_3} + C^{(i)}_{j_1,j_2,j_3}\right) \\
&\equiv g_0 + \sum_{j_k\in\partial i/j} C^{(i)}_{j_1,j_2} g_{j_1,j_2} + \sum_{j_k\in\partial i/j} C^{(i)}_{j_1,j_2,j_3} g_{j_1,j_2,j_3} + \cdots, \qquad \text{(A.2)}
\end{aligned}
$$

where g_0 is obtained in the Bethe approximation and explicitly written as $\sum_{\mathcal{A}:\mathrm{odd}} t_{\mathcal{A}}^{(i)} \prod_{j\in\mathcal{A}} M_j^{(i)}$. We summarize all the terms only with the cavity field except for the components appearing in the multipoint cavity correlation $C^{(i)}_{j_1,\cdots,j_k}$ in a function $g_{j_1,\cdots,j_k}$. The other two quantities are given in a similar way as

$$\begin{aligned}\sum_{\mathcal{A}:\mathrm{even}} t_{\mathcal{A}} \tilde{C}^{(i)}_{\mathcal{A}\cup j} &= M_j^{(i)} f_0 + \sum_{j_k\in\partial i/j} C^{(i)}_{j_1,j_2} f_{j_1,j_2} \\ &+ \sum_{j_k\in\partial i/j} C^{(i)}_{j,j_1} \tilde{f}_{j,j_1} + \sum_{j_k\in\partial i/j} C^{(i)}_{j_1,j_2,j_3} f_{j_1,j_2,j_3} + \sum_{j_k\in\partial i/j} C^{(i)}_{j,j_1,j_2} \tilde{f}_{j,j_1,j_2} + \cdots,\end{aligned} \tag{A.3}$$

and

$$\begin{aligned}\sum_{\mathcal{A}:\mathrm{odd}} t_{\mathcal{A}} \tilde{C}^{(i)}_{\mathcal{A}\cup j} &= M_j^{(i)} g_0 + \sum_{j_k\in\partial i/j} C^{(i)}_{j_1,j_2} g_{j_1,j_2} \\ &+ \sum_{j_k\in\partial i/j} C^{(i)}_{j,j_1} \tilde{g}_{j,j_1} + \sum_{j_k\in\partial i/j} C^{(i)}_{j_1,j_2,j_3} g_{j_1,j_2,j_3} + \sum_{j_k\in\partial i/j} \sum C^{(i)}_{j,j_1,j_2} \tilde{g}_{j,j_1,j_2},\end{aligned} \tag{A.4}$$

where we use two functions $\tilde{f}_{j_1,\cdots,j_k}$ and $\tilde{g}_{j_1,\cdots,j_k}$ to write the correction of all the terms only with the cavity field except for the components appearing in the multipoint cavity correlation $C^{(i)}_{j_1,\cdots,j_k}$.

For convenience, let us evaluate the ratios of these four quantities. Then we assume that the multipoint cavity correlations should be small and expand them up to the first order.

$$\begin{aligned}\frac{\sum_{\mathcal{A}:\mathrm{odd}} t_{\mathcal{A}} \tilde{C}^{(i)}_{\mathcal{A}}}{\sum_{\mathcal{A}:\mathrm{even}} t_{\mathcal{A}} \tilde{C}^{(i)}_{\mathcal{A}}} &= \frac{g_0}{f_0} + \sum_{j_k\in\partial i/j} C^{(i)}_{j_1,j_2} \frac{f_0 g_{j_1,j_2} - g_0 f_{j_1,j_2}}{f_0^2} \\ &+ \sum_{j_k\in\partial i/j} C^{(i)}_{j_1,j_2,j_3} \frac{f_0 g_{j_1,j_2,j_3} - g_0 f_{j_1,j_2,j_3}}{f_0^2} + \cdots.\end{aligned} \tag{A.5}$$

Notice that f_0 and g_0 appear in the Bethe approximation. Thus we can obtain

$$\frac{\sum_{\mathcal{A}:\mathrm{odd}} t_{\mathcal{A}} \tilde{C}^{(i)}_{\mathcal{A}}}{\sum_{\mathcal{A}:\mathrm{even}} t_{\mathcal{A}} \tilde{C}^{(i)}_{\mathcal{A}}} = T_j^{(i)} + \sum_{k=2} \sum_{j_k\in\partial i/j} C^{(i)}_{j_1,\cdots,j_k} \frac{g_{j_1\cdots,j_k} - T_j^{(i)} f_{j_1,\cdots,j_k}}{f_0}. \tag{A.6}$$

Here we define each coefficient as

$$\Gamma^{(i)}_{j,j_1,\cdots,j_k} \equiv \frac{g_{j_1,\cdots,j_k} - T_j^{(i)} f_{j_1,\cdots,j_k}}{f_0}. \tag{A.7}$$

Similarly we find

$$\frac{\sum_{\mathcal{A}:\mathrm{even}} t_{\mathcal{A}} \tilde{C}^{(i)}_{\mathcal{A}\cup j}}{\sum_{\mathcal{A}:\mathrm{even}} t_{\mathcal{A}} \tilde{C}^{(i)}_{\mathcal{A}}} = M_j^{(i)} + \sum_{k=1} \sum_{j_k\in\partial i/j} C^{(i)}_{j,j_1,\cdots,j_k} \Omega_{j,j_1,\cdots,j_k}, \tag{A.8}$$

where

$$\Omega^{(i)}_{j,j_1,\cdots,j_k} \equiv \frac{\tilde{f}_{j,j_1,\cdots,j_k}}{f_0}. \tag{A.9}$$

In addition,

$$\frac{\sum_{\mathcal{A}:\mathrm{odd}} t_{\mathcal{A}} \tilde{C}^{(i)}_{\mathcal{A}\cup j}}{\sum_{\mathcal{A}:\mathrm{even}} t_{\mathcal{A}} \tilde{C}^{(i)}_{\mathcal{A}}} = M_j^{(i)} \frac{\sum_{\mathcal{A}:\mathrm{odd}} t_{\mathcal{A}} \tilde{C}^{(i)}_{\mathcal{A}}}{\sum_{\mathcal{A}:\mathrm{even}} t_{\mathcal{A}} \tilde{C}^{(i)}_{\mathcal{A}}} + \sum_{k=1} \sum_{j_k \in \partial i/j} C^{(i)}_{j,j_1,\cdots,j_k} \Upsilon^{(i)}_{j,j_1,\cdots,j_k}, \tag{A.10}$$

where we define

$$\Upsilon^{(i)}_{j,j_1,\cdots,j_k} \equiv \frac{\tilde{g}_{j,j_1,\cdots,j_k}}{f_0}. \tag{A.11}$$

Each coefficient of the multipoint cavity correlation basically consists of the cavity field. For instance, let us evaluate $f_{j_1,j_2}/f_0$ in the coefficients

$$\frac{f_{j_1,j_2}}{f_0} = t_{ij_1} t_{ij_2} \frac{\sum_{\mathcal{B}=\mathcal{A}/j_1,j_2:\mathrm{even}} \prod_{j\in\mathcal{B}} t_{ij} M_j^{(i)}}{\sum_{\mathcal{A}:\mathrm{even}} \prod_{j\in\mathcal{A}} t_{ij} M_j^{(i)}}. \tag{A.12}$$

Here we use the following identities.

$$E_{j_1,j_2,\cdots,j_k} = E_{j_1,j_2,\cdots,j_k,j_l} + t_{il} M_l^{(i)} O_{j_1,j_2,\cdots,j_k,j_l} \tag{A.13}$$

$$O_{j_1,j_2,\cdots,j_k} = O_{j_1,j_2,\cdots,j_k,j_l} + t_{il} M_l^{(i)} E_{j_1,j_2,\cdots,j_k,j_l}, \tag{A.14}$$

where $\sum_{\mathcal{B}=\mathcal{A}/j_1,j_2,\cdots,j_k:\mathrm{even}} t_{\mathcal{B}} \prod_{j\in\mathcal{B}} M_j^{(i)} = E_{j_1,j_2,\cdots,j_k}$ and $\sum_{\mathcal{B}=\mathcal{A}/j_1,j_2,\cdots,j_k:\mathrm{odd}} t_{\mathcal{B}} \prod_{j\in\mathcal{B}} M_j^{(i)} = O_{j_1,j_2,\cdots,j_k}$. Therefore we find

$$\frac{f_{j_1,j_2}}{f_0} = t_{ij_1} t_{ij_2} \frac{E_{j_1,j_2}}{E_{j_1,j_2}(1 + t_{ij_1} t_{ij_2} M_{j_1}^{(i)} M_{j_2}^{(i)}) + O_{j_1,j_2}(t_{ij_1} M_{j_1}^{(i)} + t_{ij_2} M_{j_2}^{(i)})}. \tag{A.15}$$

Since the ratio of $O_{j_1,j_2,\cdots,j_k}/E_{j_1,j_2,\cdots,j_k}$ is equal to $T^{(i)}_{j,j_1,j_2,\cdots,j_k}$, this equality can be reduced to

$$\frac{f_{j_1,j_2}}{f_0} = \frac{t_{ij_1} t_{ij_2}}{1 + t_{ij_1} t_{ij_2} M_{j_1}^{(i)} M_{j_2}^{(i)} + T^{(i)}_{j,j_1,j_2}(t_{ij_1} M_{j_1}^{(i)} + t_{ij_2} M_{j_2}^{(i)})}. \tag{A.16}$$

In general, we can evaluate $f_{j_1,,\cdots,j_k}/f_0$ from the following recursion.

$$\frac{f_{j_1,j_2,\cdots,j_{2k},j_{2k+1}}}{f_{j_1,j_2,\cdots,j_{2k}}} = \frac{t_{ij_{2k+1}} T^{(i)}_{j,j_1,\cdots,j_{2k},j_{2k+1}}}{1 + t_{ij_{2k+1}} M^{(i)}_{j_{2k+1}} T^{(i)}_{j,j_1,\cdots,j_{2k},j_{2k+1}}} \equiv F_1(j_1, j_2, \cdots, j_{2k}, j_{2k+1}) \tag{A.17}$$

$$\frac{f_{j_1,j_2,\cdots,j_{2k}}}{f_{j_1,j_2,\cdots,j_{2k-1}}} = \frac{t_{ij_{2k}}}{T^{(i)}_{j,j_1,\cdots,j_{2k},j_{2k}} + t_{ij_{2k}} M^{(i)}_{j_{2k}}} \equiv G_1(j_1, j_2, \cdots, j_{2k-1}, j_{2k}). \tag{A.18}$$

Similarly we can evaluate the other ratios in the coefficients as

$$\frac{\tilde{f}_{j,j_1}}{f_0} = t_{ij_1} \frac{O_{j_1}}{E_\phi} = t_{ij_1} \frac{T^{(i)}_{j,j_1}}{1 + t_{ij_1} M_{j_1} T^{(i)}_{j,j_1}}, \tag{A.19}$$

and

$$\frac{g_{j_1,j_2}}{f_0} = t_{ij_1} t_{ij_2} \frac{O_{j_1,j_2}}{E_\phi} = t_{ij_1} t_{ij_2} \frac{T^{(i)}_{j,j_1,j_2}}{1 + t_{ij_1} t_{ij_2} M_{j_1}^{(i)} M_{j_2}^{(i)} + T^{(i)}_{j,j_1,j_2}(t_{ij_1} M_{j_1}^{(i)} + t_{ij_2} M_{j_2}^{(i)})} \tag{A.20}$$

$$\frac{\tilde{g}_{j,j_1}}{f_0} = t_{ij_1} \frac{E_{j_1}}{E_\phi} = t_{ij_1} \frac{1}{1 + t_{ij_1} M_{j_1} T^{(i)}_{j,j_1}}. \tag{A.21}$$

For each ratio, we can construct recursion relations similarly to the case of f. In particular, $\tilde{g}_{j,j_1,\cdots,j_{2k+1}}/\tilde{f}_{j,j_1,\cdots,j_{2k+1}} = 1/T^{(i)}_{j,j_1,\cdots,j_{2k+1}}$ and $\tilde{g}_{j,j_1,\cdots,j_{2k}}/\tilde{f}_{j,j_1,\cdots,j_{2k}} = T^{(i)}_{j,j_1,\cdots,j_{2k}}$ holds. This property ensures

$$\Omega^{(i)}_{j,j_1,\cdots,j_{2k+1}} = T^{(i)}_{j,j_1,\cdots,k_{2k+1}}\Upsilon^{(i)}_{j,j_1,\cdots,j_{2k+1}} \quad (A.22)$$
$$\Omega^{(i)}_{j,j_1,\cdots,j_{2k}} = \Upsilon^{(i)}_{j,j_1,\cdots,k_{2k}}/T^{(i)}_{j,j_1,\cdots,j_{2k}}. \quad (A.23)$$

As a result, we can find the explicit form of each coefficient. For instance,

$$\Gamma^{(i)}_{j,j_1,j_2} = \frac{t_{ij_1}t_{ij_2}(T^{(i)}_{j,j_1,j_2} - T^{(i)}_j)}{1 + t_{ij_1}t_{ij_2}M^{(i)}_{j_1}M^{(i)}_{j_2} + t_{ij_2}M^{(i)}_{j_1}T^{(i)}_{j,j_1,j_2} + t_{ij_1}M^{(i)}_{j_2}T^{(i)}_{j,j_1,j_2}} \quad (A.24)$$
$$\Omega^{(i)}_{j,j_1} = \frac{t_{ij_1}T^{(i)}_{j,j_1}}{1 + t_{ij_1}M^{(i)}_{j_1}T^{(i)}_{j,j_1}} = T^{(i)}_{j,j_1}\Upsilon^{(i)}_{j,j_1}. \quad (A.25)$$

Once we find these coefficients, we can reproduce the previous results explicitly given by Montanari and Rizzo [21]. By the present analysis, we push up their to the further improvement with multipoint correlation as

$$A^{(i)}_j = T^{(i)}_j(h_i) + \sum_{k=2}\sum_{j_k\in\partial i/j} C^{(i)}_{j_1,\cdots,j_k}\Gamma^{(i)}_{j_1,\cdots,j_k}\frac{1-t_i^2}{(1+t_iT^{(i)}_j)^2} \quad (A.26)$$
$$B^{(i)}_j = M^{(i)}_j + \sum_{k=1}\sum_{j_k\in\partial i/j} C^{(i)}_{j,j_1,\cdots,j_k}\Omega^{(i)}_{j,j_1,\cdots,j_k}. \quad (A.27)$$

We thus obtain equation (34). We reproduce the same result for the case without magnetic field and only with $C^{(i)}_{j_1,j_2}$ and $C^{(i)}_{j,j_1}$ as that given in reference [21]. Indeed the update equation can estimate closer locations of the critical point to its exact answer that those by the Bethe approximation. In order to evaluate two-point correlation to match with the given vector of the data in the case of the Inverse Ising model, we evaluate $D^{(i)}_{ij}$ from the above formulas. The result is

$$D^{(i)}_{ij} = T^{(i)}_j(h_i)M^{(i)}_j + \sum_{k=1}\sum_{j_k\in\partial i/j} C^{(i)}_{j,j_1,\cdots,j_k}\Omega^{(i)}_{j,j_1,\cdots,j_k}\frac{(T^{(i)}_j)^{d_k}}{1+t_iT^{(i)}_j}$$
$$+\sum_{k=2}\sum_{j_k\in\partial i/j} C^{(i)}_{j_1,\cdots,j_k}\Gamma^{(i)}_{j_1,\cdots,j_k}\frac{(1-t_i^2)M^{(i)}_j}{(1+t_iT^{(i)}_j)^2}. \quad (A.28)$$

and obtain equation (35) and (36). Equivalently, one may directly insert these results into equation (15) and (17).

References

[1] Bethe H A 1935 *Proceedings of the Royal Society of London. Series A - Mathematical and Physical Sciences* **150** 552
[2] Baxter R 1982 *Exactly solved models in statistical mechanics* (Academic Press)
[3] Mézard M and Parisi G 2001 *European Physical Journal* B **20** 217
[4] Mézard M and Montanari A 2009 *Information, Physics, and Computation* Oxford Graduate Texts (OUP Oxford)
[5] Peterson C and Anderson A R 1987 *Complex Systems* **1** 995
[6] Hinton G E 1989 *Neural Computation* **1** 143

[7] Kappen H and Rodrìguez F 1998 *Advances in Neural Information Processing Systems* (The MIT Press) pp 280–286
[8] Tanaka T 1998 *Phys. Rev.* E **58**(2) 2302
[9] Thouless D J, Anderson P W and Palmer R G 1977 *Philosophical Magazine* **35** 593
[10] Plefka T 1982 *Journal of Physics* A*: Mathematical and General* **15** 1971
[11] Tanaka K 2003 *IEICE TRANSACTIONS on Information and Systems* **E86-D** 1228
[12] Welling M and Teh Y W 2003 *Artificial Intelligence* **143** 19
[13] Welling M and Teh Y W 2004 *Neural Computation* **16** 197
[14] Mézard M and Mora T 2009 *Journal of Physiology-Paris* **103** 107
[15] Pearl J 1988 *Probabilistic reasoning in intelligent systems: networks of plausible inference* Representation and Reasoning Series (MORGAN KAUFMAN PUBL Incorporated)
[16] Ricci-Tersenghi F 2012 *Journal of Statistical Mechanics: Theory and Experiment* **2012** P08015
[17] Yasuda M and Tanaka K 2013 *Phys. Rev.* E **87** 012134
[18] Chertkov M and Chernyak V Y 2006 *Phys. Rev.* E **73** 065102
[19] Chertkov M and Chernyak V Y 2006 *Journal of Statistical Mechanics: Theory and Experiment* **2006** P06009
[20] Xiao J Q and Zhou H 2011 *Journal of Physics* A*: Mathematical and Theoretical* **44** 425001
[21] Montanari A and Rizzo T 2005 *Journal of Statistical Mechanics: Theory and Experiment* **2005** P10011
[22] Parisi G and Slanina F 2006 *Journal of Statistical Mechanics: Theory and Experiment* **2006** L02003
[23] Rizzo T, Wemmenhove B and Kappen H J 2007 *Phys. Rev.* E **76** 011102
[24] Barber M N, Pearson R B, Toussaint D and Richardson J L 1985 *Phys. Rev.* B **32** 1720
[25] Kramers H A and Wannier G H 1941 *Phys. Rev.* **60** 252
[26] Ohzeki M, Nishimori H and Berker A N 2008 *Phys. Rev.* E **77** 061116
[27] Ohzeki M 2009 *Phys. Rev.* E **79** 021129
[28] Nishimori H 1981 *Progress of Theoretical Physics* **66** 1169
[29] Nishimori H 2001 *Statistical physics of spin glasses and information processing: an introduction* Oxford Science Publications (Oxford University Press on Demand)

ELC International Meeting on Inference, Computation, and Spin Glasses (ICSG2013) IOP Publishing
Journal of Physics: Conference Series **473** (2013) 012006 doi:10.1088/1742-6596/473/1/012006

A generalization of improved susceptibility propagation

M Yasuda

Graduate School of Science and Engineering, Yamagata University, Yonezawa 992-8510, Yamagata pref., Japan

E-mail: muneki@yz.yamagata-u.ac.jp

Abstract. Recently, an effective susceptibility propagation method for binary Markov random fields based on a concept of diagonal consistency was proposed. This improved susceptibility propagation is a powerful method and exhibits a robust performance for various types of network structures. In this paper, a generalization of the improved susceptibility propagation using an orthonormal function expansion is proposed in which any pairwise potential functions and multi-valued random variables are acceptable. In the latter part of this paper, the proposed method is applied to a direct problem and an inverse problem on a generalized sparse prior, which is a recently proposed prior model for natural images.

1. Introduction

There is an increasing demand in the new interdisciplinary scientific fields that involve both computer sciences and statistical physics [1, 2] for techniques that can be used to evaluate local statistical quantities such as local magnetizations and local susceptibilities (covariances) of Markov random fields (MRFs) with finite sizes.

Belief propagations (BPs) are one of the most popular message-passing types of algorithms that are used to compute approximately the local magnetizations of MRFs [3]; they are equivalent to the Bethe approximations [4] used in statistical physics [5, 6]. In the last decade, a suitable message-passing technique called susceptibility propagation (SusP) (also called variational linear response in the field of information sciences) was developed to compute the local susceptibilities of MRFs [7–10]. In general, SusPs are constructed by combining BPs and linear response methods.

Recently, the author presented an effective SusP for Ising models, referred to as *improved susceptibility propagation* (I-SusP) using a concept of *diagonal consistency* [11]. Briefly, the concept of diagonal consistency is as follows. In Ising models, the second-order moment, $\langle x_i^2 \rangle$, are trivially one because $x_i \in \{+1, -1\}$. However, the variances obtained by SusPs are generally not equal to one due to higher-order effects being neglected [12]. I-SusPs are obtained by overcoming such a diagonal inconsistency problem. I-SusPs are reduced to normal SusPs on tree systems and include the adaptive Thouless-Anderson-Palmer equation [13, 14] as a special case. Almost simultaneously, a similar investigation of Ising models was presented by Raymond and Ricci-Tersenghi [15], who considered not only the diagonal consistency but also an off-diagonal consistency [16].

The I-SusP is a powerful method and exhibits a robust performance for various types of network structures. However, the original I-SusP can be applied to Ising models only. Thus, in the present paper, a generalization scheme of I-SusP using an orthonormal function expansion that was successfully used in a Bethe approximation for a general pairwise MRF [17] is proposed, in which any pairwise potential functions and multi-valued random variables are acceptable.

The remainder of this paper is organized as follows. In section 2, a pairwise MRF and its orthogonal function expansion using Chebyshev polynomials are introduced. In section 3, a Gibbs free energy for the expanded pairwise MRF derived in the previous section and its diagonal-block consistency are shown. The diagonal-block consistency plays the central role in the proposed method. In section 4, by using the diagonal-block consistency, an I-SusP based on the Bethe approximation for the expanded pairwise MRF is proposed. In section 5, the proposed method is applied to *generalized sparse priors* and is numerically verified by some results of numerical experiments. Generalized sparse priors are prior models for natural images proposed in [18]. Finally, section 6 concludes the paper with some remarks.

2. Discrete Pairwise MRF and Orthogonal Function Expansion

Let us consider an undirected graph $G = G(V, E)$ with n nodes, where $V = \{1, 2, \ldots, n\}$ is the set of all nodes and $E \subset V \times V$ is the set of all undirected links (i, j) except self-connecting links (i, i). Note that, although links (i, j) and (j, i) indicate the same undirected link, the set E has both (i, j) and (j, i) as its elements. On the graph G, let us consider a pairwise MRF with discrete random variables $\boldsymbol{x} = \{x_i \in \{0, 1, \ldots, q-1\} \mid i \in V\}$ expressed by

$$P_G(\boldsymbol{x}) \propto \exp\Big(\sum_{i \in V} \theta_i(x_i) + \frac{1}{2} \sum_{(i,j) \in E} w_{(i,j)}(x_i, x_j)\Big), \tag{1}$$

where the expressions $\{\theta_i(x_i)\}$ and $\{w_{(i,j)}(x_i, x_j)\}$ are potential functions of nodes and links, respectively. Because the graph is an undirected graph, relations $w_{(i,j)}(x_i, x_j) = w_{(j,i)}(x_j, x_i)$ are assumed. Relations $w_{(i,j)}(x_i, x_j) = w_{(i,j)}(x_j, x_i)$ are not enforced, that is, $w_{(i,j)}(x_i, x_j)$ does not have to be a symmetric function with respect to x_i and x_j.

Let us introduce an orthonormal set $\{\phi_k(x) \in \mathbb{R} \mid k = 0, 1, 2, \ldots, q-1\}$ such that

$$\sum_{x=0}^{q-1} \phi_k(x) \phi_l(x) = \delta_{k,l}, \tag{2}$$

where $\delta_{k,l}$ is the Kronecker delta. If the orthonormal set is a basis, we can expand a function $f(x)$ of an integer variable $x \in \{0, 1, \ldots, q-1\}$ by using the orthonormal set as

$$f(x) = \sum_{k=0}^{q-1} \alpha_k \phi_k(x), \quad \text{where} \quad \alpha_k := \sum_{x=0}^{q-1} f(x) \phi_k(x),$$

One possible choice of orthonormal set is the set of Chebyshev polynomials given in Appendix A. We use the Chebyshev polynomials as the orthonormal set in this paper.

Using the orthonormal set of polynomials, we can expand the potential functions $\theta_i(x_i)$ and $w_{(i,j)}(x_i, x_j)$ as

$$\theta_i(x_i) = \sum_{k=0}^{q-1} u_{[i,k]} \phi_k(x_i) \quad \text{and} \quad w_{(i,j)}(x_i, x_j) = \sum_{k,l=0}^{q-1} J_{[i,k],[j,l]} \phi_k(x_i) \phi_l(x_j), \tag{3}$$

respectively, where $u_{[i,k]}$ and $J_{[i,k],[j,l]}$ are constants given by

$$u_{[i,k]} := \sum_{x_i=0}^{q-1} \theta_i(x_i)\phi_k(x_i) \quad \text{and} \quad J_{[i,k],[j,l]} := \sum_{x_i,x_j=0}^{q-1} w_{(i,j)}(x_i,x_j)\phi_k(x_i)\phi_l(x_j), \tag{4}$$

respectively. When the functional forms of the potential functions have been specified, the values of the constants are identified uniquely via (4). By these definitions, $J_{[i,k],[j,l]} = J_{[j,l],[i,k]}$ are satisfied, but $J_{[i,k],[j,l]} \neq J_{[i,l],[j,k]}$ in general. By using the expansions in (3) and the relation $\phi_0(x) = 1/\sqrt{q}$ (see Appendix A), the MRF in (1) can be rewritten as

$$P_G(\boldsymbol{x}) = P_G(\boldsymbol{x} \mid \boldsymbol{h}, \boldsymbol{J}) = \frac{1}{Z(\boldsymbol{h}, \boldsymbol{J})} \exp\Big(-\mathcal{H}(\boldsymbol{x}; \boldsymbol{h}, \boldsymbol{J})\Big), \tag{5}$$

where

$$\mathcal{H}(\boldsymbol{x}; \boldsymbol{h}, \boldsymbol{J}) := -\sum_{k=1}^{q-1} h_{[i,k]}\phi_k(x_i) - \frac{1}{2}\sum_{(i,j)\in E}\sum_{k,l=1}^{q-1} J_{[i,k],[j,l]}\phi_k(x_i)\phi_l(x_j)$$

and the expression $Z(\boldsymbol{h}, \boldsymbol{J})$ is the partition function. The expression $h_{[i,k]}$ is defined by

$$h_{[i,k]} := u_{[i,k]} + \frac{1}{\sqrt{q}} \sum_{j\in\partial(i)} J_{[i,k],[j,0]} = u_{[i,k]} + \frac{1}{q} \sum_{j\in\partial(i)} \sum_{x_i,x_j=0}^{q-1} w_{(i,j)}(x_i,x_j)\phi_k(x_i).$$

The notation $\partial(i)$ is the set of nodes connecting to node i. In the expression in (5), the constant terms are neglected because they have no influence on the probability. This rewritten expansion of MRF is in the exponential family. Therefore, it can be found that we can obtain a distribution in the exponential family that is equivalent to the pairwise MRF in (1) by using the orthonormal expansion. In the following, we refer the distribution (5) as "the expanded pairwise MRF".

3. Gibbs Free Energy and Diagonal-Block Consistency

In this section, we formulate a Gibbs free energy (GFE) and its diagonal-block consistency for the expanded pairwise MRF. GFE representations and their diagonal-block consistencies are important components for I-SusPs provided in section 4.

3.1. Gibbs Free Energy

Let us introduce a GFE, which is a dual form of the Helmholtz free energy (HFE) defined by $F(\boldsymbol{h}, \boldsymbol{J}) := -\ln Z(\boldsymbol{h}, \boldsymbol{J})$, in the following variational way. Let us define the variational function with respect to a trial distribution $Q(\boldsymbol{x})$:

$$\mathcal{F}[Q] := \sum_{\boldsymbol{x}} \mathcal{H}(\boldsymbol{x})Q(\boldsymbol{x}) + \sum_{\boldsymbol{x}} Q(\boldsymbol{x}) \ln Q(\boldsymbol{x}), \tag{6}$$

and consider the minimization of the variational function under the restrictions

$$\sum_{\boldsymbol{x}} Q(\boldsymbol{x}) = 1, \quad \text{and} \quad \sum_{\boldsymbol{x}} \phi_k(x_i)Q(\boldsymbol{x}) = m_{[i,k]}.$$

These are the restrictions for normalizing and expectations, respectively. By using the Lagrange multipliers, the constrained minimization of $\mathcal{F}[Q]$ leads to a GFE as follows:

$$\mathcal{G}(\boldsymbol{m}) := \min_{Q} \operatorname*{extr}_{z,\boldsymbol{\gamma}} \Big\{ \mathcal{F}[Q] - z\Big(\sum_{\boldsymbol{x}} Q(\boldsymbol{x}) - 1\Big) - \sum_{i\in V} \gamma_i \Big(\sum_{\boldsymbol{x}} \phi_k(x_i)Q(\boldsymbol{x}) - m_{[i,k]}\Big)\Big\}$$

$$= -\boldsymbol{h}^{\mathrm{t}}\boldsymbol{m} + \max_{\boldsymbol{\gamma}} \Big(\boldsymbol{\gamma}^{\mathrm{t}}\boldsymbol{m} + F(\boldsymbol{\gamma}, \boldsymbol{J})\Big), \tag{7}$$

where the notation "extr" denotes the extremum with respect to the assigned parameters and the notation t is the transposition. In the GFE, we regard $\boldsymbol{m} = \{m_{[i,k]} \mid i \in V; k = 1, 2, \ldots, q-1\}$ as the independent variables. Expression $\boldsymbol{\gamma} = \{\gamma_{i,k} \mid i \in V; k = 1, 2, \ldots, q-1\}$ denote the Lagrange multipliers for the constraints for the corresponding expectations. The expression $\mathcal{G}(\boldsymbol{m})$ is the GFE of the present system and is regarded as a dual expression of the HFE by the following reasoning. Suppose that $\boldsymbol{m}^{\dagger}$ are the values of $\boldsymbol{m}$ that minimize the GFE, i.e., $\boldsymbol{m}^{\dagger} = \arg\min_{\boldsymbol{m}} \mathcal{G}(\boldsymbol{m})$, the following relations are satisfied:

$$\mathcal{G}(\boldsymbol{m}^{\dagger}) = F(\boldsymbol{h}, \boldsymbol{J}) \quad \text{and} \quad m^{\dagger}_{[i,k]} = \sum_{\boldsymbol{x}} \phi_k(x_i) P_G(\boldsymbol{x} \mid \boldsymbol{h}, \boldsymbol{J}) =: m^{\mathrm{ex}}_{[i,k]}. \tag{8}$$

This fact can be easily verified as follows. From the minimum condition of (7) with respect to $\boldsymbol{m}$, we obtain

$$-h_{[i,k]} + \gamma^{*}_{[i,k]} = 0, \tag{9}$$

where $\boldsymbol{\gamma}^{*}$ are values of $\boldsymbol{\gamma}$ that satisfy the maximum conditions in (7), i.e., $\boldsymbol{\gamma}^{*} = \arg\max_{\boldsymbol{\gamma}} \big(\boldsymbol{\gamma}^{\mathrm{t}}\boldsymbol{m} + F(\boldsymbol{\gamma}, \boldsymbol{J})\big)$, and therefore, $\boldsymbol{\gamma}^{*}$ satisfy relations

$$m_{[i,k]} = \sum_{\boldsymbol{x}} \phi_k(x_i) P_G(\boldsymbol{x} \mid \boldsymbol{\gamma}^{*}, \boldsymbol{J}). \tag{10}$$

Using (9) and (10), we arrive at the relations in (8).

3.2. Linear Response Relation and Diagonal-Block Consistency

It is known that the GFE satisfies a linear response relation that is obtained in the following way. By differentiating (10) with respect to $m_{[j,l]}$, we obtain

$$\delta_{i,j}\delta_{k,l} = \sum_{i' \in V} \sum_{r=1}^{q-1} \Big(\langle \phi_k(x_i)\phi_r(x_{i'}) \rangle_{\boldsymbol{\gamma}^{*}} - m_{[i,k]} m_{[i',r]} \Big) \frac{\partial \gamma^{*}_{[i',r]}}{\partial m_{[j,l]}}, \tag{11}$$

where notation $\langle \cdots \rangle_{\boldsymbol{\gamma}^{*}}$ denotes the expectation with respect to distribution $P_G(\boldsymbol{x} \mid \boldsymbol{\gamma}^{*}, \boldsymbol{J})$. Since $\partial^2 \mathcal{G}(\boldsymbol{m}) / \partial m_{[i,k]} \partial m_{[j,l]} = \partial \gamma^{*}_{[i,k]} / \partial m_{[j,l]}$, (11) is collectively expressed by

$$\boldsymbol{H}(\boldsymbol{m})^{-1} = \boldsymbol{\chi}(\boldsymbol{m}), \tag{12}$$

where $n(q-1) \times n(q-1)$ matrices $\boldsymbol{H}(\boldsymbol{m})$ and $\boldsymbol{\chi}(\boldsymbol{m})$ are the Hessian matrix of the GFE and the susceptibility matrix, respectively. They are defined by

$$\big[\boldsymbol{H}(\boldsymbol{m})\big]_{e(i,k),e(j,l)} := \frac{\partial^2 \mathcal{G}(\boldsymbol{m})}{\partial m_{[i,k]} \partial m_{[j,l]}}, \quad \big[\boldsymbol{\chi}(\boldsymbol{m})\big]_{e(i,k),e(j,l)} := \langle \phi_k(x_i)\phi_l(x_j) \rangle_{\boldsymbol{\gamma}^{*}} - m_{[i,k]} m_{[j,l]},$$

where $e(i,k) := (i-1)q + k$ and the notation $[\cdots]_{a,b}$ indicates the (a,b)-element of a matrix. The relation in (12) is known as one type of linear response relation and always holds for any $\boldsymbol{m}$.

By using the relations in (10) and the orthonormal set, one-variable marginal distributions of $P_G(\boldsymbol{x} \mid \boldsymbol{\gamma}^{*}, \boldsymbol{J})$, $P_i(x_i \mid \boldsymbol{\gamma}^{*}, \boldsymbol{J}) := \sum_{\boldsymbol{x} \setminus \{x_i\}} P_G(\boldsymbol{x} \mid \boldsymbol{\gamma}^{*}, \boldsymbol{J})$, can be expanded as $P_i(x_i \mid \boldsymbol{\gamma}^{*}, \boldsymbol{J}) =$

$q^{-1}+\sum_{k=1}^{q-1} m_{[i,k]}\phi_k(x_i)$. Therefore, because $\langle\phi_k(x_i)\phi_l(x_i)\rangle_{\gamma^*} = \sum_{x_i=0}^{q-1}\phi_k(x_i)\phi_l(x_i)P_i(x_i \mid \gamma^*, \boldsymbol{J})$, the diagonal-blocks in $\chi(\boldsymbol{m})$ are explicitly expressed in terms of $\boldsymbol{m}$ as

$$\big[\chi(\boldsymbol{m})\big]_{e(i,k),e(i,l)} = v_{k,l}^{(i)}(\boldsymbol{m}_i) - m_{[i,k]}m_{[i,l]}, \tag{13}$$

where $v_{k,l}^{(i)}(\boldsymbol{m}_i) := \delta_{k,l}/q + \sum_{r=1}^{q-1} m_{[i,r]}T_{k,l,r}$ and $T_{k,l,r} := \sum_{x=0}^{q-1}\phi_k(x)\phi_l(x)\phi_r(x)$. From (12) and (13), we find that the relations

$$\big[\boldsymbol{H}(\boldsymbol{m})^{-1}\big]_{e(i,k),e(i,l)} = v_{k,l}^{(i)}(\boldsymbol{m}_i) - m_{[i,k]}m_{[i,l]} \tag{14}$$

are also ensured for any $\boldsymbol{m}$. Relations in (14) are referred to as *diagonal-block consistency* in this paper.

Although the diagonal-block consistency holds in the true GFE, the diagonal-block consistency can be broken if one employs an approximate scheme such as a mean-field method [12][1]. A basic concept of I-SusP is that the diagonal-block consistency is enforced in an approximate scheme, which is always kept in the exact scheme.

4. Improved Susceptibility Propagation for Pairwise MRF

In this section, an explicit formulation of I-SusP based on the Bethe approximation for (1) is proposed. First, a general scheme of I-SusP is introduced in section 4.1 in accordance with [11]. An explicit formulation of I-SusP based on a Bethe approximation is then derived in the subsequent sections.

4.1. Scheme of Improved Susceptibility Propagation

Let us suppose an approximated GFE formed by

$$\mathcal{G}_0(\boldsymbol{m}) := -\boldsymbol{h}^{\mathrm{t}}\boldsymbol{m} + \max_{\gamma}\Big(\gamma^{\mathrm{t}}\boldsymbol{m} + F_0(\gamma, \boldsymbol{J})\Big), \tag{15}$$

and consider a minimization of the GFE. When $F_0(\gamma, \boldsymbol{J})$ is the true HFE, $\mathcal{G}_0(\boldsymbol{m})$ is equal to the true GFE.

The scheme of I-SusP can be described as follows. First, with controllable parameters $\boldsymbol{\Lambda}$, we extend an approximated GFE, $\mathcal{G}_0(\boldsymbol{m})$, by adding an extra term as

$$\mathcal{G}_1(\boldsymbol{m}, \boldsymbol{\Lambda}) := \mathcal{G}_0(\boldsymbol{m}) - \frac{1}{2}\sum_{i\in V}\sum_{k,l=1}^{q-1}\Lambda_{[i,k],[i,l]}\Big(v_{k,l}^{(i)}(\boldsymbol{m}_i) - m_{[i,k]}m_{[i,l]}\Big). \tag{16}$$

The extra term is added so that the extended approximate GFE enables to satisfy diagonal-block consistency[2]. In the I-SusP on Ising systems, it is known that this extra term eventually behaves as an Onsager reaction term [11, 15]. The goal of I-SusP is to minimize the extended approximate GFE in (16) with respect to $\boldsymbol{m}$ with searching values of $\boldsymbol{\Lambda}$ that satisfy relations

$$\big[\boldsymbol{H}_1(\boldsymbol{m}, \boldsymbol{\Lambda})^{-1}\big]_{e(i,k),e(i,l)} = v_{k,l}^{(i)}(\boldsymbol{m}_i) - m_{[i,k]}m_{[i,l]}, \tag{17}$$

where $\boldsymbol{H}_1(\boldsymbol{m}, \boldsymbol{\Lambda})$ is the Hessian matrix of $\mathcal{G}_1(\boldsymbol{m}, \boldsymbol{\Lambda})$ defined by

$$\big[\boldsymbol{H}_1(\boldsymbol{m}, \boldsymbol{\Lambda})\big]_{e(i,k),e(j,l)} := \frac{\partial^2\mathcal{G}_1(\boldsymbol{m}, \boldsymbol{\Lambda})}{\partial m_{[i,k]}\partial m_{[j,l]}}.$$

[1] Although the investigations of the breaking in [12] addressed only the Ising case, it is expected that they can be expanded to other cases involving the multi-valued case presented in this paper.
[2] In [15], this term was introduced as the Lagrange multipliers.

Relations (17) correspond to the diagonal-block consistency in (14).

One can solve the I-SusP by using the following way. When $\boldsymbol{m} = \hat{\boldsymbol{m}}(\boldsymbol{\Lambda})$, where $\hat{\boldsymbol{m}}(\boldsymbol{\Lambda}) := \arg\min_{\boldsymbol{m}} \mathcal{G}_1(\boldsymbol{m}, \boldsymbol{\Lambda})$, relations (17) can be rewritten as

$$\hat{\chi}_{[i,k],[i,l]}(\boldsymbol{\Lambda}) = v_{k,l}^{(i)}(\hat{\boldsymbol{m}}_i(\boldsymbol{\Lambda})) - \hat{m}_{[i,k]}(\boldsymbol{\Lambda})\hat{m}_{[i,l]}(\boldsymbol{\Lambda}), \tag{18}$$

where $\hat{\chi}_{[i,k],[j,l]}(\boldsymbol{\Lambda}) := \partial \hat{m}_{[i,k]}(\boldsymbol{\Lambda})/\partial h_{[j,l]}$ (see (B.2) in Appendix B). Therefore, the framework of I-SusP is summarized as follows:

STEP 1 Given $\boldsymbol{\Lambda}$, minimize $\mathcal{G}_1(\boldsymbol{m}, \boldsymbol{\Lambda})$ with respect to $\boldsymbol{m}$: $\hat{\boldsymbol{m}}(\boldsymbol{\Lambda}) = \arg\min_{\boldsymbol{m}} \mathcal{G}_1(\boldsymbol{m}, \boldsymbol{\Lambda})$

STEP 2 For the given $\boldsymbol{\Lambda}$ and $\hat{\boldsymbol{m}}(\boldsymbol{\Lambda})$, compute $\hat{\chi}_{[i,k],[j,l]}(\boldsymbol{\Lambda}) = \partial \hat{m}_{[i,k]}(\boldsymbol{\Lambda})/\partial h_{[j,l]}$.

STEP 3 Solve (18) with respect to $\boldsymbol{\Lambda}$.

STEP 4 Repeat from STEP1 to STEP3 until convergence.

Equation (18) is termed *the diagonal-block matching equation* in this paper, because it leads to the matching of the diagonal-blocks of the susceptibilities obtained by an approximation with those obtained by an exact evaluation. Note that, because the parameters $\boldsymbol{\Lambda}$ are treated as independent variables in (18), the derivatives of $\boldsymbol{\Lambda}$ with respect to $\boldsymbol{h}$ are zero in the right hand side of (18). An alternative interpretation of the diagonal-block consistency in (17) and the diagonal-block equation was provided in Appendix C.

When $\mathcal{G}_0(\boldsymbol{m}) = \mathcal{G}(\boldsymbol{m})$, parameters $\boldsymbol{\Lambda}$ automatically vanish because the diagonal-block consistency always holds in the true GFE (see Appendix C). However, if $\mathcal{G}_0(\boldsymbol{m})$ is an approximate GFE that breaks the diagonal-block consistency, parameters $\boldsymbol{\Lambda}$ remain to enforce the diagonal-block consistency.

4.2. Minimization of Extended Bethe-Gibbs Free Energy

In this section, we construct the extended GFE, $\mathcal{G}_1(\boldsymbol{m}, \boldsymbol{\Lambda})$, by using a Bethe free energy, and subsequently, we derive a message-passing equation for minimization $\mathcal{G}_1(\boldsymbol{m}, \boldsymbol{\Lambda})$ with respect to $\boldsymbol{m}$.

In the context of the Bethe approximation based on the cluster variational method presented in [17], the HFE, $F(\boldsymbol{h}, \boldsymbol{J})$, is approximated by $F(\boldsymbol{h}, \boldsymbol{J}) \approx F_0(\boldsymbol{h}, \boldsymbol{J}) = \min_{\boldsymbol{z}} F_B(\boldsymbol{z} \mid \boldsymbol{h}, \boldsymbol{J})$, where

$$\begin{aligned} &F_B(\boldsymbol{z} \mid \boldsymbol{h}, \boldsymbol{J}) \\ &:= \min_{\boldsymbol{\xi}} \Big(-\boldsymbol{h}^{\mathrm{t}}\boldsymbol{z} - \frac{1}{2}\sum_{(i,j)\in E}\sum_{k,l=1}^{q-1} J_{[i,k],[j,l]}\xi_{[i,k],[j,l]} + \sum_{i\in V}(1-|\partial(i)|)\sum_{x_i=0}^{q-1}\mathcal{P}_i(x_i \mid \boldsymbol{z}_i)\ln \mathcal{P}_i(x_i \mid \boldsymbol{z}_i) \\ &+ \frac{1}{2}\sum_{(i,j)\in E}\sum_{x_i,x_j=0}^{q-1}\mathcal{P}_{(i,j)}(x_i, x_j \mid \boldsymbol{z}_i, \boldsymbol{z}_j)\ln \mathcal{P}_{(i,j)}(x_i, x_j \mid \boldsymbol{z}_i, \boldsymbol{z}_j)\Big) \end{aligned} \tag{19}$$

is the Bethe free energy for (5) expressed by the moment representation. Marginal distributions $\mathcal{P}_i(x_i \mid \boldsymbol{z}_i)$ and $\mathcal{P}_{(i,j)}(x_i, x_j \mid \boldsymbol{z}_i, \boldsymbol{z}_j)$ are defined by $\mathcal{P}_i(x_i \mid \boldsymbol{z}_i) = q^{-1} + \sum_{k=1}^{q-1} z_{[i,k]}\phi_k(x_i)$ and

$$\mathcal{P}_{(i,j)}(x_i, x_j \mid \boldsymbol{z}_i, \boldsymbol{z}_j) := \frac{1}{q^2} + \frac{1}{q}\sum_{k=1}^{q-1}\big(z_{[i,k]}\phi_k(x_i) + z_{[j,k]}\phi_k(x_j)\big) + \sum_{k,l=1}^{q-1}\xi_{[i,k],[j,l]}\phi_k(x_i)\phi_l(x_j),$$

respectively [17]. It is noteworthy that the minimization of the Bethe free energy with respect to $\boldsymbol{z}$ can be done without any constraints, because, by the properties of the present orthogonal set and by the fact that $\phi_0(x)$ is a constant, marginal distributions $\mathcal{P}_i(x_i \mid \boldsymbol{z}_i)$ and $\mathcal{P}_{(i,j)}(x_i, x_j \mid \boldsymbol{z}_i, \boldsymbol{z}_j)$ always satisfy the normalization and reducibility conditions, i.e., $\sum_{x_i=0}^{q-1}\mathcal{P}_i(x_i \mid \boldsymbol{z}_i) = 1$,

$\sum_{x_i,x_j=0}^{q-1} \mathcal{P}_{(i,j)}(x_i, x_j \mid \boldsymbol{z}_i, \boldsymbol{z}_j) = 1$, and $\sum_{x_j=0}^{q-1} \mathcal{P}_{(i,j)}(x_i, x_j \mid \boldsymbol{z}_i, \boldsymbol{z}_j) = \mathcal{P}_i(x_i \mid \boldsymbol{z}_i)$. These can be verified by using the relation

$$\sum_{x=0}^{q-1} \phi_k(x) = \sqrt{q}\delta_{k,0}. \tag{20}$$

This is one of merits of the orthonormal expansion employed in this paper.

By employing the Bethe free energy as $F_0(\boldsymbol{\gamma}, \boldsymbol{J})$ in (15), expression $\mathcal{G}_0(\boldsymbol{m})$, namely the Bethe-Gibbs free energy, is expressed as

$$\mathcal{G}_0(\boldsymbol{m}) = -\boldsymbol{h}^{\mathrm{t}}\boldsymbol{m} + \max_{\boldsymbol{\gamma}} \big(\boldsymbol{\gamma}^{\mathrm{t}}\boldsymbol{m} + \min_{\boldsymbol{z}} F_B(\boldsymbol{z} \mid \boldsymbol{\gamma}, \boldsymbol{J})\big).$$

The maximum condition of the right hand side of this equation with respect to $\boldsymbol{\gamma}$ leads to $\boldsymbol{m} = \boldsymbol{z}^*$, where $\boldsymbol{z}^*$ are the values of $\boldsymbol{z}$ that minimize $F_B(\boldsymbol{z} \mid \boldsymbol{\gamma}, \boldsymbol{J})$. By using this relation we can reach $\mathcal{G}_0(\boldsymbol{m}) = F_B(\boldsymbol{m} \mid \boldsymbol{h}, \boldsymbol{J})$. Therefore, the extended Bethe-Gibbs free energy corresponding to (16) is expressed as

$$\mathcal{G}_1(\boldsymbol{m}, \boldsymbol{\Lambda}) = F_B(\boldsymbol{m} \mid \boldsymbol{h}, \boldsymbol{J}) - \frac{1}{2}\sum_{i\in V}\sum_{k,l=1}^{q-1} \Lambda_{[i,k],[i,l]}\Big(v_{k,l}^{(i)}(\boldsymbol{m}_i) - m_{[i,k]}m_{[i,l]}\Big) \tag{21}$$

in this case.

In the following, let us derive a message-passing equation to minimize $\mathcal{G}_1(\boldsymbol{m}, \boldsymbol{\Lambda})$ with respect to $\boldsymbol{m}$. The minimum conditions of (21) with respect to $\boldsymbol{m}$ yield

$$-H_{[i,k]} + \big(1 - |\partial(i)|\big)\sum_{x_i=0}^{q-1} \phi_k(x_i)\ln \mathcal{P}_i(x_i \mid \boldsymbol{m}_i) + \frac{1}{q}\sum_{j\in\partial(i)}\sum_{x_i,x_j=0}^{q-1} \phi_k(x_i)\ln \mathcal{P}_{(i,j)}(x_i, x_j \mid \boldsymbol{m}_i, \boldsymbol{m}_j) = 0, \tag{22}$$

where

$$H_{[i,k]} := h_{[i,k]} + \frac{1}{2}\sum_{a,b=1}^{q-1} \Lambda_{[i,a],[i,b]}T_{a,b,k} - \sum_{a=1}^{q-1} \Lambda_{[i,k],[i,a]}m_{[i,a]}.$$

The minimum conditions in $F_B(\boldsymbol{m} \mid \boldsymbol{h}, \boldsymbol{J})$ with respect to $\boldsymbol{\xi}$ lead to

$$-J_{[i,k],[j,l]} + \sum_{x_i,x_j=0}^{q-1} \phi_k(x_i)\phi_l(x_j)\ln \mathcal{P}_{(i,j)}(x_i, x_j \mid \boldsymbol{m}_i, \boldsymbol{m}_j) = 0. \tag{23}$$

By combining (23) with (22), we arrive at the message-passing equation expressed by

$$\mathcal{M}_{j\to i}^{(k)} = \sum_{x_i=0}^{q-1} \phi_k(x_i)\ln\sum_{x_j=0}^{q-1} \exp\Big(\mathcal{H}_{j\to i}(x_i, x_j)\Big), \tag{24}$$

where $\mathcal{H}_{j\to i}(x_i, x_j) := \sum_{a=1}^{q-1}\big(H_{[j,a]} + \sum_{r\in\partial(j)\setminus\{i\}} \mathcal{M}_{r\to j}^{(a)}\big)\phi_a(x_j) + \sum_{a,b=1}^{q-1} J_{[i,a],[j,b]}\phi_a(x_i)\phi_b(x_j)$. The expression $\mathcal{M}_{j\to i}^{(k)}$ denotes a message from node j to node i.

By using solutions to the message-passing equation, the values of $\boldsymbol{m}$ that minimize the extended Bethe free energy (21) are expressed by

$$m_{[i,k]} = \frac{1}{Z_i}\sum_{x_i=0}^{q-1}\phi_k(x_i)\exp\Big\{\sum_{a=1}^{q-1}\Big(H_{[i,a]} + \sum_{j\in\partial(i)}\mathcal{M}_{j\to i}^{(a)}\Big)\phi_a(x_i)\Big\}, \tag{25}$$

where expression Z_i is the normalizing constant defined by

$$Z_i := \sum_{x_i=0}^{q-1}\exp\Big\{\sum_{a=1}^{q-1}\Big(H_{[i,a]} + \sum_{j\in\partial(i)}\mathcal{M}_{j\to i}^{(a)}\Big)\phi_a(x_i)\Big\}.$$

Equations (24) and (25) are obtained by similar manipulations presented in [17]. The details of these derivation are described in Appendix D. By solving (24) and (25), we obtain the values of $\boldsymbol{m}$ that minimize (21) for a given $\boldsymbol{\Lambda}$. Note that solutions to (24) and (25) are equivalent to those to the usual BP if $\boldsymbol{\Lambda} = \boldsymbol{0}$ are fixed.

4.3. Improved Susceptibility Propagation using Bethe Approximation

In this section, we derive the the diagonal-block equation corresponding to (18) for (21).

By differentiating (25) with respect to $h_{[j,l]}$, we obtain

$$\hat{\chi}_{[i,k],[j,l]} = \sum_{a=1}^{q-1}\Big(\delta_{i,j}\delta_{a,l} - \sum_{b=1}^{q-1}\Lambda_{[i,a],[i,b]}\hat{\chi}_{[i,b],[j,l]} + \sum_{r\in\partial(i)}\frac{\partial\mathcal{M}_{r\to i}^{(a)}}{\partial h_{[j,l]}}\Big)\big(v_{k,a}^{(i)}(\boldsymbol{m}_i) - m_{[i,k]}m_{[i,a]}\big), \tag{26}$$

where $\hat{\chi}_{[i,k],[j,l]} = \partial m_{[i,k]}/\partial h_{[j,l]}$. The derivatives of messages are obtained by solving

$$\frac{\partial\mathcal{M}_{j\to i}^{(k)}}{\partial h_{[u,v]}} = \sum_{a=1}^{q-1}\Big(\delta_{j,u}\delta_{a,v} - \sum_{b=1}^{q-1}\Lambda_{[j,a],[j,b]}\hat{\chi}_{[j,b],[u,v]} + \sum_{r\in\partial(j)\setminus\{i\}}\frac{\partial\mathcal{M}_{r\to j}^{(a)}}{\partial h_{[u,v]}}\Big)$$
$$\times\sum_{x_i=0}^{q-1}\phi_k(x_i)\frac{\sum_{x_j=0}^{q-1}\phi_a(x_j)\exp\big(\mathcal{H}_{j\to i}(x_i,x_j)\big)}{\sum_{x_j=0}^{q-1}\exp\big(\mathcal{H}_{j\to i}(x_i,x_j)\big)}. \tag{27}$$

Equation (27) is obtained by differentiating (24) with respect to $h_{[u,v]}$. By letting $i = j$ in (26) and by using the relation in (18), i.e., $\hat{\chi}_{[i,k],[i,l]} = v_{k,l}^{(i)}(\boldsymbol{m}_i) - m_{[i,k]}m_{[i,l]}$, we have

$$\sum_{a,b=1}^{q-1}\Lambda_{[i,a],[i,b]}\big[\boldsymbol{\nu}_i\big]_{b,l}\big[\boldsymbol{\nu}_i\big]_{k,a} = \sum_{a=1}^{q-1}\big[\boldsymbol{\nu}_i\big]_{k,a}\sum_{j\in\partial(i)}\frac{\partial\mathcal{M}_{j\to i}^{(a)}}{\partial h_{[i,l]}}, \tag{28}$$

where $(q-1)\times(q-1)$ matrix $\boldsymbol{\nu}_i$ is defined by $\big[\boldsymbol{\nu}_i\big]_{k,l} := v_{k,l}^{(i)}(\boldsymbol{m}_i) - m_{[i,k]}m_{[i,l]}$. In the I-SusP based on the Bethe approximation, (28) corresponds to the diagonal-block matching equation in (18). Relations in (28) are collectively expressed as $\boldsymbol{\nu}_i\boldsymbol{\Lambda}_i\boldsymbol{\nu}_i = \boldsymbol{K}_i$, where $(q-1)\times(q-1)$ matrices $\boldsymbol{\Lambda}_i$ and $\boldsymbol{K}_i$ are defined by

$$\big[\boldsymbol{\Lambda}_i\big]_{k,l} := \Lambda_{[i,k],[i,l]} \quad\text{and}\quad \big[\boldsymbol{K}_i\big]_{k,l} := \sum_{a=1}^{q-1}\big(v_{k,a}^{(i)}(\boldsymbol{m}_i) - m_{[i,k]}m_{[i,a]}\big)\sum_{j\in\partial(i)}\frac{\partial\mathcal{M}_{j\to i}^{(a)}}{\partial h_{[i,l]}},$$

respectively. Therefore, we obtain

$$\boldsymbol{\Lambda}_i = \boldsymbol{\nu}_i^{-1} \boldsymbol{K}_i \boldsymbol{\nu}_i^{-1}. \tag{29}$$

By this equation, we can obtain the values of $\boldsymbol{\Lambda}$ that satisfy the diagonal-block matching equation in (28).

By simultaneously solving (24)–(27), and (29), we obtain solutions to the proposed I-SusP based on the Bethe approximation. It is noteworthy that the I-SusP is reduced to a usual SusP if $\boldsymbol{\Lambda} = \boldsymbol{0}$ are fixed.

5. Application to Generalized Sparse Priors for Natural Images

An image prior model using MRF, called the generalized sparse prior, which takes the form

$$\theta_i(x_i) = 0, \quad w_{(i,j)}(x_i, x_j) = -\alpha_{ij}|x_i - x_j|^p \tag{30}$$

was proposed [18], where $\alpha_{ij} = \alpha_{ji}$. This image prior includes some conventional image priors as special cases, because it is reduced to Q-Ising image priors when $p = 2$ and is reduced to Q-Potts priors when $p \to 0$. By changing the value of p, we can adjust the smoothness, the flatness, and the appearance frequency of edges in images. It is believed that values of p near 0.5 are appropriate for natural images.

In the following, let us consider a direct problem and an inverse problem on the generalized sparse prior on an $L \times L$ grid system, in which interaction parameters $\boldsymbol{\alpha}$ are independently drawn from Gaussian $p(\alpha_{ij} \mid \sigma) := (\sqrt{2\pi\sigma^2})^{-1} \exp(-\alpha_{ij}^2/2\sigma^2)$ and the flatness parameter p is set to 0.5.

5.1. Direct Problem on Generalized Sparse Priors

Given $\boldsymbol{\alpha} = \{\alpha_{ij}\}$ and p, in order to compute statistical quantities on a generalized sparse prior, let us apply the I-SusP proposed in the previous section to the generalized sparse priors.

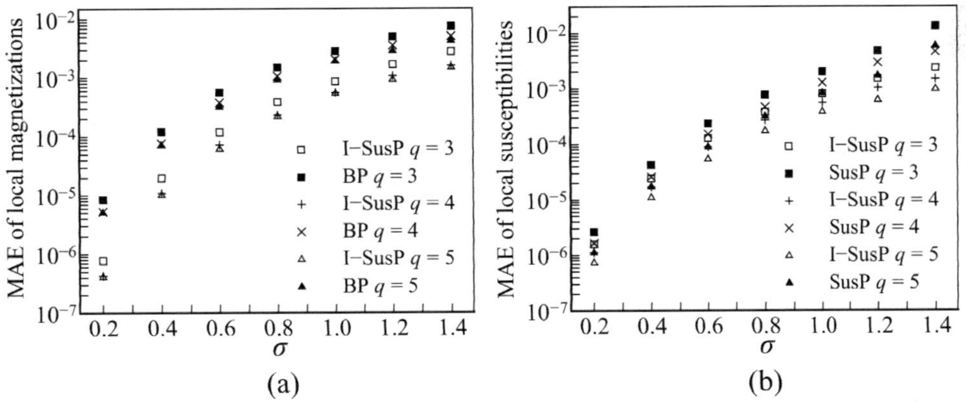

Figure 1. Plots of results of the direct problem on the generalized sparse priors when $L = 3$. (a): Comparison of local magnetizations obtained by the I-SusP with ones obtained by the BP. (b): Comparison local susceptibilities obtained by the I-SusP with those obtained by the BP. Each plot is averaged over 1000 trials.

In figure 1, mean absolute errors (MAEs) of magnetizations and susceptibilities between true values and values obtained by approximate methods against σ are plotted when $L = 3$. We see that the proposed I-SusP overcomes the normal SusP.

5.2. Inverse Problem using Observations Generated from Generalized Sparse Priors

In this section, let us solve an inverse problem, the so called inverse Ising problem, using the I-SusP without specifying the functional forms of potential functions $\{\theta_i(x_i)\}$ and $\{w_{(i,j)}(x_i, x_j)\}$. That is, we consider a non-parametric learning in this section.

Suppose that M data points, $\mathcal{D} = \{\mathbf{x}^{(\mu)} \mid \mu = 1, 2, \ldots, M\}$, are given. The aim of the inverse problem on which we focus in this section is to find the values of parameters $\boldsymbol{h}$ and $\boldsymbol{J}$ that maximize the log-likelihood function

$$l_{\mathcal{D}}(\boldsymbol{h}, \boldsymbol{J}) := \frac{1}{M}\sum_{\mu=1}^{M} \ln P_G(\mathbf{x}^{(\mu)} \mid \boldsymbol{h}, \boldsymbol{J}). \tag{31}$$

The derivatives of the log-likelihood function with respect to $\boldsymbol{h}$ and $\boldsymbol{J}$ are

$$\Delta_{h_{[i,k]}}(\boldsymbol{h}, \boldsymbol{J}) := \frac{\partial l_{\mathcal{D}}(\boldsymbol{h}, \boldsymbol{J})}{\partial h_{[i,k]}} = \langle \phi_k(x_i)\rangle_{\mathcal{D}} - \sum_{\boldsymbol{x}} \phi_k(x_i) P_G(\boldsymbol{x} \mid \boldsymbol{h}, \boldsymbol{J})$$

and

$$\Delta_{J_{[i,k],[j,l]}}(\boldsymbol{h}, \boldsymbol{J}) := \frac{\partial l_{\mathcal{D}}(\boldsymbol{h}, \boldsymbol{J})}{\partial J_{[i,k],[j,l]}} = \langle \phi_k(x_i)\phi_l(x_j)\rangle_{\mathcal{D}} - \sum_{\boldsymbol{x}} \phi_k(x_i)\phi_l(x_j) P_G(\boldsymbol{x} \mid \boldsymbol{h}, \boldsymbol{J}),$$

respectively, where $\langle \cdots \rangle_{\mathcal{D}}$ is a sample average of given data points. Because the log-likelihood function (31) is a concave function with respect to the parameters, a maximum point of the log-likelihood can be found by a gradient ascent method:

$$h_{[i,k]}^{\mathrm{new}} = h_{[i,k]}^{\mathrm{old}} + \eta \Delta_{h_{[i,k]}}(\boldsymbol{h}^{\mathrm{old}}, \boldsymbol{J}^{\mathrm{old}}), \tag{32}$$

$$J_{[i,k],[j,l]}^{\mathrm{new}} = J_{[i,k],[j,l]}^{\mathrm{old}} + \eta \Delta_{J_{[i,k],[j,l]}}(\boldsymbol{h}^{\mathrm{old}}, \boldsymbol{J}^{\mathrm{old}}), \tag{33}$$

where η is a small positive constant.

To apply the I-SusP to the inverse problem, the derivatives are approximated by

$$\Delta_{h_{[i,k]}}(\boldsymbol{h}, \boldsymbol{J}) \approx \Delta_{h_{[i,k]}}^{\mathrm{app}}(\boldsymbol{h}, \boldsymbol{J}) := \langle \phi_k(x_i)\rangle_{\mathcal{D}} - m_{i,k}, \tag{34}$$

$$\Delta_{J_{[i,k],[j,l]}}(\boldsymbol{h}, \boldsymbol{J}) \approx \Delta_{J_{[i,k],[j,l]}}^{\mathrm{app}}(\boldsymbol{h}, \boldsymbol{J}) := \langle \phi_k(x_i)\phi_l(x_j)\rangle_{\mathcal{D}} - \left(\bar{\chi}_{[i,k],[j,l]} + m_{i,k} m_{j,l}\right), \tag{35}$$

where $\boldsymbol{m}$ and $\bar{\boldsymbol{\chi}}$ are solutions to the I-SusP. By solving (24)–(27), (29), and (32) and (33) with the approximations in (34) and (35), we can solve the inverse problem by using the I-SusP for the given data set $\mathcal{D}$. If $\boldsymbol{\Lambda} = \boldsymbol{0}$ are fixed, we can obtain solutions based on the SusP. In the following, results of the inverse problem using data points generated from the generalized sparse priors are shown.

In figure 2, MAEs between true biases $\boldsymbol{h}$ and interactions $\boldsymbol{J}$ and those obtained by approximate methods are plotted for $q = 3$ and $q = 4$ when $L = 3$. Here, instead of sampled expectations, true statistical values of the generalized sparse priors are used, namely, we consider an identical case where $M \to \infty$. The results of the BP in these plots are obtained by using a non-iterative method proposed in [17]. The results obtained by the SusP are poor and nearly the same as those obtained by the BP in the region of large σ. In contrast, the results obtained by the I-SusP are improved for the whole region.

Results of the inverse problem when $L = 8$ and $q = 3$ for $\sigma = 0.4$ and $\sigma = 0.8$ are shown in table 1. In these experiments, $M = 10000$ data points, that are generated by using the Markov chain Monte Carlo (MCMC) method on the generalized sparse priors, are used. We can see that the I-SusP gives the best results.

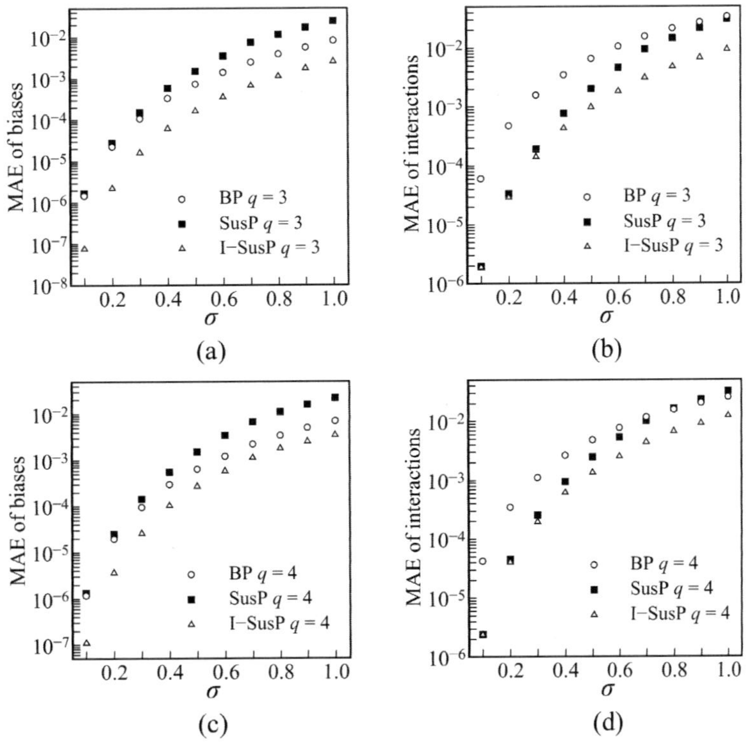

Figure 2. Plots of results of the inverse problem on the generalized sparse priors when $L = 3$. (a) and (c): MAEs of biases obtained by the BP, the SusP, and the I-SusP when $q = 3$ and $q = 4$, respectively. (b) and (d): MAEs of interactions obtained by the BP, the SusP, and the I-SusP when $q = 3$ and $q = 4$, respectively. Each plot is averaged over 1000 trials.

Table 1. Results of the inverse problem on the generalized sparse priors when $L = 8$ and $q = 3$. $M = 10000$ data points generated by using MCMC method on the generalized sparse prior are used. Each value is averaged over 30 trials.

	σ	MAE of biases	MAE of interactions
BP	0.4	0.0153	0.0262
	0.8	0.0197	0.0417
SusP	0.4	0.0153	0.0248
	0.8	0.0295	0.0365
I-SusP	0.4	0.0152	0.0246
	0.8	0.0184	0.0278

6. Conclusion and Discussion

In this paper, the generalization scheme of I-SusP using the orthogonal expansion was proposed in which any pairwise potential functions and multi-valued random variables are acceptable. Moreover, the proposed I-SusP is applied to the direct and the inverse problems on generalized sparse priors for natural images. The constraints of diagonal-block consistencies improved the performances of the algorithms and yielded results that surpass those obtained by conventional methods.

Although the orthogonal expansion using Chebyshev polynomials employed in this paper should not be the only possible choice for a generalization, it enables us to reach moment representations of free energies as presented in (19) and can make subsequent manipulations clear. This could help us in future developments, especially in inverse problems, as it helped us in inverse problems of the normal BP level [17].

In the usual SusP scheme, we first solve the BP ((24) and (25) with $\mathbf{\Lambda} = \mathbf{0}$), and obtain solutions to the BP. These solutions are the values of the local magnetizations and the messages. Subsequently, to obtain the susceptibilities, we solve the SusP ((26) and (27) with $\mathbf{\Lambda} = \mathbf{0}$) by using the solutions of the BP. Therefore, in the conventional scheme shown in figure 3 (a), the results of the SusP have no effect on the BP. In contrast, the scheme of I-SusP shown in

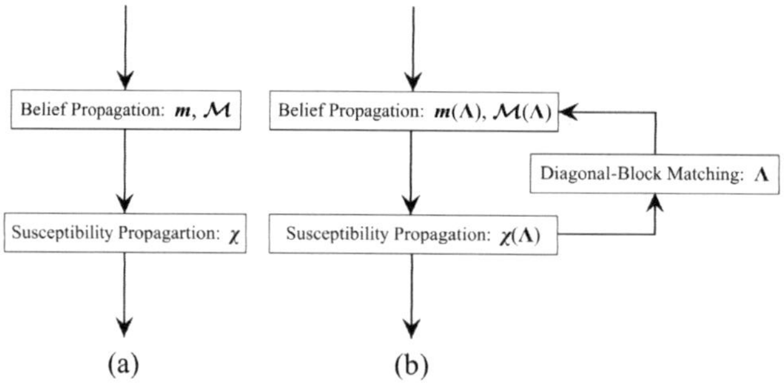

Figure 3. Illustration of schemes of (a) normal SusP and (b) I-SusP.

figure 3 (b) includes feedback from the SusP to the BP through parameters $\mathbf{\Lambda}$. This feedback marks a significant difference between the scheme of conventional SusP and the scheme of I-SusP. It is noteworthy that the computational cost of solving the I-SusP is the same as that of solving the usual SusP, because the diagonal-block matching equation in (29) does not contribute to the order of the total computational cost.

One of the most interesting future applications is a direct and an inverse problem in the deep Boltzmann machine [19, 20], which is a Boltzmann machine with a hierarchical structure and one of the most important learning models in the recent machine learning field.

Acknowledgments
This work was partly supported by Grants-In-Aid (No. 24700220) for Scientific Research from the Ministry of Education, Culture, Sports, Science and Technology, Japan.

Appendix A. Chebyshev Polynomials

One possible choice of orthonormal set is given by [21]

$$\phi_k(x) := (-1)^k \psi_k(x) \sqrt{\frac{(2k+1)[(q-1)!]^2}{(q+k)!(q-k-1)!}} \tag{A.1}$$

for $x = 0, 1, \ldots, q-1$ and $k = 0, 1, \ldots, q-1$, where $\{\psi_k(x) \mid k = 0, 1, \ldots, q-1\}$ are the Chebyshev polynomials in the discrete range given by

$$\psi_k(x) := \sum_{l=0}^{k} (-1)^l \binom{k}{l} \binom{k+l}{l} \frac{x!(q-l-1)!}{\Gamma(x-l+1)(q-1)!}, \tag{A.2}$$

where $\Gamma(x)$ is the gamma function. The Chebyshev polynomials (A.2) are also given by the recursion formula

$$(k+1)(q-1-k)\psi_{k+1}(x) = -(2x-q+1)(2k+1)\psi_k(x) - k(q+k)\psi_{k-1}(x),$$

starting from $\psi_0(x) = 1$ and $\psi_1(x) = 1 - 2x/(q-1)$. From (A.1) and (A.2), we have

$$\phi_0(x) = 1/\sqrt{q}, \phi_1(x) = \sqrt{3}(2x-q+1)/\sqrt{q(q^2-1)},$$

and so on.

Appendix B. Linear Response Relation for Approximate GFEs

For $\boldsymbol{m} = \{m_{[i,k]} \mid i \in V; k = 1, 2, \ldots, q-1\}$, let us consider a function expressed in the form

$$\hat{\mathcal{G}}(\boldsymbol{m}) = -\boldsymbol{h}^{\mathrm{t}}\boldsymbol{m} + \hat{g}(\boldsymbol{m}).$$

It is to be noted that this expression includes all the GFEs ($\mathcal{G}(\boldsymbol{m})$, $\mathcal{G}_0(\boldsymbol{m})$, and $\mathcal{G}_1(\boldsymbol{m})$) presented in this paper. At a minimum of the function, from its extremal condition, relations

$$-h_{[i,k]} + \left.\frac{\partial \hat{g}(\boldsymbol{m})}{\partial m_{[i,k]}}\right|_{\boldsymbol{m}=\boldsymbol{m}^\star} = 0 \tag{B.1}$$

hold, where $\boldsymbol{m}^\star := \arg\min_{\boldsymbol{m}} \hat{\mathcal{G}}(\boldsymbol{m})$. By differentiating (B.1) with respect to $h_{[j,l]}$, we obtain

$$\delta_{i,j} = \sum_{r\in V}\sum_{a=1}^{q-1} \left.\frac{\partial^2 \hat{g}(\boldsymbol{m})}{\partial m_{[i,k]}\partial m_{[r,a]}}\right|_{\boldsymbol{m}=\boldsymbol{m}^\star} \frac{\partial m^\star_{[r,a]}}{\partial h_{[j,l]}} = \sum_{r\in V}\sum_{a=1}^{q-1} \left.\frac{\partial^2 \hat{\mathcal{G}}(\boldsymbol{m})}{\partial m_{[i,k]}\partial m_{[r,a]}}\right|_{\boldsymbol{m}=\boldsymbol{m}^\star} \frac{\partial m^\star_{[r,a]}}{\partial h_{[j,l]}}.$$

This expression indicates that the relation

$$\hat{\boldsymbol{H}}^{-1} = \boldsymbol{\chi}^\star \tag{B.2}$$

holds, where matrix $\hat{\boldsymbol{H}}$ denotes the Hessian matrix of $\hat{\mathcal{G}}(\boldsymbol{m})$ defined by

$$[\hat{\boldsymbol{H}}(\boldsymbol{m})]_{e(i,k),e(j,l)} := \left.\frac{\partial^2 \hat{\mathcal{G}}(\boldsymbol{m})}{\partial m_{[i,k]}\partial m_{[j,l]}}\right|_{\boldsymbol{m}=\boldsymbol{m}^\star},$$

and matrix $\boldsymbol{\chi}^\star$ is defined by $[\boldsymbol{\chi}^\star]_{e(i,k),e(j,l)} := \partial m^\star_{[i,k]}/\partial h_{[j,l]}$.

Appendix C. Variational Interpretation of Diagonal-Block Equation

The diagonal-block matching equation in (18) is introduced to overcome the diagonal-block inconsistency problem. Here, we provide an alternative interpretation of the equation.

First, we define a measure of closeness between two positive definite symmetric matrices, $\boldsymbol{A}$ and $\boldsymbol{B}$, in terms of the Kullback-Leibler divergence as

$$D(\boldsymbol{A}||\boldsymbol{B}) := \int_{-\infty}^{\infty} \mathrm{d}\boldsymbol{x}\, \mathcal{N}(\boldsymbol{x} \mid \boldsymbol{A}) \ln \frac{\mathcal{N}(\boldsymbol{x} \mid \boldsymbol{A})}{\mathcal{N}(\boldsymbol{x} \mid \boldsymbol{B})},$$

where $\mathcal{N}(\boldsymbol{x} \mid \boldsymbol{\Sigma}) \propto \exp(-2^{-1}\boldsymbol{x}^{\mathrm{t}}\boldsymbol{\Sigma}\boldsymbol{x})$ represents the multivariate zero-mean Gaussian with the covariance matrix $\boldsymbol{\Sigma}^{-1}$. From the properties of the Kullback-Leibler divergence, $D(\boldsymbol{A}||\boldsymbol{B}) \geq 0$ and $D(\boldsymbol{A}||\boldsymbol{B}) = 0$ if and only if $\boldsymbol{A} = \boldsymbol{B}$ is ensured. Let us consider the quantity

$D(\boldsymbol{H}(\boldsymbol{m})||\boldsymbol{H}_1(\boldsymbol{m},\boldsymbol{\Lambda}))$, where $\boldsymbol{H}(\boldsymbol{m})$ is the Hessian matrix of true GFE defined in section 3.2 and $\boldsymbol{H}_1(\boldsymbol{m},\boldsymbol{\Lambda})$ is the Hessian matrix of $\mathcal{G}_1(\boldsymbol{m},\boldsymbol{\Lambda})$ defined in (4.1). Minimizing $D(\boldsymbol{H}(\boldsymbol{m})||\boldsymbol{H}_1(\boldsymbol{m},\boldsymbol{\Lambda}))$ corresponds to minimizing the distance [3] between the true Hessian matrix and Hessian matrix $\boldsymbol{H}_1(\boldsymbol{m},\boldsymbol{\Lambda})$.

The minimum conditions of $D(\boldsymbol{H}(\boldsymbol{m})||\boldsymbol{H}_1(\boldsymbol{m},\boldsymbol{\Lambda}))$ with respect to $\boldsymbol{\Lambda}$ are

$$v_{k,l}^{(i)}(\boldsymbol{m}_i) - m_{[i,k]}m_{[i,l]} = \left[\boldsymbol{H}_1(\boldsymbol{m},\boldsymbol{\Lambda})^{-1}\right]_{e(i,k),e(i,l)}, \tag{C.1}$$

where we use the fact that $\left[\boldsymbol{H}(\boldsymbol{m},\boldsymbol{\Lambda})^{-1}\right]_{e(i,k),e(i,l)} = v_{k,l}^{(i)}(\boldsymbol{m}_i) - m_{[i,k]}m_{[i,l]}$ holds for any $\boldsymbol{m}$ in (14). The minimum conditions in (C.1) correspond to (17) for any $\boldsymbol{m}$. When $\boldsymbol{m} = \hat{\boldsymbol{m}}(\boldsymbol{\Lambda})$, upon using (B.2), (C.1) yields (18). Therefore, we can reinterpret the diagonal-block matching equation as the condition of minimization of distance between the true Hessian matrix and its approximation in terms of the Kullback-Leibler divergence at $\boldsymbol{m} = \hat{\boldsymbol{m}}(\boldsymbol{\Lambda})$.

When $\mathcal{G}_0(\boldsymbol{m}) = \mathcal{G}(\boldsymbol{m})$, minimization of $D(\boldsymbol{H}(\boldsymbol{m})||\boldsymbol{H}_1(\boldsymbol{m},\boldsymbol{\Lambda}))$ obviously leads to $\boldsymbol{\Lambda} = \boldsymbol{0}$ because $\boldsymbol{H}_1(\boldsymbol{m},\boldsymbol{\Lambda}) = \boldsymbol{H}(\boldsymbol{m},\boldsymbol{\Lambda}) + \boldsymbol{\Lambda}$, where $\boldsymbol{\Lambda}$ is the diagonal-block matrix defined by $[\boldsymbol{\Lambda}]_{e(i,k),e(j,l)} := \delta_{i,j}\Lambda_{[i,k],[i,l]}$.

Appendix D. Derivation of Message-Passing Equation in (24)

As mentioned in section 4.2, the one-variable and the two-variable marginal distributions are expressed as

$$\mathcal{P}_i(x_i \mid \boldsymbol{m}_i) = \frac{1}{q} + \sum_{k=1}^{q-1} m_{[i,k]}\phi_k(x_i), \tag{D.1}$$

$$\mathcal{P}_{(i,j)}(x_i, x_j \mid \boldsymbol{m}_i, \boldsymbol{m}_j) = \frac{1}{q^2} + \frac{1}{q}\sum_{k=1}^{q-1}\big(m_{[i,k]}\phi_k(x_i) + m_{[j,k]}\phi_k(x_j)\big) + \sum_{k,l=1}^{q-1}\xi_{[i,k],[j,l]}\phi_k(x_i)\phi_l(x_j). \tag{D.2}$$

Since relations $\sum_{x_j=0}^{q-1}\mathcal{P}_{(i,j)}(x_i,x_j \mid \boldsymbol{m}_i,\boldsymbol{m}_j) = \mathcal{P}_i(x_i \mid \boldsymbol{m}_i)$ always hold, relations

$$\sum_{x_i=0}^{q-1}\phi_k(x_i)\ln\mathcal{P}_i(x_i \mid \boldsymbol{m}_i) = \sum_{x_i=0}^{q-1}\phi_k(x_i)\ln\sum_{x_j=0}^{q-1}\mathcal{P}_{(i,j)}(x_i,x_j \mid \boldsymbol{m}_i,\boldsymbol{m}_j) \tag{D.3}$$

also hold.

We can express (22) as

$$-H_{[i,k]} + \sum_{x_i=0}^{q-1}\phi_k(x_i)\ln\mathcal{P}_i(x_i \mid \boldsymbol{m}_i) - \sum_{j\in\partial(i)}\mathcal{M}_{j\to i}^{(k)} = 0, \tag{D.4}$$

where

$$\mathcal{M}_{j\to i}^{(k)} := \sum_{x_i=0}^{q-1}\phi_k(x_i)\ln\mathcal{P}_i(x_i \mid \boldsymbol{m}_i) - \frac{1}{q}\sum_{x_i,x_j=0}^{q-1}\phi_k(x_i)\ln\mathcal{P}_{(i,j)}(x_i,x_j \mid \boldsymbol{m}_i,\boldsymbol{m}_j). \tag{D.5}$$

[3] This measure is generally not a distance in a precise mathematical sense, because $D(\boldsymbol{A}||\boldsymbol{B}) \neq D(\boldsymbol{B}||\boldsymbol{A})$.

Without loss of generalities, we can rewrite $\mathcal{P}_i(x_i \mid \boldsymbol{m}_i)$ as

$$\mathcal{P}_i(x_i \mid \boldsymbol{m}_i) = \exp\big(\ln \mathcal{P}_i(x_i \mid \boldsymbol{m}_i)\big) \propto \exp\Big(\sum_{k=1}^{q-1} c_{[i,k]}\phi_k(x_i)\Big), \tag{D.6}$$

where $c_{[i,k]} := \sum_{x_i=0}^{Q-1} \phi_k(x_i) \ln \mathcal{P}_i(x_i \mid \boldsymbol{m}_i)$ and relation $\phi_0(x_i) = 1/\sqrt{q}$ is used. Substituting (D.6) in (D.4) and using (20), we get

$$c_{[i,k]} = H_{[i,k]} + \sum_{j\in\partial(i)} \mathcal{M}^{(k)}_{j\to i}. \tag{D.7}$$

Therefore, from (D.6) and (D.7), $\mathcal{P}_i(x_i \mid \boldsymbol{m}_i)$ is expressed as

$$\mathcal{P}_i(x_i \mid \boldsymbol{m}_i) = \frac{1}{Z_i}\exp\Big\{\sum_{k=1}^{Q-1}\Big(H_{[i,k]} + \sum_{j\in\partial(i)} \mathcal{M}^{(k)}_{j\to i}\Big)\phi_k(x_i)\Big\}, \tag{D.8}$$

where Z_i is the normalization constant. On the other hand, from (D.1) and (20), relations

$$\sum_{x_i=0}^{q-1} \phi_k(x_i)\mathcal{P}_i(x_i \mid \boldsymbol{m}_i) = m_{[i,k]} \tag{D.9}$$

should always hold. Combining (D.8) with (D.9), we arrive at (25).

In the following, we derive an alternative expression of the two-variable marginal distribution in a similar way to above derivation. Without loss of generalities, we can rewrite $\mathcal{P}_{(i,j)}(x_i, x_j \mid \boldsymbol{m}_i, \boldsymbol{m}_j)$ as

$$\begin{aligned}\mathcal{P}_{(i,j)}(x_i, x_j \mid \boldsymbol{m}_i, \boldsymbol{m}_j) &= \exp\big(\ln \mathcal{P}_{(i,j)}(x_i, x_j \mid \boldsymbol{m}_i, \boldsymbol{m}_j)\big)\\ &\propto \exp\Big(\frac{1}{\sqrt{q}}\sum_{k=1}^{q-1} d_{[i,k],[j,0]}\phi_k(x_i) + \frac{1}{\sqrt{q}}\sum_{l=1}^{q-1} d_{[i,0],[j,l]}\phi_l(x_j) + \sum_{k,l=1}^{q-1} d_{[i,k],[j,l]}\phi_k(x_i)\phi_l(x_j)\Big),\end{aligned} \tag{D.10}$$

where $d_{[i,k],[j,l]} := \sum_{x_i,x_j=0}^{q-1} \phi_k(x_i)\phi_l(x_j) \ln \mathcal{P}_{(i,j)}(x_i, x_j \mid \boldsymbol{m}_i, \boldsymbol{m}_j)$. From (23), we find

$$d_{[i,k],[j,l]} = J_{[i,k],[j,l]} \tag{D.11}$$

for $1 \le k \le q-1$ and $1 \le l \le q-1$. Using (D.4), (D.5), and the definition of $d_{[i,k],[j,l]}$, we get

$$\begin{aligned}\mathcal{M}^{(k)}_{j\to i} &= \sum_{x_i=0}^{q-1} \phi_k(x_i)\ln \mathcal{P}_i(x_i \mid \boldsymbol{m}_i) - \frac{1}{\sqrt{q}}\sum_{x_i,x_j=0}^{q-1} \phi_k(x_i)\phi_0(x_j)\ln \mathcal{P}_{(i,j)}(x_i, x_j \mid \boldsymbol{m}_i, \boldsymbol{m}_j)\\ &= H_{[i,k]} + \sum_{j\in\partial(i)} \mathcal{M}^{(k)}_{j\to i} - \frac{1}{\sqrt{q}} d_{[i,k],[j,0]}.\end{aligned} \tag{D.12}$$

Similarly, we can obtain

$$\mathcal{M}^{(l)}_{i\to j} = H_{[j,l]} + \sum_{i\in\partial(j)} \mathcal{M}^{(l)}_{i\to j} - \frac{1}{\sqrt{q}} d_{[i,0],[j,l]}. \tag{D.13}$$

From (D.10)–(D.13), we find

$$\mathcal{P}_{(i,j)}(x_i, x_j \mid \boldsymbol{m}_i, \boldsymbol{m}_j) = \frac{1}{Z_{(i,j)}} \exp\Big\{ \sum_{k=1}^{q-1} \Big(H_{[i,k]} + \sum_{r\in\partial(i)\setminus\{j\}} \mathcal{M}^{(k)}_{r\to i} \Big) \phi_k(x_i) + \sum_{l=1}^{q-1} \Big(H_{[j,l]} + \sum_{r\in\partial(j)\setminus\{i\}} \mathcal{M}^{(l)}_{r\to j} \Big) \phi_l(x_j) + \sum_{k=1}^{Q-1} \sum_{l=l}^{Q-1} J^{(k,l)}_{(i,j)} \phi_k(x_i)\phi_l(x_j) \Big\}, \tag{D.14}$$

where $Z_{(i,j)}$ is the normalization constant.

Substituting (D.8) and (D.14) in (D.3) and using (2) and (20), we can reach the message-passing equation in (24).

References

[1] Opper M and Saad D (eds) 2001 *Advanced Mean Field Methods—Theory and Practice* (MIT Press)
[2] Mézard M and Montanari A 2009 *Information, Physics and Computation* (Oxford University Press)
[3] Pearl J 1988 *Probabilistic Reasoning in Intelligent Systems: Networks of Plausible Inference (2nd ed.)* (San Francisco, CA: Morgan Kaufmann)
[4] Bethe H A 1935 *Proceedings of the Royal Society of London. Series A: Mathematical and Physical Sciences* **150** 552–75
[5] Kabashima Y and Saad D 1998 *Europhys. Lett.* **44** 668–74
[6] Yedidia J S, Freeman W T and Weiss Y 2001 *Advances in Neural Information Processing Systems* vol 13 pp 689–95
[7] Tanaka K 2003 *IEICE Trans. on Information and Systems* **E86-D** 1228–42
[8] Welling M and Teh Y W 2003 *Artifcial Intelligence* **143** 19–50
[9] Welling M and Teh Y W 2004 *Neural Computation* **16** 197–221
[10] Mézard M and Mora T 2009 *Journal of Physiology-Paris* **103** 107–13
[11] Yasuda M and Tanaka K 2013 *Phys. Rev.* E **87** 012134
[12] Yasuda M and Tanaka K 2007 *J. Phys. A: Math. and Theor.* **40** 9993–10007
[13] Opper M and Winther O 2001 *Phys. Rev. Lett.* **86** 3695–9
[14] Opper M and Winther O 2001 *Phys. Rev.* E **64** 056131
[15] Raymond J and Ricci-Tersenghi F 2013 *In IEEE ICC'13 - Workshop on Networking across disciplines: Communication Networks, Complex Systems and Statistical Physics (NETSTAT) (ICC'13 - IEEE ICC'13 Workshop NETSTAT)*
[16] Raymond J and Ricci-Tersenghi F 2013 *Phys. Rev.* E **87** 05211
[17] Yasuda M, Kataoka S and Tanaka K 2012 *J. Phys. Soc. Jpn.* **81** 044801
[18] Tanaka K, Yasuda M and Titterington D M 2012 *J. Phys. Soc. Jpn.* **81** 114802
[19] Salakhutdinov R and Hinton G E 2009 *Proceedings of the 12th International Conference on Artificial Intelligence and Statistics (AISTATS 2009)* (Clearwater Beach, FL) pp 448–55
[20] Salakhutdinov R and Hinton G E 2012 *Neural Computation* **24** 1967–2006
[21] Morita T 1993 *J. Phys. Soc. Jpn.* **62** 4218–23

Loop calculus and bootstrap-belief propagation for perfect matchings on arbitrary graphs

M Chertkov[1], A Gelfand[1,2] and J Shin[3]

Theoretical Division & Center for Nonlinear Studies, Los Alamos National Laboratory, Los Alamos, NM 87545, USA
Department of Computer Science, University of California, Irvine, Irvine, CA 92697-3435, USA
Mathematical Sciences Department, IBM T. J. Watson Research, Yorktown Heights, NY 10598, USA

E-mail: chertkov@lanl.gov

Abstract. This manuscript discusses computation of the Partition Function (PF) and the Minimum Weight Perfect Matching (MWPM) on arbitrary, non-bipartite graphs. We present two novel problem formulations - one for computing the PF of a Perfect Matching (PM) and one for finding MWPMs - that build upon the inter-related Bethe Free Energy (BFE), Belief Propagation (BP), Loop Calculus (LC), Integer Linear Programming and Linear Programming frameworks. First, we describe an extension of the LC framework to the PM problem. The resulting formulas, coined (fractional) Bootstrap-BP, express the PF of the original model via the BFE of an alternative PM problem. We then study the zero-temperature version of this Bootstrap-BP formula for approximately solving the MWPM problem. We do so by leveraging the Bootstrap-BP formula to construct a sequence of MWPM problems, where each new problem in the sequence is formed by contracting odd-sized cycles (or blossoms) from the previous problem. This Bootstrap-and-Contract procedure converges reliably and generates an empirically tight upper bound for the MWPM. We conclude by discussing the relationship between our iterative procedure and the famous Blossom Algorithm of Edmonds '65 and demonstrate the performance of the Bootstrap-and-Contract approach on a variety of weighted PM problems.

1. Introduction

Belief Propagation (BP) is an iterative message passing algorithm with physics roots in H. Bethe [1] and R. Peierls' [2] work on "melting of lattices", information theory roots in decoding of Gallager's Low Density Parity Check (LDPC) codes [3], and artificial intelligence roots dating to J. Pearl [4] (who actually coined the term). For decades the three fields developed BP and the homonymous Bethe-Peierls approximation independently. Only recently, under the Bethe Free Energy (BFE) framework of Yedidia, Freeman and Weiss in [5, 6], have the parallel developments been united.

BP is a naturally distributed and easy to implement algorithm for iteratively solving the system of BP equations - a set of equations that also characterize the fixed points of the BFE function [5, 6]. The so-called BP/BFE approach can be used to compute the Partition Function (PF) or find the Maximum Likelihood (ML) assignment of a Graphical Model (GM), where a GM defines a joint probability over a set of random variables via local probability measures that factorize according to some underlying graph structure. BP/BFE provides an exact answer to the PF and ML tasks if the underlying graph is a tree. BP/BFE is a heuristic on graphs with loops, but it typically provides an accurate approximation.

Two recent developments have altered the paradigm that BP/BFE is only "good" for tree or tree-like GMs. First, it was shown that BP can exactly recover the ML assignment for certain binary GMs with dense, loopy graph structures. For example, BP can be used to find a Perfect Matching (PM) on a bipartite graph [7, 8, 9]. Interestingly, all models of this special type can be formulated as Linear Programming (LP) problems that have an integral optimal solution, though this condition is not sufficient in general [9, 10, 11].

The Loop Calculus (LC) was developed at the same time and provides a way to express the PF of a general binary GM as a finite (but still exponential in the size of the graph) series [12, 13]. Each term in the series is related to a generalized loop of the underlying graph (which is a subgraph where all vertices have degree two or higher) and is expressed explicitly using the solution to the BP equations. In some special cases, e.g. for PMs over bipartite graphs where the partition function is a permanent [14], the multiplicative mismatch between the PF of the original problem and the respective BFE/BP expression is itself the PF of a new GM defined on the same graph. In such cases, the LC can be used to improve upon the BP estimate for the PF (when the BP estimate is not tight) by re-summing a number of important LC terms [15, 16]. As suggested in [17, 18, 19, 20], the BP approximation to the ML problem can be improved by analyzing higher order terms of the LC (sometimes re-summing as in [20]) and using this information to modify the original GM. The hope is that when BP is run on the new GM it will be exact or, at least, improve upon its initial approximation.

In this manuscript we continue to work on improving BP, focusing on Minimum Weight Perfect Matching (MWPM) problems in general (non-bipartite) graphs. This choice is not arbitrary. MWPM is a special problem in Matching Theory which can be solved in polynomial time by Edmonds' famous Blossom Algorithm (BA) [21] (see [22] for details of BlossomV, the state-of-the-art implementation of the BA). The most recent development in this classical sub-field of Computer Science is due to Chandrasekaran, Végh and Vempala [23] who proposed a polynomial time "cutting-plane" procedure to finding a MWPM. In contrast to Edmonds' primal-dual algorithm, their approach carefully constructs a sequence of linear inequalities (or cut constraints) that tighten the LP relaxation, while ensuring that the added "cuts" maintain half-integrality of the LP.

While Edmonds' BA is efficient, it is not a distributed algorithm. Several recent results motivate the use of distributed, message passing implementations of BP to solve the MWPM problem. First, standard max-product BP was shown to be convergent and correct when the optimal solution to the LP relaxation of the weighted matching problem is integral. This was shown for PMs on bipartite graphs in [7, 8, 9] and extended to general weighted matchings in [11]. And in [20], the MWPM was shown to be an example of a problem where the LP optima and iterative BP solution are equivalent (when the latter converges).

Unfortunately, the initial LP relaxation is not tight in general. The work in [20] considered situations when the optimal solution is non-integral and proposed running BP on a modified version of the weighted matching GM. In particular, higher order factors enforcing Edmonds' blossom (cut) constraints were added to the original GM and a finite temperature version of BP (with proper annealing and damping) was used to improve convergence to the LP optima. A provably accurate alternative to [20], discussed in [24], would involve: (a) adaptively selecting the blossoms/constraints used to modify the GM; (b) maintaining half-integrality of the underlying LP at every iteration; and (c) proving convergence and correctness using the computational tree approach of [9, 11].

This manuscript develops a new theoretical and algorithmic approach towards constructing efficient and distributed algorithms for resolving the MWPM problem. This new approach, which is complementary to both [20] and [24], is based on extending the LC paradigm of [12, 13, 14, 25] to PMs over an arbitrary graph at a finite temperature, T, and careful analysis of the ML, $T \to 0$, limit. Analysis of the $T \to 0$ limit led to the development of an algorithm for (approximately) solving the MWPM problem - the BP-Bootstrap-and-Contract procedure - which we evaluated empirically.

The material in the manuscript is organized as follows. We provide some background on GMs and PM problems in section 2; the BP/BFE framework is briefly reviewed in section 3 and in Appendix A. The main results involving the LC are covered in section 4. Finally, the Bootstrap-BP and fractional Bootstrap-BP [25] formulas are discussed in section 5 and Appendix B. The ML, $T \to 0$, versions of the Bootstrap-BP formulas are discussed and illustrated on a small example in section 6. The Bootstrap-

and-Contract algorithm is presented in section 7 and evaluated empirically in section 8. Finally, we summarize and discuss open questions in section 9.

2. Graphical Model of Perfect Matching

Consider an arbitrary undirected graph, $\mathcal{G} = (\mathcal{V}, \mathcal{E})$, with associated edge weights $w = (\exp(-\varepsilon_{ij}/T)|\{i,j\} \in \mathcal{E})$, where T is a positive temperature parameter term and ε_{ij} is the energy of a Perfect Matching (PM), or dimer[1], associated with the edge $\{i,j\}$. One problem of interest is to evaluate the PF of the PM over the tuple $(\mathcal{V}, \mathcal{E}, w)$:

$$Z(w) = \sum_{\sigma\in\Sigma} \exp\left(-\frac{1}{T}\sum_{\{i,j\}\in\mathcal{E}} \varepsilon_{ij}\sigma_{ij}\right) = \sum_{\sigma\in\Sigma}(w_{ij})^{\sigma_{ij}} = \sum_{\sigma\in\Sigma} W(\sigma), \tag{1}$$

where $\Sigma = \left\{\sigma \,\middle|\, \forall\{i,j\} : \sigma_{ij} = \{0,1\};\ \forall i : \sum_{j:\{i,j\}\in\mathcal{E}} \sigma_{ij} = 1\right\}$ is the set of PMs in $\mathcal{G}$, where a PM is a subset of the edges in $\mathcal{E}$ such that an edge is adjacent to each vertex in $\mathcal{V}$. Note that $\mathcal{P}(\sigma) = W(\sigma)/Z(w)$ is thus the probability of the PM σ.

The total weight, E, of the Minimum Weight Perfect Matching (MWPM) arises by taking the zero temperature limit, $T \to 0$, of the PF

$$E = -\lim_{T\to 0} T\log Z = \min_{\sigma\in\Sigma} \sum_{\{i,j\}\in\mathcal{E}} \varepsilon_{ij}\sigma_{ij}. \tag{2}$$

The right-hand side (rhs) of (2) is an Integer Linear Programming (ILP) problem. The LP relaxation of (2) is

$$E_0 = \min_{x\in X} \sum_{\{i,j\}\in\mathcal{E}} \varepsilon_{ij}x_{ij}, \tag{3}$$

where $X = \left\{x_{ij} \,\middle|\, \forall\{i,j\} : x_{ij} \in [0,1];\ \forall i : \sum_{j:\{i,j\}\in\mathcal{E}} x_{ij} = 1\right\}$ is the so-called PM polytope. Since (3) constitutes a relaxation of (2), $E_0 \le E$. As shown in [26] the optimal x in (3) is half integer.

If the graph is bipartite then (a) computing the PF in (1) is equivalent to computing the permanent of the matrix of w-weights; (b) the PM polytope, X is equal to the convex hull of Σ; (c) $E_0 = E$, i.e. the LP-relaxation is gapless; and (d) the optimal x in (3) is integral (assuming the formulation in (2) is not degenerate).

3. Bethe Free Energy, Belief Propagation and LP-BP

For a generic Graphical Model (GM) over $\mathcal{G}$ assigning (un-normalized) weight w to state σ, one defines the exact variational [2] function

$$\mathcal{F}\{b\} \equiv T\sum_{\sigma} b(\sigma)\ln\frac{b(\sigma)}{W(\sigma)}. \tag{4}$$

The belief, $b(\sigma)$ is understood as a proxy to the probability $\mathcal{P}(\sigma)$ because under the normalization condition $\sum_{\sigma\in\Sigma} b(\sigma) = 1$, the Gibbs function is convex and it achieves its only minimum at $b(\sigma) = \mathcal{P}(\sigma)$ and $\mathcal{F}\{\mathcal{P}\} = -T\ln Z$.

Belief Propagation (BP) is an iterative method for computing beliefs that is exact when the underlying GM is a tree. As shown in [6], the BP fixed point equations can be derived as a relaxation of the constrained Gibbs function (4). We briefly review the concepts of [6] with application to the PM problem. The material generalizes the description of [14], which was limited to bipartite graphs (permanents).

[1] The two terms, used respectively in Statistical Physics and Computer Science, are equivalent.

[2] It is referred to as the Gibbs and Kullback-Leibler function in statistical physics and statistics, respectively.

The BP approximate belief of state $\sigma \in \Sigma$ is

$$b(\sigma) \approx b_{\text{BP}}(\sigma) = \frac{\prod_i b_i(\sigma_i)}{\prod_{\{i,j\}\in\mathcal{E}} b_{ij}(\sigma_{ij})}, \tag{5}$$

where $b_i(\sigma_i)$ ($b_{ij}(\sigma_{ij})$) are vertex (edge) beliefs related to each other by

$$\forall\{i,j\} \in \mathcal{E}: \quad b_{ij}(\sigma_{ij}) = \sum_{\sigma_i\backslash\sigma_{ij}} b_i(\sigma_i) = \sum_{\sigma_j\backslash\sigma_{ij}} b_j(\sigma_j), \tag{6}$$

and where $\sigma_i = (\sigma_{ij}|j : \{i,j\} \in \mathcal{E})$ is a vector of edge beliefs. The beliefs should also satisfy the normalization conditions:

$$\forall\{i,j\} \in E: \quad b_{ij}(1) + b_{ij}(0) = 1. \tag{7}$$

Substituting (5) into (4) and utilizing (6) and (7) one arrives at the following Bethe Free Energy function (BFE)

$$\mathcal{F}_{\text{BP}}(\beta) \equiv E - TS, \quad E \equiv \sum_{\{i,j\}} b_{ij}(1)\varepsilon_{ij}, \tag{8}$$

$$S \equiv \sum_{\{i,j\}\in\mathcal{E}} \sum_{\sigma_{ij}} b_{ij}(\sigma_{ij}) \ln b_{ij}(\sigma_{ij}) - \sum_{i\in\mathcal{V}} \sum_{\sigma_i} b_i(\sigma_i). \tag{9}$$

Equations (6), (7), (8) and (9) can be greatly simplified. One can express (6) and (7) solely in terms of $\beta_{ij} \equiv b_{ij}(1)$ satisfying the following Doubly Stochastic (DS) constraints

$$\forall\{i,j\} \in \mathcal{E} : 0 \le \beta_{ij} \le 1; \quad \forall i : \sum_j \beta_{ij} = 1. \tag{10}$$

We will use $\beta \in$ DS to indicate that β is DS, i.e. lies in the DS polytope defined by (10). The entropy (9) becomes

$$S(\beta) = \sum_{\{i,j\}\in\mathcal{E}} \left((1-\beta_{ij})\ln(1-\beta_{ij}) - \beta_{ij}\ln\beta_{ij}\right). \tag{11}$$

Therefore, the BFE/BP approximation to the partition function of the PM model over $\mathcal{G}$ becomes

$$\begin{aligned} -\log Z_{\text{BP}} &= \frac{1}{T} \min_{\beta\in\text{DS}} \mathcal{F}_{\text{BP}}(\beta) \\ &= \min_{\beta\in\text{DS}} \sum_{\{i,j\}\in\mathcal{E}} \left(\beta_{ij}\left(\frac{\varepsilon_{ij}}{T} + \log\beta_{ij}\right) - (1-\beta_{ij})\log(1-\beta_{ij})\right). \end{aligned} \tag{12}$$

The MWPM can also be estimated within the BFE/BP approach as:

$$E_{\text{BP}} = \min_{\beta\in\text{DS}} \beta_{ij}\varepsilon_{ij}. \tag{13}$$

Notice that (13) is an LP, that will be called BPLP because of its relation to the BP concept. As shown in [9], $E_{\text{BP}} = E_0$, i.e. LP $\equiv$ BPLP, where LP is defined by (3). One also finds that the optimal β in (13) is half-integral.

For a most general GM the BFE function is bounded from below but non-convex. However, as shown in [27] for the PM over bi-partite graph, $\mathcal{F}_{\text{BP}}(\beta)$ is a convex function over $\beta \in$ DS. As argued in Appendix A the BFE convexity proof of [27] extends straightforwardly to PM GM over arbitrary $\mathcal{G}$.

To analyze the minima of the constrained BFE in (12) we incorporate Lagrange multipliers μ_i enforcing the constraints in (10). Looking for a stationary point of the respective Lagrange function over the β variables, one arrives at the following set of quadratic equations for edge variables, β_{ij}

$$\forall \{i,j\} \in \mathcal{E}: \quad \beta_{ij}(1-\beta_{ij}) = \exp\left(\frac{\mu_i + \mu_j - \varepsilon_{ij}}{T}\right). \tag{14}$$

Equations (14) combined with the conditions (10) constitute the so-called Belief Propagation (BP) equations.

To find a solution of BP equations (10) and (14) one relies on an iterative procedure. For a description of a set of iterative BP algorithms convergent to a minimum of the BFE for the perfect matching problem we refer the interested reader to [28, 29, 30].

4. Loop Series

As shown in [12, 13], the exact partition function of a generic GM can be expressed in terms of a Loop Series (LS), where each term is computed explicitly using beliefs found by solving the BP equations.

Adapting this general result to the PM problem one derives

$$Z = Z_{\text{BP}} \cdot z, \quad z \equiv 1 + \sum_C r_C, \quad r_C = \prod_{i \in C} \psi_{i;C}, \tag{15}$$

$$\psi_{i;C} \equiv \frac{\sum_{\vec{\sigma}_i} b_i(\sigma_i) \prod_{j:\{i,j\}\in C}(m_{ij} - 2\sigma_{ij} + 1)}{\prod_j^{\{i,j\}\in C} \sqrt{1-(m_{ij})^2}} = (1-q_i) \prod_{j \in C:\{i,j\}\in C} \sqrt{\frac{\beta_{ij}}{1-\beta_{ij}}}, \tag{16}$$

$$m_{ij} \equiv \sum_{\sigma_{ij}} (2\sigma_{ij} - 1) b_{ij}(\sigma_{ij}) = 2\beta_{ij} - 1, \quad q_i \equiv \sum_{j \in C}^{\{i,j\}\in C} 1, \tag{17}$$

where once again $\{b\}$ (or $\{\beta\}$) are solutions to the BP equations (10) and (14) and C stands for an arbitrary generalized loop, defined as a subgraph of the complete bipartite graph with all its vertices having a degree larger than 1. The q_i in (16) are the C-dependent degrees, i.e. $q_i = \sum_{j|\{i,j\}\in C} 1$.

Let us clarify derivation of the compact expressions on the rhs of (16):

$$\sum_{\sigma_i} b_i(\sigma_i) \prod_{j:\{i,j\}\in C} (m_{ij} - 2\sigma_{ij} + 1) = \sum_{k:(i,k)\notin C} \beta_{ik} \prod_{j:\{i,j\}\in C} (m_{ij}+1)$$
$$+ \sum_{k:(i,k)\in C} \beta_{ik}(m_{ik}-1) \prod_{j \neq k}^{\{i,j\}\in C} (m_{ij}+1) = 2^{q_i}(1-q_i) \prod_{j:\{i,j\}\in C} \beta_{ij}, \tag{18}$$

where we used relations (17) and also the double stochasticity of β.

Combining (15) and (16) one derives

$$r_C = \left(\prod_{i \in C}(1-q_i)\right) \prod_{\{i,j\}\in C} \frac{\beta_{ij}}{1-\beta_{ij}}. \tag{19}$$

In (19), loops comprised of an even (odd) number of vertices give positive (negative) contributions to r_C. Equation (19) was stated in [14] for the bipartite case. From the above derivation, we can see that the result is valid for any graph $\mathcal{G}$.

5. Bootstrap-BP and Fractional Bootstrap-BP

Another important result for PMs is the following re-summation of the LS

$$\frac{Z(w)}{Z_{\text{BP}}(w)} = \frac{Z(\tilde{\beta})}{\prod_{\{i,j\}\in\mathcal{E}}(1-\beta_{ij})}, \tag{20}$$

where $\tilde{\beta} = (\beta_{ij}(1-\beta_{ij})|\{i,j\} \in \mathcal{E})$. Equation (20) was stated in [14] only for the case of permanent - PM over binary graph, even though direct verification confirms (see Appendix B) that the formula is generic, applicable to PM over an arbitrary graph [3].

We call (20) the Bootstrap-BP equation because the rhs of (20) provides a quantitative measure of the relative accuracy of BP.

Extending the approach of [25] from bipartite to general graphs, one arrives at the following fractional generalization of (20)

$$\frac{Z(w)}{Z_\gamma(w)} = Z(\tilde{\beta}_\gamma) \prod_{\{i,j\}\in\mathcal{E}} (1-\beta_{ij})^\gamma, \tag{21}$$

$$-\log Z_\gamma(w) = \min_{\beta\in\text{DS}} \sum_{\{i,j\}\in\mathcal{E}} \left(\beta_{ij}\left(\frac{\varepsilon_{ij}}{T} + \log\beta_{ij}\right) + \gamma(1-\beta_{ij})\log(1-\beta_{ij})\right), \tag{22}$$

where $\tilde{\beta}_\gamma = (\beta_{ij}(1-\beta_{ij})^{-\gamma}|\{i,j\} \in \mathcal{E})$ and the doubly stochastic β_{ij}'s entering (21) and (22) come from solving the following γ-version of (14)

$$\forall\{i,j\} \in \mathcal{E}: \quad \beta_{ij}(1-\beta_{ij})^{-\gamma} = \exp\left(\frac{\mu_i + \mu_j - \varepsilon_{ij}}{T}\right), \tag{23}$$

γ can be any number in $[-1, 1]$, with $\gamma = -1$ corresponding to the BP case. Notice, that the Loop Series representation for the partition function correction only applies to the BP case. Its generalization to arbitrary γ is not known to the authors.

6. Finding MWPMs via BP: An Illustration

For an arbitrary graph, the BPLP = LP relaxation of the MWPM problem is known to be half-integral [11]. When the solution to the relaxation is integral, one has a certificate of optimality and $E = E_0$. When the solution is half integral, $E_0 < E$, simply because BPLP is a relaxation of the original MWPM problem. The Bootstrap-BP recasting of the LS provides a compact expression relating E and E_0 in terms of some new partition function. The question explored in the remainder of this section is the following:

- Can the (fractional) Bootstrap-BP formula in (21) be used to close the gap between E_0 and E?

The rhs of (21) involves the PF of a new GM defined on graph $\mathcal{G}$. Our hope is that the new GM is easier to deal with than the original GM. Before checking to see if this is true, we first analyze the $T \to 0$ limit of (21).

The fractional parameter γ in (21) can be chosen arbitrary from the interval $[-1, 1]$. However, in order to avoid resolving a fictitious singularity [4] it is convenient to consider $\gamma = 0$. Then in the $T \to 0$ limit one derives from (21) and (23):

$$E = E_0 + E_1, \quad E_1 = \min_{\sigma\in\Sigma} \sum_{\{i,j\}\in\mathcal{E}} (\varepsilon_{ij} - \mu_i - \mu_j)\sigma_{ij} \tag{24}$$

[3] We have been informed by one of the Referees that the general result is also stated in the thesis of Y. Watanabe [31]

[4] The algebraic factor on the rhs of (21) is singular in the $T \to 0$ limit at $\gamma < 0$ when some β_{ij} approaches unity. However, the singularity is fictitious, as the PF on the rhs of (21) will turn to 0 (in the limit). Since $Z(w) \sim \exp(-E/T)$ and $Z_\gamma(w) \sim \exp(-E_0/T)$ and $E_0 < E$ in the case of interest, we expect to get zero (in the limit) for the rhs of (21).

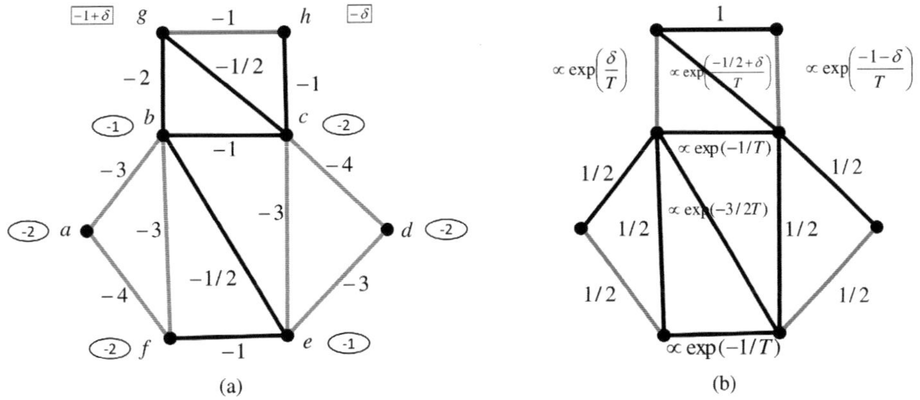

Figure 1. Illustrative 8-node example analysed with the $\gamma = 0$ version of the bootstrap-BP approach. (a) Numbers next to edges are ε_{ij}. Edges shown in green and red output $\beta_{ij} = 1/2$ and $\beta_{ij} = 1$ respectively in BP-LP. MWPM is $\{a, f\}, \{e, d\}, \{g, b\}, \{h, c\}$. Numbers shown in ovals and squares correspond to the leading asymptotic estimates for the Lagrangian multipliers, μ_i. Freedom in choosing the parameter δ, entering μ_i adjusted to $\beta_{ij} = 1$ does not affect the $E - E_0$ correction we are looking for in here. (b) Expressions next to the edges are respective β_{ij}. Edges marked blue give the leading order PM contribution to $Z(\tilde{\beta})$.

where E and E_0 are as in (2) and (3), and μ are dual variables (often called chemical potentials in statistical physics) that can be reconstructed from the optimal solution of the BPLP (3) as follows:

- One first identifies any non-intersecting half-integral cycles C of $\mathcal{G}$, i.e. odd-length cycles where $\beta_{ij} = 1/2$ for each edge $\{i, j\} \in C$ and $\{\beta_{ij}\}$ is the solution of BPLP [5].
- One computes Lagrangian multipliers, μ_i, along each half-integral cycle, resolving, separately for every cycle the set of linear equations, $\mu_i + \mu_j = \varepsilon_{ij}$.
- For every edge, $\{i, j\}$ with $\beta_{ij} = 1$ one computes the Lagrangian multipliers associated with the vertices i and j, $\mu_i \leftarrow \varepsilon_{ij}/2 + \delta_{ij}$ and $\mu_j \leftarrow \varepsilon_{ij}/2 - \delta_{ij}$, where δ_{ij} can be chosen arbitrarily, for example $\delta_{ij} \leftarrow \varepsilon_{ij}/2$. [6]

To demonstrate utility of the bootstrap BPLP formulas (3) and (24) let us consider the 8-node example shown in figure 1. In this example, the MWPM is $\{a, f\}, \{e, d\}, \{g, b\}, \{h, c\}$ and, therefore, $E = -4 - 3 - 2 - 1 = -10$. The solution to the BPLP is fractional. All edges on the two cycles, $\{a, b, f\}$ and $\{c, d, e\}$ have $\beta = 1/2$, edge $\{g, h\}$ has $\beta_{gh} = 1$ and all other edges in the graph have $\beta = 0$. Therefore, $E_0 = (-4 - 3 - 3)/2 + (-4 - 3 - 3)/2 - 1 = -11$, which is clearly less than E. All dual variables, μ_i, associated with the half-integral cycles (shown in green in figure 1 (a)) are determined unambiguously. The dual variables for edge $\{g, h\}$ must satisfy $\mu_g + \mu_h + 1 = 0$ and we choose $\mu_g - \mu_h = 2\delta$. In principle, the freedom can be fixed taking into account the DB vertex constraints on all the β-corrections. However, this is not necessary because the δ's simply cancel when computing the correction $E_1 = \delta + 1 - \delta = 1 = E_0 - E$ for the PM contribution to $Z(\beta)$ marked in blue in figure 1 (b).

A number of things are worth noting from this simple example. First, the Bootstrap-BP formula (20) allowed us to reduce the original MWPM problem to an auxiliary MWPM problem on the same graph. At first, the new MWPM problem does not appear to be easier to solve than the original MWPM problem.

[5] If the BPLP is fractional, then at least two such cycles exist. This is guaranteed by half-integrality of BPLP and because we are solving a PM problem.

[6] The choice of δ does not affect the optimal σ defining E_1 in (24).

However, this example suggests a direction to further simplify the auxiliary problem by transforming it to a MWPM problem on a strictly smaller graph. Indeed, for the example considered one can *contract* each of the "green, half-integer" cycles to individual nodes. This suggestion is inspired by the following observations [21, 22, 23]:

- Define a *blossom* to be an odd-sized cycle and choose our blossoms to be the odd-sized cycles with $\beta_{ij} = 1/2$ – marked green in figure 1 (a). Such blossoms contribute zero weight to the energy correction E_1.
- Edmond's "outer" constraint associated with any of the "half-integer" (green) blossom states that *at least one edge* from the cut (outer) edges should be in the optimal MWPM solution of the original problem. Note that for any of the "half-integer" blossoms in the example of figure 1 exactly one cut edge (and not more than one!) contributes to the optimal MWPM of the original problem.

In the following section we describe an algorithm that repeatedly applies this contraction concept to identify a sequence of corrections to the MWPM energy E.

7. Algorithm for finding MWPM sequentially

Algorithm: BP-Bootstrap-and-Contract

Initialization: The master system is defined as a non-degenerate MWPM over the $(\mathcal{V}, \mathcal{E}, w)$ tuple. (Add small random corrections to w to ensure non-degeneracy.) Set $E = 0$.

Iterate:

(1) Solve the LP = BPLP, (3) and (13) over $(\mathcal{V}, \mathcal{E}, w)$. (For efficient distributed implementation use the algorithm of [9, 11].) β and E_0 are the outputs. Set, $E \mathrel{+}= E_0$.

(2) If solution of the LP = BPLP is integer: **Stop**.

(3) Otherwise, modify the master system following the procedure:

- **(a)** Identify non-intersecting half-integral cycles of $\mathcal{G}$, i.e. cycles along which all edges are half-integral, $\beta_{ij} = 1/2$, in the solution of the LP = BPLP. (Existence of at least two such cycles is guaranteed.)
- **(b)** Compute the Lagrangian multipliers, μ_i, along the half-integer cycles, resolving, separately for every cycle the set of linear equations, $\mu_i + \mu_j = \varepsilon_{ij}$.
- **(c)** For every edge, $\{i, j\}$, which is in PM according to the LP = BPLP, i.e. $\beta_{ij} = 1$, compute the Lagrangian multipliers associated with the vertices i and j, $\mu_i \leftarrow \varepsilon_{ij}/2 + \delta_{ij}$ and $\mu_j \leftarrow \varepsilon_{ij}/2 - \delta_{ij}$, where δ_{ij} can be chosen arbitrarily, for example $\delta_{ij} \leftarrow \varepsilon_{ij}/2$. [6]
- **(d)** Update weights, $\forall\{i, j\} : \varepsilon_{ij} \leftarrow \varepsilon_{ij} - \mu_i - \mu_j$.
- **(e)** Update the $(\mathcal{V}, \mathcal{E}, w)$ tuple according to the following rules: Contract all vertices correspondent to a half-integer cycle into one new vertex. If multiple edges have appeared in the result of the contraction, combine them in one edge choosing the new weight equal to be the largest weight of the combined edges. Repeat it for all the half-integer cycles.
- **(f)** Add small random correction to w to remove possible (future) degeneracy.

[6] Note that the steps (b) and (c) are necessary for determining the weight of edge corresponding to $\beta_{ij} = 0$.

As already noted, this algorithm is suboptimal because the optimal MWPM may select more than one "cut edges" for the green cycles. In spite of this potential sub-optimality, the Bootstrap-and-Contract

Table 1. Evaluation of Bootstrap-and-Contract on random complete graph instances for two levels of sparsity. N and M indicate the number of nodes and average number of edges in each of the 100 instances. *# Cor.* is the number of instances (out of 100) in which Bootstrap-and-Contract found the weight of the optimal matching. *# LPs* indicates the average number of LPs solved by Bootstrap-and-Contract. The bracketed numbers indicate the minimum and maximum number of LPs solved across the 100 instances. *# Blos.* is the average number of blossoms collapsed in each MWPM problem.

50 % sparse				90 % sparse			
N / M	# Cor.	# LPs	# Blos.	N / M	# Cor.	# LPs	# Blos.
100 / 1961	100	1.4[1,4]	0.9[0,6]	100 /475	100	1.4[1,4]	0.8[0,8]
250 / 12284	100	1.6[1,5]	1.2[0,8]	250 / 2973	100	1.6[1,3]	1.2[0,4]
500 / 49159	100	1.8[1,6]	1.7[0,12]	500 / 11898	100	1.7[1,4]	1.5[0,6]
1000 / 196696	100	1.9[1,5]	1.9[0,8]	1000 / 47576	99	1.8[1,4]	1.8[0,8]

Table 2. Evaluation of Bootstrap-and-Contract on random Triangulation instances.

N	M	# Cor.	# LPs	# Blos.
100	285	95	5.7[2,13]	16.0[4,32]
250	733	83	10.6[4,28]	45.8[18,82]
500	1481	75	16.4[5,36]	92.1[44,132]
1000	2979	60	28.8[9,70]	200.3[140,294]
5000	14975	5	78.5[24,202]	996.0[796,1212]

Table 3. Evaluation of Bootstrap-and-Contract on a chain of triangles connected by a single edge or 3 edges.

Singly-Connected				3-Connected			
K	# Cor.	# LPs	# Blos.	K	# Cor.	# LPs	# Blos.
100	100	10.0[3,17]	59.2[36,72]	100	100	19.6[4,39]	60.3[18,102]
200	100	16.4[5,29]	128.1[86,156]	200	100	34.3[5,71]	130.2[42,192]
500	100	34.4[11,68]	336.6[262,390]	500	100	63.2[19,143]	355.8[200,466]
1000	100	58.8[19,111]	704.8[582,802]	1000	100	119.7[20,223]	793.8[490,956]

procedure outputs an energy E which is an upper bound on the value of the MWPM (proof of this statement will be presented in a future work). Once the upper bound is derived, one can use the following heuristics to find the PM corresponding to the output. Assume that all the ϵ weights in the original PM formulation are integer and solve the following auxiliary LP,

$$\min \sum_{\{i,j\}\in\mathcal{E}} \varepsilon_{ij}x_{ij}, \text{ s.t. } \sum_{\{i,j\}\in\mathcal{E}} \varepsilon_{ij}x_{ij} \le E,\ \forall i\in\mathcal{V}: \sum_{j:\{i,j\}\in\mathcal{E}} x_{ij}=1, \quad \forall\{i,j\}\in\mathcal{E}:\ x_{ij}\ge 0.$$

If the LP outputs an integral solution, we stop. Otherwise, we run the same LP with $E-1$ instead of E, and repeat the procedure till an integral solution is found.

8. Experiments

We conducted a set of experiments to evaluate the performance of the Bootstrap-and-Contract procedure. The Bootstrap-and-Contract algorithm solves a series of LP problems, $LP^{(0)}, ..., LP^{(S)}$, where $LP^{(S)}$

is the first LP with an integral solution. $LP^{(t+1)}$, the LP in iteration $t+1$, is constructed from $LP^{(t)}$ by formulating a new weighted PM problem.

Since the LP relaxation to the PM problem is half-integral [26], we are assured that the set of edges for which $\beta_{ij}^{(t)} = 1/2$ will form non-intersecting cycles of odd length (note that $\beta_{ij}^{(t)}$ denotes the belief on edge $\{i, j\}$ in the solution to $LP^{(t)}$). In addition, each odd-sized half-integral cycle identifies a constraint that can be used to tighten the relaxation of $LP^{(t)}$. Such constraints are referred to as *blossom* constraints in the matching literature [21] and we will continue to adopt that name herein. Contracting all vertices correspondent to an odd-sized half-integral cycle has the effect of adding the blossom constraint to the LP. Thus, by contracting blossoms in each iteration, we hope to construct a series $LP^{(0)}$, ..., $LP^{(S)}$ of increasingly tight LP relaxations to the PM problem.

Unfortunately, the procedure of greedily collapsing blossoms is not guaranteed to produce a tight LP for the MWPM problem. This means that the Bootstrap-and-Contract procedure will not necessarily recover the weight E of the optimal perfect matching. We therefore rely on experimentation to gauge the efficacy of the proposed Bootstrap-and-Contract heuristic.

We synthetically generated several types of weighted graph problems and compared the weight of the matching found by Bootstrap-and-Contract to the weight of the optimal matching. The weight of the optimal perfect matching was found using Kolmogorov's BlossomV algorithm - the state-of-the-art algorithm for solving weighted matching problems [22].

We conducted experiments on three problem types: 1) Sparse complete graph instances; 2) Triangulation instances; and 3) Chained triangle instances. The sparse complete graph instances were generated by taking a complete graph on N nodes and eliminating edges with probability $p = \{0.5, 0.9\}$. [7] Edge weights were assigned an integral weight drawn from a discrete uniform distribution in $[1, 2^{20}]$. The triangulation instances were generated by randomly placing points in the $2^{20} \times 2^{20}$ square and computing a Delaunay triangulation on this set of points. The edge weights were set to the rounded Euclidean distance between two points. The chained triangle instances were used in the First DIMACS Implementation Challenge [32]. Based on the algorithms t.f and tt.f by N. Ritchey and B. Mattingly, we generate a sequence of K triangles. The chain of triangles is formed by connecting neighboring triangles via a single edge or by three edges. While certainly not exhaustive, this set of problems encompasses an interesting range of graph types.

The results from running Bootstrap-and-Contract on sparse complete graph instances are shown in table 1. For each number of nodes $N \in \{100, 250, 500, 1000\}$, we generated 100 random instances. We report the average number of edges, M, for each setting of N and *# Cor.*, the number of instances (out of 100) for which BP-Bootstrap found the optimal weight matching. In addition, we report the average and [minimum, maximum] number of LPs solved by BP-Bootstrap as well as the number of blossoms contracted by the Bootstrap-and-Contract procedure. From this table it seems clear that Bootstrap-and-Contract procedure is accurate on dense complete graphs. A reason for the success of the method on this class of graphs is that $LP^{(0)}$, the initial LP relaxation, is tight a large percentage of the time. This is indicated by the fact that on average less than two LPs need to be solved by Bootstrap-and-Contract.

In contrast, the Bootstrap-and-Contract method performs rather poorly on the random Triangulation instances shown in table 2. The greedy blossom contracting approach does not appear to be successful on this class of problems. In fact, when we examined the operation of Kolmogorov's BlossomV algorithm on these instances we found that BlossomV frequently needed to backtrack and expand blossoms that had been collapsed - an operation not supported by our greedy heuristic.

Finally, the results from running Bootstrap-and-Contract on the chained triangle instances are shown in table 3. The Bootstrap-and-Contract method performs well on these problems. Interestingly, we see that the greedy heuristic is effective even though these instances require solving many LPs and collapsing many blossoms.

[7] In doing so, we ensure that the graph contains a perfect matching by ensuring that the graph remains edge-connected.

9. Questions yet to be addressed in this project

We conclude by describing what we believe are the two most important challenges needing to be resolved in our "Matching" using BP project.

- As proven in [33] for the bipartite (permanent) case: $Z \geq Z_{\rm BP}$. As noticed in [25], the proof of [33] is in fact a direct corollary of (20) combined with a combinatorial result of Schrijver from [34]. Indeed, it was proven in [34] (see Corollary 1c) that $\text{Perm}(\beta *. (1-\beta)) \geq \prod_{\{i,j\}\in\mathcal{E}}(1-\beta_{ij})$ (where $*.$ stands for the element-by-element product of matrices), which is essentially the statement that the rhs of (20) in the bipartite case is equal or larger than unity. The proof of Schrijver is indirect and the author mentions in passing "... we have tried to find a direct proof of it, based on continuity and differentiability, and did not succeed". Is it possible to find such a direct proof based on the combination of the Loop Calculus (15) and (19) and the Bootstrap-BP (20) formulas in the bipartite case?
- In general, and as explained above, our Bootstrap-and-Contract algorithm outputs an upper bound on the MWPM. In our empirical study, we observed that when our algorithm fails to produce an optimal matching, the BlossomV algorithm needs to backtrack and expand one or more contracted blossoms. A remaining question is how to incorporate the undo/expand step into our procedure. In particular, can one do so in a distributed manner, via a message passing (or perhaps just greedy) procedure?

Appendix A. Convexity of the Bethe Free Energy

This Appendix generalizes the main result of [27], where convexity of the BFE was proven for the case of PM GM over the bi-partite graph, to the case of PM GM over a general graph. Indeed, according to Theorem 20 of [27], $\sum_{j:\{i,j\}\in\mathcal{E}} \log\beta_{ij} - (1-\beta_{ij})\log(1-\beta_{ij})$ is convex within the polytope defined by, $\sum_{j:\{i,j\}\in\mathcal{E}}\beta_{ij} = 1$ and $\beta_{ij}\geq 0$, for all $\{i,j\}\in\mathcal{E}$ and regardless of the global structure of the graph $\mathcal{G}$. Therefore, $\mathcal{F}_{\rm BP}$ (12) is convex within the domain DS simply because it is represented as a sum of functions which are all convex within the domain.

Appendix B. Derivation of the bootstrap formulas

In this Appendix we derive (20) and (21) following the logic of [14], see also Appendixes B.1 and D.1 of [25]. For the purpose of generality we will consider arbitrary (fractional) γ and arbitrary graph $\mathcal{G}$.

Our algebraic derivation consists of comparing two formulas. First, considering vectors of weights corresponding to the two sides of (23) and equating the respective partition functions one derives

$$Z(\tilde{\beta}_\gamma) = \sum_{\sigma\in\Sigma}\exp\left(\sum_{\{i,j\}\in\mathcal{E}}\sigma_{ij}\frac{\mu_i+\mu_j-\varepsilon_{ij}}{T}\right) = Z(w)\exp\left(\frac{\sum_{i\in\mathcal{V}}\mu_i}{T}\right), \tag{B.1}$$

where we have used, $\sum_{\{i,j\}\in\mathcal{E}}\sigma_{ij}(\mu_i+\mu_j) = \sum_{i\in\mathcal{V}}\mu_i\sum_{j:\{i,j\}\in\mathcal{E}}\sigma_{ij} = \sum_{i\in\mathcal{V}}\mu_i$. Second, let us apply log to both sides of (23) and then sum the results over all the edges of $\mathcal{G}$ weighted with β_{ij}

$$\sum_{\{i,j\}\in\mathcal{E}}\beta_{ij}\left(\log\beta_{ij} - \gamma\log(1-\beta_{ij})\right) = \frac{1}{T}\left(\sum_{i\in\mathcal{V}}\mu_i - \sum_{\{i,j\}\in\mathcal{E}}\sigma_{ij}\varepsilon_{ij}\right), \tag{B.2}$$

where we have used the double stochasticity of the optimal β through, $\sum_{\{i,j\}\in\mathcal{E}}\beta_{ij}(\mu_i+\mu_j) = \sum_{i\in\mathcal{V}}\mu_i\sum_{j:\{i,j\}\in\mathcal{E}}\beta_{ij} = \sum_{i\in\mathcal{V}}\mu_i$. Then, combining (B.1) and (B.2) with (22) we arrive at (21). ■

References

[1] Bethe H A 1935 *Proceedings of Royal Society of London* A **150** 552–75
[2] Peierls H A 1936 *Proceedings of Cambridge Philosophical Society* **32** 477–81
[3] Gallager R G 1963 *Low density parity check codes* (Cambridge, MA: MIT Press)
[4] Pearl J 1988 *Probabilistic Reasoning in Intelligent Systems: Networks of Plausible Inference* (San Francisco: Morgan Kaufmann Publishers, Inc.)
[5] Yedidia J S, Freeman W T and Weiss Y 2001 Generalized belief propagation *Advances in Neural Information Processing Systems* vol 13 (Cambridge, MA: MIT Press) pp 689–95
[6] Yedidia J S, Freeman W T and Weiss Y 2005 *IEEE Transactions on Information Theory* **51** 2282–312
[7] Bayati M, Shah D and Sharma M 2005 Maximum weight matching via max-product belief propagation *Proceedings of the 2005 IEEE International Symposium on Information Theory* (Adelaide, Australia) pp 1763–7
[8] Bayati M, Shah D and Sharma M 2006 A simpler max-product maximum weight matching algorithm and the auction algorithm *Proceedings of the 2006 IEEE International Symposium on Information Theory* (Seattle, WA) pp 557–61
[9] Bayati M, Shah D and Sharma M 2008 *IEEE Transactions on Information Theory* **54** 1241–51
[10] Chertkov M 2008 *Journal of Statistical Mechanics: Theory and Experiment* **2008** P10016 (*Preprint* arXiv:0801.0341[cond-mat.stat-mech]) URL http://iopscience.iop.org/1742-5468/2008/10/P10016
[11] Sanghavi S, Malioutov D and Willsky A 2011 *IEEE Transactions on Information Theory* **57** 2203–12
[12] Chertkov M and Chernyak V 2006 *Physical Review* E **73** 065102(R)
[13] Chertkov M and Chernyak V 2006 *Journal of Statistical Mechanics: Theory and Experiment* **2006** P06009
[14] Watanabe Y and Chertkov M 2010 *Journal of Physics A: Mathematical and Theoretical* **43** 242002 URL http://stacks.iop.org/1751-8121/43/i=24/a=242002
[15] Gómez V, Mooij J M and Kappen H J 2007 *Journal for Machine Learning Research* **8** 1987–2016
[16] Gomez V, Kappen H J and Chertkov M 2010 *Journal of Machine Learning Research* **11** 1273–96 URL http://arxiv.org/abs/0901.0786
[17] Chertkov M and Chernyak V Y 2006 Loop calculus helps to improve belief propagation and linear programming decodings of low-density-parity-check codes (*Preprint* arXiv:cs/0609154[cs.IT])
[18] Kudekar S, Johnson J K and Chertkov M 2011 Linear programming based detectors for two-dimensional intersymbol interference channels *Proceedings of the 2011 IEEE International Symposium on Information Theory* (St. Petersburg, Russia) pp 2999–3003 (*Preprint* arXiv:1102.5386[cs.IT])
[19] Kudekar S, Johnson J K and Chertkov M 2011 Improved linear programming decoding using frustrated cycles (*Preprint* arXiv:1105.4665[cs.IT])
[20] Gelfand A, Chertkov M and Shin J 2013 Belief propagation for linear programming (*Preprint* arXiv:1305.4130[cs.DS]) URL http://arxiv.org/abs/1305.4130
[21] Edmonds J 1965 *Canadian Journal of Mathematics* **17** 449–67 URL http://www.cs.princeton.edu/introcs/papers/edmonds.pdf
[22] Kolmogorov V 2009 *Mathematical Programming Computation* **1** 43–67 URL http://www.springerlink.com/index/E251L2WWW0258265.pdf
[23] Chandrasekaran K, Végh L A and Vempala S 2012 The cutting plane method is polynomial for perfect matchings *Proceedings of the 2012 IEEE 53rd Annual Symposium on Foundations of Computer Science* (New Brunswick, NJ) pp 571–580 (*Preprint* arXiv:1207.5813[cs.DS]) URL http://arxiv.org/abs/1207.5813
[24] Chertkov M, Gelfand A and Shin J 2013 Graphical transformation for belief propagation: Maximum weight matchings and odd cycles (*Preprint* arXiv:1306.1167[cs.DS])
[25] Chertkov M and Yedidia A B 2013 *Journal of Machine Learning Research* **14** 2029–66
[26] Schrijver A 2003 *Combinatorial Optimization* (Springer-Verlag)
[27] Vontobel P O 2013 *IEEE Transactions on Information Theory* **59** 1866–901
[28] Chertkov M, Kroc L and Vergassola M 2008 Belief propagation and beyond for particle tracking (*Preprint* arXiv:0806.1199[cs.IT]) URL http://arxiv.org/abs/0806.1199
[29] Chertkov M, Kroc L, Krzakala F, Vergassola M and Zdeborova L 2010 *Proceedings of National Academy of Sciences* **107**(17) 7663–8 (*Preprint* arXiv:0909.4256[physics.data-an]) URL http://arxiv.org/abs/0909.4256
[30] Huang B and Jebara T 2009 Approximating the permanent with belief propagation (*Preprint* arXiv:0908.1769[cs.LG]) URL http://arxiv.org/abs/0908.1769
[31] Watanabe Y 2010 *Ph.D. Thesis* http://arxiv.org/abs/1004.4942
[32] Johnson D S and McGeoch C C 1993 *Network Flows and Matching: First DIMACS Implementation Challenge* (Providence, RI: American Mathematical Society) generators available at ftp://dimacs.rutgers.edu/pub/netflow/generators/matching
[33] Gurvits L 2011 Unleashing the power of Schrijver's permanental inequality with the help of the Bethe Approximation (*Preprint* arXiv:1106.2844[math.CO]) URL http://arxiv.org/abs/1106.2844

Journal of Physics: Conference Series **473** (2013) 012007 doi:10.1088/1742-6596/473/1/012007

[34] Schrijver A 1998 *Journal of Combinatorial Theory. Series B* **72** 122–35

ELC International Meeting on Inference, Computation, and Spin Glasses (ICSG2013) IOP Publishing
Journal of Physics: Conference Series **473** (2013) 012008 doi:10.1088/1742-6596/473/1/012008

Statistical inference of co-movements of stocks during a financial crisis

Takero Ibuki†, Shunsuke Higano‡, Sei Suzuki§, Jun-ichi Inoue£ and Anirban Chakraborti¶

† Data Mining Group, Service & Solution Development Department, Research and Development Center, NTT DOCOMO, INC., 3-6 Hikarino-oka, Yokosuka-shi, Kanagawa 239-8536, Japan
‡ Hokkaido Prefectural Police, N2-W7, Chuo-Ku, Sapporo 060-8520, Japan
§ Department of Basic Sciences, Saitama Medical University, 38 Morohongo, Moroyama, Saitama 350-0495, Japan
£ Graduate School of Information Science and Technology, Hokkaido University, N14-W9, Kita-Ku, Sapporo 060-0814, Japan
¶ Laboratoire de Mathématiques Appliquées aux Systèmes, École Centrale Paris, 92290 Châtenay-Malabry, France

E-mail: † takerou.ibuki.xt@nttdocomo.com, ‡ higano@complex.ist.hokudai.ac.jp, £ j_inoue@ist.hokudai.ac.jp, jinoue@cb4.so-net.ne.jp, § sei01@saitama-med.ac.jp, ¶ anirban.chakraborti@ecp.fr

Abstract. In order to figure out and to forecast the emergence phenomena of social systems, we propose several probabilistic models for the analysis of financial markets, especially around a crisis. We first attempt to visualize the collective behaviour of markets during a financial crisis through cross-correlations between typical Japanese daily stocks by making use of multi-dimensional scaling. We find that all the two-dimensional points (stocks) shrink into a single small region when a economic crisis takes place. By using the properties of cross-correlations in financial markets especially during a crisis, we next propose a theoretical framework to predict several time-series simultaneously. Our model system is basically described by a variant of the multi-layered Ising model with random fields as non-stationary time series. Hyper-parameters appearing in the probabilistic model are estimated by means of minimizing the 'cumulative error' in the past market history. The justification and validity of our approaches are numerically examined for several empirical data sets.

1. Introduction

One of central modern issues in quantitative finance is to determine to what extent the market is 'efficient'; crudely, whether there is an equal chance that the stock is under or over value at any time point. From the view point of statistics, the market is regarded as efficient when the market price is an unbiased estimate, in other words, when the price can be greater or less than the true value as long as the deviation is completely random.

Recently, as huge high-frequency financial data sets can be stored and analysed, the so-called 'stylized (empirical) facts' [1, 2, 3, 4] such as heavy tails of returns, volatility clustering, gain/loss asymmetry *etc.* have been found, in particular, in the research field of econophysics [5, 6, 7, 8, 9, 10, 11]. At the same time, several empirical facts provide an evidence to show that there exist some 'seasonal effects' in financial market (the so-called 'market anomaly'). In fact, it is well-known that buying stocks at the end of year and selling them at the beginning of the next year is sometimes less risky (the so-called *January effects* [12]). Hence, it is now partially accepted that the financial market is 'weakly' efficient or it is sometimes inefficient to some extent and in certain time scale.

Actually, there are several evidences to show the market inefficiency during financial crisis. At financial crisis, traders are more likely to behave according to the 'mood' (atmosphere) in society (financial market) and they incline to take rather 'irrational' strategies in some sense. Thus, this collective behavior might cause some market anomaly.

In the literature of behavioral economics [13], a concept of the so-called *information cascade* (or *Herding effect*) is well-known as a result of such human (traders') collective behaviour. One of the key measurements to understand such financial cascade is 'correlation' between ingredients in the societies. For instance, cross-correlations between stocks, traders are quite important to figure out the human collective phenomena. As the correlation could be found in various scale-lengths, from macroscopic stock price level to microscopic trader's level, the cascade also might be observed 'hierarchically' in such various scales from prices of several stocks to ways (strategies) of trader's decision making. Actually, we sometimes encounter the problem to find unusual structure in correlated time series observed from multi-dimensional information channels. Such time series obtained from multi-channel measurement have been widely provided in both natural and social sciences. Hence, it is now quite important for us to carry out empirical data analysis extensively to solve various modern and serious problems around us.

After the earthquake on 11th March 2011, Japanese NIKKEI stock market quickly responded to the crisis and quite a lot of traders sold their stocks of companies whose branches or plants are located in that disaster stricken area. As the result, the Nikkei stock average suddenly drops after the crisis [14]. However, it is impossible for us to mention the co-movement of stocks having

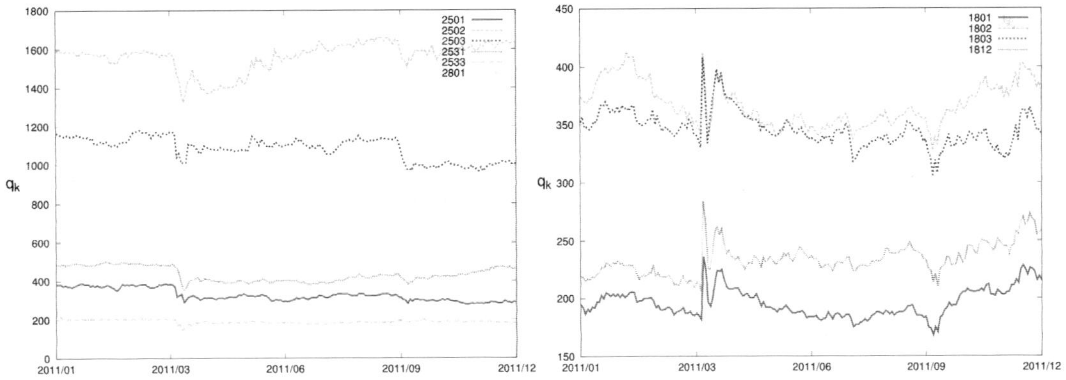

Figure 1. The prices of several major stocks in food industries (left) and construction industries (right) as a function of time (these are daily data sets). Each number shown at each line caption denotes the ID for each company: (*i.e.* 2501: Sapporo Breweries, 2502: Asahi Breweries, 2503: Kirin Holdings, 2531: Takara Holdings, 2533: Oenon Holdings, 2801: Kikkoman Corporation, 1801: Taisei Corporation, 1802: Obayashi Corporation, 1803: Shimizu Corporation, 1812: Kajima Corporation. These IDs can be checked at the web site [23]) (COLOR ONLINE)

correlation (with the majority bulk component including themselves) or anti-correlation during the crisis. In Figure 1, we plot the prices of several major stocks in food business (left) and construction business (right) as a function of time (these are daily data sets). From this figure, we find that The price of stocks for the same type of business behaves as correlated time series, whereas for different types of business, say, food industries and construction business, they have apparent any-correlations especially during the crisis.

Hence, it might be quite important for us to make an attempt to bring out more 'microscopic' useful information, which is never obtained from the averaged macroscopic quantities such as stock average, about the market. As a candidate of such 'microscopic information', we can use the (linear) correlation coefficient based on the two-body interactions between stocks [14, 15, 16, 17, 18, 19, 20, 21]. To make out the mechanism of financial crisis, it might be helpful for us to visualize such correlations in stocks and compare the dynamical behaviour of the correlation before and after crisis.

In this paper, in order to show and explain the hierarchical information cascade, we visualize the correlation of each stock in two-dimension. We specify each location of K stocks from a given set of the $K(K-1)/2$ distances by making use of the so-called multi-dimensional scaling (MDS) [22]. We also propose a theoretical framework to predict several time-series simultaneously by using cross-correlations in financial markets. The justification of this assumption is numerically checked for the empirical Japanese stock data, for instance, those around 11 March 2011, and for foreign currency exchange rates around Greek crisis in spring 2010.

This paper is organized as follows. In section 2, we explain our tools of analysis, namely, the correlation coefficient and multi-dimensional scaling. In section 3, our forecasting model for stock prices is introduced. In section 4, we examine our model for empirical data sets. The last section gives several remarks.

2. Linear correlation coefficient and multi-dimensional scaling

We utilize the linear correlation coefficient to measure the strength of correlation between stocks [15, 16, 17, 18, 19, 20]. The correlation coefficient (Pearson estimator) is calculated as follows.

Let us define $p_t^{(i)} (\geq 0)$ as a price of stock i at time t. Then, we evaluate the return of the price $p_t^{(i)}$ in terms of the logarithmic measurement as

$$\Delta r_i(t) \equiv \log p_t^{(i)} - \log p_{t-1}^{(i)}. \tag{1}$$

For the above logarithmically rescaled return, we calculate the moving average over a time window (interval) with width M as

$$\overline{\Delta r_i(t)} \equiv \frac{1}{M} \sum_{l=t-M+1}^{t} \Delta r_i(l) \tag{2}$$

for stock i, and we also evaluate the two-body correlation between stocks i, j by the following definition

$$\overline{\Delta r_i(t) \Delta r_j(t)} \equiv \frac{1}{M} \sum_{l=t-M+1}^{t} \Delta r_i(l) \Delta r_j(l). \tag{3}$$

Then, the linear correlation coefficient is given by

$$c_{ij}(t) = \frac{\overline{\Delta r_i(t) \Delta r_j(t)} - (\overline{\Delta r_i(t)})(\overline{\Delta r_j(t)})}{\sqrt{[\overline{(\Delta r_i(t))^2} - (\overline{\Delta r_i(t)})^2][\overline{(\Delta r_j(t))^2} - (\overline{\Delta r_j(t)})^2]}}. \tag{4}$$

2.1. Distribution of linear coefficient

As empirical data set, we pick up 200 stocks including the so-called TOPIX (TOkyo stock Price IndeX) Core30, which consists of typical 30 stock indices being picked up from the view point of 'current price' or 'liquidity' from the Japanese Nikkei stock market [23]. It should be kept in mind that the data itself is not provided as 'tick-by-tick' intra-day data, the minimal time interval of the data is one day (the closing price is given in the data set).

In order to investigate the statistical properties, we evaluate the distribution of the correlation coefficient $P(c)$ and plot the result in Figure 2. From this figure, we clearly find that the

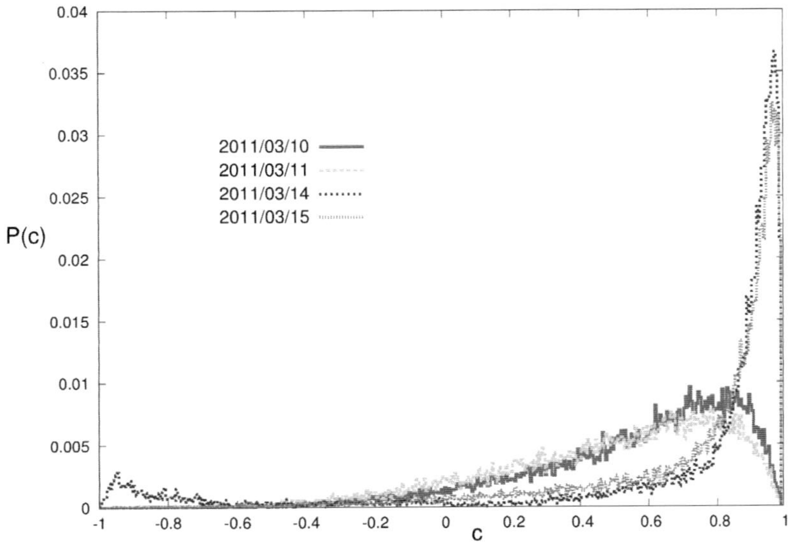

Figure 2. The distribution $P(c)$ of linear correlation coefficients. The result for before crisis (10th, 11th March 2011) and after crisis (14th, 15th March 2011) are shown. It should be noted that 12th and 13th were weekend and the market was closed. We find that a single peak before the crisis splits into the 'correlated' (with the majority bulk component including themselves) and 'anti-correlated' clusters on 14th. However, the anti-correlated cluster disappears on the next day. (COLOR ONLINE)

distribution is skewed and possesses a single peak at $c > 0$ before the crisis. Namely, most of the 200 stocks are mutually correlated. On the other hand, just after the crisis, say, 14th March 2012, the single peak splits into two components and the bulk in which some pairs of two stocks posses the negative correlation appears. Thus, one can grasp the macroscopic aspect of the collective behaviour of the 200 stocks for both before and after the crisis. However, much more microscopic properties of the 200 stocks are unfortunately hided behind the distribution $P(c)$.

To reveal such hidden microscopic aspects of the collective behaviour of the 200 stocks, we shall next attempt to visualize the relationship between these stocks by specifying the location of each stock in two-dimensional space.

2.2. The multi-dimensional scaling: From correlation to distance

To make a plot of the location of each stock, one needs the information about the Euclidean distance between arbitrary two stocks. As we saw in the previous subsection, the correlation coefficient $c_{ij}(t)$ might posses some useful information about the relationship between arbitrary two stocks i, j, however, it should be noticed that the correlation coefficient (4) satisfies $-1 \leq c_{ij}(t) \leq 1$, and apparently it cannot be treated as a 'distance'. Hence, here we transform the correlation coefficient $c_{ij}(t)$ into the distance $d_{ij}(t)$ between the stocks i, j as

$$d_{ij}(t) = \sqrt{\frac{1 - c_{ij}(t)}{2}}. \tag{5}$$

We should bear in mind that the above distance satisfies $0 \le d_{ij}(t) \le 1$ and defines the metric space in multi-dimension [14].

Obviously, once we obtain the location vectors $\boldsymbol{X}_i, \boldsymbol{X}_j$ for the stocks i, j , one can easily calculate the distance between them as $\|\boldsymbol{X}_i - \boldsymbol{X}_j\|$. However, the inverse process, namely, to specify the location vectors $\boldsymbol{X}_i, \boldsymbol{X}_j$ for a given distance $\|\boldsymbol{X}_i - \boldsymbol{X}_j\|$ is not so easy when the number of the stocks N increases. To carry out the inverse process systematically, we can use the well-known method named as *multi-dimensional scaling (MDS)* (see *e.g.* [22]). In following, we explain the procedure.

Let us first specify the location of an arbitrary stock i at time t by means of a P-dimensional vector $\boldsymbol{X}_i^{(t)}$ in the following way (Hereafter, we consider the case of $P = 2$ especially):

$$\boldsymbol{X}_i(t) \equiv (x_{i1}(t), x_{i2}(t), \cdots, x_{iP}(t)), \ i = 1, \cdots, K. \tag{6}$$

Naturally, the Euclidean distance between arbitrary two stocks i, j is now given by

$$d_{ij}(t) = \sqrt{\sum_{m=1}^{P} (x_{im}(t) - x_{jm}(t))^2}. \tag{7}$$

Hence, the inner product of location vectors of stocks i and j is also calculated as

$$2\{\boldsymbol{X}_i(t) \cdot \boldsymbol{X}_j(t)\} = 2z_{ij}(t) = \frac{1}{K}\sum_{i=1}^{K} d_{ij}(t)^2 + \frac{1}{K}\sum_{j=1}^{K} d_{ij}(t)^2 - \frac{1}{K^2}\sum_{i,j=1}^{K} d_{ij}(t)^2 - d_{ij}(t)^2 \tag{8}$$

where we should notice that we chose the origin of axis as the 'center of mass' $\boldsymbol{X}_c(t)$ for K stocks points, that is to say,

$$\boldsymbol{X}_c(t) \equiv \frac{1}{K}\sum_{i=1}^{K} \boldsymbol{X}_i(t) = \mathbf{0} \tag{9}$$

to specify an arbitrary stock (vector) at each time t. This equation (9) implies that the center of mass is a time-independent vector and it is definitely fixed at the origin for all time $t \ge 0$.

Then, in order to look for the locations $\boldsymbol{X}_i(t)$, $i = 1 \cdots, K$ which generates a set of distances $\{d_{ij}(t)\}$ consistently, we should minimize the following energy function (cost):

$$E_t = \sum_{i,j=1}^{K} \left(z_{ij}(t) - \sum_{m=1}^{P} x_{im}(t)x_{jm}(t) \right)^2 \tag{10}$$

with respect to $\boldsymbol{X}_i(t)$, $i = 1, \cdots, K$.

Thus, our problem to find the best possible locations for stocks is now rewritten in terms of an optimization problem to look for the solution $\boldsymbol{X}_i(t)$, $i = 1, \cdots, K$ that minimizes the energy function E_t at each time step t. We plot the result in Figure 3 for the same data set used in the plot of Figure 2. From these panels, we clearly find that after the crisis, the scattered plots actually shrink into a small limited region centered at the origin (the center of mass) as we expected before. Isolated several dots separating from main clusters denote the price of 'building industry'. These isolated points anti-correlated with the bulk are construction business, such as ID 1934: Yurtec corporation, which is a construction industry in Tohoku area, ID 1826: Sata construction Co. Ltd. Apparently we recognize that the large amount of demand is expected in such industries even after the crisis (mega earthquakes).

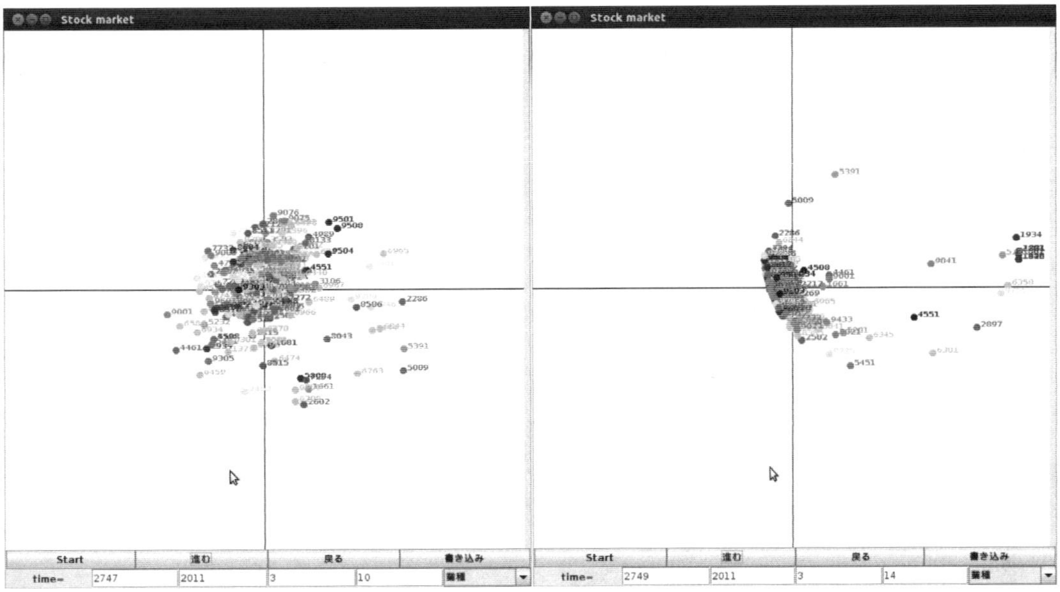

Figure 3. The result of the MDS. We pick up 200 stocks including the so-called TOPIX Core30 and the Nikkei stock average as empirical data set. The results on 10th (left) and 14th (right) March 2011 are shown. Different colors indicate different types of business. The numbers accompanying the dots show company IDs. The isolated points anti-correlated with the bulk are construction business, such as **1934**: Yurtec Corporation, which is a construction industry in Tohoku area, **1826**: Sata Construction Co. Ltd. We set the width of time window to evaluate the correlation coefficient as $M = 7$ (days). (COLOR ONLINE)

3. Prediction by using cross-correlation

It is necessary for us to investigate the human collective behaviour of the social agents such as traders in order to construct persistent systems in earthquake disasters. In financial systems, the price of each commodity as a macroscopic quantity is determined by huge number of trader's making decisions. To predict the price efficiently, a lot of mathematical tools such as AR model and its extensions called as ARCH model or GARCH model [6], or Kalman filter and its various variants have been proposed. Recently, besides these rather traditional models, several physics inspired models have been also introduced by several authors [24, 25, 26]. However, these models apparently lack the microscopic view point. Namely, in these models, the predicted price is not constructed by the result from microscopic trader's decision making.

Among these studies of prediction of stock prices, Kaizoji [27] attempted to represent 'buying' and 'selling' signals posted by traders by 'Ising spin'. He proposed a model in which the return of the price is determined by the 'magnetization' of the Ising system. He carried out computer simulations and concluded that there are interesting relationship between financial phenomena such as 'bubble' or 'crush' and physically collective phenomena which are referred to as 'phase transition' in the Ising magnetic system.

However, those studies seem to be not yet extensive and there exist several open questions to be clarified. For instance, dynamics of macroscopic quantities which might specify the collective behaviour of traders should be revealed more extensively. And as we saw in the previous sections, just after crisis, several stocks are strongly correlated. Hence, we might use the cross-correlation to predict several prices simultaneously by modifying the Kaizoji's model [27] by means of the cross-correlations. Therefore, from this section, we are focusing on the prediction of prices of several stocks simultaneously by using the cross-correlation in stock markets.

3.1. A link from microscopic making decisions to macroscopic prices

In order to investigate the effect of the cross-correlation on the prediction of several stock-prices, we first explain the model proposed by Kaizoji [27] as a basic model.

Let us define $p_t^{(k)}$ as the price of commodity k at time t. Then, the return, which is defined as the difference between prices at successive two time steps t and $t+1$ is given by

$$\Delta p_t^{(k)} \equiv p_{t+1}^{(k)} - p_t^{(k)} = m_t^{(k)}, \quad k = 1, \cdots, K. \tag{11}$$

In order to construct the return $m_t^{(k)}$ from the microscopic view point, we assume that each trader $(i = 1, \cdots, N_k)$ possessing the stock k buys or sells $v_{it}^{(k)}$-volumes at each time step t. Then, let us call the group of buyers as $\mathcal{A}_+(k,t)$, whereas the group of sellers is referred to as $\mathcal{A}_-(k,t)$. Hence, the total volumes of buying and selling are explicitly given by

$$\psi_+(k,t) \equiv \sum_{i \in \mathcal{A}_+(k,t)} v_{it}^{(k)}, \ \psi_-(k,t) \equiv \sum_{i \in \mathcal{A}_-(k,t)} v_{it}^{(k)}, \tag{12}$$

respectively. We should keep in mind that the total number of traders dealing with the stock k is conserved, namely, the condition:

$$\mathcal{A}_+(k,t) + \mathcal{A}_-(k,t) = N_k \tag{13}$$

should be satisfied.

Then, the return $m_t^{(k)}$ is naturally defined by means of (12)

$$m_t = \lambda(\psi_+(k,t) - \psi_-(k,t)) \tag{14}$$

where λ is a positive constant. Namely, when the volume of buyers is greater than that of sellers, $\psi_+(k,t) > \psi_-(k,t)$, the return becomes positive $m_t^{(k)} > 0$. As the result, the price of the commodity k should be increased at the next time step as $\Delta p_t^{(k)} = m_t^{(k)}$.

3.1.1. Ising spin representation The above microscopic observation and the set-up might be naturally accepted, however, here we shall make the situation much more simpler. Namely, we omit the information about the volume by setting $v_{it}^{(k)} = 1\,(\forall_{i,t,k})$. The making decision of each trader $(i = 1, \cdots, N_k)$ is now obtained simply by an Ising spin:

$$S_i^{(k)}(t) = \begin{cases} +1 & \text{(the trader } i \text{ buys the stock } k \text{ at time } t) \\ -1 & \text{(the trader } i \text{ sells the stock } k \text{ at time } t) \end{cases} \tag{15}$$

The return is also simplified as

$$m_t^{(k)} = \lambda(\psi_+(k,t) - \psi_-(k,t)) = \lambda \sum_{i=1}^{N_k} v_{it}^{(k)} S_i^{(k)}(t) \tag{16}$$

where we set $\lambda = N_k^{-1}$ to make the return:

$$m_t^{(k)} = \frac{1}{N_k} \sum_{i=1}^{N_k} S_i^{(k)}(t) \tag{17}$$

satisfying $|m_t^{(t)}| \le 1$. Thus, m_t^k corresponds to the 'magnetization' in statistical physics, and the update rule of the price for the stock k is governed in terms of the magnetization $m_t^{(k)}$ as

$$\Delta p_t^{(k)} = m_t^{(k)}, \ k = 1, \cdots, K. \tag{18}$$

3.2. The energy function

We next introduce the energy function of the system.

$$E_t(\boldsymbol{S}^{(k)}) = -\frac{J_t^{(k)}}{N_k}\sum_{ij} S_i^{(k)}S_j^{(k)} - h_t^{(k)}\sum_i \sigma_\tau^{(k)}(t)S_i^{(k)} - \gamma_t^{(k)}\sum_i \left(\frac{1}{K}\sum_{\mu\neq k}^{K} c_{k\mu}(t)m_t^{(\mu)}\right)S_i^{(k)} \quad (19)$$

where we omitted the time dependence in $S_i^{(k)}$ for simplicity. The first term in the right hand side of (19) induces human collective behaviour, namely, each agent inclines to take the same decision as the others to decrease the total energy. The effect of this first term on the minimization of total energy might be recognized as the so-called *Keynes's beauty contest*. It means that traders tend to make the same decision as the others, in particular, during a crisis. Namely, when a big negative news (mood) such as earthquake in Japan is broadcasted, almost all of the traders might sell their own stocks as the others also sell without any rationality. In other words, the first term causes the human collective behaviour of the traders, which is sometimes refereed to as information cascade in the literature of behavioral economics, for $J_t^{(k)} \geq h_t^{(k)}, \gamma_t$. Actually, we find that the lower bound of the first term in the right hand side of (19) is evaluated as

$$-\frac{J_t^{(k)}}{N_k}\sum_{ij} S_i^{(k)}S_j^{(k)} \geq -N_k J_t^{(k)} \quad (20)$$

where the equal sign is satisfied if and only if $S_i^{(k)} = 1\ (\forall_i)$ or $S_i^{(k)} = -1\ (\forall_i)$ holds.

The second term in (19) represents the cross-correlation between the decisions of traders and market (historical) information. In this paper, we choose the 'trends' :

$$\sigma_\tau^{(k)}(t) \equiv \frac{p_t - p_{t-\tau}}{\tau} \quad (21)$$

for such information. This means that the total energy should decrease when each trader posts the sign $S_i^{(k)} = \pm 1$ to the market so as to make the product $\sigma_\tau^{(k)}(t)S_i^{(k)}$ definitely positive. In other words, each trader buys the commodity k if the price tends to increase over the past τ-time steps, whereas the trader sells the stock vice versa. Actually, the second term is definitely minimized as

$$-h_t^{(k)}\sum_i \sigma_\tau^{(k)}(t)S_i^{(k)} \geq -N_k h_t^{(k)}|\sigma_\tau^{(k)}(t)| \quad (22)$$

where the equality should be satisfied when $\mathrm{sgn}[S_i^{(k)}] = \mathrm{sgn}[\sigma_\tau^{(k)}(t)]$ holds.

The third term, which does not appear in the references [27], comes from correlation between the trader i possessing the commodity k and a 'typical trader' (a mean-field) $m_t^{(\mu)}$ possessing stocks $\mu\ (\neq k)$. A factor $c_{k\mu}(t)$ appearing in the third term stands for the correlation coefficient and we should remember that it is given by (4) explicitly. Hence, when the stocks k and μ are correlated in terms of the positive coefficient, namely, $c_{k\mu}(t) > 0$ from time $t-M+1$ to time t, the trader possessing the stock k inclines to take the same decision as the typical trader dealing with the stock μ, that is, $m_t^{(\mu)}$.

From the above argument, it might be useful for us to investigate to what extent those traders behave collectively through the values of macroscopic hyper-parameters $J_t^{(k)}, h_t^{(k)}$ and $\gamma_t^{(k)}$. Namely, in the realistic stock market, there might exist a possibility that human behaviour turns out to be collective and 'irrational' in some sense when these parameters change the values as $h_t^{(k)}/J_t^{(k)} \to 0, \gamma_t^{(k)}/J_t^{(k)} \to 0$ and $J_t^{(k)} \to 1$ (a critical point of infinite range ferromagnetic Ising model).

3.3. The Boltzmann-Gibbs distribution

It should be noticed that the state vectors of the agents: $\boldsymbol{S}^{(k)} = (S_1, \cdots, S_{N_k})$, $k = 1, \cdots, K$ are determined so as to minimize the energy function (19) from the argument in the previous subsection. For most of the cases, the solution should be unique. However, in realistic financial markets, the decisions by agents should be much more 'diverse'. Thus, here we consider statistical ensemble of traders for each commodity $\boldsymbol{S}^{(k)}$ and define the distribution of the ensemble by $P(\boldsymbol{S}^{(k)})$. Then, we shall look for the suitable distribution which maximizes the so-called Shannon's entropy

$$H = -\sum_{\boldsymbol{S}^{(k)}} P(\boldsymbol{S}^{(k)}) \log P(\boldsymbol{S}^{(k)}) \tag{23}$$

under two distinct constraints:

$$\sum_{\boldsymbol{S}^{(k)}} P(\boldsymbol{S}^{(k)}) = 1, \ \sum_{\boldsymbol{S}^{(k)}} P(\boldsymbol{S}^{(k)}) E(\boldsymbol{S}^{(k)}) = E_k. \tag{24}$$

Namely, according to Jaynes [28], we choose the distribution which minimizes the following functional $f_k\{P(\boldsymbol{S}^{(k)})\}$:

$$\begin{aligned} f_k\{P(\boldsymbol{S}^{(k)})\} &= -\sum_{\boldsymbol{S}^{(k)}} P(\boldsymbol{S}^{(k)}) \log P(\boldsymbol{S}^{(k)}) - \lambda_1^{(k)} \left(\sum_{\boldsymbol{S}^{(k)}} P(\boldsymbol{S}^{(k)}) - 1 \right) \\ &- \lambda_2^{(k)} \left(\sum_{\boldsymbol{S}^{(k)}} P(\boldsymbol{S}^{(k)}) E_k(\boldsymbol{S}^{(k)}) - E_k \right) \end{aligned} \tag{25}$$

where $\lambda_1^{(k)}, \lambda_2^{(k)}$ are Lagrange's multipliers. After some easy algebra, we immediately notice that $\lambda_1^{(k)}$ gives a normalization constant of $P(\boldsymbol{S}^{(k)})$ and $\lambda_2^{(k)} \equiv \beta$ is a control parameter for 'thermal fluctuation' in the system. Thus, we have the solution as

$$P(\boldsymbol{S}^{(k)}) = \frac{\exp[-\beta E(\boldsymbol{S}^{(k)})]}{\sum_{\boldsymbol{S}^{(k)}} \exp[-\beta E(\boldsymbol{S}^{(k)})]} \tag{26}$$

where β stands for the inverse-temperature, and hereafter we set $\beta = 1$. We also defined the sum with respect to $\boldsymbol{S}^{(k)}$ by

$$\sum_{\boldsymbol{S}^{(k)}} (\cdots) \equiv \sum_{S_1^{(k)}=\pm 1} \cdots \sum_{S_N^{(k)}=\pm 1} (\cdots). \tag{27}$$

3.4. A Bayesian interpretation

It might be helpful for us to interpret the above probability distribution (26) from the view point of Bayesian inference. Actually, one can derive the above Boltzmann-Gibbs form $P(\boldsymbol{S}^{(k)})$ by means of a posterior distribution in Bayesian statistics. Let us assume that the price change of stock might be caused by $\sigma_\tau^{(k)}(t)$ due to the microscopic decision making $S_i^{(k)} = \pm 1$ for trader i. Then, the following conditional probability

$$P(\sigma_\tau^{(k)}(t) | \boldsymbol{S}^{(k)}) \propto \exp\left(h_t^{(k)} \sum_{i=1}^{N} \sigma_\tau^{(k)}(t) S_i^{(k)} \right) \tag{28}$$

denotes a likelihood of the result $\sigma_\tau^{(k)}(t)$ caused by the decision making of traders $\boldsymbol{S}^{(k)}$. Therefore, in order to forecast the decision of traders $\boldsymbol{S}^{(k)}$ for a given market data $\sigma_\tau^{(k)}(t)$, we construct the posterior $P(\boldsymbol{S}^{(k)} | \sigma_\tau^{(k)}(t))$ by means of the Bayesian formula

$$P(\boldsymbol{S}^{(k)} | \sigma_\tau^{(k)}(t)) \propto P(\sigma_\tau^{(k)}(t) | \boldsymbol{S}^{(k)}) Q(\boldsymbol{S}^{(k)} | \{m_t^{(\mu \neq k)}\}) \tag{29}$$

where $Q(\boldsymbol{S}^{(k)}|\{m_t^{(\mu\neq k)}\})$ stands for the prior distribution. For a given set of mean-fields from the other Ising layers $\{m_t^{(\mu\neq k)}\}$, it is naturally accepted to choose the prior as

$$Q(\boldsymbol{S}^{(k)}|\{m_t^{(\mu\neq k)}\}) \propto \exp\left[\frac{J_t^{(k)}}{N_k}\sum_{ij} S_i^{(k)}S_j^{(k)} + \gamma_t^{(k)}\sum_i \left(\frac{1}{K}\sum_{\mu\neq k}^{K} c_{k\mu}(t)m_t^{(\mu)}\right)S_i^{(k)}\right] \tag{30}$$

Then, we have the posterior

$$\begin{aligned} & P(\boldsymbol{S}^{(k)}|\sigma_\tau^{(k)}(t)) \\ = & \frac{\exp[\frac{J_t^{(k)}}{N_k}\sum_{ij} S_i^{(k)}S_j^{(k)} + h_t^{(k)}\sum_i \sigma_\tau^{(k)}(t)S_i^{(k)} + \gamma_t^{(k)}\sum_i(\frac{1}{K}\sum_{\mu\neq k}^K c_{k\mu}(t)m_t^{(\mu)})S_i^{(k)}]}{\sum_{\boldsymbol{S}^{(k)}}\exp[\frac{J_t^{(k)}}{N_k}\sum_{ij} S_i^{(k)}S_j^{(k)} + h_t^{(k)}\sum_i \sigma_\tau^{(k)}(t)S_i^{(k)} + \gamma_t^{(k)}\sum_i(\frac{1}{K}\sum_{\mu\neq k}^K c_{k\mu}(t)m_t^{(\mu)})S_i^{(k)}]}. \end{aligned} \tag{31}$$

Thus, our model system is described by multi-layered Ising model in which arbitrary two layers (stocks) are coupled through the mean-fields.

3.5. *The mean-field equation for instantaneous return*

As it is well-known, when ingredients of the system are 'fully-connected', the partition function Z (the numerator of (31)) is evaluated at the saddle point as

$$Z \simeq \exp\left[N_k \Phi(m^{(k)} : J_t^{(k)}, h_t^{(k)}, \gamma_t^{(k)}, \{m_t^{(\mu)}\})\right] \tag{32}$$

in the limit of $N_k \to \infty$ with 'free energy density'

$$\Phi = -\frac{J_t^{(k)}}{2}(m^{(k)})^2 + \log\cosh\left(J_t^{(k)}m^{(k)} + h_t^{(k)}\sigma_\tau^{(k)}(t) + \frac{\gamma_t^{(k)}}{K}\sum_{\mu\neq k}^{K} c_{k\mu}(t)m_t^{(\mu)}\right). \tag{33}$$

Thus, the saddle point equation $\partial\Phi/\partial m^{(k)} = 0$ yields

$$m^{(k)} = \tanh\left(J_t^{(k)}m^{(k)} + h_t^{(k)}\sigma_\tau^{(k)}(t) + \frac{\gamma_t^{(k)}}{K}\sum_{\mu\neq k}^{K} c_{k\mu}(t)m_t^{(\mu)}\right). \tag{34}$$

The second and third terms appearing in $\tanh(\cdots)$ are an external field from the market history and mean-fields from the other Ising layers, respectively. For a given non-stationary field $\sigma_t^{(k)}(t)$ and mean-field $\{m_t^{(\mu)}\}$, it could not be expected that the equilibrium solution for each layer $m^{(k)}$ exists. Hence, here we assume that such non-equilibrium effects could be built-in by dealing with the following time-dependent (naive) mean-field equation for the 'instantaneous' returns.

$$m_t^{(k)} = \tanh\left(J_{t-1}^{(k)}m_{t-1}^{(k)} + h_{t-1}^{(k)}\sigma_\tau^{(k)}(t-1) + \frac{\gamma_{t-1}^{(k)}}{K}\sum_{\mu\neq k}^{K} c_{k\mu}(t-1)m_{t-1}^{(\mu)}\right),\quad k=1,\cdots,K \tag{35}$$

By solving the above non-linear maps numerically for a given set of past real market history $\{\sigma_\tau^{(k)}(t-1)\}$, one can forecast the price of each stock by means of $\Delta p_t = m_t^{(k)},\ k=1,\cdots,K$ simultaneously.

3.6. Hyper-parameter estimation from non-stationary time-series

In order to use the mean-field equation (35) for forecasting the stock prices, systematic estimation for the so-called hyper-parameters $(J_t^{(k)}, h_t^{(k)}, \gamma_t^{(k)})$ appearing in the right hand side of (35) is needed. In the literature of probabilistic information processing, say, in Bayesian image restoration [29], one cannot use the mean-square error as a cost function because it needs the 'true (original) image' to be constructed. Therefore, we usually use the marginal likelihood as the cost function to estimate the hyper-parameters. However, fortunately in our present model system, one can utilize the past market history, let us to say, the 'true returns' in the past $\Delta q_l^{(k)} \equiv q_l^{(k)} - q_{l-1}^{(k)}$, $l = 1, \cdots, t-1$, and it means that the 'cumulative error' could be defined as a cumulative mean-square error between the true observable and the estimate by

$$\begin{aligned}\mathcal{E}_k &\equiv \frac{1}{2}\sum_{l=1}^{t-1}\left(\overline{\Delta q_l^{(k)}} - \overline{\Delta p_l^{(k)}}\right)^2 \\ &= \frac{1}{2}\sum_{l=1}^{t-1}\left[\overline{\Delta q_l^{(k)}} - \tanh\left(J_{t-1}^{(k)}\overline{\Delta q_{l-1}^{(k)}} + h_{t-1}^{(k)}\sigma_{\tau}^{(k)}(t-1) + \frac{\gamma_{t-1}^{(k)}}{K}\sum_{\mu=1}^{K} c_{k\mu}(t-1)\overline{\Delta q_{l-1}^{(\mu)}}\right)\right]^2 \quad (36)\end{aligned}$$

where we defined the 'forecasted return' $\Delta p_l^{(k)} \equiv p_l^{(k)} - p_{l-1}^{(k)}$, $l = 1, \cdots, t-1$ and the time-averages

$$\overline{\Delta q_l^{(k)}} \equiv \frac{1}{M}\sum_{i=l-M+1}^{l}(q_{i+1}^{(k)} - q_i^{(k)}),\ \overline{\Delta p_l^{(k)}} \equiv \frac{1}{M}\sum_{i=l-M+1}^{l}(p_{i+1}^{(k)} - p_i^{(k)}) \quad (37)$$

for the time window with width M. To obtain the last line in the above equation (36), we used

$$\overline{\Delta p_l^{(k)}} \simeq m_l^{(k)} = \tanh\left(J_{t-1}^{(k)} m_{l-1}^{(k)} + h_{t-1}^{(k)}\sigma_{\tau}^{(k)}(t-1) + \frac{\gamma_{t-1}^{(k)}}{K}\sum_{\mu\neq k}^{K} c_{k\mu}(t-1) m_{l-1}^{(\mu)}\right) \quad (38)$$

and replaced the $m_{l-1}^{(k)}$, $k = 1, \cdots, K$ appearing in $\tanh(\cdots)$ by the corresponding observables $\overline{\Delta q_{l-1}^{(k)}}$, $k = 1, \cdots, K$ because one can actually use these values before forecasting.

Hence, we should infer these hyper-parameters $(J_t^{(k)}, h_t^{(k)}, \gamma_t^{(k)})$ from the past data set in the financial market by the gradient descent learning

$$J_t^{(k)} = J_{t-1}^{(k)} - \eta\frac{\partial \mathcal{E}_k}{\partial J_{t-1}^{(k)}},\ h_t^{(k)} = h_{t-1}^{(k)} - \eta\frac{\partial \mathcal{E}_k}{\partial h_{t-1}^{(k)}},\ \gamma_t^{(k)} = \gamma_{t-1}^{(k)} - \eta\frac{\partial \mathcal{E}_k}{\partial \gamma_{t-1}^{(k)}}, \quad (39)$$

where η is a learning rate. In the next section, we examine the above forecasting framework for empirical (intra-day) data sets in which a crisis appears.

4. Performance evaluation of forecasting for empirical high-frequency data sets

In this section, we check the usefulness of our prediction model with cross-correlation in stocks. As a simple examination, for the case of three different time-series $K = 3$, we check the accuracy of our prediction procedure. In section 2, we visualized 200 stocks which are given as daily data during a crisis. However, the number of data points is not enough for our forecasting procedure. Hence, here we pick up EUR/AUD ($k = 1$), EUR/CAD ($k = 2$), EUR/JPY ($k = 3$) exchange rates (EUR: Euro, CAD: Canadian dollar, AUD: Australian dollar, JPY: Japanese yen), which are given as high-frequency tick-by-tick data, from 27th April 2010 to 13th May 2010 as real values $q_t^{(k)}$. We plot those three true time series in Figure 4 as solid lines. We observe that these time series posses a crisis which corresponds to Greek crisis in spring 2010. In the left panels of Figure 4, the resulting prices $p_t^{(k)}$ predicted by our model are shown. We set the time window size as $M = \tau = 100$ and the learning rate as $\eta = 0.01$. Of course, we can choose the leaning rate as 'adaptive one' like $\eta = \eta(t)$, however, in this paper we set the value to a positive constant. From these panels, we find that our prediction procedure works well and the error is only a few percent of the average value of the rate.

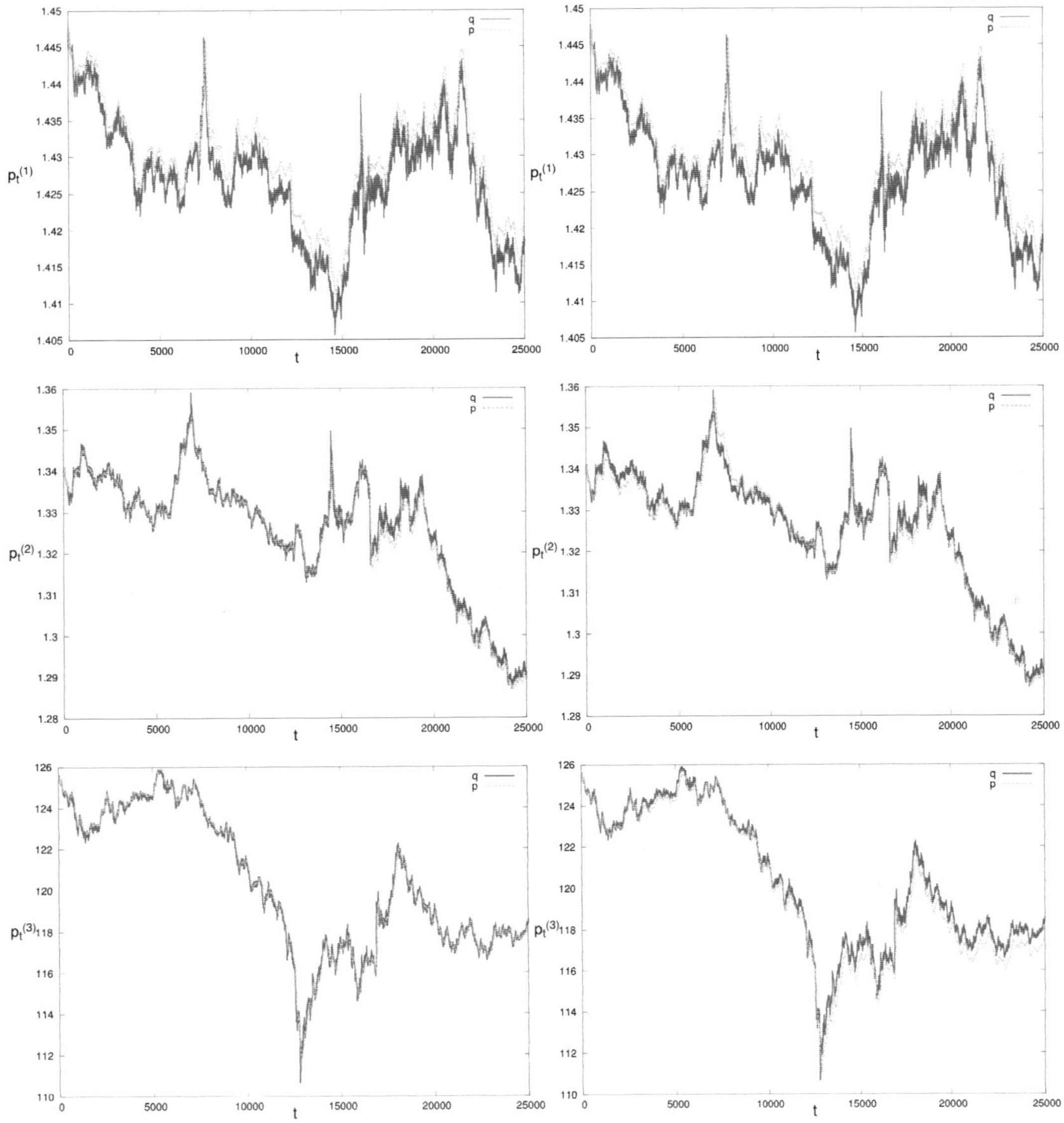

Figure 4. From the top to the bottom, EUR/AUD ($k = 1$), real value $q_t^{(k)}$ and its prediction $p_t^{(k)}$ are plotted for EUR/CAD ($k = 2$), EUR/JPY ($k = 3$) exchange rates from 27th April 2010 to 13th May 2010. The panels in the left column are obtained by 'stock-wise optimization', whereas the panels in the right are calculated by 'pair-wise optimization' for the correlation strengths. We set the width of time window and the learning rate as $M = \tau = 100, \eta = 0.01$. (COLOR ONLINE)

4.1. Pair-wise optimization for hyper-parameters of correlation strengths

We next examine the following slight modification

$$\frac{\gamma_t^{(k)}}{K}\sum_{\mu\neq k}^{K} c_{k\mu}(t-1)\overline{\Delta q_{l-1}^{(\mu)}} \to \frac{1}{K}\sum_{\mu\neq k}^{K}\gamma_t^{(k\mu)} c_{k\mu}(t-1)\overline{\Delta q_{l-1}^{(\mu)}} \tag{40}$$

in (35)-(36). Namely, the strength of correlation between Ising layers is evaluated independently for each pair of layers. Due to this modification, the learning equation for $\gamma_t^{(k)}$ in (39) should be corrected as

$$\gamma_t^{(k\mu)} = \gamma_{t-1}^{(k\mu)} - \eta\,\frac{\partial \mathcal{E}_k}{\partial \gamma_{t-1}^{(k\mu)}},\ \mu\neq k. \tag{41}$$

From now on, the forecasting model with the above modification (40)(41) is referred to as 'pair-wise optimization', whereas the original version (35)(36) is called as 'stock-wise optimization' for hyper-parameters for the correlation strengths.

We plot the result in the right panels in Figure 4. From these panels, we are confirmed that the pair-wise optimization makes the result worse against our expectation. Especially, for large t regime, the gap between the stock-wise optimization and the pair-wise optimization becomes large. From the result, we should consider the optimal number of hyper-parameters to fit the model to the time-series without over-fitting.

4.2. Flows of hyper-parameters during a crisis

Finally, we consider the dynamical behaviour of hyper-parameters during the crisis. We show the results in Figure 5. These panel show the dynamics of hyper-parameters $(J_t^{(k)}, h_t^{(k)}, \gamma_t^{(k)})$ for

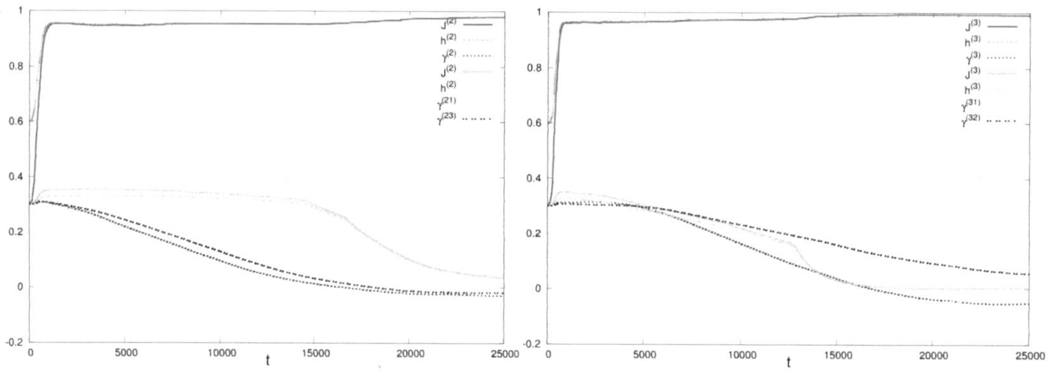

Figure 5. Dynamics of macroscopic parameters $(J_t^{(k)}, h_t^{(k)}, \gamma_t^{(k)})$ for 'stock-wise optimization' (The top three line captions) and $(J_t^{(k)}, h_t^{(k)}, \gamma_t^{(k\mu)})$ for 'pair-wise optimization' (The bottom four line captions). We show the result for $k=2$ (left panel) and $k=3$ (right). (COLOR ONLINE)

the stock-wise optimization and and $(J_t^{(k)}, h_t^{(k)}, \gamma_t^{(k\mu)})$ for the pair-wise optimization.

From this panel, we clearly find that for both cases ($k = 2,3$) of exchange rates in our forecasting model described by stock-wise optimization, the hyper-parameters $J^{(k)}, h^{(k)}$ converge to $h^{(k)} = 0$ and $J^{(k)} = 1$, respectively. Then, the $\gamma_t^{(k)}$ converges to zero for $k = 2$, which corresponds to the critical point of order-disorder phase transition in the infinite-range ferromagnetic Ising model. However, for $k = 3$ it converges to slightly negative value across the zero. Hence for $k = 3$ the stock-wise optimization does not detect the critical behaviour, and the deviation from the critical point becomes large as t goes on. On the other hand, the model with the pair-wise optimization also detects the critical behaviour for $k = 2$, however, the behavior of the hyper-parameters does not reflect the critical point (crisis) for $k = 3$.

We can partially figure out these results in the original time series shown in Figure 4. From this figure, we can observe that the exchange rate $k = 3$ was recovered from the crisis relatively faster than $k = 2$, whereas it was still in a crisis for the exchange rate $k = 2$ even in large t regime. Actually, in this period, Japanese yen was strong, and at the crisis, this tendency of strong yen was enhanced. However, the term of extreme strong yen is not so long and the value of Euro against yen was recovered quickly. On the other hand, continuous drops of Euro against Canadian dollar was serious after the crisis and it could not escape from a 'crush domain' even at $t = 250000$ (ticks). In this sense, the deviation from the critical point should be also observed through our model and actually it was confirmed in Figure 5.

5. Discussions and concluding remarks

There are many ways to visualize the co-movements of stocks, and using MDS is one of them. When the MDS studies were performed with daily data [35], we found that it was easier to visualize or detect specific sectors, strongly correlated pairs and market events. It was suggested that this type of plots using daily data may be used in designing strategies of pairs trade or identifying clusters or detecting market trends. It was also shown in Ref. [35], that we could follow the evolution of the market and/or trace the movements of particular companies (e.g., Lehmann Brothers) with respect to the rest of the market, before or after a market crisis.

In this paper, in order to show and forecast some cascade in financial systems, we visualized the correlation of each pair of stocks in two-dimension using MDS. We also proposed a theoretical framework based on the multi-layered Ising model to predict several time-series simultaneously by using cross-correlations in financial markets. Actually, in this paper, we showed that the knowledge of statistical mechanics of information (see *e.g.* [30]) could be applicable to the research topics outside of information processing, namely, quantitative finance or economics.

Finally, we would like to mention several remarks concerning our future direction.

5.1. On the time window size

In our model system, we set $M = \tau = 100$ for the width of time window to evaluate several statistics in our forecasting model (see (21) and (37)). However, we should chose these lengths more carefully. Recently, Livan, Inoue and Scalas [31] examined the effect of non-stationarity of time series on the portfolio optimization by using several statistical test including some knowledge of the random matrix theory [32], and they found that the longer time window does not always give the better estimate of the true value of the portfolio. This implies that there might exist some optimal size of window to construct the forecasting model.

5.2. The turnover

The turnover, namely, the total number of volume being dealt with for a stock at time t:

$$a_k(t) \equiv \psi_+(k,t) + \psi_-(k,t) = \sum_{i=1}^{N_k} v_{it}^{(k)} \tag{42}$$

might be a good indicator for the financial crisis (see also (12)). Actually, in Figure 6, we show the empirical plot of the turnover for several stocks during the crisis in Japan 2011. From this figure, we find that the turnover takes a sharp peak around the shock (11th March 2011), which was remarkably observed in stocks of construction industries (see the right panel of Figure 6). In our preliminary study [33], we made a model to estimate the turnover by means of three state Ising model in which each spin can take zero ('staying') besides ± 1 for 'selling' or 'buying'.

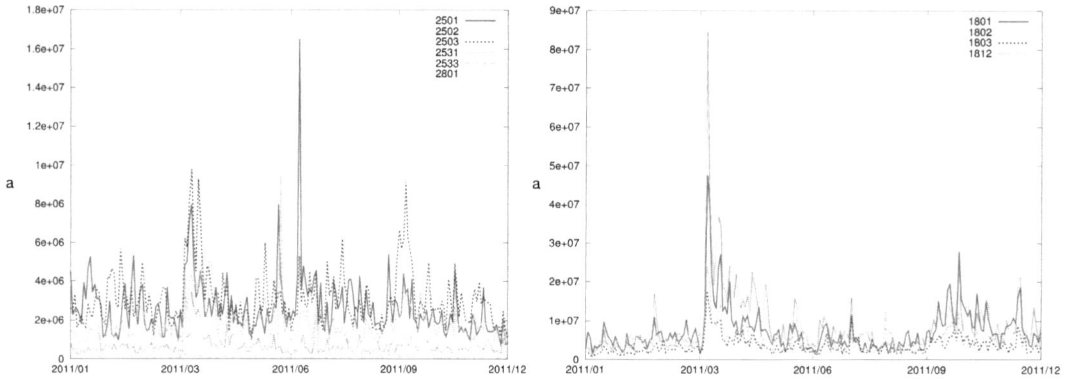

Figure 6. The turnover as a function of time (day). The number of caption denotes the ID for each company: (*i.e.* 2501: Sapporo Breweries, 2502: Asahi Breweries, 2503: Kirin Holdings, 2531: Takara Holdings, 2533: Oenon Holdings, 2801: Kikkoman Corporation, 1801: Taisei Corporation, 1802: Obayashi Corporation, 1803: Shimizu Corporation, 1812: Kajima Corporation. These IDs can be checked at the web site [23]) (COLOR ONLINE)

5.3. Suitable estimator for non-synchronous time series

The data sets used in the forecasting examination are not daily data but high-frequency data. Those are 'non-synchronous' time series, namely, there is no one-to-one correspondence in time axis in arbitrary two stocks. In this paper, for evaluating, say, $\Delta r_i(t)\Delta r_j(t^{'})$ where $t \neq t^{'}$, we chose the $t^{'}$ simply by

$$t^{'} = \mathrm{argmin}_t \|t - t^{'}\| \tag{43}$$

and of course, it might posses some evaluation error. In this sense (namely, 'strict sense'), the Pearson estimator is not suitable to evaluate the cross-correlation, and we should use another way, say, the so-called *Hayashi-Yoshida estimator* [34, 35].

5.4. The inverse-Ising problem

In our forecasting model, we assumed that the traders are fully-connected. However, the graph topology is important for us to consider the communities in the markets. Hence, the procedure to estimate the adjacency matrix from the empirical data (behavior of traders) should be done. Definitely, it is formulated as an 'inverse Ising problem'.

The studies related to the above four issues are now on-going and we will report the results at the meeting if we obtain the preliminary.

Acknowledgment

The authors would like to thank Enrico Scalas, Giacomo Livan, Frédéric Abergel for fruitful discussion and useful comments. SS and JI thank Saha Institute of Nuclear Physics for their support during our stay in Kolkata. One of the authors (JI) thanks Basque Center for Applied Mathematics, École Centrale Paris for their warm hospitality. JI also acknowledges the financial support by Grant-in-Aid for Scientific Research (C) of Japan Society for the Promotion of Science, No. 22500195 (2010-2012) and No.25330278 (2013-2015). AC is grateful to Hokkaido University for support during his stay in Sapporo.

References

[1] Cont R 2001 *Quantitative Finance* **1** 223
[2] Chakraborti A, Muni Toke I, Patriarca M and Abergel F 2011 *Quantitative Finance* **11** 991

[3] Chakraborti A, Muni Toke I, Patriarca M and Abergel F 2011 *Quantitative Finance* **11** 1013
[4] Chakraborti A, Patriarca M and Santhanam M S 2007 *Econophysics of Markets and Business Networks* (Milan: Springer) p 51
[5] Bouchaud J P and Potters M 2000 *Theory of Financial Risk and Derivative Pricing* (Cambridge: Cambridge University Press)
[6] Mantegna R N and Stanley H E 2000 *An Introduction to Econophysics* (Cambridge: Cambridge University Press)
[7] Garibaldi U and Scalas E 2010 *Finitary Probabilistic Methods in Econophysics* (Cambridge: Cambridge University Press)
[8] Aoyama H, Fujiwara Y, Ikeda Y, Iyetomi H, Souma W and Yoshikawa H 2011 *Econophysics and Companies: Statistical Life and Death in Complex Business Networks* (Cambridge: Cambridge University Press)
[9] Sinha S, Chatterjee A, Chakraborti A and Chakrabarti B K 2011 *Econophysics: An Introduction* (Berlin: Wiley-VCH)
[10] Chakrabarti B K, Chakraborti A, Chakravarty S R and Chatterjee A 2013 *Econophysics of Income and Wealth Distributions* (Cambridge: Cambridge University Press)
[11] Eds. Chakrabarti B K, Chakraborti A and Chatterjee A 2006 *Econophysics and Sociophysics: Trends and Perspectives* (Weinheim: Wiley-VCH)
[12] Keim D B 1983 *J. Financial Economics* **12** 13
[13] Kahneman D and Tversky A 1979 *Econometrica* **47** 263
[14] Ibuki T, Suzuki S and Inoue J 2012 *Econophysics of Systemic Risk and Network Dynamics* (Milan: Springer) p 239
[15] Mantegna R N 1999 *Eur. Phys. J.* B **11** 193
[16] Onnela J-P, Chakraborti A, Kaski K, Kertesz J and Kanto A 2003 *Phys. Rev.* E **68** 056110
[17] Chakraborti A 2006 *Econophysics of Stock and other Markets* (Milan: Springer) p 13
[18] Onnela J-P, Chakraborti A, Kaski K, Kertesz J and Kanto A 2003 *Physica Scripta* T **106** 48
[19] Onnela J-P, Chakraborti A, Kaski K and Kertesz J 2003 *Physica* A **324** 247
[20] Onnela J-P, Chakraborti A, Kaski K and Kertesz J 2002 *Eur. Phys. J.* B **30** 285
[21] Ibuki T, Higano S, Suzuki S and Inoue J 2012 *ASE Human Journal* **1** 74
[22] Borg I and Groenen P 2005 *Modern Multidimensional Scaling: theory and applications* (New York: Springer)
[23] http://finance.yahoo.co.jp/
[24] Bouchaud J P and Cont R 1998 *Eur. Phys. J.* B **6** 543
[25] Masukawa J 2002 *Journal of the Japan Society for Simulation Technology* **21** 92
[26] Watanabe K, Takayasu H and Takayasu M 2009 *Phys. Rev.* E **80** 056110
[27] Kaizoji T 2000 *Physica* A **287** 493
[28] Jaynes T E 1957 *Physical Review* **106** 620
[29] Inoue J and Tanaka K 2002 *Phys. Rev.* E **65** 016125
[30] Nishimori H 2001 *Statistical Physics of Spin Glasses and Information Processing: An Introduction* (Oxford: Oxford University Press)
[31] Livan G, Inoue J and Scalas E 2012 *J. Stat. Mech.: Theory and Experiment* P07025
[32] Livan G, Alfarano S and Scalas E 2011 *Phys. Rev.* E **84** 016113
[33] Murota M and Inoue J 2013 *Econophysics of Agent-based Models* (Milan: Springer) in press
[34] Hayashi T and Yoshida N 2005 *Bernoulli* **11** 359
[35] Tilak G, Széll T, Chicheportiche R and Chakraborti A 2012 *Econophysics of Systemic Risk and Network Dynamics* (Milan: Springer) p 77

Single-trial estimation of stimulus and spike-history effects on time-varying ensemble spiking activity of multiple neurons: a simulation study

Hideaki Shimazaki
RIKEN Brain Science Institute, Wako, Saitama, Japan
E-mail: shimazaki@brain.riken.jp

Abstract. Neurons in cortical circuits exhibit coordinated spiking activity, and can produce correlated synchronous spikes during behavior and cognition. We recently developed a method for estimating the dynamics of correlated ensemble activity by combining a model of simultaneous neuronal interactions (e.g., a spin-glass model) with a state-space method (Shimazaki et al. 2012 PLoS Comput Biol 8 e1002385). This method allows us to estimate stimulus-evoked dynamics of neuronal interactions which is reproducible in repeated trials under identical experimental conditions. However, the method may not be suitable for detecting stimulus responses if the neuronal dynamics exhibits significant variability across trials. In addition, the previous model does not include effects of past spiking activity of the neurons on the current state of ensemble activity. In this study, we develop a parametric method for simultaneously estimating the stimulus and spike-history effects on the ensemble activity from single-trial data even if the neurons exhibit dynamics that is largely unrelated to these effects. For this goal, we model ensemble neuronal activity as a latent process and include the stimulus and spike-history effects as exogenous inputs to the latent process. We develop an expectation-maximization algorithm that simultaneously achieves estimation of the latent process, stimulus responses, and spike-history effects. The proposed method is useful to analyze an interaction of internal cortical states and sensory evoked activity.

1. Introduction

Neurons in the brain make synaptic contacts to each other and form specific signaling networks. A typical cortical neuron receives synaptic inputs from $3000 - 10000$ other neurons, and makes synaptic contacts to several thousands of other neurons. They send and receive signals using pulsed electrical discharges known as action potentials, or spikes. Therefore, individual neurons in a circuit can be activated in a coordinated manner when relevant information is processed. In particular, nearly simultaneous spiking activity of multiple neurons (synchronous spikes) occurs dynamically in relation to a stimulus presented to an animal, the animal's behavior, and the internal state of the brain (attention and expectation) [1–5].

Recently, it was reported that a model of synchronous spiking activity that accounts for spike rates of individual neurons and interactions between pairs of neurons can explain $\sim 90\%$ of the synchronous spiking activity of a small subset (~ 10) of retinal ganglion cells [6, 7] and cortical neurons [8] *in vitro*. This model is known as a maximum entropy model or an Ising/spin-glass model in statistical physics. However, since the model assumes stationary data, it is not directly applicable to non-stationary data recorded from awake behaving animals. In these data sets, spike-rates of individual neurons and even interactions among them may vary across time.

In order to analyze time-dependent synchronous activity of neurons, we recently developed a method for estimating the dynamics of correlations between neurons by combining the model of neuronal interactions (e.g., the Ising/spin-glass model) with a state-space method [9, 10]. In classical neurophysiological experiments, neuronal activity is repeatedly recorded under identical experimental conditions in order to obtain reproducible features in the spiking activity across the 'trials'. Typically, neurophysiologists estimate average time-varying firing rates of individual neurons in response to a stimulus from the repeated trials [11,12]. In the same fashion, the state-space method in [9, 10] aims to estimate the dynamics of the neuronal interactions, including higher-order interactions, that occurs repeatedly upon the onset of externally triggered events. When this method is applied to three neurons recorded simultaneously from the primary motor cortex of a monkey engaged in a delayed motor task (data from [2]), it was revealed that these neurons dynamically organized into a group characterized by the presence of a higher-order (triple-wise) interaction, depending on the behavioral demands to the monkey [12].

However, neurophysiological studies in the past decades revealed that spiking activity of individual neurons is subject to large variability across trials due to structured ongoing activity of the networks that arises internally to the brain [13–15]. In these conditions, the method developed in [12] would not efficiently detect the stimulus responses because a signal-to-noise ratio may be small even in the trial-averaged activity. Although statistical methods for detecting responses of individual neurons from single-trial data have been investigated [16–19], no methods are available for estimating synchronous responses of multiple neurons to a stimulus in a single trial when these neurons are subject to the activity that is largely unrelated to the stimulus.

In the analysis of single-trial data, it is critical to consider dependency of the current activity of neurons on the past history of their activity. A neuron undergoes an inactivation period known as a refractory period after it generates an action potential. Therefore, a model neuron significantly improves its goodness-of-fit to data if it captures this biophysical property [20, 21]. In addition, estimating the dependency of the current activity level of a neuron on past spiking history of another neuron allows us to construct effective connectivity of the network within an observed set of neurons [22, 23]. Including the spike-history effects in the models of synchronous ensemble activity is thus an important topic, and investigated also in [24] in the framework of a continuous-time point process theory.

In this study we construct a method for simultaneously estimating the stimulus and spike-history effects on ensemble spiking activity when the activity of these neurons is dominated by ongoing activity. For this goal, we extend the previously developed state-space model of neuronal interactions: We model the ongoing activity, i.e., time-varying spike rates and interactions, of neurons as a latent process, and include the stimulus and spike-history effects on the activity as exogenous inputs to the latent process. We develop an expectation-maximization (EM) algorithm for this model, which efficiently combines construction of a posterior density of the latent process and estimation of the parameters for stimulus and spike-history effects. The method is tested using simulated spiking activity of 3 neurons with known underlying architecture. We provide an approximation method for determining inclusion of these exogenous inputs in the model and a surrogate method to test significance of the estimated parameters.

2. Methods

In this study, we analyze spike sequences simultaneously obtained from N neurons. From these spike sequences, we construct binary spike patterns at discrete time steps by dividing the sequences into disjoint time bins with an equal width of Δ ms (in total, T bins). The width Δ determines a permissible range of synchronous activity of neurons in this analysis. We let X_i^t be a binary variable of the i-th neuron ($i = 1, 2, \ldots, N$) in the t-th time bin ($t = 1, 2, \ldots, T$).

Here a time bin containing '1' indicates that one or more spikes exist in the time bin whereas '0' indicates that no spike exists in the time bin. The binary pattern of N neurons at time bin t is denoted as $\mathbf{X}_t = [X_1^t, X_2^t, \ldots, X_N^t]'$. The prime indicates the transposition operation to the vector. The entire observation of the discretized ensemble spiking activity is represented as $\mathbf{X}_{1:T} = [\mathbf{X}_1, \mathbf{X}_2, \ldots, \mathbf{X}_T]$.

2.1. The model of time-varying simultaneous interactions of neurons

We analyze the ensemble spike patterns using time-dependent formulation of a joint probability mass function for binary random variables. Let x_i be a binary variable, namely $x_i = \{0, 1\}$. The joint probability mass function of N-tuple binary variables, $\mathbf{x} = [x_1, x_2, \ldots, x_N]$, at time bin t ($t = 1, 2, \ldots, T$) can be written in an exponential form as

$$p(\mathbf{x}|\boldsymbol{\theta}_t) = \exp\left[\sum_i \theta_i^t x_i + \sum_{i<j} \theta_{ij}^t x_i x_j + \cdots + \theta_{1\cdots N}^t x_1 \cdots x_N - \psi(\boldsymbol{\theta}_t)\right]. \tag{1}$$

Here $\boldsymbol{\theta}_t = [\theta_1^t, \theta_2^t, \ldots, \theta_{12}^t, \theta_{13}^t, \ldots, \theta_{1\cdots N}^t]'$ summarizes the time-dependent canonical parameters of the exponential family distribution. The canonical parameters for the interaction terms, e.g., θ_{ij}^t $(i, j = 1, \ldots, N)$, represent time-dependent instantaneous interactions at time bin t among the neurons denoted in its subscript. $\psi(\boldsymbol{\theta}_t)$ is a log normalization parameter to satisfy $\sum p(\mathbf{x}|\boldsymbol{\theta}_t) = 1$.

Using a feature vector that captures simultaneous spiking activities of subsets of the neurons, $\mathbf{f} = [f_1, f_2, \ldots, f_{12}, f_{13}, \ldots, f_{1\cdots N}]'$, where

$$\begin{aligned} f_i(\mathbf{x}) &= x_i, & i &= 1, \cdots, N \\ f_{ij}(\mathbf{x}) &= x_i x_j, & i &< j \\ &\vdots \\ f_{1\cdots N}(\mathbf{x}) &= x_1 \cdots x_N, \end{aligned}$$

the probability mass function (Eq. 1) is compactly written as $p(\mathbf{x}|\boldsymbol{\theta}_t) = \exp\left[\boldsymbol{\theta}_t' \mathbf{f}(\mathbf{x}) - \psi(\boldsymbol{\theta}_t)\right]$. The expected occurrence rates of simultaneous spikes of multiple neurons is given by a vector $\boldsymbol{\eta}_t = E\left[\mathbf{f}(\mathbf{x})|\boldsymbol{\theta}_t\right]$, where expectation is performed using $p(\mathbf{x}|\boldsymbol{\theta}_t)$.

Eq. 1 specifies the probabilities of all 2^N spike patterns by using $2^N - 1$ parameters. One reasonable approach to reduce the number of parameters is to select and fix interesting features in the spiking activity, and construct a probability model that maximizes entropy. For example, maximization of entropy of the spike patterns given the low-order features, $\mathbf{f} = [f_1, f_2, \ldots, f_N, f_{12}, f_{13}, \ldots, f_{N-1,N}]'$, yields a spin-glass model that is similar to Eq. 1, but does not include interactions higher than the second order. While it is important to explore a characteristic feature vector to neuronal ensembles, here we note that the method developed in this study does not depend on the choice of the vector, $\mathbf{f}$. Below, we denote d as the number of elements in the vector, $\mathbf{f}$.

Given the observed ensemble spiking activity $\mathbf{X}_{1:T}$, the likelihood function of $\boldsymbol{\theta}_{1:T} = [\boldsymbol{\theta}_1, \boldsymbol{\theta}_2, \ldots, \boldsymbol{\theta}_T]$ is given as

$$p(\mathbf{X}_{1:T}|\boldsymbol{\theta}_{1:T}) = \prod_{t=1}^{T} \exp[\boldsymbol{\theta}_t' \mathbf{f}(\mathbf{X}_t) - \psi(\boldsymbol{\theta}_t)], \tag{2}$$

assuming conditional independence across the time bins. Eq.2 constitutes an observation equation of our state-space model.

2.2. Inclusion of stimulus and spike-history effects in the state model

The main focus of attention in this study is modeling of a process for the time-dependent canonical parameters, $\boldsymbol{\theta}_t$, in Eq. 1. We model their evolution as a first-order auto-regressive (AR) model. The effects of the stimulus and spike history are included as exogenous inputs to the AR model (an ARX model). In its full expression, the state model is written as

$$\boldsymbol{\theta}_t = \mathbf{F}\boldsymbol{\theta}_{t-1} + \mathbf{G}\mathbf{S}_t + \sum_{i=1}^{p} \mathbf{H}_i \mathbf{X}_{t-i} + \boldsymbol{\xi}_t, \tag{3}$$

for $t = 2, \ldots, T$. Here the matrix $\mathbf{F}$ ($d \times d$ matrix) is the first order auto-regressive parameter. $\boldsymbol{\xi}_t$ ($d \times 1$ matrix) is a random vector independently drawn from a zero-mean multivariate normal distribution with covariance matrix $\mathbf{Q}$ ($d \times d$ matrix) at each time bin. The state process starts with an initial value $\boldsymbol{\theta}_1$ that follows a normal distribution with mean $\boldsymbol{\mu}$ ($d \times 1$ matrix) and covariance matrix $\boldsymbol{\Sigma}$ ($d \times d$ matrix), namely $\boldsymbol{\theta}_1 \sim \mathcal{N}(\boldsymbol{\mu}, \boldsymbol{\Sigma})$. Below, we describe details of the exogenous terms.

The second term represents responses to external signals, or stimuli, $\mathbf{S}_t$, which are observed concurrently with the spike sequences. The vector $\mathbf{S}_t$ is a column vector of n_s external signals at time bin t. The each element is the value of an external signal at time bin t. If an external signal is represented as a sequence of discrete events, we denote the corresponding element of $\mathbf{S}_t$ by '1' if an event occurred within time bin t and '0' otherwise. Multiplying $\mathbf{S}_t$ by the matrix $\mathbf{G}$ ($d \times n_s$ matrix) produces weighted linear combinations of the external signals at time bin t.

The third term represents the effects of spiking activity during the previous p time bins, $\mathbf{X}_{t-i}$ ($i = 1, \ldots, p$), on the current activity. The matrix $\mathbf{H}_i$ ($d \times N$ matrix) represents the spike-history effects of spiking activity in the previous time bin $t - i$ on the state at time bin t. The spike-history effects are collectively denoted as $\mathbf{H} \equiv [\mathbf{H}_1, \mathbf{H}_2, \ldots, \mathbf{H}_p]$ ($d \times Np$ matrix).

Eq. 3 constitutes a prior density of the latent process in our state-space model. We denote the set of parameters in the prior distribution, called hyper-parameters, as $\mathbf{w} \equiv [\mathbf{F}, \mathbf{G}, \mathbf{H}, \mathbf{Q}, \boldsymbol{\mu}, \boldsymbol{\Sigma}]$. In this study, we refer to $\mathbf{w}$ as a parameter. In addition, we simplify Eq. 3 as $\boldsymbol{\theta}_t = \mathbf{F}\boldsymbol{\theta}_{t-1} + \mathbf{U}\mathbf{u}_t + \boldsymbol{\xi}_t$, where $\mathbf{u}_t$ is a single column vector constructed by stacking the stimulus vector and spike-history vectors at time bin t in a row, i.e., $\mathbf{u}_t = [\mathbf{S}_t; \mathbf{X}_{t-1}; \mathbf{X}_{t-2}; \ldots; \mathbf{X}_{t-p}]$. Similarly, we define a matrix $\mathbf{U}$ as $\mathbf{U} = [\mathbf{G}, \mathbf{H}]$. With this simplification, the prior density defined in Eq. 3 is written as $p(\boldsymbol{\theta}_{1:T}|\mathbf{w}) = p(\boldsymbol{\theta}_1|\boldsymbol{\mu}, \boldsymbol{\Sigma}) \prod_{t=2}^{T} p(\boldsymbol{\theta}_t|\boldsymbol{\theta}_{t-1}, \mathbf{F}, \mathbf{U}, \mathbf{Q})$, where the transition probability, $p(\boldsymbol{\theta}_t|\boldsymbol{\theta}_{t-1}, \mathbf{F}, \mathbf{U}, \mathbf{Q})$, is given as a normal distribution with mean $\mathbf{F}\boldsymbol{\theta}_{t-1} + \mathbf{U}\mathbf{u}_t$ and covariance matrix $\mathbf{Q}$.

3. Estimation of stimulus responses and spike-history effects

We estimate the parameter, $\mathbf{w}$, based on the principle of maximizing a (log) marginal likelihood function. Namely, we select the parameter that maximizes

$$l(\mathbf{w}) = \log \int p(\mathbf{X}_{1:T}, \boldsymbol{\theta}_{1:T}|\mathbf{w})\, d\boldsymbol{\theta}_{1:T}. \tag{4}$$

For this goal, we use the expectation-maximization (EM) algorithm [25–27]. In this method, we iteratively obtain the optimal parameter $\mathbf{w}^*$ that maximizes the lower bound of the above log marginal likelihood. This alternative function, known as the expected complete data log-likelihood (a.k.a., q-function), is computed as

$$\begin{aligned} q(\mathbf{w}^*|\mathbf{w}) &\equiv E\left[\log p\left(\mathbf{X}_{1:T}, \boldsymbol{\theta}_{1:T}|\mathbf{w}^*\right)|\mathbf{X}_{1:T}, \mathbf{w}\right] \\ &= \sum_{t=1}^{T}\left(E\boldsymbol{\theta}_t'\mathbf{f}\left(\mathbf{X}_t\right) - E\psi\left(\boldsymbol{\theta}_t\right)\right) - \frac{d}{2}\log 2\pi - \frac{1}{2}\log\det\boldsymbol{\Sigma}^* \\ &- \frac{1}{2}E[(\boldsymbol{\theta}_1-\boldsymbol{\mu}^*)'\boldsymbol{\Sigma}^{*-1}(\boldsymbol{\theta}_1-\boldsymbol{\mu}^*)] - \frac{(T-1)\,d}{2}\log 2\pi - \frac{(T-1)}{2}\log\det\mathbf{Q}^* \\ &- \frac{1}{2}\sum_{t=2}^{T}E[(\boldsymbol{\theta}_t - \mathbf{F}^*\boldsymbol{\theta}_{t-1} - \mathbf{U}^*\mathbf{u}_t)'\mathbf{Q}^{*-1}(\boldsymbol{\theta}_t - \mathbf{F}^*\boldsymbol{\theta}_{t-1} - \mathbf{U}^*\mathbf{u}_t)]. \end{aligned} \tag{5}$$

The expectation, $E[\cdot|\mathbf{X}_{1:T}, \mathbf{w}]$, in Eq. 5 is performed using the smoother posterior density of the state obtained by a nominal parameter, $\mathbf{w}$, namely

$$p\left(\boldsymbol{\theta}_{1:T}|\mathbf{X}_{1:T}, \mathbf{w}\right) = \frac{p\left(\mathbf{X}_{1:T}|\boldsymbol{\theta}_{1:T}\right)p\left(\boldsymbol{\theta}_{1:T}|\mathbf{w}\right)}{p\left(\mathbf{X}_{1:T}|\mathbf{w}\right)}. \tag{6}$$

In particular, Eq. 5 can be computed using the following expected values by the posterior density: The smoother mean $\boldsymbol{\theta}_{t|T} = E\left[\boldsymbol{\theta}_t|\mathbf{X}_{1:T}, \mathbf{w}\right]$, the smoother covariance matrix $W_{t|T} = E[(\boldsymbol{\theta}_t - \boldsymbol{\theta}_{t|T})(\boldsymbol{\theta}_t - \boldsymbol{\theta}_{t|T})'|\mathbf{X}_{1:T}, \mathbf{w}]$, and the lag-one covariance matrix, $W_{t,t-1|T} = E[(\boldsymbol{\theta}_t - \boldsymbol{\theta}_{t|T})(\boldsymbol{\theta}_{t-1} - \boldsymbol{\theta}_{t-1|T})'|\mathbf{X}_{1:T}, \mathbf{w}]$. These values are obtained using the approximate recursive Bayesian filtering/smoothing algorithm developed in [9, 10] (See Appendix A and Eqs. A.5, A.6, and A.7 therein).

In the EM-algorithm, we obtain the parameter that maximizes the q-function by alternating the expectation (E) and maximization (M) steps. In the E-step, we obtain the above expected values in Eq. 5 by the approximate recursive Bayesian filtering/smoothing algorithm using a fixed $\mathbf{w}$ (Appendix A). In the M-step, we obtain the parameter, $\mathbf{w}^*$, that maximizes Eq. 5. The resulting $\mathbf{w}^*$ is then used in the next E-step. Below, we derive an algorithm for optimizing the parameter at the M-step.

For the state model that includes the auto-regressive parameter and stimulus and/or spike-history effects, these parameters are estimated simultaneously. From $\frac{\partial}{\partial\mathbf{F}^*}q(\mathbf{w}^*|\mathbf{w}) = \mathbf{0}$, we obtain

$$\mathbf{F}^*\sum_{t=2}^{T}\left(\mathbf{W}_{t-1,t|T} + \boldsymbol{\theta}_{t-1|T}\boldsymbol{\theta}_{t-1|T}'\right) + \mathbf{U}^*\sum_{t=2}^{T}\boldsymbol{u}_t\boldsymbol{\theta}_{t-1|T}' = \sum_{t=2}^{T}\left(\mathbf{W}_{t-1,t|T} + \boldsymbol{\theta}_{t|T}\boldsymbol{\theta}_{t-1|T}'\right). \tag{7}$$

Here, $\boldsymbol{\theta}_{t|T}$, $\mathbf{W}_{t|T}$, and $\mathbf{W}_{t-1,t|T}$ are the smoother mean and covariance, and the lag-one covariance matrix given by Eqs. A.5, A.6, and A.7, respectively. Similarly, from $\frac{\partial}{\partial\mathbf{U}^*}q(\mathbf{w}^*|\mathbf{w}) = \mathbf{0}$, we obtain

$$\mathbf{F}^*\sum_{t=2}^{T}\boldsymbol{\theta}_{t-1|T}\boldsymbol{u}_t' + \mathbf{U}^*\sum_{t=2}^{T}\boldsymbol{u}_t\boldsymbol{u}_t' = \sum_{t=2}^{T}\boldsymbol{\theta}_{t|T}\boldsymbol{u}_t'. \tag{8}$$

Hence, the simultaneous update rule for $\mathbf{F}^*$ and $\mathbf{U}^*$ is given as

$$\begin{aligned} \left[\begin{array}{cc} \mathbf{F}^* & \mathbf{U}^* \end{array}\right] &= \left[\begin{array}{cc} \sum_{t=2}^{T}\left(\mathbf{W}_{t-1,t|T} + \boldsymbol{\theta}_{t|T}\boldsymbol{\theta}_{t-1|T}'\right) & \sum_{t=2}^{T}\boldsymbol{\theta}_{t|T}\boldsymbol{u}_t' \end{array}\right] \\ &\left[\begin{array}{cc} \sum_{t=2}^{T}\left(\mathbf{W}_{t-1,t|T} + \boldsymbol{\theta}_{t-1|T}\boldsymbol{\theta}_{t-1|T}'\right) & \sum_{t=2}^{T}\boldsymbol{\theta}_{t-1|T}\boldsymbol{u}_t' \\ \sum_{t=2}^{T}\boldsymbol{u}_t\boldsymbol{\theta}_{t-1|T}' & \sum_{t=2}^{T}\boldsymbol{u}_t\boldsymbol{u}_t' \end{array}\right]^{-1}. \end{aligned} \tag{9}$$

Here the inverse matrix on the r.h.s. is obtained by using the blockwise inversion formula:

$$\begin{bmatrix} A & B \\ C & D \end{bmatrix}^{-1} = \begin{bmatrix} A^{-1} + A^{-1}B\left(D - CA^{-1}B\right)^{-1}CA^{-1} & -A^{-1}B\left(D - CA^{-1}B\right)^{-1} \\ -\left(D - CA^{-1}B\right)^{-1}CA^{-1} & \left(D - CA^{-1}B\right)^{-1} \end{bmatrix}.$$

The covariance matrix, $\mathbf{Q}$, can be optimized separately. From $\frac{\partial}{\partial \mathbf{Q}^*} q\left(\mathbf{w}^*|\mathbf{w}\right) = \mathbf{0}$, the update rule of $\mathbf{Q}$ is obtained as

$$\begin{aligned} \mathbf{Q}^* =& \frac{1}{T-1}\sum_{t=2}^{T}[\mathbf{W}_{t|T} - \mathbf{W}_{t-1,t|T}\mathbf{F}' - \mathbf{F}\mathbf{W}'_{t-1,t|T} + \mathbf{F}\mathbf{W}_{t-1|T}\mathbf{F}'] \\ &+ \frac{1}{T-1}\sum_{t=2}^{T}\left(\boldsymbol{\theta}_{t|T} - \mathbf{F}\boldsymbol{\theta}_{t-1|T} - \mathbf{U}\mathbf{u}_t\right)\left(\boldsymbol{\theta}_{t|T} - \mathbf{F}\boldsymbol{\theta}_{t-1|T} - \mathbf{U}\mathbf{u}_t\right)'. \end{aligned} \tag{10}$$

Finally, the mean of the initial distribution is updated with $\boldsymbol{\mu}^* = \boldsymbol{\theta}_{1|T}$ from $\frac{\partial}{\partial \boldsymbol{\mu}^*} q\left(\mathbf{w}^*|\mathbf{w}\right) = \mathbf{0}$. The covariance matrix $\boldsymbol{\Sigma}$ for the initial parameters is fixed in this optimization.

4. Results

4.1. Simulation of a network of 3 neurons

In order to test the method, we simulate spiking activity of 3 neurons that possess specific characteristics in spike generation and connectivity as follows (See Fig. 1A). (1) The instantaneous firing rate of each neuron model depends on its own spike history in order to reproduce refractoriness in neuronal spiking activity. To achieve this, we adopt a renewal point process model whose instantaneous inter-spike interval (ISI) distribution is given by an inverse Gaussian distribution as a model of the stochastic spiking activity. (2) The firing rate of each neuron model varies across time in order to reproduce the ongoing activity. For that purpose, spike times of each neuron are generated from the renewal process by adding inhomogeneity to the underlying rate using the time-rescaling method described in [20]. The underlying rate of the inhomogeneous renewal point process model is modulated using a sinusoidal function (frequency: 1 Hz; mean and amplitude: 30 spikes/s). This rate modulation is common to the 3 neurons. (3) The neurons are activated by externally triggered stimulus inputs. To realize the stimulus responses, we deterministically induce spikes at predetermined timings of the stimuli. We consider two stimuli, one (Stimulus 1) that induces a spike in Neuron 1, and the other (Stimulus 2) that induces synchronous spikes in Neuron 2 and Neuron 3. The timings of external stimuli are not related to the sinusoidal time-varying rate, but randomly selected in the observation period (On average each stimulus happens once in 1 second). (4) There is feedforward circuitry in the network. We assume that Neuron 1 makes excitatory synaptic contacts to Neuron 2 and Neuron 3. To realize this, 5ms after a spike occurs in Neuron 1, we induce simultaneous spikes in Neuron 2 and Neuron 3 with a probability 0.5.

We simulate spike sequences with a length of 30 seconds using 1 ms resolution for numerical time steps (An example of a short period (1 s) is shown in Fig. 1B). Figure 1C displays the instantaneous spike-rates (conditional intensity functions of point processes) of Neuron 1 (Top, red line) and Neuron 2 & 3 (Bottom, green and blue lines) underlying the spiking activity in Fig. 1B. The black lines indicate sinusoidal rate modulation common to all neurons. In addition, spikes are induced in Neuron 1 at the onsets of Stimulus 1 (magenta triangles). Similarly, simultaneous spikes of Neuron 2 and Neuron 3 are generated at the onsets of Stimulus 2 (cyan triangles). In the traces of instantaneous spike-rates in Fig. 1C, instantaneous increases caused by the stimuli and synaptic inputs are not displayed. The instantaneous spike-rate of a neuron is reset to zero whenever a spike is induced in that neuron.

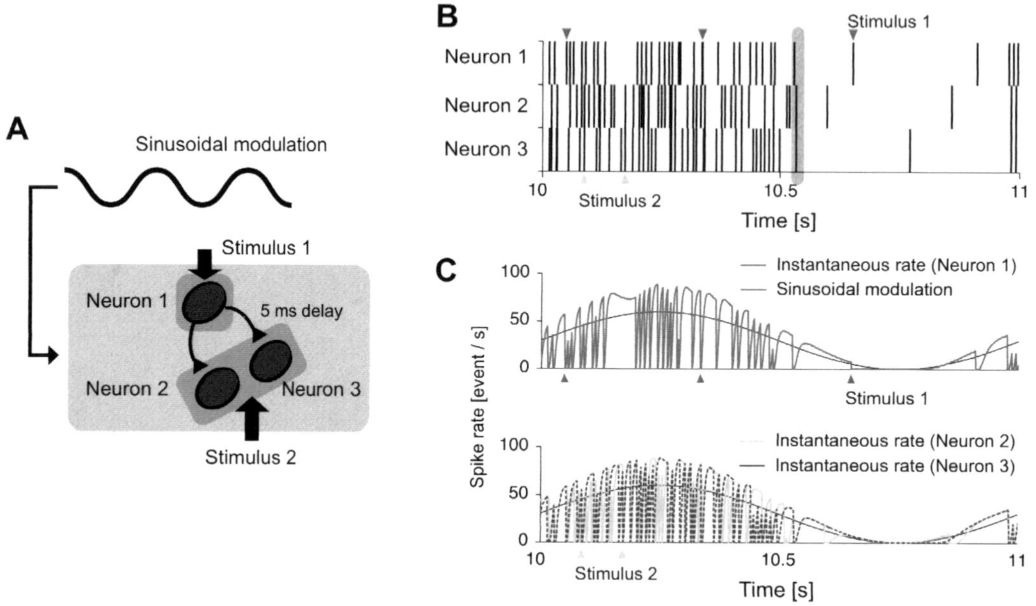

Figure 1. (A) Schematic diagram of a simulated network of 3 neurons. Neuron 1 makes excitatory synaptic contacts to Neuron 2 and 3. Stimulus 1 excites Neuron 1 whereas Stimulus 2 excites Neuron 2 and 3 simultaneously. In addition, all neurons receive sinusoidal rate modulation. (B) Simulated spiking activity of the network. A short period (1 s) of the total 30 s length is shown. The magenta and cyan triangles represent occurrence times of Stimulus 1 and 2, respectively. The gray bar highlights simultaneous spikes of Neuron 2 and 3 that are causally induced 5 ms after a spike occurs in Neuron 1. (C) Instantaneous spike rates. (Top) The red trace is the instantaneous spike rate of Neuron 1 simulated as an inhomogeneous renewal point process whos instantaneous inter-spike interval is given by the inverse Gaussian distribution ($f\left(t;\kappa\right)=\sqrt{\frac{\kappa}{2\pi t^3}}\exp\left[-\frac{\kappa}{2t}\left(t-1\right)^2\right]$ for $x>0$, 0 for $x<0$, $\kappa=1.8$ for all neurons). The inhomogeneous rate is modulated by the sinusoidal function (black solid line, frequency: 1 Hz; mean rate and amplitude: 30 spikes/s). (Bottom) Instantaneous spike rates of Neuron 2 and 3 (solid green line and dashed blue line, respectively).

4.2. Selection of a state model

We analyze the simulated ensemble activity by the proposed state-space model. For this goal, we first construct binary spike patterns, $\mathbf{X}_{1:T}$, from the simulated spike sequences of 30 seconds (Note: spike times are recorded in 1 ms resolution) by discretizing them using disjoint time bins with 2 ms width. We then apply state-space models to the binary data. The observation model used here contains interactions up to the second order (a pairwise interaction model):

$$p(\mathbf{x}|\boldsymbol{\theta}_t)=\exp\left[\theta_1^t x_1+\theta_2^t x_2+\theta_3^t x_3+\theta_{12}^t x_1 x_2+\theta_{13}^t x_1 x_3+\theta_{23}^t x_2 x_3-\psi(\boldsymbol{\theta}_t)\right]. \quad (11)$$

For the state model, we consider 5 different models that include a set of different components in Eq. 3. We select a model based on the framework of model selection in order to avoid over-fitting of a model to the data. Details of each state model are described as follows.

The first state model assumes $\mathbf{F}=\mathbf{I}$, where $\mathbf{I}$ is an identity matrix, and does not include any of exogenous inputs. In this model, we optimize only the covariance matrix $\mathbf{Q}$. The first model is

Journal of Physics: Conference Series **473** (2013) 012009 doi:10.1088/1742-6596/473/1/012009

denoted as [**Q**]. The second state model, denoted as [**Q**, **F**], is the first-order auto-regressive model. In this model, we optimize both the covariance matrix **Q** and the auto-regressive parameter **F**. The third model, denoted as [**Q**, **F**, **G**], additionally includes the stimulus term as exogenous inputs (Stimulus 1 and Stimulus 2). Both the matrix **F** and **G** are optimized simultaneously in addition to **Q**. The fourth model includes both stimulus and spike-history terms. In this model, the state model includes the history of spiking activity up to the last 6 time bins ($p = 6$). All parameters **F**, **G**, and **H** are optimized simultaneously in addition to **Q**. This model is denoted as [**Q**, **F**, **G**, **H**6]. The structure of the fifth model is the same as the fourth model, but contains the history of spiking activity up to the last 12 time bins ($p = 12$). The last model is denoted as [**Q**, **F**, **G**, **H**12].

In order to select the most predictive model among them, we select the state-space model that minimizes the Akaike (Bayesian) information criterion (AIC) [28],

$$\mathrm{AIC} = -2l\left(\mathbf{w}^*\right) + 2\dim \mathbf{w}^*, \tag{12}$$

where $\mathbf{w}^*$ is the optimized parameter in the Methods section. The (marginal) likelihood function in Eq. 12 is obtained by a log-quadratic approximation, i.e, the Laplace method [10] (See Appendix B for the complete equation). Figure 2B displays decreases in AICs (ΔAIC) of the last four models from the AIC of the first model, [**Q**]. The larger the ΔAIC is, the better the state-space model is expected to predict unseen data. For these data sets, inclusion of exogenous inputs, in particular the spike history, significantly decreases the AIC. From this result, we select the state model that includes the stimulus response term and the spike-history terms up to the previous 6 time bins.

4.3. Parameter estimation

We now look at the estimated parameters of the model selected by the AIC, namely [**Q**, **F**, **G**, **H**6]. Due to the limitation in the space, we do not display the estimated dynamics of the canonical parameters, $\boldsymbol{\theta}_t$, by the recursive Bayesian method (See [9, 10] for the detailed analysis on dynamics of $\boldsymbol{\theta}_t$ by this method). The estimated parameters, **G** and **H**, are summarized in Fig. 3.

Here, in order to test the significance of the estimated parameters, we construct confidence bounds of the estimates, using a surrogate method. In this approach, we apply the same state-space model, [**Q**, **F**, **G**, **H**6], to the surrogate data set for the exogenous inputs. In the surrogate data set, the onset times of external signals (Stimulus 1 and Stimulus 2) are randomized in the

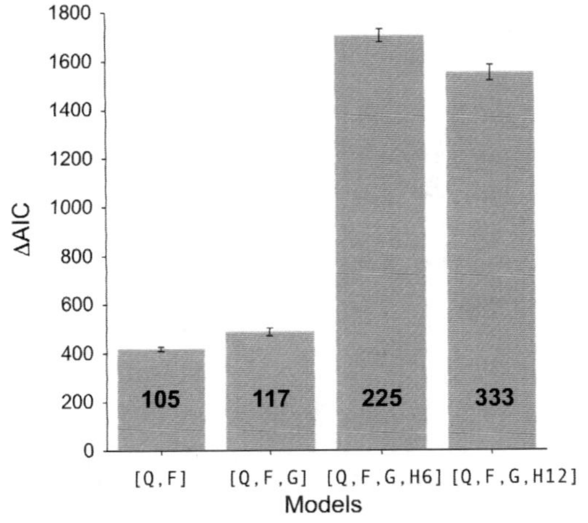

Figure 2. Comparison of state models by the Akaike Information Criterion (AIC). The state-space models with the following five different state models are comapared: [**Q**], [**Q**, **F**],[**Q**,**F**,**G**], [**Q**,**F**, **G**, **H**6], and [**Q**,**F**, **G**, **H**12] (See details of the models for main text). The reduction of the AIC of the last four models from the AIC of the model [**Q**] (ΔAIC) was repeatedly computed for 10 times. The height of the bar indicates the average ΔAIC. The error bar indicates $\pm$ 2 S.E. The numbers marked on each bar are dimensions of the models (The number of free parameters in the state model).

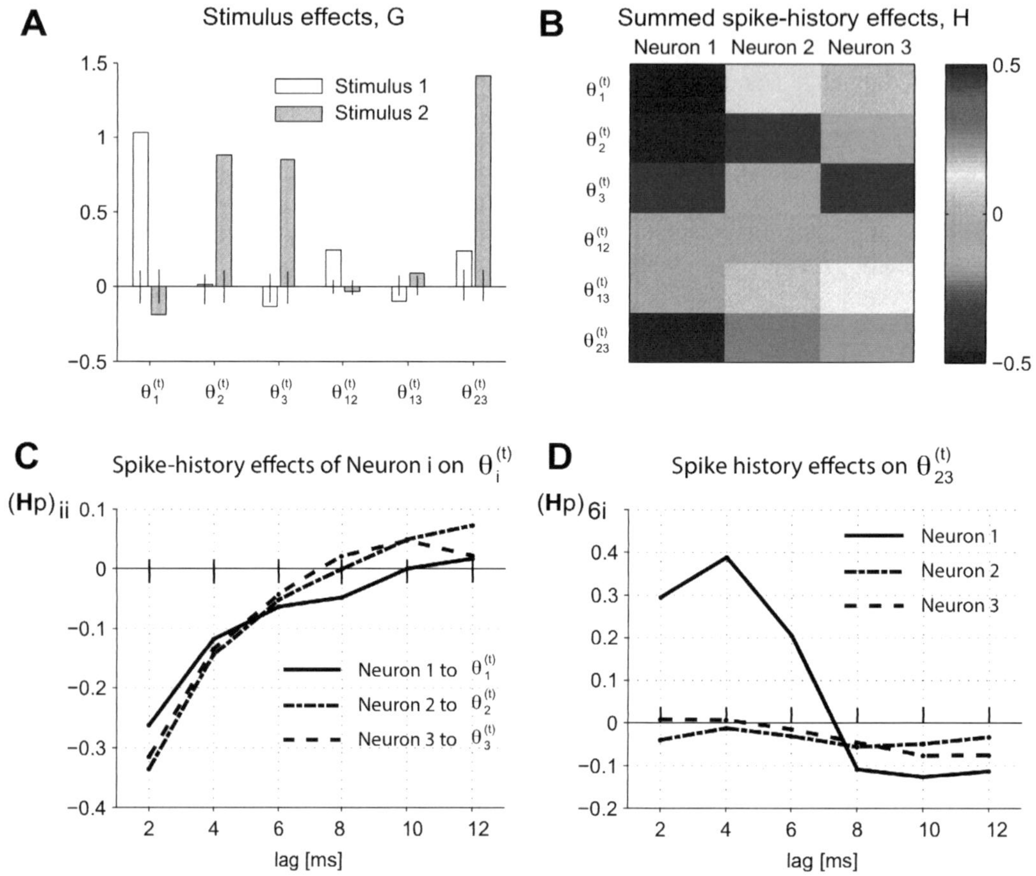

Figure 3. Parameter estimation of the state-space model. (A) Effects of Stimulus 1 and 2 on the canonical paramters (the first and second column of $\mathbf{G}$). The vertical ticks on abscissa indicate the 95% confidence bounds for each parameters obtained by the surrogate method. (B) Summed spike-history effects. The matrices of spike-history effects, $\mathbf{H}_p$, are summed over the time-lags and shown using color. (C) The effect of a spike occurrence in Neuron i at p time bins before the tth bin on $\theta_i^{(t)}$ $(i = 1, 2, 3)$. (D) The effect of a spike occurence in Neuron i $(i = 1, 2, 3)$ on the interaction parameter $\theta_{23}^{(t)}$.

observation period. Similarly, we randomly select $p = 6$ bins from the past spiking activity to obtain surrogate spike history, instead of selecting the last consecutive 6 bins from time bin t. Thus the estimated parameters, $\mathbf{G}$ and $\mathbf{H}$, are not related to the structure specified in the Section 4.1. We repeatedly applied the state-space model to the surrogate data (1000 times) to obtain the 95% confidence bound for the parameter estimation (vertical ticks in Fig. 3A, C and D).

Figure 3A displays effects of the two stimuli, $\mathbf{G}$, on the respective elements in $\boldsymbol{\theta}_t$. First, Stimulus 1 significantly increases $\theta_1^{(t)}$ whereas changes in the pairwise interactions by Stimulus 1 are relatively small, indicating that Stimulus 1 induces spikes in Neuron 1. On the contrary, Stimulus 2 increases to $\theta_2^{(t)}$, $\theta_3^{(t)}$, and $\theta_{23}^{(t)}$. In particular, the increase in the interaction parameter

$\theta_{23}^{(t)}$ by Stimulus 2 indicates that the presence of Stimulus 2 induces excess simultaneous spikes in Neuron 2 and Neuron 3 more often than the chance coincidence expected for the two neurons.

The spike-history effects are summarized as a summed matrix, $\sum_{i=1}^{p} \mathbf{H}_i$, shown in Fig. 3B. Two major effects are observed. First, the spike history of Neuron i significantly decreases $\theta_i^{(t)}$ $(i = 1, 2, 3)$ (See diagonal of the first 3×3 matrix in Fig. 3B). Figure 3C displays the contribution of a spike in Neuron i during the previous p time bins to the parameter $\theta_i^{(t)}$. These components primarily, albeit not exclusively, capture the renewal property of the simulated neuron models. Second, the spike history of Neuron 1 increases $\theta_2^{(t)}$, $\theta_3^{(t)}$, and $\theta_{23}^{(t)}$ (See the first column in Fig. 3B), indicating that spike interactions from Neuron 1 to Neuron 2 & 3. In particular, the increase in $\theta_{23}^{(t)}$ due to the spikes in Neuron 1 during previous 1-3 time bins (Fig. 3D) indicates that the inputs from Neuron 1 induces excess synchronous spikes in the other two neurons with approximately 2-6 ms delay.

5. Conclusion

We developed a parametric method for estimating stimulus responses and spike-history effects on the simultaneous spiking activity of multiple neurons when the ensemble themselves exhibit ongoing activity. The method was tested by simulated multiple neuronal spiking activity with known underlying architecture. We provided two methods to corroborate the fitted models. First, based on the result in the preceding paper, we provided an approximate equation for the log marginal likelihood (see Appendix B), which was used to select the most predictive state-space model. Second, we provided a method for obtaining confidence bounds of the estimated parameters based on a surrogate approach.

Example spike sequences simulated in this study are overly simplified. Therefore, the method needs be tested using real neuronal spike data, e.g., from cultured neurons whose underlying circuit is identified by electrophysiological studies. In practical applications, it is recommended to utilize basis functions such as raised cosine bumps used in [23] in the exogenous terms in order to capture the stimulus and spike-history effects with a fewer parameters. In addition, an appropriate bin size must be selected in order to obtain a meaningful result in the analysis of real data. Since the bin size determines a permissible range of synchronous activity, a physiological interpretation of the result depends on the choice of the bin size. It is thus recommended to present results based on multiple different bin sizes in order to confirm a specific hypothesis in a study as shown in [10, 29]. Methods to overcome an artifact due to the disjoint binning are discussed in [30–34]. In future, inclusion of such advanced methods will allow us to detect near-synchronous responses without sacrificing temporal resolution of the analysis.

Given that applicability of the method is confirmed in real data, the proposed method is useful to investigate how ensemble activity of multiple neurons in a local circuit changes configurations of their simultaneous responses (synchronous responses) to different stimuli applied to an animal. Further, it would be interesting to see different effects of the same stimulus on the ensemble activity when an animal undergoes different cortical states.

Acknowledgments

The present study is based on the modeling framework developed in [9, 10]. The author acknowledges Prof. Shun-ichi Amari, Prof. Emery N. Brown, and Prof. Sonja Grün for their support in construction of the original model. The author also thanks to Dr. Christopher L. Buckley and Dr. Erin Munro for critical reading of the manuscript.

Appendix A. Construction of a posterior density by the recursive Bayesian filtering/smoothing algorithm

A posterior density of the time-varying $\boldsymbol{\theta}_t$, which specifies the joint probability mass function of spike patterns at time bin t, are obtained by a non-linear recursive Bayesian estimation method developed in [9, 10]. The method allows us to find a maximum a posteriori (MAP) estimate of $\boldsymbol{\theta}_t$ and its uncertainty, namely the most probable paths of time-varying canonical parameters $\boldsymbol{\theta}_t$ and their confidence bounds given the observed simultaneous activity of multiple neurons. The estimation procedure completes by a forward recursion to construct a filter posterior density and then by a backward recursion to construct a smoother posterior density. In this approach, the posterior densities are approximated as a multivariate normal probability density function.

In the forward filtering step, we first compute mean and covariance of one step prediction density:

$$\boldsymbol{\theta}_{t|t-1} = \mathbf{F}\boldsymbol{\theta}_{t-1|t-1} + \mathbf{U}\mathbf{u}_t, \tag{A.1}$$

$$\mathbf{W}_{t|t-1} = \mathbf{F}W_{t-1|t-1}\mathbf{F}' + \mathbf{Q}. \tag{A.2}$$

Then, a mean vector and covariance matrix of the filter posterior density, which is approximated as a normal density, is given as

$$\boldsymbol{\theta}_{t|t} = \boldsymbol{\theta}_{t|t-1} + n\mathbf{W}_{t|t-1}(\mathbf{y}_t - \boldsymbol{\eta}_{t|t}), \tag{A.3}$$

$$\mathbf{W}_{t|t}^{-1} = \mathbf{W}_{t|t-1}^{-1} + n\mathbf{J}_{t|t}, \tag{A.4}$$

where $\boldsymbol{\eta}_{t|t} = E\left[\mathbf{f}\left(\mathbf{x}\right)|\mathbf{X}_{1:t}, \mathbf{w}\right]$ is the simultaneous spike rates at time bin t expected from the joint probability mass function, Eq. 1, specified by $\boldsymbol{\theta}_{t|t}$. Thus Eq. A.3 is a non-linear equation. We solve Eq. A.3 by a Newton-Raphson method. It can be shown that the solution is unique. The matrix $\mathbf{J}_{t|t}$ is a Fisher information matrix of Eq. 1 evaluated at $\boldsymbol{\theta}_{t|t}$.

Finally, we compute mean and covariance of a smoother posterior density as

$$\boldsymbol{\theta}_{t|T} = \boldsymbol{\theta}_{t|t} + \mathbf{A}_t\left(\boldsymbol{\theta}_{t+1|T} - \boldsymbol{\theta}_{t+1|t}\right), \tag{A.5}$$

$$\mathbf{W}_{t|T} = \mathbf{W}_{t|t} + \mathbf{A}_t\left(\mathbf{W}_{t+1|T} - \mathbf{W}_{t+1|t}\right)\mathbf{A}_t'. \tag{A.6}$$

with $\mathbf{A}_t = \mathbf{W}_{t|t}\mathbf{F}'\mathbf{W}_{t+1|t}^{-1}$ for $t = T, T-1, \ldots, 2, 1$. Namely, we start computing Eqs. A.5 and A.6 in a backward manner, using $\boldsymbol{\theta}_{T|T}$ and $\mathbf{W}_{T|T}$ obtained in the filtering method at the initial step. The lag-one covariance smoother, $W_{t-1,t|T}$, is obtained using the method of De Jong and Mackinnon [35]:

$$\mathbf{W}_{t-1,t|T} \equiv E[(\boldsymbol{\theta}_{t-1} - \boldsymbol{\theta}_{t-1|T})(\boldsymbol{\theta}_t - \boldsymbol{\theta}_{t|T})'|\, y_{1:T}] = \mathbf{A}_{t-1}\mathbf{W}_{t|T}. \tag{A.7}$$

Appendix B. Approximate marginal likelihood function

The approximated formula of the log marginal likelihood (Eq. 4) was obtained in [10] as

$$l(\mathbf{w}) \approx \sum_{t=1}^{T} n\left(\mathbf{y}_t'\boldsymbol{\theta}_{t|t} - \psi\left(\boldsymbol{\theta}_{t|t}\right)\right) + \frac{1}{2}\sum_{t=1}^{T}\left(\log\det W_{t|t} - \log\det W_{t|t-1}\right)$$
$$- \frac{1}{2}\sum_{t=1}^{T} tr\left[\mathbf{W}_{t|t-1}^{-1}\left(\boldsymbol{\theta}_{t|t} - \boldsymbol{\theta}_{t|t-1}\right)\left(\boldsymbol{\theta}_{t|t} - \boldsymbol{\theta}_{t|t-1}\right)'\right]. \tag{B.1}$$

Here we briefly provide the derivation (See [10] for details). The log marginal likelihood is written as

$$l(\mathbf{w}) = \sum_{t=1}^{T} \log p(\mathbf{y}_t|\mathbf{y}_{1:t-1}, \mathbf{w}) = \sum_{t=1}^{T} \log \int p(\mathbf{y}_t|\boldsymbol{\theta}_t)p(\boldsymbol{\theta}_t|\mathbf{y}_{1:t-1}, \mathbf{w})d\boldsymbol{\theta}_t. \tag{B.2}$$

The integral in the above equation is approximated as

$$\int p(\mathbf{y}_t|\boldsymbol{\theta}_t)p(\boldsymbol{\theta}_t|\mathbf{y}_{1:t-1},\mathbf{w})d\boldsymbol{\theta}_t = \frac{1}{\sqrt{(2\pi)^d|W_{t|t-1}|}}\int \exp\left[q(\boldsymbol{\theta}_t)\right]d\boldsymbol{\theta}_t \approx \frac{\sqrt{(2\pi)^d|W_{t|t}|}}{\sqrt{(2\pi)^d|W_{t|t-1}|}}\exp\left[q\left(\boldsymbol{\theta}_{t|t}\right)\right], \tag{B.3}$$

where $q(\boldsymbol{\theta}_t) = n\left(\mathbf{y}_t'\boldsymbol{\theta}_t - \psi\left(\boldsymbol{\theta}_t\right)\right) - \frac{1}{2}\left(\boldsymbol{\theta}_t - \boldsymbol{\theta}_{t|t-1}\right)'\mathbf{W}_{t|t-1}^{-1}\left(\boldsymbol{\theta}_t - \boldsymbol{\theta}_{t|t-1}\right)$. To obtain the second approximate equality, we used the Laplace approximation: the integral in Eq. B.3 is given as $\int \exp\left[q\left(\boldsymbol{\theta}_t\right)\right]d\boldsymbol{\theta}_t \approx \sqrt{(2\pi)^d|W_{t|t}|}\exp\left[q\left(\boldsymbol{\theta}_{t|t}\right)\right]$. Here we note that a solution of $q(\boldsymbol{\theta}_t) = 0$ is equivalent to the filter MAP estimate, $\boldsymbol{\theta}_{t|t}$. By applying Eq. B.3 to Eq. B.2, we obtain Eq. B.1.

References

[1] Vaadia E, Haalman I, Abeles M, Bergman H, Prut Y, Slovin H and Aertsen A 1995 *Nature* **373** 515–518
[2] Riehle A, Grün S, Diesmann M and Aertsen A 1997 *Science* **278** 1950–1953
[3] Steinmetz P N, Roy A, Fitzgerald P J, Hsiao S S, Johnson K O and Niebur E 2000 *Nature* **404** 187–190
[4] Fujisawa S, Amarasingham A, Harrison M T and Buzsáki G 2008 *Nat Neurosci* **11** 823–833
[5] Ito H, Maldonado P E and Gray C M 2010 *J Neurophysiol* **104** 3276–3292
[6] Schneidman E, Berry M J, Segev R and Bialek W 2006 *Nature* **440** 1007–1012
[7] Shlens J, Field G D, Gauthier J L, Grivich M I, Petrusca D, Sher A, Litke A M and Chichilnisky E J 2006 *J Neurosci* **26** 8254–8266
[8] Tang A, Jackson D, Hobbs J, Chen W, Smith J L, Patel H, Prieto A, Petrusca D, Grivich M I, Sher A, Hottowy P, Dabrowski W, Litke A M and Beggs J M 2008 *J Neurosci* **28** 505–518
[9] Shimazaki H, Amari S I, Brown E N and Grün S 2009 *Proc. IEEE ICASSP2009* pp 3501–3504
[10] Shimazaki H, Amari S I, Brown E N and Grün S 2012 *PLoS Comput Biol* **8** e1002385
[11] Shimazaki H and Shinomoto S 2007 *Neural Comput* **19** 1503–1527
[12] Shimazaki H and Shinomoto S 2010 *J Comput Neurosci* **29** 171–182
[13] Arieli A, Sterkin A, Grinvald A and Aertsen A 1996 *Science* **273** 1868–1871
[14] Tsodyks M, Kenet T, Grinvald A and Arieli A 1999 *Science* **286** 1943–1946
[15] Kenet T, Bibitchkov D, Tsodyks M, Grinvald A and Arieli A 2003 *Nature* **425** 954–956
[16] Nawrot M, Aertsen A and Rotter S 1999 *Journal of Neuroscience Methods* **94** 81–92
[17] Cunningham J, Yu B, Shenoy K, Sahani M, Platt J, Koller D, Singer Y and Roweis S 2008 *Advances in Neural Information Processing Systems* **20** 329–336
[18] Czanner G, Eden U T, Wirth S, Yanike M, Suzuki W A and Brown E N 2008 *J Neurophysiol* **99** 2672–2693
[19] Yu B M, Cunningham J P, Santhanam G, Ryu S I, Shenoy K V and Sahani M 2009 *J Neurophysiol* **102** 614–635
[20] Brown E N, Barbieri R, Ventura V, Kass R E and Frank L M 2001 *Neural Comput* **14** 325–346
[21] Barbieri R, Quirk M C, Frank L M, Wilson M A and Brown E N 2001 *J Neurosci Methods* **105** 25–37
[22] Truccolo W, Eden U T, Fellows M R, Donoghue J P and Brown E N 2005 *J Neurophysiol* **93** 1074–1089
[23] Pillow J W, Shlens J, Paninski L, Sher A, Litke A M, Chichilnisky E J and Simoncelli E P 2008 *Nature* **454** 995–999
[24] Kass R E, Kelly R C and Loh W L 2011 *Ann Appl Stat* **5** 1262–1292
[25] Dempster A P, Laird N M and Rubin D B 1977 *J Roy Stat Soc B Met* **39** 1–38 ISSN 00359246
[26] Shumway R and Stoffer D 1982 *J Time Ser Anal* **3** 253–264
[27] Smith A C and Brown E N 2003 *Neural Comput* **15** 965–991
[28] Akaike H 1980 *Bayesian Statistics* ed Bernardo J M, Groot M H D, Lindley D V and Smith A F M (Valencia, Spain: University Press) pp 143–166
[29] Riehle A, Grammont F, Diesmann M and Grün S 2000 *J Physiol Paris* **94** 569–582
[30] Grün S, Diesmann M, Grammont F, Riehle A and Aertsen A 1999 *J Neurosci Methods* **94** 67–79
[31] Pipa G, Wheeler D W, Singer W and Nikolić D 2008 *J Comput Neurosci* **25** 64–88
[32] Grün S 2009 *J Neurophysiol* **101** 1126–1140
[33] Hayashi T and Yoshida N 2005 *Bernoulli* **11** 359–379
[34] Chakraborti A, Toke I M, Patriarca M and Abergel F 2011 *Quant Finance* **11** 991–1012
[35] De Jong P and Mackinnon M J 1988 *Biometrika* **75** 601–602

Inference of Hopfield-Potts patterns from covariation in protein families: calculation and statistical error bars

Simona Cocco[1], Rémi Monasson[2] and Martin Weigt[3]
[1] Laboratoire de Physique Statistique de l'Ecole Normale Supérieure - UMR 8550, associé au CNRS et à l'Université Pierre et Marie Curie, 24 rue Lhomond, 75005 Paris,France
[2] Laboratoire de Physique Théorique de l'Ecole Normale Supérieure - UMR 8549, associé au CNRS et à l'Université Pierre et Marie Curie, 24 rue Lhomond, 75005 Paris, France
[3] Laboratoire de Génomique des Microorganismes - UMR 7238, Université Pierre et Marie Curie, 15 rue de l'Ecole de Médecine, 75006 Paris, France

E-mail: monasson@lpt.ens.fr

Abstract. We consider the Hopfield-Potts model for the covariation between residues in protein families recently introduced in Cocco, Monasson, Weigt (2013). The patterns of the model are inferred from the data within a new gauge, more symmetric in the residues. We compute the statistical error bars on the pattern components. Results are illustrated on real data for a response regulator receiver domain (Pfam ID PF00072) family.

1. Introduction

A protein family is a set of protein sequences which are evolutionary related. While the amino-acid sequences show great diversity they generally give rise to a common fold. Extracting structural information about this fold from the statistical features of the distribution over the sequences is an important question [1]. Maximum entropy (MaxEnt) modelling [2, 3], which defines least constrained distributions capable of reproducing some statistical features are useful in this regard. When the statistical features of interest are local residue conservation and residue-residue correlations MaxEnt predicts that the sequences are extracted from the equilibrium distribution of a Potts model, whose interaction parameters have to be inferred from the data [4, 5]. This approach, combined with a mean-field approximation for the computation of the parameters, was successful to predict sites which are in contact in the 3-D structure of the protein [6], se also [7, 8, 9] for related approaches, and [10, 11, 12, 13, 14, 15, 16, 17, 18] for applications to use such sequence information to guide tertiary and quaternary structure prediction, and [16] for an example how predictions can be used to rationally design mutagenesis experiments influencing a protein's functionality.

However, given the length of the sequences, $L \sim$ few hundreds, and the number of possible values for each residue on a site, $q = 21$ (20 amino-acids and the gap symbol in the alignment), the number of statistical couplings to be inferred scales as $(Lq)^2$ and can reach millions. To drastically reduce the number of parameters in the model and avoids overfitting we have recently introduced the Hopfield-Potts model [19] for the study of covariation between residues in protein families. The Hopfield-Potts model is a special case of the Potts model in which the coupling matrix between the residues has a low rank, p, a priori much smaller than Lq. The non-zero modes of the coupling matrix, called patterns, define specific directions in the sequence space. These patterns show some similarities with position-specific scoring matrices, defining ideal sequence motifs which real sequences in the protein family are likely to have, but instead of encoding independent-site amino-acid preferences, they include statistical couplings between sequence positions. Some of the motifs are enforced through 'attractive' patterns assigning positive statistical scores to the sequence aligned along them. In addition, in distinction to the original Hopfield model [20], the model also include 'repulsive' patterns, giving statistical negative scores to all sequences but the ones including a few strongly coevolving residues. Repulsive patterns are very useful, since they are often localized on sites in contact on the 3D fold, see [19] for more details. The Hopfield-Potts model thus offers an efficient way to infer contacts, with much less parameters than the usual Potts model.

In this paper we pursue the study of the Hopfield-Potts along two directions. As one and only one residue can be present on any given site, the Potts model has local gauge invariance, which was broken in a specific way in our previous publication [19]. Here we propose another choice for the gauge, which is more symmetric in the residues, and repeat the calculation of the most likely patterns given the protein multi-sequence alignment. In addition we show how the statistical error bars on the patterns can be computed within the statistical physics framework. Knowledge of the error bars allows us to derive a new criterion for pattern selection and for the number of patterns, p, to be considered. Our analytical results are then illustrated on real genomic data for one protein family.

The paper is organized as follows. In Section 2, we propose a brief reminder on the covariation analysis of protein families with the Maximum Entropy approach leading to the definition of a Potts model for the protein sequences in one family. The Hopfield-Potts model is introduced and the different gauges of the models are discussed in Section 3. In Section 4 we calculate the expressions for the most likely patterns. Error bars are computed in Section 5. Some applications to real protein data are shown in Section 6, while conclusions are given in Section 7. Readers not interested in the details of the statistical mechanics calculations can access the main results for the pattern values and their error bars in Sections 4.1 and 5.1.

2. Reminder on the statistical physics approach to covariation

2.1. Co-evolution and amino-acid correlations

The multi-sequence alignment (MSA) is denoted as $A = \{a_i^m | i = 1, ..., L,\ m = 1, ..., M\}$ with the index i running over the L columns in each row of the alignment (sites), and m over the M sequences, which are the rows in the MSA. The amino-acids a_i^m are assumed to be represented by natural numbers $1, ..., q$ with $q = 21$. This number includes the 20 standard amino acids, and the alignment gap.

In our approach, we do not use the data directly, but we summarize them by the amino-acid occupancies in single MSA columns and pairs of columns,

$$f_i(a) = \frac{1}{M} \sum_{m=1}^{M} \delta_{a,a_i^m} \ , \quad f_{ij}(a,b) = \frac{1}{M} \sum_{m=1}^{M} \delta_{a,a_i^m} \delta_{b,a_j^m} \ , \tag{1}$$

with $i, j = 1, ..., L$ and $a, b = 1, ...q$. Note that, to lower the effect of finite and dependent sampling of phylogenetically related biological sequences, in [6] a pseudocount and a reweighting scheme were used. The latter assigns lower the contribution of very close, almost repeated sequences in the above equations, and to increase the one of isolated ones. We describe how to include pseudocount and reweighting in Section 6.1.

If the two columns i and j were independently occupied by amino-acids, the joint distribution $f_{ij}(a, b)$ would factorize into the product $f_i(a)f_j(b)$. We therefore measure correlations of the amino-acid occupancies of columns i and j via the covariance matrix

$$C_{ij}(a,b) = f_{ij}(a,b) - f_i(a)f_j(b) \ . \tag{2}$$

Note that finite sampling may introduce a spurious non-zero correlation between two residues, even if those residues do not covary.

2.2. *Maximum entropy modeling*

The presence of a strong correlation between residues does not entail that they are in direct contact on the 3D fold of the protein. The reason is the following [4, 5]: When i is in contact with j, and j is in contact with k, also i and k will show correlation. It is thus important to distinguish between *direct interaction* and *indirect correlation*, and to infer networks of direct couplings which generate the empirically observed correlations.

Disentangling direct and indirect correlations is only possible, if the statistics of full-length protein sequences is considered. It can be done by constructing a statistical model $P(a_1, ..., a_L)$ describing the probability of observing a particular amino-acid sequence $a_1, ..., a_L$. Due to the limited amount of available data, we require this model to reproduce empirical frequency counts for single MSA columns and column pairs,

$$f_i(a_i) = \sum_{\{a_k|k\neq i\}} P(a_1, ..., a_L) \ , \quad f_{ij}(a_i, a_j) = \sum_{\{a_k|k\neq i,j\}} P(a_1, ..., a_L) \ , \tag{3}$$

i.e. marginal distributions of $P(a_1, ..., a_L)$ are required to coincide with the empirical counts. Beyond this coherence, we aim at the *least constrained* statistical description. To this aim we apply the *maximum-entropy principle* [2, 3], which consists in maximizing the entropy $H[P] = -\sum_{a_1,...,a_L} P(a_1, ..., a_L) \log P(a_1, ..., a_L)$, under the constraints in Eqn. (3). If the constraints are imposed via Lagrange multipliers, we readily find the analytical form

$$P(a_1, ..., a_2) = \frac{1}{\mathcal{Z}(\{e_{ij}(a,b), h_i(a)\})} \exp\left\{\frac{1}{2}\sum_{i,j} e_{ij}(a_i, a_j) + \sum_i h_i(a_i)\right\} \ , \tag{4}$$

where $\mathcal{Z}$ is a normalization constant for P:

$$\mathcal{Z}(\{e_{ij}(a,b), h_i(a)\}) = \sum_{a_1,...,a_L} \exp\left\{\frac{1}{2}\sum_{i,j} e_{ij}(a_i, a_j) + \sum_i h_i(a_i)\right\} \ . \tag{5}$$

Thus, the maximum-entropy model takes the form of a (generalized) q-states Potts model. The parameters $e_{ij}(a,b)$ denote the direct couplings between MSA columns, and the $h_i(a)$ local fields (biases) in single sites. Values for these parameters have to be determined such that Eqs. (3) are satisfied. Applied to amino-acid sequence data and combined with a mean-field calculation of the couplings e, this approach was called *direct coupling analysis* (DCA) [21, 6].

2.3. *A Bayesian perspective*

The maximum-entropy approach of the last section can be rephrased in a Bayesian framework. Assume the model to be given by Eqn. (4), and assume the sequences in the alignment A to be independently sampled. We thus find the probability of the alignment to be

$$P\left[A|\{e_{ij}(a,b), h_i(a)\}\right] = \prod_{m=1}^{M} P(a_1^m, ..., a_L^m) \ . \tag{6}$$

Plugging in Eqn. (4) we find the log-likelihood of the MSA A,

$$\begin{aligned}\mathcal{L}\left[\{e_{ij}(a,b),h_i(a)\}|A\right] &= \frac{1}{M}\log P\left[A|\{e_{ij}(a,b),h_i(a)\}\right] \\ &= \frac{1}{2}\sum_{i,j}\sum_{a,b}e_{ij}(a,b)f_{ij}(a,b)+\sum_{i,a}h_i(a)f_i(a)-\log\mathcal{Z}(\{e_{ij}(a,b),h_i(a)\})\end{aligned} \tag{7}$$

One can readily see that the interaction and field parameters maximizing $\mathcal{L}$ fulfill Eqs. (3).

3. The Hopfield-Potts model

The Potts model in (4) is defined from $O\big((Lq)^2\big)$ parameters to be inferred from the data. To reduce the risk of overfitting we now introduce the Hopfield-Potts model, a special case of the Potts model, where the number of parameters to be inferred is much lower.

3.1. Definition of the model and dimensional reduction

The main idea is to express the matrix $e_{ij}(a,b)$ in terms of $p \ll Lq$ patterns $\{\xi_i^\mu(a), \mu = 1,...,p\}$, and to write

$$e_{ij}(a,b) = \frac{1}{L}\sum_{\mu=1}^{p}\xi_i^\mu(a)\xi_j^\mu(b) \tag{8}$$

with $i,j = 1,...,L$ being the site indices, and $a,b = 1,...,q$ being amino acids, or Potts states. Note that this matrix, for linearly independent patterns, has rank p, and it depends only on pLq parameters. By defining the log-score of a sequence $(a_1,...,a_L)$ for one pattern ξ^μ as

$$S(a_1,...,a_L|\xi^\mu) = \left[\sum_{i=1}^{L}\xi_i^\mu(a_i)\right]^2 . \tag{9}$$

the probability (4) of that amino-acid sequence can be written as an the exponential of the sum of log-scores along a number p of patterns through :

$$P(a_1,\ldots,a_L) = \frac{1}{\mathcal{Z}}\exp\left\{\frac{1}{2L}\sum_{\mu=1}^{p}S(a_1,\ldots,a_L|\xi^\mu)\right\} \tag{10}$$

An important remark is that patterns are not necessarily real valued. We will find both real and imaginary-valued patterns. The first case correspond to *attractive patterns*, which lead to a positive sign on the log-score $S(a_1,...,a_L|\xi^\mu)$, the second case to *repulsive patterns* which lead to a negative sign on the log-score $S(a_1,...,a_L|\xi^\mu)$. Consequently for the probability $P(a_1,...,a_L)$ to be large, the absolute value of the scores $S(a_1,...,a_L|\boldsymbol{\xi})$ for attractive patterns must be large, whereas for repulsive patterns it must be small (close to zero).

Following closely the derivations for the Ising-case in [22], we will show how mean-field theory can be used to derive an expression for the patterns $\xi_i^\mu(a)$ in terms of the empirical 1- and 2-point frequencies (1).

3.2. Gauge fixing for the Potts states

Due to the presence on each site i of a single amino-acid a in the Potts model defined in Eqn. (4) changes of the couplings $e_{ij}(a,b) \to e_{ij}(a,b) + g_i(a)$ can be compensated by corresponding changes of the field $h_i(a) \to h_i(a) - g_i(a)$. In a previous work [19] we removed this gauge invariance by specializing to couplings matrices $e_{ij}(a,b)$ where the qth row $(a = q)$ and column

$(b = q)$ are equal to zero for every pairs of site $i < j$. Within the Hopfield-Potts model this amounted to enforce $\xi_i^\mu(q) = 0$ for all sites i and patterns μ.

Hereafter we consider another, natural choice for the gauge, expressed as a linear constraint over the pattern components:

$$\sum_a f_i(a)\ \xi_i^\mu(a) = 0\ , \tag{11}$$

for all sites i and all patterns μ. This gauge is more symmetric over the Potts states than the gauge adopted in [19].

3.3. Gauge fixing in the pattern space

It is important to realize that Hopfield's expression for the couplings in terms of patterns introduces another gauge, in the pattern space. The representation given in Eqn. (8) is indeed not unique. The expression is invariant with respect to a rotation in pattern space, *i.e.* with respect to multiplying all patterns with an orthogonal $p \times p$-matrix $\mathcal{O}$. More precisely, it is easy to check that the couplings $e_{ij}(a,b)$ defined by Eqn. (8) are unchanged under the transformation

$$\xi_i^\mu(a) \to \sum_{\nu=1}^{p} \mathcal{O}^{\mu,\nu}\ \xi_i^\nu(a)\ , \tag{12}$$

for fixed a and i.

To have a unique expression for the patterns, we have to eliminate this arbitrariness in analogy to the gauge fixation discussed in the previous Section for the Potts states. A simple and convenient prescription is to impose that the patterns are orthogonal for the following dot product:

$$\sum_{i,a} f_i(a)\ \xi_i^\mu(a)\ \xi_i^\nu(a) = 0 \qquad (\mu \neq \nu)\ . \tag{13}$$

The number of distinct orthogonality conditions, $p(p-1)/2$, is equal to the number of generators of the p-dimensional orthogonal group[1], entailing that the solution to (13) is generally unique (up to a discrete permutation symmetry of the patterns).

4. Calculation of the Hopfield-Potts patterns

4.1. Expressions of the most likely patterns

We define from the multiple sequence alignment the (Lq)–dimensional Pearson correlation matrix,

$$\Gamma_{ij}(a,b) = \frac{c_{ij}(a,b)}{\sqrt{f_i(a)\, f_j(b)}} \tag{14}$$

These matrix has (at least) L zero eigenvalues, because, on each site i, the probabilities of the $q = 21$ amino-acid or gap symbols sum up to 1:

$$\sum_{a=1}^{q} f_i(a) = 1 \Rightarrow \sum_{b=1}^{q} \Gamma_{ij}(a,b)\ \sqrt{f_j(b)} = 0\ , \forall\ i, a. \tag{15}$$

We look for the non-zero eigenmodes:

$$\sum_{j,b} \Gamma_{ij}(a,b)\ v_{jb}^\mu = \lambda^\mu\ v_{ia}^\mu\ , \tag{16}$$

[1] Strictly speaking the gauge group is not the orthogonal group $O(p)$ but the indefinite orthogonal group $O(p_+, p_-)$ where p_+ and p_- are, respectively, the numbers of real- and imaginary-valued patterns. The number of generators of this group is nevertheless equal to $p(p-1)/2$, with $p = p_+ + p_-$.

where the eigenvectors define an orthonormal basis of the $L(q-1)$–dimensional space orthogonal to the null space of Γ:

$$\sum_a v^\mu_{ia}\sqrt{f_i(a)} = 0 \quad \forall i\ , \qquad \sum_{i,a} v^\mu_{ia}\, v^\nu_{ia} = 0 \quad \forall \mu \neq \nu\ , \qquad \sum_{i,a} (v^\mu_{ia})^2 = L \quad \forall \mu\ . \tag{17}$$

It is easy to check that

$$\sum_{\mu=1}^{L(q-1)} \lambda^\mu = \sum_{i=1}^{L}\sum_{a=1}^{q} \Gamma_{ii}(a,a) = \sum_{i=1}^{L}\sum_{a=1}^{q} \big(1 - f_i(a)\big) = L(q-1)\ . \tag{18}$$

Therefore, the average value of the eigenvalues λ^μ of the matrix Γ (after having removed the L zero modes) is equal to unity.

The statistical mechanics of the inverse Hopfield-Potts model, presented in the next Section, allows us to derive the following expressions for the most likely patterns given the data, *i.e.* the MSA A,

$$\xi^\mu_i(a) = \sqrt{1 - \frac{1}{\lambda^\mu}}\ \frac{v^\mu_{i,a}}{\sqrt{f_i(a)}}\ . \tag{19}$$

By virtue of (17) the gauge conditions (11) are fulfilled. The pattern $\xi^\mu(a)$ is therefore simply written as a function of the eigenvector of the Pearson correlation matrix v^μ with our choice of the gauge. Expression (19) for the patterns in turn defines the Lq–dimensional coupling matrix e through Eqn. (8). Note that the prefactor $\sqrt{1-1/\lambda_\mu}$ is real for $\lambda_\mu > 1$, vanishes for $\lambda_\mu = 1$, and becomes imaginary for $\lambda_\mu < 1$. According to the discussion above, large eigenvalues (> 1) therefore correspond to attractive patterns, and small eigenvalues (< 1) to repulsive patterns. We will discuss how eigenvalues should be selected in Section 4.3.

4.2. Statistical mechanics derivation

We now present the detailed derivation of the results presented above. Our aim is to calculate the partition function $\mathcal{Z}(\{e_{ij}(a,b), h_i(a)\})$, defined in Eqn. (5), in the mean-field approximation. The sum over the sequence can be performed using Hubbard-Stratonovich transformations for each μ,

$$\begin{aligned}\mathcal{Z} &= \int \prod_{\mu=1}^{p} \frac{dx^\mu}{\sqrt{2\pi/L}} \sum_{\{a_i|i=1,\dots,L\}} \exp\left\{ -\frac{L}{2}\sum_{\mu=1}^{p}(x^\mu)^2 + \sum_{i=1}^{L}\left(h_i(a_i) + \sum_{\mu=1}^{p} x^\mu \xi^\mu_i(a_i)\right)\right\} \qquad (20)\\ &= \int \prod_{\mu=1}^{p} \frac{dx^\mu}{\sqrt{2\pi/L}} \exp\left\{ -\frac{L}{2}\sum_{\mu=1}^{p}(x^\mu)^2 + \sum_{i=1}^{L}\log\left(\sum_{a=1}^{q}\exp\left[h_i(a) + \sum_{\mu=1}^{p} x^\mu\xi^\mu_i(a)\right]\right)\right\}\ .\end{aligned}$$

The leading contribution of the x^μ-integrations can be determined by the saddle-point approximation. The saddle points x^μ_0 satisfy the equations

$$x^\mu_0 = \frac{1}{L}\sum_{ia} \xi^\mu_i(a) T_i(a) \quad \text{with} \quad T_i(a) = \frac{\exp\left[h_i(a) + \sum_{\mu=1}^{p} x^\mu_0 \xi^\mu_i(a)\right]}{\sum_{b=1}^{q}\exp\left[h_i(b) + \sum_{\mu=1}^{p} x^\mu_0\xi^\mu_i(b)\right]}\ . \tag{21}$$

As the next step, we determine the Gaussian corrections to this saddle point. To this aim, we need to calculate the second derivatives (in $x^\mu = x^\mu_0$):

$$\frac{\partial^2}{\partial x^\mu \partial x^\nu}\sum_i \log\left(\sum_a \exp\left[h_i(a) + \sum_{\mu=1}^{p} x^\mu \xi^\mu_i(a)\right]\right) = \sum_{i,a,b} \xi^\mu_i(a)\xi^\nu_i(b)\left[T_i(a)\delta_{a,b} - T_i(a)T_i(b)\right]\ . \tag{22}$$

Carrying out the Gaussian integrations we find the following estimate for the log-likelihood of the MSA, see Eqn. (7),

$$\begin{aligned}\mathcal{L}[\{\xi_i^\mu(a),h_i(a)\}|A] &= \frac{1}{2L}\sum_{\mu,i,j,a,b}\xi_i^\mu(a)\,\xi_j^\mu(b)\,f_{ij}(a,b)+\sum_{ia}h_i(a)\,f_i(a)+\frac{1}{2}\log\det G\\ &+\frac{L}{2}\sum_{\mu=1}^{p}(x_0^\mu)^2-\sum_{i=1}^{L}\log\left(\sum_{a=1}^{q}\exp\left[h_i(a)+\sum_{\mu=1}^{p}x_0^\mu\,\xi_i^\mu(a)\right]\right)\ , \end{aligned}\tag{23}$$

up to an irrelevant additive constant. The entries of the matrix G are defined through

$$G_{\mu\nu}=\delta_{\mu\nu}-\frac{1}{L}\sum_i\sum_{a,b}\xi_i^\mu(a)\,\xi_i^\nu(b)\,[T_i(a)\delta_{ab}-T_i(a)T_i(b)]\ .\tag{24}$$

Maximizing $\mathcal{L}$ over the fields $h_i(a)$ we readily obtain

$$T_i(a)=f_i(a)\ ,\tag{25}$$

where the $T_i(a)$'s have been defined in (21). According to the Potts gauge (11) and the saddle-point equations (21) we have $x_0^\mu=0$ for all μ's. Furthermore the pattern gauge condition (13) implies that G defined above is a diagonal matrix. After some elementary algebra we are left with the simpler expression for the log-likelihood of the data,

$$\begin{aligned}\mathcal{L}[\{\xi_i^\mu(a)\}|A] &= \sum_{ia}f_i(a)\log f_i(a)+\frac{1}{2L}\sum_{\mu,i,j,a,b}\xi_i^\mu(a)\,\xi_j^\mu(b)\,f_{ij}(a,b)\\ &+\frac{1}{2}\sum_\mu\log\left(1-\frac{1}{L}\sum_{i,a}f_i(a)\,\xi_i^\mu(a)^2\right)\ .\end{aligned}\tag{26}$$

We find the trivial result that, for $p=0$ (no pattern and, thus, no coupling between the sites), the likelihood is the negative of the sum of all single-column entropies. In the presence of patterns ($p\geq 1$) the values of the latter are found by optimizing $\mathcal{L}$. We write

$$\begin{aligned}\frac{\partial\mathcal{L}}{\partial\xi_i^\mu(a)} &= \sum_{j,b}f_{ij}(a,b)\,\xi_j^\mu(b)-\frac{1}{G^\mu}\,f_i(a)\,\xi_i^\mu(a)\\ &= \sqrt{f_i(a)}\left(\sum_{j,b}\Gamma_{ij}(a,b)\,\sqrt{f_j(b)}\,\xi_j^\mu(b)-\frac{1}{G^\mu}\,\sqrt{f_i(a)}\,\xi_i^\mu(a)\right)\ ,\end{aligned}\tag{27}$$

where the entries of matrix Γ have been defined in (14), and

$$G^\mu=1-\frac{1}{L}\sum_{i,a}f_i(a)\,\xi_i^\mu(a)^2\ .\tag{28}$$

As a consequence the optimal patterns $\xi_i^\mu(a)$ are in one-to-one correspondence with the eigenvectors v_{ia}^μ of Γ. More precisely, for all μ,i,a, $\sqrt{f_i(a)}\,\xi_i^\mu(a)=C^\mu\,v_{ia}^\mu$. The proportionality constant C^μ is such that the eigenvalue λ^μ coincides with $1/G^\mu$. According to Eqn. (28) we find

$$G^\mu=\frac{1}{\lambda^\mu}=1-(C^\mu)^2\ .\tag{29}$$

Hence C^μ is real-valued if $\lambda^\mu>1$ and imaginary-valued if $\lambda^\mu<1$. The resulting expression for the patterns is given in (19). We check a posteriori that the patterns fulfill the gauge conditions (11) and (13).

4.3. Pattern selection: maximum likelihood criterion
The final expression for the maximal likelihood is therefore

$$\mathcal{L}[A] = \sum_{i,a} f_i(a) \log f_i(a) + \sum_{\mu=1}^{p} \Delta\mathcal{L}(\lambda_\mu) \quad \text{with} \quad \Delta\mathcal{L}(\lambda) = \frac{1}{2}(\lambda - 1 - \log \lambda) \ . \tag{30}$$

If the number p of patterns is imposed we have to select the p eigenvalues λ_μ giving the largest contributions $\Delta\mathcal{L}(\lambda_\mu)$ to the likelihood. Large contributions arrive from both the largest and the smallest eigenvalues, whereas eigenvalues close to one contribute little. We thus have to determine a threshold value θ for the log-likelihood such that there are exactly p patterns with larger, and $L(q-1)-p$ patterns with smaller (and thus neglected) contributions to the likelihood. This amounts to finding the two positive real solutions $\ell_\pm$ ($\ell_- < \ell_+$) of the equation

$$\Delta\mathcal{L}(\ell_\pm) = \theta \tag{31}$$

and to include only patterns with $\lambda_\mu < \ell_-$ or $\lambda_\mu > \ell_+$. We will call hereafter p_+ and, respectively, p_- the number of eigenvalues λ_μ larger than ℓ_+, respectively smaller than ℓ_-. We obviously have $p_+ + p_- = p$.

5. Statistical error bars on the inferred patterns

5.1. Expressions for the error bars
Let $\mathcal{H}$ be the $(p+1)L(q-1)$–dimensional Hessian matrix of minus the log-likelihood (26), computed in the inferred patterns and fields,

$$\mathcal{H} \equiv \begin{pmatrix} \mathcal{H}_{ij}^{\mu,\nu}(a,b) & \mathcal{H}_{ij}^{\mu,0}(a,b) \\ \mathcal{H}_{ij}^{0,\nu}(a,b) & \mathcal{H}_{ij}^{0,0}(a,b) \end{pmatrix} = \begin{pmatrix} -\frac{\partial^2 \mathcal{L}}{\partial \xi_i^\mu(a) \partial \xi_j^\nu(b)} & -\frac{\partial^2 \mathcal{L}}{\partial \xi_i^\mu(a) \partial h_j(b)} \\ -\frac{\partial^2 \mathcal{L}}{\partial h_i(a) \partial \xi_j^\nu(b)} & -\frac{\partial^2 \mathcal{L}}{\partial h_i(a) \partial h_j(b)} \end{pmatrix} . \tag{32}$$

Asymptotically, for a large number M of sequences in the data base, the posterior probability of patterns and fields is Gaussian, with covariance matrix equal to the inverse of $M\ \mathcal{H}$. Let us call $\delta\xi_i^\mu(a)$ and $\delta h_i(a)$ the deviations of the patterns and of the fields with respect to their most likely values. Then those deviations are normally distributed, with zero averages, and covariances given by

$$\begin{aligned} \langle \delta\xi_i^\mu(a)\, \delta\xi_j^\nu(b) \rangle &= \frac{1}{M}(\mathcal{H}^{-1})_{ij}^{\mu,\nu}(a,b) \ , \quad \langle \delta h_i(a)\, \delta h_j(b) \rangle = \frac{1}{M}(\mathcal{H}^{-1})_{ij}^{0,0}(a,b) \ , \\ \langle \delta\xi_i^\mu(a)\, \delta h_j(b) \rangle &= \frac{1}{M}(\mathcal{H}^{-1})_{ij}^{\mu,0}(a,b) \ , \end{aligned} \tag{33}$$

where $\langle \cdot \rangle$ denotes the average over the posterior distribution. In the following we call error bars on the patterns and on the fields the standard deviations of their components. For instance the error bar on the component (i,a) of pattern μ is given by $\sqrt{\langle \delta\xi_i^\mu(a)^2 \rangle} = \sqrt{\frac{1}{M}(\mathcal{H}^{-1})_{ii}^{\mu,\mu}(a,a)}$.

We show in the next Section that the error bars on the pattern components are given by

$$\langle \delta\xi_i^\mu(a)^2 \rangle = \frac{1}{M}\left[\frac{(v_{ia}^\mu)^2}{2\lambda_\mu(\lambda_\mu - 1)} + \sum_{A \le p\ \&\ A \ne \mu} \frac{(v_{ia}^A)^2 (\lambda_\mu - 1)\lambda_A}{(\lambda_\mu |\lambda_A - 1| + \lambda_A |\lambda_\mu - 1|)^2} + \sum_{A > p} \frac{(v_{ia}^A)^2}{\lambda_\mu - \lambda_A} \right] . \tag{34}$$

Note that the second sum in (34) runs over all the eigenvectors of Γ with eigenvalues different from zero and smaller than the ones corresponding to the inferred patterns ($A > p$), while the first sum runs over the top p eigenvectors (except eigenvector μ).

The above expression is correct when all selected patterns are attractive. Assume now that, say, $p_- \geq 1$ repulsive patterns (corresponding to the smallest p_- eigenvalues) and p_+ attractive patterns are retained. Expression (34) is still valid for an attractive pattern, *i.e.* such that $\mu \leq p_+$, upon changing condition ($A \leq p$ & $A \neq \mu$) into ($A \leq p_+$ & $A \neq \mu$, or $A \geq L(q-1)-p_-$) and condition ($A > p$) into ($p_+ < A < L(q-1) - p_-$). For a repulsive pattern, *i.e.* such that $\mu > L(q-1) - p_-$, formula (34) holds upon changing condition $A \leq p$ into ($A \leq p_+$, or $A \geq L(q-1) - p_-$ & $A \neq \mu$) and condition ($A > p$) into ($p_+ < A < L(q-1) - p_-$). Note that for repulsive patterns the components $\xi_i^\mu(a)$ are imaginary, and the covariances $\langle \delta\xi_i^\mu(a)^2 \rangle$ are negative real numbers.

5.2. Calculation of the Hessian matrix of the log-likelihood

The calculation of the inverse of $\mathcal{H}$ was done in [22] for the Ising case ($q = 2$). We now briefly review how it can be extended to the Potts case, and limit ourselves to the calculation of the error bars on the pattern components only. We start by differentiating the log-likelihood (23) twice with respect to the patterns ξ:

$$\begin{aligned}\mathcal{H}_{ij}^{\mu,\nu}(a,b) &= \frac{\delta^{\mu\nu}}{L}\Big[- C_{ij}(a,b) + \lambda_\mu \delta_{ij}\big(f_i(a)\delta^{ab} - f_i(a)f_i(b)\big) + \frac{\lambda_\mu}{L} f_i(a) f_j(b) \sum_\alpha \lambda_\alpha\, \xi_i^\alpha(a)\xi_j^\alpha(b)\Big] \\ &+ \frac{1}{L^2}\lambda_\mu\,\lambda_\nu\,\xi_i^\nu(a)\,\xi_j^\mu(b)\ . \end{aligned} \tag{35}$$

The sum over the index α above runs over the p selected patterns. We define a set of $p(q-1)L$ vectors, $\vec{e}^{(A,\omega)}$, with A taking values from 1 to $L(q-1)$ and ω from 1 to p, and whose components on the canonical basis are given by

$$\big(\vec{e}^{(A,\omega)}\big)_{i,a}^{\mu} = \delta_{\mu,\omega}\, \frac{v_{i,a}^A}{\sqrt{f_i(a)}}\ . \tag{36}$$

The vectors $\vec{e}^{(A,\omega)}$ form a basis of the subspace orthogonal to the L-dimensional null subspace of Γ. We order the vectors such that the first p vectors, with $A = 1, 2, \ldots, p$ corresponds to the p selected patterns, see Eqn. 19. In the new basis, the entries of the matrix $\tilde{\mathcal{H}}$ defined in (35) are given by:

$$\tilde{\mathcal{H}}_{(A,\omega),(B,\tau)} = \delta_{\omega,\tau}\delta_{A,B}\Big[-\lambda_A + \lambda_\omega + \lambda_\omega(\lambda_A - 1)\delta_{A\leq p}\Big] + \sqrt{\lambda_\omega\lambda_\tau(\lambda_\omega - 1)(\lambda_\tau - 1)}\delta_{\omega,B}\delta_{\tau,A}\ . \tag{37}$$

Here, we have used $\delta_{A\leq p} = 1$ is A is smaller or equal to p, and 0 otherwise. Matrix $\tilde{\mathcal{H}}$ can now be easily diagonalized:

- For $A > p$, $\tilde{\mathcal{H}}_{(A,\omega),(B,\tau)} = (\lambda_\omega - \lambda_A)\delta_{\omega,\tau}\delta_{A,B}$, which is diagonal. Hence, we have already found $((q-1)L - p) \times p$ eigenvalues $\lambda_\omega - \lambda_A$, with $A = p+1, p+2, \ldots, p(q-1)L$ and $\omega = 1, 2, \ldots, p$.
- For $1 \leq A \leq p$, $\tilde{\mathcal{H}}_{(A,\omega),(B,\tau)} = 2\lambda_\tau(\lambda_\tau - 1)\delta_{\omega,\tau}\delta_{A,B}$, which is diagonal. Hence, we have p eigenvalues $2\lambda_\mu(\lambda_\mu - 1)$, with $\mu = 1, \ldots, p$.
- There are $\frac{1}{2}p(p-1)$ eigenvalues $2\lambda_\mu\lambda_\nu - \lambda_\mu - \lambda_\nu$, with $1 \leq \mu < \nu \leq p$. The corresponding (non-normalized) eigenvectors are $\delta_{B,\mu}\delta_{\tau,\nu}\sqrt{1 - 1/\lambda_\nu} + \delta_{B,\nu}\delta_{\tau,\mu}\sqrt{1 - 1/\lambda_\mu}$.
- The remaining $\frac{1}{2}p(p-1)$ eigenvalues vanish. They correspond to rotations of the patterns (in the μ-space) which leave the couplings and the log-likelihood unchanged. Their number coincides with the number of generators of the p-dimensional rotation group, see Section 3.3.

We may now write the expression for the pseudo-inverse of $\tilde{\mathcal{H}}$ in the $\vec{e}$–basis,

$$\begin{aligned}[\tilde{\mathcal{H}}^{-1}]_{(A,\omega),(B,\tau)} &= \frac{\delta^{\omega,\tau}\delta^{A,B}\delta_{A>p}}{\lambda_\omega - \lambda_A} + \delta^{\omega,\tau}\delta^{A,B}(1-\delta^{A,\omega})\delta_{A\le p}\frac{(\lambda_\omega - 1)\lambda_A}{(2\lambda_\omega\lambda_A - \lambda_A - \lambda_\omega)^2} \\ &+ \frac{\delta^{\omega,\tau}\delta^{\omega,A}\delta^{\omega,B}}{2\lambda_\omega(\lambda_\omega - 1)} + \delta^{\tau,A}\delta^{\omega,B}(1-\delta^{\tau,\omega})\frac{\sqrt{\lambda_\omega\lambda_\tau(\lambda_\omega-1)(\lambda_\tau-1)}}{(2\lambda_\omega\lambda_\tau - \lambda_\omega - \lambda_\tau)^2} .\end{aligned} \tag{38}$$

Using expression (33) for the covariance of the pattern fluctuations we obtain

$$\langle \delta\xi_i^\mu(a)\, \delta\xi_j^\nu(b)\rangle = \frac{L}{M}\,[\tilde{H}^{-1}]_{ij}^{\mu,\nu}(a,b) = \frac{L}{M}\sum_{A,\omega,B,\tau}[\tilde{\mathcal{H}}^{-1}]_{(A,\omega),(B,\tau)}\left(\vec{e}^{(A,\omega)}\right)_{i,a}^{\mu}\left(\vec{e}^{(B,\tau)}\right)_{j,b}^{\nu} . \tag{39}$$

The resulting expression for the diagonal elements of the covariance matrix is given in Eqn. (34).

5.3. Pattern selection: Bayesian criterion

Knowledge of the uncertainties over the patterns allows us to define a Bayesian criterion for pattern selection. Informally speaking, patterns whose components have strong deviations around their most likely values, that is, of the order of the pattern components themselves cannot be considered as reliable and should be discarded. Therefore, for each pattern $\boldsymbol{\xi}$, we consider the ratio of the squared fluctuations to the squared norm of the pattern,

$$\rho(\boldsymbol{\xi}) = \frac{\sum_{ia}\langle(\delta\xi_i(a))^2\rangle}{\sum_{ia}(\xi_i(a))^2} . \tag{40}$$

This ratio is positive for attractive and for repulsive patterns. We will decide that the pattern is reliable if the ratio ρ is smaller than some arbitrary error threshold, say, 1 or 2/3 [22]. Exemplary results for one protein family (response regulator domain Pfam ID PF00014) are given in Supplementary Information of [19], Fig. 13. Note that the error bars depend on the error threshold itself, smaller error thresholds lead to increased errors of the selected patterns. As a consequence, pattern selection according to the uncertainty of patterns is a self-consistent criterion, which can be solved in a iterative way.

5.4. Error bars on the couplings

We now turn to the derivation of the error bars over the couplings, see Eqn. (8),

$$e_{ij}(a,b) = \frac{1}{L}\sum_\mu \xi_i^\mu(a)\,\xi_j^\mu(b) , \tag{41}$$

where the index μ takes the values $1, 2, \ldots, p_+$ and $L(q-1)-p_-, \ldots, L(q-1)$. Approximating the distribution of the patterns as a Gaussian law with covariance matrix $\mathcal{H}^{-1)}/M$, centered in the inferred patterns $\boldsymbol{\xi}^\mu$, we find the variances of the couplings:

$$\begin{aligned}\langle e_{ij}(a,b)^2\rangle - \langle e_{ij}(a,b)\rangle^2 &= \frac{1}{ML^2}\sum_{\mu,\nu}\Big\{\xi_i^\mu(a)\,\xi_i^\nu(a)(\mathcal{H}^{-1})_{jb,jb}^{\mu\nu} + \xi_j^\mu(b)\,\xi_j^\nu(b)(\mathcal{H}^{-1})_{ia,ia}^{\mu\nu} \\ &+ \xi_i^\mu(a)\,\xi_j^\nu(b)(\mathcal{H}^{-1})_{ia,jb}^{\mu\nu} + \xi_i^\nu(a)\,\xi_j^\mu(b)(\mathcal{H}^{-1})_{ia,jb}^{\mu\nu} \\ &+ \frac{1}{M}(\mathcal{H}^{-1})_{ia,jb}^{\mu\nu}(\mathcal{H}^{-1})_{jb,ia}^{\mu\nu} + \frac{1}{M}(\mathcal{H}^{-1})_{ia,ia}^{\mu\nu}(\mathcal{H}^{-1})_{jb,jb}^{\mu\nu}\Big\} .\end{aligned} \tag{42}$$

Notice that the mean values of the couplings,

$$\langle e_{ij}(a,b)\rangle \;=\; \frac{1}{L}\sum_{\mu}\left(\xi_i^{\mu}(a)\,\xi_j^{\mu}(b)+\frac{1}{M}(\mathcal{H}^{-1})^{\mu\mu}_{ia,jb}\right)\,, \tag{43}$$

are slightly shifted with respect to their typical (most likely) values given by Eqn. (8). The relative error bar on the couplings can be defined through

$$\epsilon = \frac{\sum_{ij,ab}(\langle e_{ij}(a,b)^2\rangle - \langle e_{ij}(a,b)\rangle^2)}{\sum_{ij,ab}\langle e_{ij}(a,b)\rangle^2}\ . \tag{44}$$

The error ϵ is a function of the number p of retained patterns, or, equivalently, of the thresholds $\ell_{\pm}$ over the selected small/large eigenvalues.

6. Application to protein data

In [19], the Hopfield-Potts approach in its maximum-likelihood formulation has been applied to a number of protein families, in order to establish the connection between the theoretical inference approach, and the information contained in multiple-sequence alignment of large protein families. Here we are going to summarize some of the findings of [19], and we are going to compare them to the Bayesian approach, where patterns are selected according to their relative errors. We will do this concentrating on one single protein family, more specifically the *response regulator* domain (Pfam ID PF00072 [23]) which is given by an alignment of $M = 62,074$ sequences of length $L = 112$. The X-ray crystal structure used to test our results has PDB ID 1nxw [24].

Using this example, we will describe in the following what has to be considered when working with real protein sequence data, and what are the characteristics and the biological meaning of the selected patterns

6.1. *Data preprocessing*

The theoretical setup in the first part of this article assumes the sequence data to be identically and independently distributed with respect to some statistical model, which is a priori unknown but shall be reconstructed from the data. Real sequence data are, however, phylogenetically related. In addition they are not frequent enough to well sample rare amino-acid combinations, which results in a rank-deficient covariance matrix (beyond the L zero eigenvalues due to the non-independence of the frequency count of $a = q$ from the values for $a = 1, ..., q-1$, cf. the discussion about the gauge invariance of the model). To be able to provide good inference results we follow two heuristic steps of data preprocessing [6]:

- *Pseudocount regularization:* The maximum rank $L(q-1)$ of the matrix is re-established by introducing a so-called pseudocount ν, which formally corresponds to the addition of the (the average over) completely random amino-acid sequences to the MSA.
- *Reweighting:* To reduce the bias due to uneven sampling, we assign a reduced statistical weight to sequences in densely sampled regions in sequence space. More precisely, for each sequence $m \in \{1, ..., M\}$ a weight

$$w_m = ||\,\{n \in \{1, ..., M\} \mid d_H(n,m) < xL\}\,||^{-1} \tag{45}$$

is calculated which equals the inverse of the number of sequences n within Hamming distance $d_H(n,m) < xL$ from m, with x being an arbitrary but fixed number from the interval $(0,1)$. Here, $d_H(n,m)$ denotes the Hamming distance between rows n and m of the MSA. The total weight

$$M_{eff} = \sum_{m=1}^{M} w_m \tag{46}$$

can be seen as the effective number of independent sequences, and it is used in evaluating the error bars instead of the original sequence number M.

With these two modifications, frequency counts become

$$
\begin{aligned}
f_i(a) &= \frac{1}{M_{eff}+\nu}\left[\frac{\nu}{q}+\sum_{m=1}^{M} w_m\, \delta_{a,a_i^m}\right] \\
f_{ij}(a,b) &= \frac{1}{M_{eff}+\nu}\left[\frac{\nu}{q^2}+\sum_{m=1}^{M} w_m\, \delta_{a,a_i^m}\, \delta_{b,a_j^m}\right]
\end{aligned} \tag{47}
$$

Values $\nu = M_{eff}$ and $x = 0.2$ were found to work well in mean-field DCA across many families [6], we use the same values. The modified frequency counts are used to determine covariance and Pearson correlation matrices, the remaining part of the Hopfield-Potts model learning is performed as explained in the previous Sections.

6.2. Spectral density and pattern features

The upper panel of Fig. 1 displays the spectral density (only the $L(q-1) = 2240$ non-zero eigenvalues are shown). It is characterized by a pronounced peak around $\lambda = 1$. Non-zero eigenvalues range from $\lambda_{min} \simeq 0.132$ to $\lambda_{max} \simeq 45.1$.

The middle panel of the same figure shows the contributions of all patterns to the log-likelihood of the Hopfield-Potts model, given by Eqn. (30), vs. their eigenvalues. Maximum-likelihood inference consists of deciding a log-likelihood threshold (i.e. a horizontal line in the plot) such that there are exactly p diamonds above, and $L(q-1)-p$ below this line. Whereas the largest contributions come from the attractive patterns corresponding to large eigenvalues, already for $p = 10$ the first repulsive pattern (smallest eigenvalue) is selected.

The lower panel of Fig. 1 gives the weighted inverse participation ratio (IPR)

$$
\mathrm{IPR}(\xi) = \frac{\sum_{ia} f_i(a)^2\, [\xi_i(a)]^4}{\left(\sum_{ia} f_i(a)\, [\xi_i(a)]^2\right)^2} \tag{48}
$$

of patterns ξ as a function of the corresponding eigenvalue. The IPR can take values between 1 (for a pattern which has only one single non-zero component) and $1/Lq \simeq 0.00042$ (for a completely distributed pattern of constant entries). This range is in practice reduced due to the gauge, which does not allow a single component to be non-zero, or all components to have the same sign. The presence of the $f_i(a)$ factor in Eqn. (48) allows us to weigh the pattern components in the IPR according to the frequencies of the amino-acids in the alignment. We observe that strongly localized patterns correspond to eigenvalues which are either small (i.e. having a large contribution to the likelihood) or close to one (i.e. not contributing to the likelihood). Attractive patterns are extended.

The upper row of Fig. 2 shows the three most repulsive patterns, which correspond to the smallest eigenvalues. These patterns are found to be very localized in a few sites, which are in contact in the three-dimensional protein fold. It is important to notice that, due to the negative sign of the score function (9) for imaginary-valued patterns, repulsive patterns enforce the presence of particular combination of amino-acids on specific sites in a tight manner. Take for example the first repulsive pattern in Fig. 2: the score function is zero only if both of the two amino-acids, which correspond to the two large (positive and negative) components of the pattern are present, or if none of them are present. The score function is, on the contrary, negative if only one of the above amino-acid is present. As is described in more detail in [19] we have verified that this strong constraint is due to specific chemical bonds or physical interactions between the two amino-acids.

On the contrary, the three most attractive patterns, displayed in the lower row of Fig. 2, are extended. As discussed in [19], the components of the first attractive pattern are strongly correlated to residue conservation. The other two patterns show strong peaks in the values of a corresponding to the alignment gap symbol. They have the largest entries, in absolute value, close to the two ends of the protein sequence. These two patterns actually represent an artifact of the alignment procedure: It is easier to extend a gap than to open a new one after an aligned amino acid. This leads, to correlations between the occurrence of gaps in close-by positions, in particular at the beginning and at the end of each sequence.

6.3. Errors vs. likelihood

Whereas pattern selection in [19] uses a maximum-likelihood approach, the original theoretical proposal of Hopfield-model learning in the inverse Ising problem [22] uses a Bayesian approach. In this Bayesian approach, error bars for patterns are estimated, and only patterns up to some maximum relative error are considered. As in maximum-likelihood pattern selection, this leads to the inclusion of patterns in both tails of the spectrum, i.e. to attractive and repulsive patterns corresponding to large resp. small eigenvalues. As is explained above, here we have generalized the expressions for the errors on patterns and couplings to the Hopfield-Potts case, in order to compare the two selection criteria.

The results are given in Fig. 3: The main figure contains the error bars of the patterns as a function of the associated eigenvalue. Note that the error bars in Eqn. (34) depend explicitly on the selected patterns, *i.e.* for a given error threshold (or a given pattern number p), the included patterns have to be determined self-consistently. In general, the selection of a smaller number of patterns increases slightly the error bars.

In the insert of Fig. 3, we plot the log-likelihood contribution $\Delta\mathcal{L}(\lambda)$ of patterns against the error bars (determined without error threshold). We find an impressive collapse of the data over several orders of magnitude for both the attractive and the repulsive patterns on a single joint curve. Some minor fluctuations are observed in the tail of the most repulsive patterns.

As a consequence, we find that for all but very small numbers p of selected patterns, the maximum-likelihood and the Bayesian criterion almost coincide. Due to the increased computational complexity of the Bayesian criterion (error bar calculation and self-consistent selection), we therefore stick to the maximum-likelihood criterion of [19].

Figure 3 contains also a parametric plot of the relative error ϵ of the Hopfield-Potts couplings, cf. Eqn. (44), as a function of both ℓ_- (left branch) and ℓ_+ (right branch), the latter being determined via maximum-likelihood selection. A strong accumulation of the individual pattern errors into coupling errors can be observed, underlining the importance of dimensional reduction for obtaining high-quality couplings.

6.4. Contact prediction by the Hopfield-Potts model

The output of the Hopfield-Potts modeling is a $q \times q$ matrix for each pair (i, j) of sites, i.e. of residue positions in the aligned protein sequences. To predict contacts, we have to say which of these matrices are "large", i.e. stand for strong couplings, and which are small. A single scaler score has to be assigned to each coupling matrix. To this end, we follow [25]. We first introduce the Frobenius norm

$$F_{ij} = \sqrt{\sum_{a,b=1}^{q} \tilde{e}_{ij}(a,b)^2} \tag{49}$$

of the transformed coupling matrices

$$\tilde{e}_{ij}(a,b) = e_{ij}(a,b) - e_{ij}(\cdot,b) - e_{ij}(a,\cdot) + e_{ij}(\cdot,\cdot) \tag{50}$$

with the dot denoting an average over all amino acids and the gap in the concerned position. This transformation leads to a gauge where the sum over each column or row of the coupling matrix for each given (i,j) vanishes; this gauge minimizes the Frobenius norm with respect to the gauge freedom. Intuitively, this gauge puts "as much as possible" of the statistical modeling into the local field parameters, and "as little as necessary" into the couplings. This score is adjusted by the heuristic *average product correction* (APC) [26],

$$F_{ij}^{APC} = F_{ij} - \frac{F_{\cdot j}F_{i\cdot}}{F_{\cdot\cdot}} \ , \tag{51}$$

now with the dot indicating an average over all sites $1, ..., L$.

Once this is done, position pairs $(i,j), 1 \leq i < j \leq L$, are ranked according to their F^{APC} values. The highest-ranking ones are expected to be in contact. Our capacity to actually predict residue contacts is assessed in Fig. 4, where predictions are compared with the native contact map of the response regulator protein. We observe that most 'islands' of tertiary contacts are identified by one or more of the 50 top F^{APC} values. Remarkably the quality of the prediction with only $p = 150$ patterns is very similar to the one obtained from all $p = L(q-1) = 2240$ patterns. Hence most patterns do not actually convey much information about contacts, and can be discarded in the inference.

We show in Fig. 5 the success rate of the prediction, defined as the fraction of true positive contacts among the top x scores F^{APC} as a function of x. Only contacts between far away residues (by more than 4 sites) on the chain are taken into account. We observe that with all $p = L(q-1)$ patterns the first 58 scores correspond to true contacts (black line), while inference with only $p = 150$ patterns make a first false positive prediction after 34 predicted contacts (blue line). Strikingly, however, the Hopfield-Potts model with the $p = 150$ most repulsive patterns only shows remarkable performances and beats the all-pattern model for success rates $\sim 85-90\%$, corresponding to $\simeq 130-140$ predicted contacts (red line). On the contrary, the Hopfield-Potts with the same number $p = 150$ of attractive patterns only shows much poorer performances (green line).

7. Conclusion

In this paper we have shown, in a detailed way, how the inverse Hopfield-Potts model can be solved within the mean field approximation. In general this inverse model is useful to analyze the correlated activity of a population of variables with the aim of extracting an interaction network between them. The Hopfield-Potts model is expected to be efficient for large data sets, which correspond to networks with many nodes and interactions, and to be robust against undersampling.

We have here applied the inverse Hopfield Potts approach to the correlated mutations found in different sequences of a protein family; the interaction network between the sequence sites allows us to make predictions about the sites in contact in the tridimensional structure of the protein. We have focused on the analysis of the protein family PF00072, which has 112 sites (the variables of our interaction network), each one carrying 20 possible amino-acids plus the gap symbol. This defines a large network with a total of $(112\times21)2/2 \simeq 3\,10^6$ interaction parameters to be inferred from a sampled data set made of only 62,000 sequences. The inverse Hopfield-Potts approach offers several interesting features: i) The coupling matrix is expressed as a function of its principal modes, called patterns, and allows for a strong dimensional reduction compared to the usual Potts model; ii) The inferred patterns can be simply related to the eigenvectors of the Pearson correlation matrix of the data set, see (19); iii) The log-likelihood of the data given the patterns has a simple expression (30) and gives a natural way to select the best patterns; iv) An alternative and equivalent way to select patterns is the evaluation of their statistical error bars, see inset of Fig. 3. Moreover here we have shown how to calculate the statistical errors on the couplings from the statistical errors on the patterns; v) Among the patterns which contribute most to the log-likelihood many low-eigenvalues modes of the Pearson Correlation Matrix are present. The importance of these modes is new with respect to the classical PCA approach in which only large-eigenvalues modes are selected. Low- eigenvalues modes, which we call the repulsive patterns, are often localized on sites in contact on the 3D fold of the protein. Remarkably, for the PF00072 family studied here, the quality of the prediction with only $p = 150$ repulsive patterns is very similar to the one obtained from all $p = L(q-1) = 2240$ patterns as shown in the contact map of Fig. 4. A more exhaustive analysis on 15 protein families can be found in [19]. We have also discussed in detail the two gauge invariances of the Hopfield-Potts model. The first one, typical of the Potts model, results from the conservation of the probability of observing one of the possible symbols on a site, while the second gauge invariance is typical of the Hopfield model, and is due to the invariance of the couplings under rotations in the pattern space. We have chosen here the gauges in a different way with respect to [19]. The choice of the Hopfield gauge is irrelevant as far as couplings are concerned when all $p = L(q-1)$ patterns are selected, but it does affect the couplings when only a limited number of patterns are selected. Moreover the Hopfield gauge choice could a priori change the sites in which the patterns are localized. However, it is interesting to remark that the two choices of the gauge do not alter the main findings, i.e. the localization of repulsive patterns on the same sites and the accurate prediction of the contacts from a limited number of repulsive patterns. It would be interesting to have a better comprehension of how the gauge changes the representation of the patterns and if there is an optimal choice for the gauge.

It would also be useful to have a statistical mechanics justification for the pre-processing of data, through the reweighting procedure and the introduction of a large pseudo-count, and for the use of the average product correction to have a better estimator of contacts. Finally, let us mention that it is in principle possible to calculate corrections to the mean field approximation, as was done in [22]. However, in the case of small pattern amplitudes (of the order $1/\sqrt{L}$), a very good sampling is needed for the corrections to be effective. Calculating corrections to the large and localized pattern components encountered here remains an open problem.

References

[1] de Juan D, Pazos F and Valencia A 2013 *Nature Reviews Genetics* **14** 249
[2] Jaynes E T 1957 *Physical Review Series II* **106** 620
[3] Jaynes E T 1957 *Physical Review Series II* **108** 171
[4] Lapedes A S, Giraud B G, Liu L and Stormo G D 1999 *Lecture Notes-Monograph Series: Statistics in Molecular Biology and Genetics* **33** 236
[5] Weigt M, White R A, Szurmant H, Hoch J A and Hwa T 2009 *Proc. Natl. Acad. Sci. U. S. A.* **106** 67
[6] Morcos F, Pagnani A, Lunt B, Bertolino A, Marks D S, Sander C, Zecchina R, Onuchic J N, Hwa T and Weigt M 2011 *Proc. Natl. Acad. Sci. U. S. A.* **108** E1293
[7] Burger L and van Nimwegen E 2010 *PLoS Comput. Biol.* **6** E1000633
[8] Balakrishnan S, Kamisetty H, Carbonell J G, Lee S I and Langmead C J 2011 *Proteins: Struct., Funct., Bioinf.* **79** 1061
[9] Jones D T, Buchan D W A, Cozzetto D and Pontil M 2012 *Bioinformatics* **28** 184
[10] Schug A, Weigt M, Onuchic J N, Hwa T and Szurmant H 2009 *Proc Natl Acad Sci USA* **106** 22124
[11] Marks D S, Colwell L J, Sheridan R P, Hopf T A, Pagnani A, Zecchina R and Sander C 2011 *PLoS ONE* **6** e28766
[12] Sadowski M I, Maksimiak K and Taylor W R 2011 *Computational Biology and Chemistry* **35** 323
[13] Sulkowska J I, F Morcos F, Weigt M, Hwa T and Onuchic J N 2012 *Proc. Natl. Acad. Sci.* **109** 10340
[14] Nugent T and Jones D T 2012 *Proceedings of the National Academy of Sciences* **109** E1540
[15] Hopf T A, Colwell L J, Sheridan R, Rost B, Sander C and Marks D S 2012 *Cell* **149** 1607

[16] Dago A E, Schug A, Procaccini A, Hoch J A, Weigt M and Szurmant H 2012 *Proc Natl Acad Sci USA* **109** 10148
[17] Marks D S, Hopf T A and Sander C 2012 *Nature Biotechnology* **30** 1072
[18] Taylor W R, Hamilton R S and Sadowski M I 2013 *Current Opinion in Structural Biology* –
[19] Cocco S, Monasson R and Weigt M 2013 *PLoS Comput. Biol.* **9** E1003176
[20] Hopfield J J 1982 *Proc. Natl. Acad. Sci.* **79** 2554
[21] Lunt B, Szurmant H, Procaccini A, Hoch J, Hwa T and Weigt M 2010 *Methods in Enzymology* **471** 43
[22] Cocco S, Monasson R and Sessak V 2011 *Physical Review* E **83** 051123
[23] Punta M, Coggill P C, Eberhardt R Y, Mistry J, Tate J G, Boursnell C, Pang N, Forslund K, Ceric G, Clements J, Heger A, Holm L, Sonnhammer E L L, Eddy S R, Bateman A and Finn R D 2012 *Nucleic Acids Res.* **40** D290
[24] Bent C J, Isaacs N W, Mitchell T J and Riboldi-Tunnicliffe A 2004 *J. Bacteriol.* **186** 2872
[25] Ekeberg M, Lövkvist C, Lan Y, Weigt M and E A 2013 *Phys. Rev.* E **87** 012707
[26] Dunn S D, Wahl L M and Gloor G B 2008 *Bioinformatics* **24** 333

8. Figures and figure captions

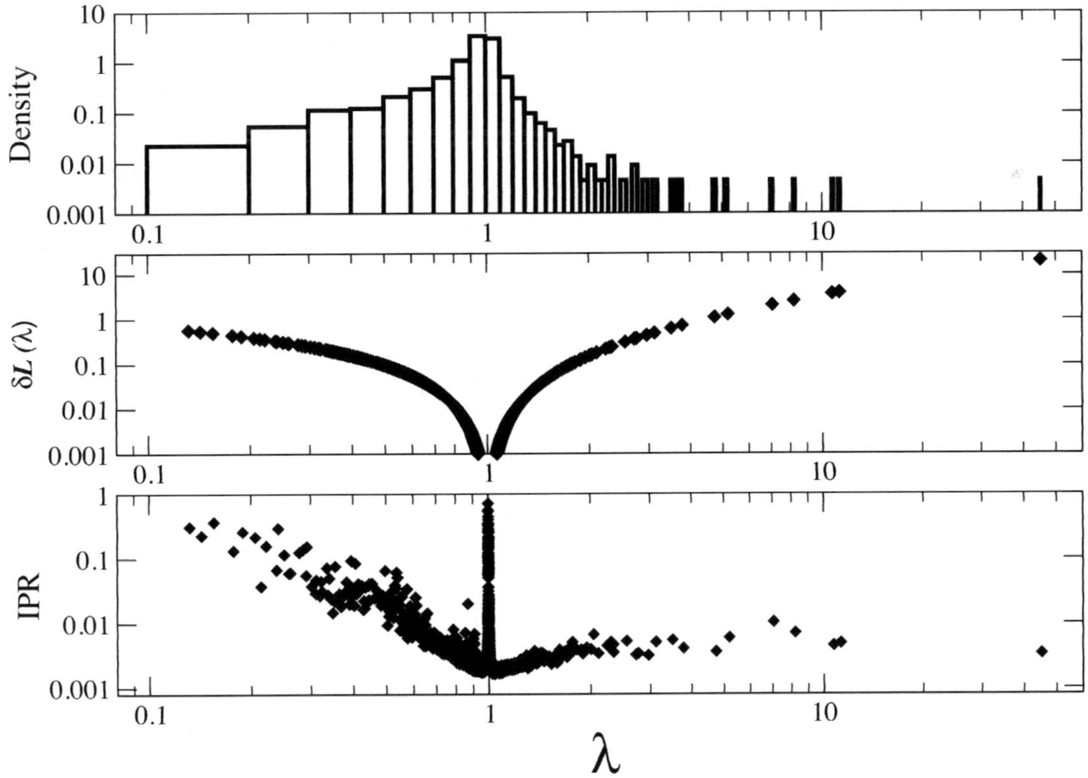

Figure 1. Spectral density (top), contributions to the log-likelihood (middle) and inverse participation ratios of the patterns (bottom) for the family PF00072. See text for description.

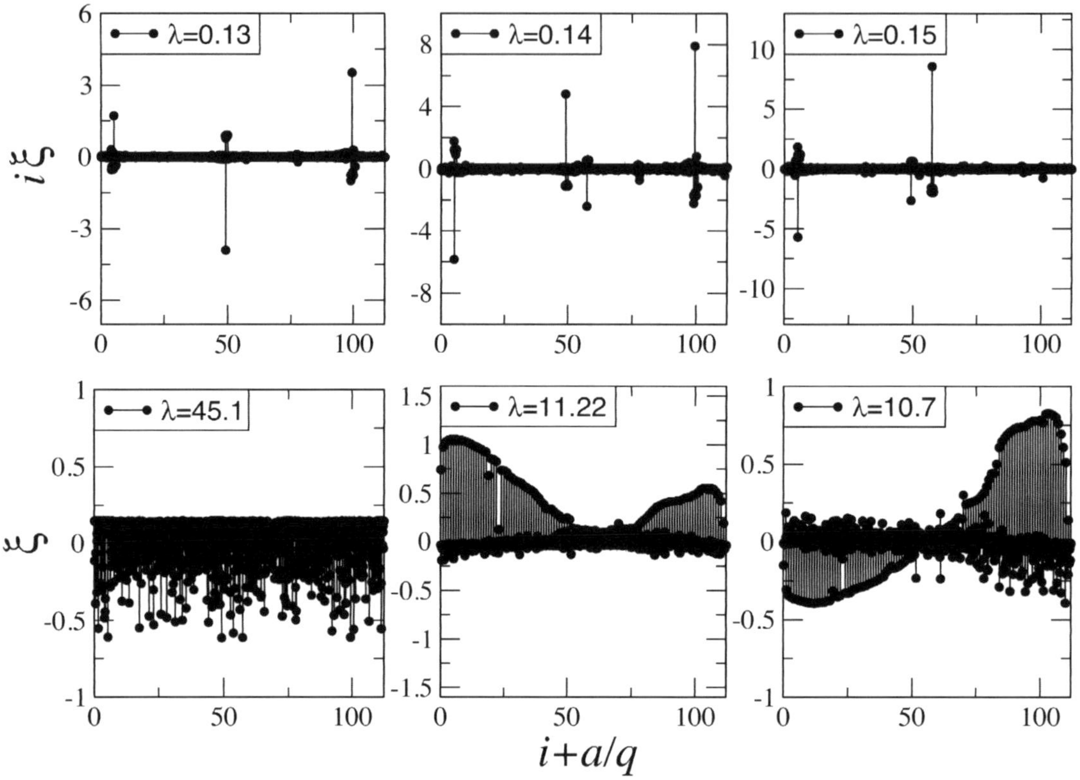

Figure 2. Patterns for the family PF00072. Top: the three patterns corresponding to the three lowest eigenvalues; as pattern components are purely imaginary we plot $i\xi$ instead of ξ. Bottom: the three patterns corresponding to the three largest eigenvalues. The residue coordinate is a rationale number $x = i + a/q$, where i is the site index and a is the amino-acid index.

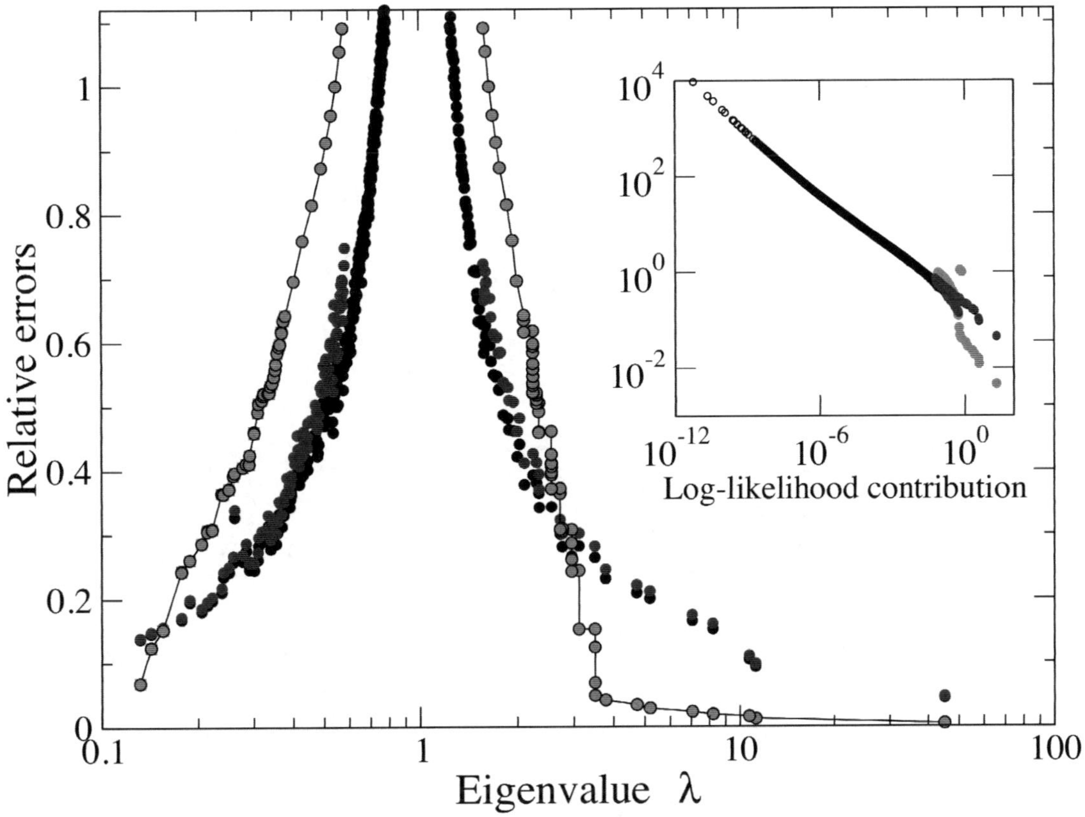

Figure 3. Bayesian pattern selection for the family PF00072. Relative error $\sqrt{\rho^\mu}$, see Eqn. (40), as a function of the eigenvalues for $p = 150$ selected patterns according to the log-likelihood contributions (blue dots) and for all $p = 2240$ patterns (black dots). The number M of sequences in formula (44) for the error is replaced with the number of independent sequences, $M_{eff} = 29408$, estimated from the reweighting procedure. Magenta dots give the relative error $\sqrt{\epsilon}$ of the Hopfield-Potts couplings as a function of the eigenvalue ℓ_- of the maximal selected repulsive pattern (left branch) and of the eigenvalue ℓ_+ minimal selected attractive pattern (right branch). The relative error on the couplings is displayed up to a maximum of $p = 120$ selected patterns. The inset shows a parametric plot of the error for each pattern as a function of the pattern contribution $\Delta\mathcal{L}(\lambda)$; same color code as for the main figure. The relative error on the couplings as a function of $\Delta\mathcal{L}(\lambda)$ is also given in the inset (magenta dots) up to 120 selected patterns.

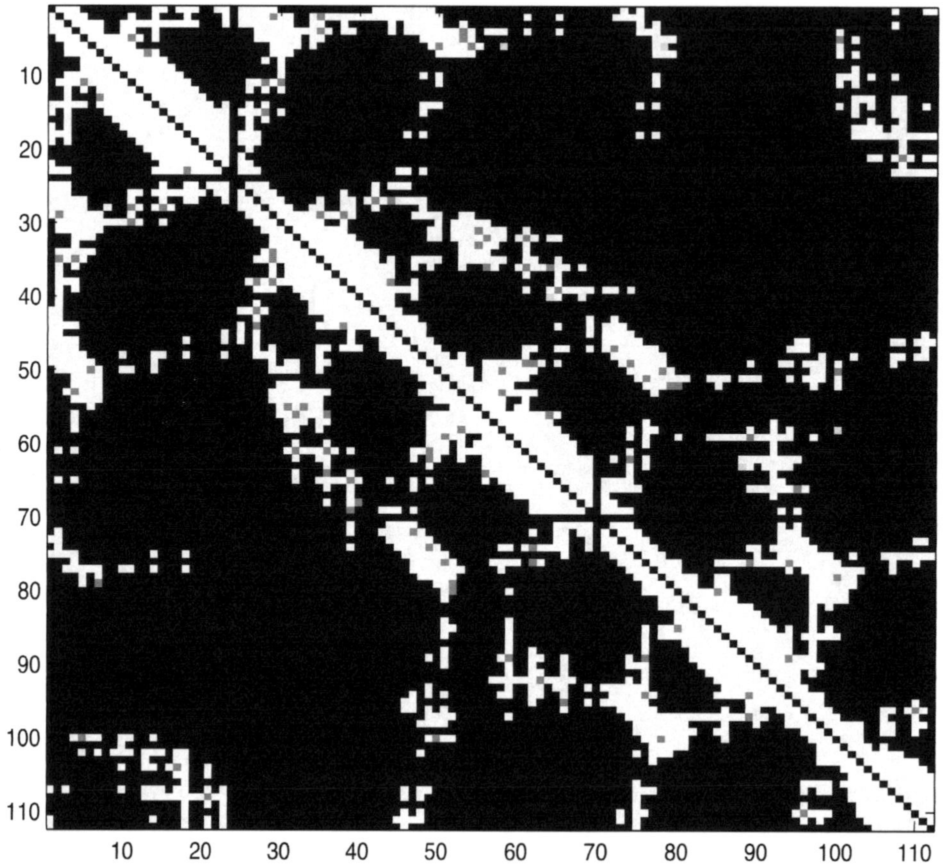

Figure 4. Contact map of the family PF00072. Yellow squares show residues i and j in contact (distance between Calpha smaller than 8Å); For the sake of clarity we have removed contacts between residues nearby on the chain, *i.e.* such that $|i-j| \leq 4$. Red and green dots are contact predictions from the Hopfield-Potts model with $p = 150$ patterns (top right corner) and with all $P = L(q-1) = 2240$ patterns (bottom left corner). Predictions are made from the top 50 Frobenius norms, red dots indicate true positives and green dots locate false positives.

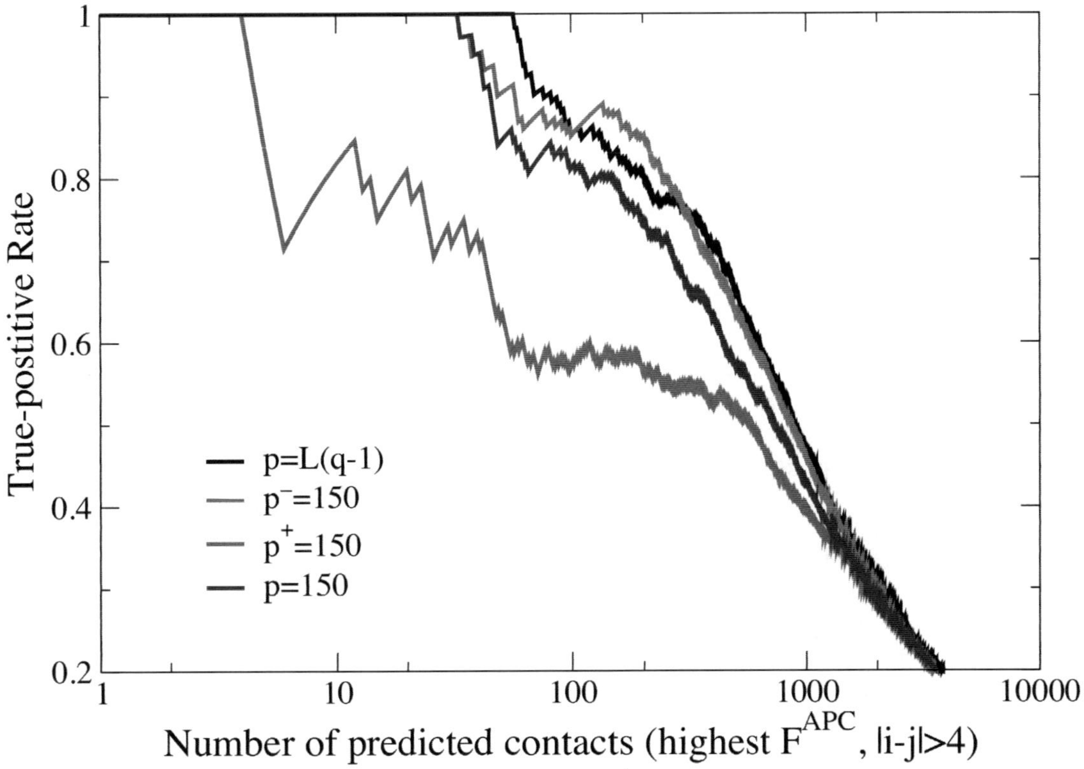

Figure 5. True positive rates for the family PF00072. Fraction of true contacts as a function of the number of top scores F^{APC} retained for the $p = 150$ patterns with largest contributions to the log-likelihood ($p_+ = 39, p_- = 111$, blue curve), for the most 150 repulsive patterns (red curve), for the most 150 attractive patterns (green curve), and for all $p = 2240$ patterns (black curve). Only contacts between sites i, j such that $|i - j| > 4$ on the chain are considered to compute the fraction of true positive.

Thermal, quantum and simulated quantum annealing: analytical comparisons for simple models

V Bapst and G Semerjian

LPTENS, Unité Mixte de Recherche (UMR 8549) du CNRS et de l'ENS, associée à l'UPMC Univ Paris 06, 24 Rue Lhomond, 75231 Paris Cedex 05, France.

E-mail: victor.bapst@polytechnique.edu, guilhem@lpt.ens.fr

Abstract. We study various annealing dynamics, both classical and quantum, for simple mean-field models and explain how to describe their behavior in the thermodynamic limit in terms of differential equations. In particular we emphasize the differences between quantum annealing (i.e. evolution with Schrödinger equation) and simulated quantum annealing (i.e. annealing of a Quantum Monte Carlo simulation).

1. Introduction

Combinatorial optimization is a branch of computer science concerned with the task of finding the minimum of a cost function $C(\underline{\sigma})$ of discrete variables $\underline{\sigma} = (\sigma_1, \dots, \sigma_N)$. The computational complexity theory [1] classifies the difficulty of such a task, depending on the problem type (the "shape" of the function), in terms of the time and memory requirements of algorithms that find optimal configurations of the variables (or approximations with a controlled accuracy). If the cost function is viewed as an energy, optimal configurations correspond to groundstates and this analogy leads to physics-inspired annealing procedures aiming at discovering the groundstate of a given system. Thermal (simulated) annealing [2] in particular exploits thermal fluctuations to achieve this goal: in a Monte Carlo numerical experiment a fictitious temperature can be slowly decreased down to very low temperatures. If this decrease is slow enough the system remains close to equilibrium at all times, and at the end of this annealing, where the temperature is very low, the typical configurations encountered will be the groundstates of the system. Instead of thermal fluctuations, or in addition to them, one can envision the use of quantum fluctuations to accelerate the exploration of the configuration space towards the discovery of the groundstate. This idea appeared with several variants and names [3, 4, 5, 6, 7], in particular quantum annealing and quantum adiabatic algorithm (see [8, 9, 10, 11] for reviews). In one of these variants one considers the evolution of an isolated system at zero temperature, thus described by the Schrödinger equation, with a quantum Hamiltonian interpolating during the course of the evolution between a simple Hamiltonian with an easy to prepare groundstate (typically a transverse field) and the Hamiltonian encoding the cost function one wants to minimize. If this interpolation is slow enough (i.e. if the adiabaticity condition is fulfilled), a system initially prepared in the groundstate of the simple Hamiltonian will end up in the sought-for groundstate of the cost function. This procedure, that we shall call quantum annealing in the following, could be implemented in an ideal quantum computer (with a perfect coherence), but its simulation on a classical computer is of course very costly (the dimension of the quantum

Hilbert space growing exponentially with the number of variables). Another variant, that we shall call simulated quantum annealing in the following (also known as Path-Integral Monte Carlo Quantum Annealing in [8, 10] or Quantum Monte Carlo Annealing in [9]), amounts to perform a Quantum Monte Carlo numerical experiment, based on an imaginary-time path integral representation of the quantum partition function, at positive temperature, in which some parameters, in particular the transverse field, are slowly modified during the simulation. Recent claims that such a procedure can be implemented experimentally with a rather large number of variables can be found in [12].

A satisfactory understanding of these various annealing procedures would require being able to answer at least the following two questions: i) how long should be the annealing to reach the groundstate? ii) if the annealing is too fast, what is the residual energy (i.e. the difference between the energy of the configuration reached at the end of the annealing and the one of the groundstate) as a function of the annealing time? These questions, and in particular the behavior of these quantities in the thermodynamic limit, have notably been studied in the context of mean-field spin-glasses [13], that are relevant as typical instances of random combinatorial optimization problems [14]. However, even in this mean-field setting, only partial answers to questions i) and ii) are known. Consider for instance the thermal annealing case; mean-field spin-glasses do exhibit a so-called dynamic transition temperature $T_{\rm d}$, below which metastable pure states proliferate and drive the system out-of-equilibrium if the annealing is not slow enough to cross the free energy barriers between the pure states [15]. In many of these models these barriers are extensive, hence the time required for an annealing to bring the system in its groundstate is exponentially large in the size of the system. Even if the computation of $T_{\rm d}$ is now well-understood, the rate of extensive growth of the barriers, hence the answer to question i), is generically unkown. Question ii) can be more easily adressed in a particular limit, i.e. for annealing durations which go to infinity after the thermodynamic limit has been taken, see in particular [16, 17, 18] for recent works on this question with the so-called state following method.

The current knowledge for the analysis of the quantum annealing procedure is similar and slightly less advanced: from the equilibrium analysis of quantum mean-field spin glasses, via the replica method for fully-connected models or the cavity one for diluted models (see [11] for a recent review of these works), the location of the quantum phase transition where the gap between the groundstate and the first excited one closes in the thermodynamic limit can be computed (similarly to $T_{\rm d}$ for the classical case), yet the quantitative behavior of the gap is known analytically only in some simple cases [19, 20, 21]. The equivalent of the state following method in the quantum case has not yet been developed. For what concerns the simulated quantum annealing, the dynamic transition of the path-integral model representing the quantum problem at finite temperature has been computed; as demonstrated in [11] this is the point at which a QMC annealing falls out-of-equilibrium, but again no prediction is available for its long (but subexponential) time behaviour. Let us also mention the existence of worst-case bounds on the annealing schedules of both quantum and simulated quantum annealing that ensure the convergence to the groundstate in an infinite time (see [10] for a review of these results) and of several numerical experiments comparing the efficiency of the various annealing procedures (for instance [22, 23], and [8, 9] and references therein).

The strategy followed in this paper consists in studying simple models on which these questions can be investigated analytically, namely non-disordered fully-connected models. The thermal annealing of these models is rather simple to analyze, see [24] for early results. We previously considered in great details the quantum annealing procedure (i.e. the real-time Schrödinger evolution) [21], here we show that the simulated quantum annealing case is also tractable (which was previously discovered in [25]) and we emphasize the difference between the last two dynamics.

The rest of this paper is organized as follows. In Sec. 2 we give explicit definitions of the various annealing dynamics and of the models to be studied. Then in Sec. 3 we present the evolution equations that govern the dynamics in the thermodynamic limit, and study their long-time behavior; this unveils a crucial difference in the behavior of the quantum annealing and the simulated quantum annealing (even at very low temperature). We finally come to our conclusions in Sec. 4, sketching some extensions of these results to more complicated models and other time regimes. For reason of space limitations we refrain from giving details of the computations that shall be presented elsewhere [26].

2. Definitions

2.1. *Various annealing dynamics*

We consider models whose classical configurations are given by the values of N Ising spins, denoted $\underline{\sigma} = (\sigma_1, \ldots, \sigma_N)$. In the quantum case we consider the 2^N-dimensional Hilbert space spanned by the vectors denoted $\{|\underline{\sigma}\rangle\}$, that form the so-called computational basis. We write $\widehat{\sigma}_i^z$ and $\widehat{\sigma}_i^x$ for the Pauli matrices acting on the i-th spin according to $\widehat{\sigma}_i^z|\underline{\sigma}\rangle = \sigma_i|\underline{\sigma}\rangle$ and $\widehat{\sigma}_i^x|\underline{\sigma}\rangle = |\underline{\sigma}^{(i)}\rangle$, where $\underline{\sigma}^{(i)}$ is the configuration obtained from $\underline{\sigma}$ by flipping the i-th spin.

In the various annealing procedures we denote $\mathcal{T}$ the total physical time of the evolution, which is parametrized by a reduced time parameter $s \in [0,1]$. All parameters are allowed to evolve with s, i.e. $\beta(s)$ is the time varying inverse temperature, and in the most general case we consider a time-dependent quantum Hamiltonian written as

$$\widehat{H}(s) = \sum_{\underline{\sigma}} E(\underline{\sigma}; s)|\underline{\sigma}\rangle\langle\underline{\sigma}| - \Gamma(s)\sum_{i=1}^{N} \widehat{\sigma}_i^x \ . \tag{1}$$

The "classical" (i.e. diagonal in the computational basis) energy $E(\underline{\sigma}; s)$ can a priori vary in time, as well as the intensity $\Gamma(s)$ of the transverse field.

Let us now define the dynamics to be studied.

2.1.1. *Thermal annealing*

This first case is purely classical ($\Gamma(s) = 0$), with a dynamics made of a Markov stochastic process in the configuration space, described by a continuous time master equation for the probability $P(\underline{\sigma}; s)$ to observe the system in the configuration $\underline{\sigma}$ at (reduced) time s:

$$\frac{1}{\mathcal{T}}\frac{\mathrm{d}}{\mathrm{d}s}P(\underline{\sigma}; s) = \sum_{\underline{\sigma}'} W(\underline{\sigma}, \underline{\sigma}'; s)P(\underline{\sigma}'; s) \ . \tag{2}$$

The jump rates from configuration $\underline{\sigma}'$ to $\underline{\sigma}$ (with diagonal terms chosen to ensure the conservation of probabilities) are assumed to verify the detailed balance condition with respect to the instantaneous Gibbs-Boltzmann distribution, i.e.

$$W(\underline{\sigma}, \underline{\sigma}'; s)e^{-\beta(s)E(\underline{\sigma}';s)} = W(\underline{\sigma}', \underline{\sigma}; s)e^{-\beta(s)E(\underline{\sigma};s)} \ . \tag{3}$$

In the usual form of simulated annealing $E(\underline{\sigma}; s)$ is independent of s and equal to the cost function $C(\underline{\sigma})$ to be minimized, and $\beta(s)$ increases from 0 to ∞ as s goes from 0 to 1. Note that β and E always appear through the combination βE; to simplify the discussion below we shall assume that β is fixed to some arbitrary value and that all the time-dependence of the annealing schedule is included in $E(\underline{\sigma}; s)$. We also assume that the initial condition at $s = 0$ is at equilibrium.

We shall further assume that the dynamics flips one spin at a time, W is hence non-zero only if $\underline{\sigma}$ and $\underline{\sigma}'$ are at Hamming distance at most 1. For definiteness we further specialize this choice to

$$W(\underline{\sigma}^{(i)}, \underline{\sigma}; s) = \frac{1}{2}\left(1 - \tanh\left(\beta\frac{E(\underline{\sigma}^{(i)}; s) - E(\underline{\sigma}; s)}{2}\right)\right) \ . \tag{4}$$

This form of the transition rate corresponds to the Glauber (or heat bath) rule, and can be expressed in terms of the effective magnetic field h_i acting on spin i, defined as

$$h_i(\underline{\sigma}; s) = \sigma_i \frac{E(\underline{\sigma}^{(i)}; s) - E(\underline{\sigma}; s)}{2} . \tag{5}$$

This expression is actually independent of σ_i.

2.1.2. Quantum annealing This case corresponds to the evolution of an isolated system. Its wave function $|\psi(s)\rangle$ thus follows Schrödinger's equation,

$$\frac{i}{T} \frac{\mathrm{d}}{\mathrm{d}s} |\psi(s)\rangle = \widehat{H}(s) |\psi(s)\rangle . \tag{6}$$

We shall assume that the interpolation starts from a purely transverse field, $E(\underline{\sigma}; s = 0) = 0$, with $|\psi(s = 0)\rangle$ the groundstate of the transverse field (i.e. the uniform superposition of the 2^N states of the computational basis), and that at the end of the annealing the transverse field is turned off and the classical energy corresponds to the cost function to be optimized: $\Gamma(s = 1) = 0$ and $E(\underline{\sigma}; s = 1) = C(\underline{\sigma})$.

2.1.3. Simulated quantum annealing Quantum models can be represented as classical models with an additional imaginary time dimension (using for instance the Suzuki-Trotter identity [27]), of length corresponding to the inverse temperature. In the case of Ising spins in a transverse field one is thus led to consider a configuration space of trajectories, that we shall denote $\underline{\boldsymbol{\sigma}} = (\boldsymbol{\sigma}_1, \ldots, \boldsymbol{\sigma}_N)$, with the convention that bold font symbols stand for imaginary-time dependent quantities, here $\boldsymbol{\sigma}_i = \{\sigma_i(t) : t \in [0, \beta]\}$ with $\sigma_i(t) = \pm 1$ is the imaginary time trajectory of an Ising spin. The quantum partition function and the thermodynamic average can then be computed by introducing the following equilibrium measure on the space of trajectory configurations (for a classical energy $E(\underline{\sigma})$, a transverse field Γ and an inverse temperature β):

$$P_{\mathrm{eq}}(\underline{\boldsymbol{\sigma}}) = \frac{1}{Z} \prod_{i=1}^{N} \Gamma^{|\boldsymbol{\sigma}_i|} e^{-\int_0^\beta E(\underline{\sigma}(t)) \mathrm{d}t} , \tag{7}$$

where $|\boldsymbol{\sigma}_i|$ denotes the number of spin flips in the trajectory of the i-th spin, and we assume periodic boundary conditions in the imaginary time direction, $\underline{\sigma}(t = 0) = \underline{\sigma}(t = \beta)$. A Quantum Monte Carlo simulation proceeds by a stochastic exploration of the space of configurations $\underline{\boldsymbol{\sigma}}$. One can use for instance a quantum version of the heat-bath algorithm, where elementary moves correspond to the update of one spin trajectory $\boldsymbol{\sigma}_i$ (for the whole imaginary time axis), redrawn from the equilibrium measure conditioned on the current state of all trajectories on sites distinct from i. What we call simulated quantum annealing is precisely this type of stochastic process for $\underline{\boldsymbol{\sigma}}$, described by the master equation,

$$\frac{1}{T} \frac{\mathrm{d}}{\mathrm{d}s} P(\underline{\boldsymbol{\sigma}}; s) = \sum_{\underline{\boldsymbol{\sigma}}'} W(\underline{\boldsymbol{\sigma}}, \underline{\boldsymbol{\sigma}}'; s) P(\underline{\boldsymbol{\sigma}}'; s) , \tag{8}$$

where the transition rates are those of heat-bath updates, respecting detailed balance with respect to the instantaneous equilibrium distribution (7) with the current values of $E(\cdot; s)$ and $\Gamma(s)$ (for simplicity of notation we assume the inverse temperature $\beta(s)$ to be held constant). The initial condition will be considered to be equilibrated at $s = 0$, i.e. in a purely transverse field.

2.2. Mean-field fully-connected models

The models we shall consider are generalizations of the (quantum) Curie-Weiss model (also related to the Lipkin-Meshkov-Glick model [28]), with their energy depending on the configurations only through the total magnetization:

$$E(\underline{\sigma}; s) = N g \left(\frac{1}{N} \sum_{i=1}^{N} \sigma_i ; s \right) , \tag{9}$$

$g(m; s)$ being the energy density for a configuration of magnetization density m at reduced time s. The effective field (see the definition in Eq. (5)) created by such a configuration then reads

$$h(m; s) = -\frac{\partial}{\partial m} g(m; s) , \tag{10}$$

at the leading order in the large N limit.

As the transverse field $\Gamma(s)$ is also conjugated to the total (transverse) magnetization, such quantum models are invariant under any permutation of the spins (or in more physical terms their Hamiltonians commute with the total spin), which greatly simplifies their analysis.

3. Evolution equations

3.1. Differential equations

As we shall see now the high level of symmetry in the definition of these models allows one to reduce the study of their dynamics, in the thermodynamic limit $N \to \infty$ with the annealing time $\mathcal{T}$ kept finite, to simple differential equations, for the three type of dynamics introduced above and arbitrary time dependencies of their parameters.

3.1.1. Thermal annealing If the initial condition $P(\underline{\sigma}; s = 0)$ of the master equation (2) is invariant under the permutation of the sites (i.e. only depends on the magnetization of the initial condition), then this property is conserved along the dynamical evolution (even for finite N), allowing one to deduce a master equation for $P(m; s)$. In addition in the thermodynamic limit this probability distribution concentrates around its most likely value $m(s)$ (if the initial condition is not too wide), the latter being solution of a simple, deterministic, differential equation:

$$\frac{1}{\mathcal{T}} \frac{\mathrm{d}m}{\mathrm{d}s} = \tanh(\beta h(m(s); s)) - m(s) . \tag{11}$$

This expression, that can be easily obtained after a short computation, has a natural interpretation: when a spin is updated its average magnetization corresponds to that of a single spin at equilibrium at inverse temperature β, in an effective magnetic field created by the rest of the configuration and thus equal to $h(m(s); s)$.

3.1.2. Quantum annealing The Schrödinger evolution of such mean-field spin models was studied in details in [21] (see also [29, 30] for similar treatments of particle models), where it was shown that the thermodynamic limit leads to semi-classical Hamiltonian equation of motions for the typical magnetization $m(s)$ and its conjugated momentum $\widetilde{m}(s)$:

$$\begin{cases} \frac{1}{\mathcal{T}} \frac{\mathrm{d}}{\mathrm{d}s} m(s) = \frac{\partial}{\partial \widetilde{m}} \mathcal{H}(m(s), \widetilde{m}(s); s) \\ \frac{1}{\mathcal{T}} \frac{\mathrm{d}}{\mathrm{d}s} \widetilde{m}(s) = -\frac{\partial}{\partial m} \mathcal{H}(m(s), \widetilde{m}(s); s) \end{cases} , \tag{12}$$

where the effective Hamiltonian $\mathcal{H}$ is given in terms of the parameters of the model as

$$\mathcal{H}(m, \widetilde{m}; s) = g(m; s) - \Gamma(s) \sqrt{1 - m^2} \cos(2\widetilde{m}) . \tag{13}$$

The initial condition is $(m(s = 0), \widetilde{m}(s = 0)) = (0, 0)$.

3.1.3. Simulated quantum annealing Let us finally consider the case of the annealing of a Quantum Monte Carlo simulation. Its analysis is at first sight much more complicated than the thermal annealing one, as the configurations have an additional imaginary time dependence. It turns out however that in the large N limit the magnetization density concentrates around a typical value $m(s)$, independently of the imaginary time (i.e. the static approximation of [25] becomes exact), and that the correlations along this axis are irrelevant (see [26] for a detailed derivation of this statement). This magnetization density is found to obey a simple differential equation,

$$\frac{1}{\mathcal{T}}\frac{\mathrm{d}m}{\mathrm{d}s} = \frac{h(m(s);s)}{\sqrt{h(m(s);s)^2+\Gamma(s)^2}} \tanh(\beta\sqrt{h(m(s);s)^2+\Gamma(s)^2}) - m(s) \ , \tag{14}$$

as was already obtained in [25] for a particular choice of $h(m,s)$. As it should this equation reduces to (11) when the transverse field $\Gamma(s)$ vanishes at all times. In fact the intuitive explanation given after (11) remains valid here, as the first term in (14) is precisely the average value of the z-magnetization of a quantum spin 1/2 in a longitudinal field h and transverse field Γ.

3.2. Analysis in the large annealing time limit

For any finite value of $\mathcal{T}$ the evolution equations (11), (12) and (14) can be easily integrated numerically, for any annealing schedule. One can however make further analytical progresses by considering the limit $\mathcal{T} \to \infty$ of a very slow annealing (we recall that we have already taken the thermodynamic limit $N \to \infty$). As the thermal annealing is a particular case of the simulated quantum annealing we shall not discuss explicitly the thermal case, and contrast the behavior of the quantum and simulated quantum annealing.

Consider the evolution equation (14) of the latter. In the large $\mathcal{T}$ limit one can check that $m(s)$ will be close to $m_*(s)$, where $m_*(s)$ is such that the right-hand side of (14) vanishes. This condition is easily seen to be equivalent to $m_*(s)$ being an extremum of the free energy $f(m;s)$, a standard computation (see for example [21]) leading indeed to

$$f(m;s) = \underset{\lambda}{\mathrm{ext}} \left[g(m;s) + m\lambda - \frac{1}{\beta} \ln 2\cosh(\beta\sqrt{(1-s)^2+\lambda^2}) \right] \ . \tag{15}$$

Of course the free energy can have several extrema, in that case one has to be more precise to define $m_*(s)$. The rules to remove this ambiguity are as follows. $m_*(0)$ has to be chosen as the local minimum reached by a steepest-descent from the initial condition. Then $m_*(s)$ remains stuck in the same local minimum as long as the latter exists. At all bifurcations where $m_*(s)$ loses its stability, it flows to a new local minimum following a steepest-descent search of the free energy. For future use let us write down a more explicit form of the above free energy in the zero-temperature limit:

$$f(m;s) = g(m;s) - (1-s)\sqrt{1-m^2} \ . \tag{16}$$

Let us now turn to the analysis of the slow annealing limit of the quantum annealing equations (12). These have the form of Hamiltonian equations of motion, and their large $\mathcal{T}$ limit can thus be analyzed using the theory of adiabatic invariants of classical mechanics. The initial condition $(m(s=0), \widetilde{m}(s=0)) = (0,0)$ is a local minimum of the Hamiltonian $\mathcal{H}(m,\widetilde{m};s=0)$, hence for large annealing times $\mathcal{T}$ the solution of (12) remains close to $(m(s),\widetilde{m}(s)) = (m_*(s),0)$, as long as this point corresponds to a local minimum of $\mathcal{H}$. This is actually also the minimum of the zero-temperature free energy (16), hence the quantum annealing and the zero-temperature simulated quantum annealing coincide in this limit as long as no bifurcation is encountered. However a drastic difference between the two dynamics appears as soon as the local minimum

Figure 1. Sketch of the difference of behavior between the zero temperature simulated quantum annealing (orange circle) and the quantum annealing (red circle and wavelet) when crossing a spinodal point along their annealing.

loses its stability: the semi-classical dynamics describing the quantum annealing is indeed conservative, hence a non-zero value of the momentum $\widetilde{m}$ is induced at the bifurcation, and an oscillatory behavior for $m(s)$ in its potential energy well takes place. On the contrary the zero-temperature simulated quantum annealing flows to the newly accessible local minimum. This difference of behavior is illustrated on figure 1.

4. Conclusions

In this paper we have studied various types of annealing dynamics of quantum spin models, simple enough to be reducible to unidimensional problems, their magnetization playing the role of a particle coordinate. The main message to be emphasized is the difference between a quantum annealing (an evolution with Schrödinger equation) and a low temperature simulated quantum annealing (a time-varying Quantum Monte Carlo simulation) that appears when a bifurcation occurs in the structure of the effective energy potential along the annealing (see also [31] for a discussion of this difference from another point of view). This difference persists even in the limit of zero-temperature, which can be understood physically as a system coupled to a "zero-temperature heat bath" can release part of its energy, at variance with an isolated system. Microscopic descriptions of the coupling to the environment might thus be necessary to interpret the results of experiments on quantum annealers [12].

When the groundstate of such a model encounters a first-order quantum phase transition both quantum and simulated quantum annealing remains stuck in the metastable continuation of the groundstate between the transition and the spinodal point. Right after the spinodal the energy is strictly smaller in the simulated quantum annealing case, as the coupling to the heat bath provides a mechanism for energy relaxation. This could seem to suggest that including both thermal and quantum fluctuations is the best strategy to attain the lowest possible residual energy. This conclusion is however wrong in general, as the subsequent relative evolution of the depth and width of the multiple energy wells between the spinodal and the end of the annealing can in some cases reverse this ordering [26].

The long-term goal sketched in the introduction would be to understand these annealing dynamics in disordered mean-field systems that correspond to random optimization problems. The discussion presented here in terms of effective free energy is reminiscent of the quantum and dynamical TAP equations approach of [32, 33] for disordered fully-connected models. Such models, even if obviously much more intricate than the simple ones discussed here, share with the latter the possibility of first-order phase transitions accompanied by metastable states and spinodals (thanks to their mean-field character), one can thus hope for some level of universality for the mechanism encountered at spinodals. As an intermediate step in this program we shall present in [26] further results on the quantum random subcubes model [34, 35]; in the regime where the number of cubes is finite one can exploit the permutation symmetry between sites inside a few blocks of spins, which leads to a similar analysis of the annealing dynamics with the magnetization replaced by a finite dimensional vector of block magnetizations.

Let us finally emphasize that we concentrated here on a particular time regime, namely with the limit $\mathcal{T} \to \infty$ of long annealing times taken after the thermodynamic limit $N \to \infty$. In presence of metastability this time scale is however too short to reach the groundstate energy, the adiabatic time (for quantum or simulated annealing) growing exponentially fast with the system size. It is thus also important to consider the regime where $\mathcal{T}$ and N are taken to infinity simultaneously, with $\mathcal{T} = \exp[N\tau]$. We have shown in [21] how to compute the residual energy of the quantum annealing as a function of τ via the determination of the exponentially small avoided level crossings along the metastable branches of the eigenvalues of the Hamiltonian $\widehat{H}(s)$. In the case of simulated quantum annealing one can also investigate this exponential time regime by computing the height of the free energy barriers, the adiabatic time being controlled by these barriers through an Arrhenius law [26].

References

[1] Papadimitriou C H 1994 *Computational Complexity* (Addison-Wesley)
[2] Kirkpatrick S, Gelatt Jr C D and Vecchi M 1983 *Science* **220** 671–80
[3] Apolloni B, Carvalho C and de Falco D 1989 *Stoc. Proc. Appl.* **33** 233–44
[4] Finnila A B, Gomez M A, Sebenik C, Stenson C and Doll J D 1994 *Chem. Phys. Lett.* **219** 343–8
[5] Kadowaki T and Nishimori H 1998 *Phys. Rev.* E **58** 5355–63
[6] Brooke J, Bitko D, Rosenbaum T F and Aeppli G 1999 *Science* **284** 779–81
[7] Farhi E, Goldstone J, Gutmann S, Lapan J, Lundgren A and Preda D 2001 *Science* **292** 472–5
[8] Santoro G E and Tosatti E 2006 *J. Phys. A: Math. Gen.* **39** R393
[9] Das A and Chakrabarti B K 2008 *Rev. Mod. Phys.* **80** 1061–81
[10] Morita S and Nishimori H 2008 *J. Math. Phys.* **49** 125210 .
[11] Bapst V, Foini L, Krzakala F, Semerjian G and Zamponi F 2013 *Physics Reports* **523** 127–205
[12] Boixo S *et al.* 2013 Quantum annealing with more than one hundred qubits (*Preprint* arXiv:1304.4595 [quant-ph])
[13] Mézard M and Montanari A 2009 *Information, Physics and Computation* (New York: Oxford University Press)
[14] Cheeseman P, Kanefsky B and Taylor W M 1991 Where the really hard problems are *Proc. 12th IJCAI (San Mateo CA USA)* (Morgan Kaufmann) pp 331–7
[15] Cugliandolo L F and Kurchan J 1993 *Phys. Rev. Lett.* **71** 173–6
[16] Zdeborová L and Krzakala F 2010 *Phys. Rev.* B **81** 224205
[17] Franz S and Parisi G 2013 *Journal of Statistical Mechanics: Theory and Experiment* **2013** P02003
[18] Krzakala F and Zdeborová L 2013 *this volume.*
[19] Jörg T, Krzakala F, Kurchan J and Maggs A C 2008 *Phys. Rev. Lett.* **101** 147204
[20] Farhi E, Goldstone J, Gosset D, Gutmann S and Shor P 2011 *Quantum Information & Computation* **11** 840–54
[21] Bapst V and Semerjian G 2012 *Journal of Statistical Mechanics* **2012** P06007
[22] Santoro G E, Martonak R, Tosatti E and Car R 2002 *Science* **295** 2427
[23] Stella L, Santoro G E and Tosatti E 2006 *Phys. Rev.* B **73** 144302
[24] Griffiths R B, Weng C-Y and Langer J S 1966 *Phys. Rev.* **149** 301–5
[25] Inoue J I 2010 *Journal of Physics: Conference Series* **233** 012010
[26] Bapst V and Semerjian G 2013 in preparation
[27] Suzuki M 1976 *Progress of Theoretical Physics* **56** 1454–69
[28] Lipkin H J, Meshkov N and Glick A J 1965 *Nucl. Phys.* **62** 188–98
[29] Sciolla B and Biroli G 2011 *J. Stat. Mech.* **2011** P11003
[30] Schiró M and Fabrizio M 2010 *Phys. Rev. Lett.* **105** 076401
[31] Hastings M B and Freedman M H 2013 Obstructions to classically simulating the quantum adiabatic algorithm (*Preprint* arXiv:1302.5733 [quant-ph])
[32] Biroli G 1999 *Journal of Physics A: Mathematical and General* **32** 8365–88
[33] Biroli G and Cugliandolo L F 2001 *Phys. Rev.* B **64** 014206
[34] Mora T and Zdeborová L 2008 *J. Stat. Phys.* **131** 1121–38
[35] Foini L, Semerjian G and Zamponi F 2010 *Phys. Rev. Lett.* **105** 167204

An irreversible Markov-chain Monte Carlo method with skew detailed balance conditions

K Hukushima and Y Sakai

Department of Basic Science, Graduate School of Arts and Sciences, The University of Tokyo, 3-8-1 Komaba, Meguro-ku, Tokyo 153-8902

E-mail: hukusima@phys.c.u-tokyo.ac.jp

Abstract. An irreversible Markov-chain Monte Carlo (MCMC) method based on a skew detailed balance condition is discussed. Some recent theoretical works concerned with the irreversible MCMC method are reviewed and the irreversible Metropolis-Hastings algorithm for the method is described. We apply the method to ferromagnetic Ising models in two and three dimensions. Relaxation dynamics of the order parameter and the dynamical exponent are studied in comparison to those with the conventional reversible MCMC method with the detailed balance condition. We also examine how the efficiency of exchange Monte Carlo method is affected by the combined use of the irreversible MCMC method.

1. Introduction

Since Metropolis algorithm was introduced in 1953 [1], Monte Carlo (MC) methods have been used intensively in a wide area of physics [2], biology and statistical sciences [3, 4]. Most of the MC methods in statistical physics are on the basis of the Metropolis strategy, in which a Markov chain is constructed in order for its invariant distribution to coincide with the desired distribution. There exist various improvements on the Metropolis MC algorithms, which are mainly categorized into two directions, i.e., non-local updating methods such as cluster algorithm [5, 6] and extended ensemble methods [7]. Typical examples of the latter are the multicanonical method [8], the simulated tempering [9] and the exchange MC method [10] or parallel tempering. These methods have been applied to various complex systems, e.g. protein folding problems and spin glasses, and turned out to be quite useful for simulating these systems.

These Markov-chain Monte Carlo (MCMC) methods have been developed within the framework of the detailed balance condition (DBC). It is, however, not always necessary to construct the MCMC method using DBC. In fact, some MCMC methods without DBC have recently been proposed [11, 12, 13] and have attracted a great deal of attention as an alternative direction of the development of the MCMC method. Suwa and Todo [11, 14] have proposed a general constructing way for the MCMC method without DBC and have showed that the proposed method is able to bring about several times reduction in the correlation time of Potts model [15]. Turitsyn *et al.* [12] and Fernandes and Weigel [13] have proposed another type of MCMC method without DBC separately and they have found a qualitative improvement in efficiency of the MCMC method in a mean-field Ising model. While Peskun's theorem [16] gives a useful guiding principle for constructing an efficient MCMC method with DBC, no such a principle has been established in the case of the MCMC method without DBC. Therefore, it would be worth investigating the MCMC method without DBC in some model systems.

In the present paper, we pay our attention to an irreversible MCMC method, particularly with a skew detailed balance condition (SDBC) [12]. We make a review on the method in which its basic idea and an implementation of the algorithm are described. We also present our numerical results applying the irreversible MCMC method with SDBC to Ising models in two- and three dimensions.

2. An irreversible Markov-chain Monte Carlo algorithm

2.1. Irreversible MCMC and skew detailed balance conditions

In this section, we review an irreversible Markov-chain Monte Carlo method based on the skew detailed balance condition originally proposed by Turitsyn, Chertkov and Vucelja [12] in 2011.

In the present paper, we focus our attention to N Ising spin systems for simplicity, although it is straightforward to extend the method to any discrete state models [17] such as the Potts model [15]. A state of the Ising model is specified by a vector $\boldsymbol{\sigma} = (\sigma_1, \ldots, \sigma_N)$ with $\sigma_j \in \{-1, +1\}$ for $j = 1, \ldots, N$. The target distribution $\pi(\boldsymbol{\sigma})$ for finding the state $\boldsymbol{\sigma}$ in the statistical physics is often proportional to the Boltzmann factor $\exp(-\beta\mathcal{H}(\boldsymbol{\sigma}))$ where β is an inverse temperature and $\mathcal{H}$ is the Hamiltonian to be studied. The main aims of the MCMC methods are to generate samples of the state from the target distribution $\pi(\boldsymbol{\sigma})$, and to calculate an expectation value of a function $\hat{f}$ under the target distribution, e.g. $\langle \hat{f} \rangle_\pi = \sum_{\boldsymbol{\sigma}} \pi(\boldsymbol{\sigma}) f(\boldsymbol{\sigma})$ where $f(\boldsymbol{\sigma})$ is the realization of $\hat{f}$ with the state $\boldsymbol{\sigma}$ and $\sum_{\boldsymbol{\sigma}}$ denotes the summation over 2^N states.

In order to utilize the irreversible MCMC scheme by the SDBC, the state space is doubled by introducing an additional Ising variable $\epsilon \in \{-1, +1\}$. The state in the enlarged state space is denoted by $X = (\boldsymbol{\sigma}, \epsilon) \in \{-1, +1\}^{N+1}$ and the corresponding probability distribution $\tilde{\pi}$ is assumed to be independent of the additional Ising spin ϵ:

$$\tilde{\pi}(\boldsymbol{\sigma}, \epsilon) = \tilde{\pi}(\boldsymbol{\sigma}, -\epsilon) = \frac{\pi(\boldsymbol{\sigma})}{2}. \tag{1}$$

We consider a single spin-flip update for both the original spin $\boldsymbol{\sigma}$ and the additional spin ϵ as an elementary process in the Markov chain. Let F_j be a spin-flip operator on the j-th site: $F_j\boldsymbol{\sigma} = (\sigma_1, \ldots, -\sigma_j, \ldots, \sigma_N)$ A transition rate per unit time from state $(\boldsymbol{\sigma}, \epsilon)$ to $(F_j\boldsymbol{\sigma}, \epsilon)$ is denoted as $w_j(\boldsymbol{\sigma}, \epsilon)$ and that from state $(\boldsymbol{\sigma}, \epsilon)$ to $(\boldsymbol{\sigma}, -\epsilon)$ is $\lambda(\boldsymbol{\sigma}, \epsilon)$. Using these transition rates, balance condition (BC) is expressed as

$$\sum_j w_j(F_j\boldsymbol{\sigma}, \epsilon)\tilde{\pi}(F_j\boldsymbol{\sigma}, \epsilon) - \sum_j w_j(\boldsymbol{\sigma}, \epsilon)\tilde{\pi}(\boldsymbol{\sigma}, \epsilon) + \lambda(\boldsymbol{\sigma}, -\epsilon)\tilde{\pi}(\boldsymbol{\sigma}, -\epsilon) - \lambda(\boldsymbol{\sigma}, \epsilon)\tilde{\pi}(\boldsymbol{\sigma}, \epsilon) = 0. \tag{2}$$

This ensures that $\tilde{\pi}$ is the unique invariant distribution of the Markov chain. In order to determine the transition rate $w_j(\boldsymbol{\sigma}, \epsilon)$, we impose the SDBC:

$$\tilde{\pi}(\boldsymbol{\sigma}, \epsilon)w_j(\boldsymbol{\sigma}, \epsilon) = \tilde{\pi}(F_j\boldsymbol{\sigma}, -\epsilon)w_j(F_j\boldsymbol{\sigma}, -\epsilon). \tag{3}$$

This requires that the stochastic flow from state $(\boldsymbol{\sigma}, +\epsilon)$ to $(F_j\boldsymbol{\sigma}, +\epsilon)$ is balanced out by that from $(F_j\boldsymbol{\sigma}, -\epsilon)$ to $(\boldsymbol{\sigma}, -\epsilon)$. In general, this condition breaks the detailed balance conditions (DBC): $\tilde{\pi}(\boldsymbol{\sigma}, \epsilon)w_j(\boldsymbol{\sigma}, \epsilon) = \tilde{\pi}(F_j\boldsymbol{\sigma}, \epsilon)w_j(F_j\boldsymbol{\sigma}, \epsilon)$. As a specific solution of (3), the transition rate $w_j(\boldsymbol{\sigma}, \epsilon)$ we use throughout the present work is given by

$$w_j(\boldsymbol{\sigma}, \epsilon) = \frac{1}{2}\alpha\,(1 - \sigma_j \tanh\beta h_j)\,(1 - \delta\epsilon\sigma_j), \tag{4}$$

where α is a time constant and h_j is a local field acting on the site j. The possible range of δ is -1 to 1 and DBC is recovered in (4) with $\delta = 0$. The transition rate is equivalent to the conventional heat-bath transition rate under the virtual external field ϵH with $H = \frac{1}{\beta}\,\mathrm{arctanh}\,\delta$. Thus, the additional spin ϵ represents the direction of the virtual field. While the additional spin ϵ is coupled to the local order parameter σ_j in this transition rate, one can replace it with any other linear function of σ_j such as a local energy $\sigma_j h_j$. The choice of the transition rate might affect the efficiency of the MCMC method, depending on the model system to be studied, but this has not been clarified yet.

By using SDBC, BC is rewritten as

$$\lambda(\boldsymbol{\sigma}, \epsilon) - \lambda(\boldsymbol{\sigma}, -\epsilon) = \sum_j \left(w_j(\boldsymbol{\sigma}, -\epsilon) - w_j(\boldsymbol{\sigma}, \epsilon)\right). \tag{5}$$

The explicit form of the transition rate for ϵ flip is not unique. Turitsyn *et al.* [12] have proposed the transition rate as

$$\lambda(\boldsymbol{\sigma}, \epsilon) = \max\left(0, \sum_j \left(w_j(\boldsymbol{\sigma}, -\epsilon) - w_j(\boldsymbol{\sigma}, \epsilon)\right)\right), \tag{6}$$

which is referred to as the Turitsyn-Chertkov-Vucelja (TCV) type. Another type of $\lambda(\boldsymbol{\sigma}, \epsilon)$ is also given as

$$\lambda(\boldsymbol{\sigma}, \epsilon) = \sum_j w_j(\boldsymbol{\sigma}, -\epsilon), \tag{7}$$

which is referred to as the Sakai-Hukushima 1 (SH_1) type [18]. These transition rates are available for a general class of the Ising models.

2.2. Irreversible Metropolis-Hastings algorithm

In this subsection, we explain an actual procedure in MCMC simulations which is based on Metropolis-Hastings algorithm [19]. Let $X^{(n)}$ be the state in the enlarged state space after n iterations. The irreversible MCMC method starts with an arbitrary initial state $X^{(0)}$ and iterates the following steps for $n = 1, 2, \dots$:

(a) Suppose that the current state $X^{(n)} = (\boldsymbol{\sigma}, \epsilon)$ and choose a site j at random.

(b) Accept the new state as $X^{(n+1)} = (F_j\boldsymbol{\sigma}, \epsilon)$ with the probability $w_j(\boldsymbol{\sigma}, \epsilon)$.

(c) If it is rejected, accept the ϵ flipped state as $X^{(n+1)} = (\boldsymbol{\sigma}, -\epsilon)$ with an acceptance rate

$$A(\epsilon \to -\epsilon; \boldsymbol{\sigma}) = \frac{\frac{1}{N}\lambda(\boldsymbol{\sigma}, \epsilon)}{1 - \frac{1}{N}\sum_j w_j(\boldsymbol{\sigma}, \epsilon)}. \tag{8}$$

(d) If it is also rejected, set $X^{(n+1)} = X^{(n)}$. Return to (a) and repeat the steps (a)–(d).

It is proved that these steps satisfy the BC [17]. One MC step is defined as N iterations of the steps (a)–(d). To evaluate the acceptance rate in step (c), the summation with respect to the site is necessary and its computational complexity is of the order of N. In practice, once the summation is evaluated at the initial condition, it is sufficient to update the value of the summation when the spin-flip process is accepted. The complexity for the update is of the order of 1 in statistical-mechanical models with short range interactions.

2.3. Irreversible Glauber dynamics in one-dimensional Ising model

Although the MCMC methods are powerful simulation tools, it is helpful to understand analytically dynamical properties of the Markov chain in a large probabilistic model in statistical mechanics. A kinetic Ising model [20] in one dimension is a simplest model, exactly solved in the case of Markov-chain dynamics with DBC [21]. The correlation length ξ of the one-dimensional Ising model is given by $\xi(\beta) = (-\log(\tanh \beta J))^{-1}$, which diverges at zero temperature. Time evolution of the order parameter and also its autocorrelation function exhibit an exponential decay in the model [21]. The relaxation time is obtained as $\tau \propto (1 - \tanh 2\beta J)^{-1}$. For $\beta \to \infty$, a dynamical scaling relation, $\tau \sim \xi^z$, holds with the dynamical exponent $z = 2$.

Recently, the kinetic Ising model under an irreversible Markov chain with SDBC has been studied [18]. For the SH_1 type transition rate of (7), the magnetization relaxation is solved and then the relaxation time is exactly obtained as

$$\tau \propto \frac{1}{(1+\delta^2)(1-\gamma)}, \tag{9}$$

with $\gamma = \tanh 2\beta J$, meaning that the reduction in the relaxation time from DBC is constant $1 + \delta^2$.

Another transition rate for the ϵ flip specific to the one-dimensional Ising model is proposed as

$$\lambda(\boldsymbol{\sigma}, \epsilon) = \frac{1}{2}\alpha\delta(1 - \tanh 2\beta J)(1 + \frac{\epsilon}{N}\sum_i \sigma_i), \tag{10}$$

which is called as SH_2 type [18]. The transition rate combined with (4) also satisfies BC. Under a plausible assumption which is valid in the vicinity of the equilibrium state, the time evolution of the order parameter is also solved and its relaxation time is obtained as

$$\tau \propto \frac{1}{1 - \gamma + \delta\sqrt{1-\gamma^2}}, \tag{11}$$

which implies that z is down to 1. The result analytically showed that the specific choice of the irreversible transition rate is able to change the dynamical critical phenomena drastically.

3. Numerical results

In this section we present some numerical results obtained from applications of the irreversible MCMC method with SDBC explained in the previous section to some Ising models in high dimensions. The model Hamiltonian is given as

$$\mathcal{H}(\boldsymbol{\sigma}) = -J\sum_{\langle ij \rangle} \sigma_i \sigma_j, \tag{12}$$

where spins are defined on the sites of a square lattice in two dimensions and of a cubic lattice in three dimensions. The summation is over all the nearest-neighbor pairs in the lattice and the periodic boundary condition is imposed in all directions. The strength J is set to the unit of temperature.

We measure the relaxation of order parameter

$$m(t) = \langle \hat{m} \rangle_t, \tag{13}$$

where $\hat{m} = \frac{1}{N}\sum_i \sigma_i$ and $\langle \cdots \rangle_t$ denotes a dynamical average at the MC step t starting from an initial condition. At the initial state with $t = 0$, all the spins are fixed to $+1$ and spin ϵ takes ± 1 at random. It is argued that the transition rate of the SH_1 type reduces only the time constant of the magnetization relaxation in general Ising models in comparison to that of the reversible MCMC method with DBC [18]. As a nontrivial example, we use the transition rate of the TCV type with $\alpha = 1/(1 + \delta)$ and the parameter δ in (4) is set to 0.9.

3.1. Dynamical exponent in irreversible MCMC

Recent numerical studies have suggested that an irreversible MCMC method is able to reduce the dynamical exponent z of a mean-field Ising model [12, 13]. Further, as discussed in section 2.3, it is analytically shown in the one-dimensional Ising model that a specific type of the transition rate satisfying SDBC changes the value of z from 2 as in DBC to 1. This is the changeover from diffusive to ballistic relaxation of the order parameter. However, the irreversible MCMC method does not always make such a changeover. In fact, the transition rate of the SH_1 type for the ϵ flip reduces only a finite factor in the relaxation time independent of temperature, but not the dynamical exponent.

In this subsection, we study the dynamical exponent of the Ising model in two and three dimensions by performing the irreversible MCMC method with SDBC explained above. A standard way for evaluating the value of z is to estimate the autocorrelation time τ for different sizes L at the critical temperature and to fit to the power-law formula as $\tau \sim L^z$. We use here a simpler method called non-equilibrium relaxation method [22], which is another tool for evaluating the critical exponents through a dynamical process from a non-equilibrium initial state to the equilibrium one. According to the non-equilibrium relaxation method, it is sufficient to measure a dimensionless quantity

$$f_{mm}(t) = N\left[\frac{\langle \hat{m}^2 \rangle_t}{\langle \hat{m} \rangle_t^2} - 1\right], \tag{14}$$

to determine the dynamical exponent. Assuming the dynamical scaling ansatz for $\langle \hat{m} \rangle_t$ and $\langle \hat{m}^2 \rangle_t$, the function $f_{mm}(t)$ is expected to exhibit power-law divergence in a large t limit at the critical temperate as $f_{mm}(t) \sim t^{\lambda_{mm}}$. With the help of the hyper-scaling relation, the exponent λ_{mm} is expressed as d/z with d being the spatial dimensionality.

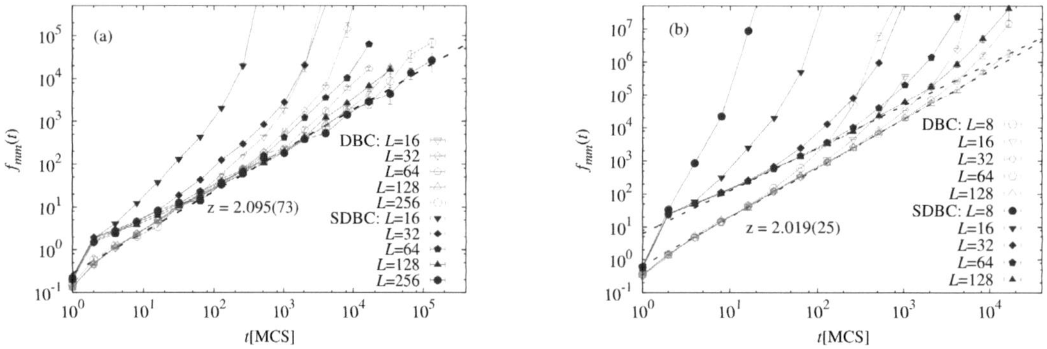

Figure 1. MC-step dependence of the magnetization correlation function $f_{mm}(t)$ at $T = T_c$ of the Ising models in two (a) and three dimensions (b). The asymptotic slope of $f_{mm}(t)$ provides an estimate of d/z where d is the spatial dimensionality and z the dynamical exponent. The data with DBC and SDBC are marked by open (red) and filled (blue) symbols, respectively.

Numerical results of $f_{mm}(t)$ for two and three dimensions are presented in figure 1 (a) and (b), respectively. The system sizes studied are from $L = 16$ to 256 in two dimensions and from 8 to 128 in three dimensions. The asymptotic slope of $f_{mm}(t)$ in the figure gives an estimate of d/z. While finite-size effect is found for both DBC and SDBC, that for SDBC is relatively large. In two dimensions, the difference between them decreases in the large t behaviour with the system size increasing and eventually two limiting functions will merge into the same curve. Fitting the slope gives $z = 2.095(75)$, which is consistent with the established estimate, $z_{2d} = 2.165(5)$, obtained by the reversible MCMC simulations [22]. In three dimensions, the envelope curve independent of the size has a different slope in the two methods within the observed time window. The estimated value from DBC simulation, $z = 2.019(25)$, is marginally consistent with the recent value $z_{3d} = 2.055(10)$ [22]. The gradual slope for SDBC gives an effective value of z larger than that with DBC. This is presumably due to a non-asymptotic effect and the two estimates would coincide with each other asymptotically. Thus, no quantitative reduction of the dynamical exponent is achieved for the Ising model in both two and three dimensions by the present irreversible MCMC method with the TCV type. This is in contrast to the mean-field Ising model [12] and the one-dimensional Ising model with the SH_2 type [18].

3.2. Order-parameter relaxation in two-dimensional Ising model

In this subsection, we see relaxation dynamics of the magnetization at and off critical temperature. We show the magnetization $m(t)$ obtained with SDBC and DBC as a function of the MC step in the two-dimensional Ising model at $T = 2.4$ above T_c (figure 2 (a)) and at $T = T_c = 2.2691...$ (figure 2 (b)). For relatively smaller L, $m(t)$ by SDBC decays faster than that by DBC at these temperatures. For example, in the case with $L = 8$, a time constant of $m(t)$ for SDBC is about 10 times shorter than that for DBC. However, the gain of SDBC decreases with increasing the size. While we still have a finite gain of SDBC at $T = 2.4$ even in the large N limit, it disappears at T_c, which is consistent with no reduction of z seen in the previous section. A similar behaviour is found below T_c and also in the three-dimensional Ising model. This indicates that the efficiency of the irreversible MCMC method with the TCV type depends on temperature and it is enhanced at off-critical temperatures.

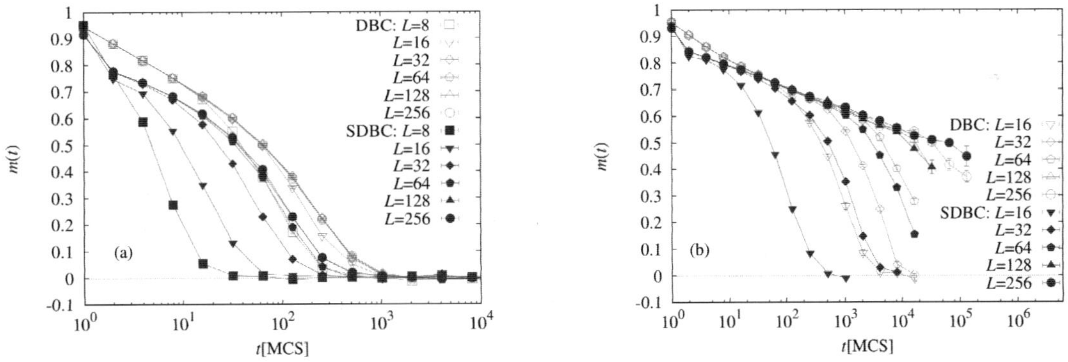

Figure 2. MC-step dependence of the averaged magnetization of the ferromagnetic Ising model in two dimensions at (a) $T/J = 2.4$ and (b) $T/J = 2.2691....$ The data with DBC and SDBC are marked by open (red) and filled (blue) symbols, respectively.

3.3. Efficiency of exchange Monte Carlo method with irreversible MCMC update

Since the extended ensemble MC methods have been applied to various problems, one may consider that the efficiency of the MC methods does not seriously depend on the local update scheme. The effect of the local update on the extended ensemble MC methods has not been extensively studied. Recently, it has been pointed out that the average round-trip time of the exchange MC method, which is an important factor in determining the efficiency, is reduced by an appropriate choice of frequency of the local update [23] and the use of efficient local update [24]. The latter case also improved significantly the performance of the Wang-Landau method [25, 26]. We therefore study the efficiency of the exchange MC method when the irreversible MCMC method is used as the local update method.

In the exchange MC method, MCMC simulations for M replicated Ising systems with different temperatures are performed by some local update method in parallel and an exchange process of states in the replicated systems is applied to every two replicas at adjacent temperatures. The model examined is the ferromagnetic Ising model in three dimensions. The number of replicas is fixed to 32 and the highest and the lowest temperatures are chosen as $4J$ and $7J$, respectively, between which the critical temperature lies. The intermediate temperatures are determined so that the acceptance ratio $p_{\rm ac}$ of the exchange process is almost independent of temperature for the largest system size attained in this work, $N = 32^3$. The same set of temperatures is used in our simulations with two different local updates and for different sizes. An elementary time scale in the exchange MC method is characterized by the round-trip time, also called the ergodic time, which is defined by the average MC steps for each replica to move from the highest to the lowest temperatures and return to the highest one.

The acceptance ratio $p_{\rm ac}$ with the conventional MCMC method with DBC and that with SDBC are shown as a function of temperature in figure 3. It is found that these two acceptance ratios completely coincide with each other for different temperatures and sizes, meaning that the local update scheme has no influence on the acceptance ratio of the exchange process. However, as shown in figure 4, there is some substantial difference in the round-trip time. Namely, the round-trip time in the MCMC method with SDBC is a few times smaller than that with DBC. This result shows that the irreversible MCMC method improves the efficiency of the exchange MC method at least for practical values of N, while it is not clear at present for an asymptotic large N.

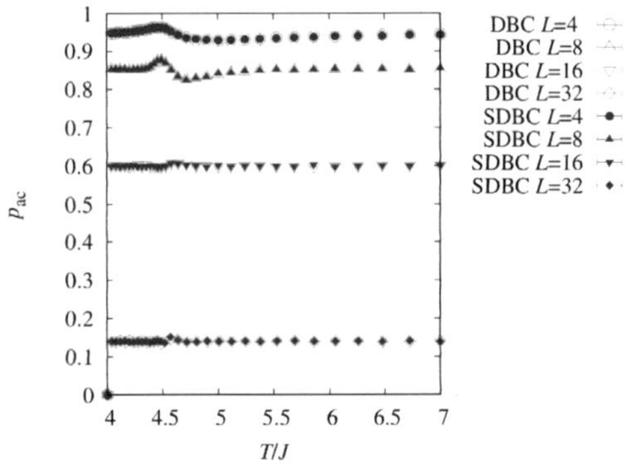

Figure 3. Temperature dependence of the acceptance ratio $p_{\rm ac}$ of the replica exchange process of the three-dimensional Ising model for $L = 4, 8, 16$ and 32. The data with DBC and SDBC are marked by open (red) and filled (blue) symbols, respectively.

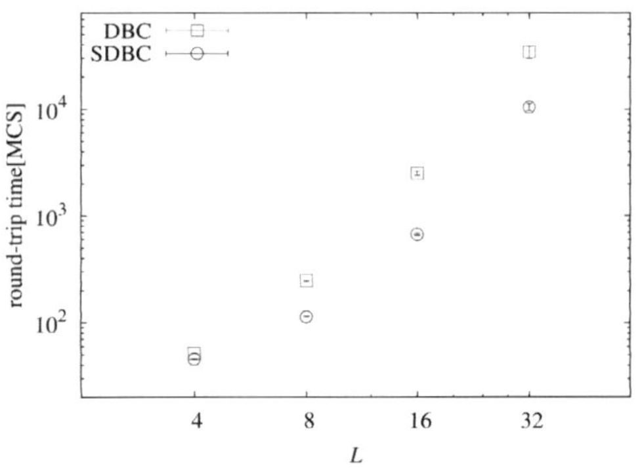

Figure 4. Linear-size L dependence of round-trip time in the exchange Monte Carlo method of the three-dimensional Ising model. The data with DBC and SDBC are denoted by squares and circles, respectively.

It may not be trivial that the round-trip time depends on the choice of the local update scheme in spite of the insensitivity of $p_{\rm ac}$. In order to understand the reason for such a behaviour, we study a typical example of the exchange process as a function of the MC step. It is clearly seen in figure 5 (a) that a bottleneck in the replica move occurs around the critical temperature T_c in the MCMC method with DBC. Although the exchange trial is accepted at around T_c, the replica coming across T_c often goes back to the former temperature in the following few MC steps. Thus, it takes a long round-trip time while the acceptance ratio $p_{\rm ac}$ is independent of temperature on average. In contrast, such a bottleneck phenomenon is not found in the MCMC method with SDBC.

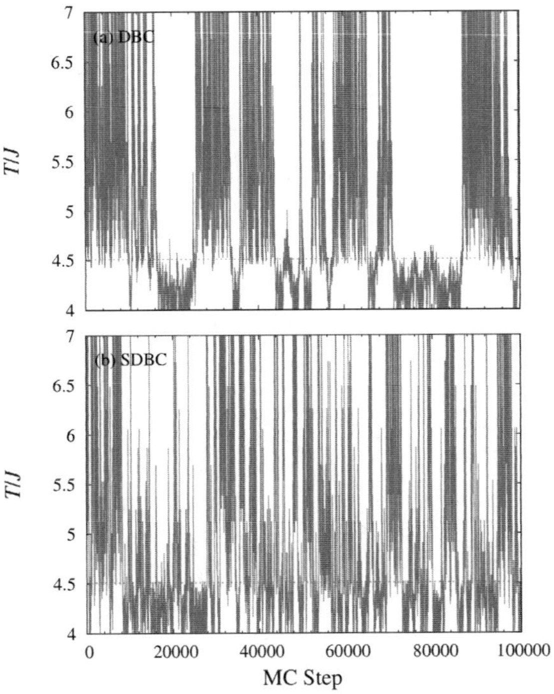

Figure 5. Typical trajectory of temperature for a specific replica as a function of Monte Carlo step in the exchange Monte Carlo method of a three-dimensional Ising ferromagnetic model. A trajectory obtained by the MC method with DBC (a) and one by the MC method with SDBC (b) are shown. The system size examined in these simulations is 16^3. The dotted line represents the critical temperature, $T_c \simeq 4.51$.

4. Summary

In the present paper, we have discussed an irreversible MCMC method with a skew detailed balance condition, which breaks the detailed balance condition used in conventional MCMC methods. The method is applied to some Ising models in dimensions more than one, which exhibit a finite-temperature phase transition. Although the method is able to reduce a correlation time in the Markov chain at off-critical temperature, unfortunately no significant reduction of the dynamical exponent is found, that is in sharp contrast to that observed in the mean-field Ising model and one-dimensional Ising chain. We also show that the efficiency of the exchange MC method is improved by the use of the irreversible MCMC method with SDBC as a local update scheme. At present, it is not well understood how the irreversible MCMC methods affect the dynamics of the Markov chain. Therefore, we consider that further theoretical studies are needed to clarify the efficiency of the irreversible MCMC methods.

Acknowledgments

The numerical simulation in the present paper has partially been performed by using the facility of the Supercomputer Center, Institute for Solid State Physics, University of Tokyo. This research was supported by the Grants-in-Aid for Scientific Research from the JSPS, Japan, No. 22340109 and No. 25610102.

References

[1] Metropolis N, Rosenbluth A W, Rosenbluth M N, Teller A H and Teller E 1953 *J. Chem. Phys.* **21** 1087–92

[2] Landau D P and Binder K 2005 *A Guide to Monte Carlo Simulations in Statistical Physics* (Cambridge: Cambridge University Press) 2nd ed
[3] Liu J S 2001 *Monte Carlo Strategies in Scientific Computing* (New York: Springer)
[4] Doucet A, De Freitas N and Gordon N (eds) 2001 *Sequential Monte Carlo Methods in Practice* (New York: Springer-Verlag)
[5] Swendsen R H and Wang J S 1987 *Phys. Rev. Lett.* **58** 86–8
[6] Wolff U 1989 *Phys. Rev. Lett.* **62** 361–4
[7] Iba Y 2001 *Int. J. Mod. Phys.* C **12** 623–56
[8] Berg B A and Neuhaus T 1992 *Phys. Rev. Lett.* **68** 9–12
[9] Marinari E and Parisi G 1992 *Europhys. Lett.* **19** 451–8
[10] Hukushima K and Nemoto K 1996 *J. Phys. Soc. Jpn.* **65** 1604–8
[11] Suwa H and Todo S 2010 *Phys. Rev. Lett.* **105** 120603
[12] Turitsyn K S, Chertkov M and Vucelja M 2011 *Physica* D **240** 410–4
[13] Fernandes H C M and Weigel M 2011 *Comput. Phys. Comm.* **182** 1856–9
[14] Todo S and Suwa H 2013 ICSG2013 proceedings
[15] Wu F Y 1982 *Rev. Mod. Phys.* **54** 235–68
[16] Peskun P H 1973 *Biometrika* **60** 607–12
[17] Sakai Y and Hukushima K in preparation.
[18] Sakai Y and Hukushima K 2013 *J. Phys. Soc. Jpn.* **82** 064003
[19] Hastings W K 1970 *Biometrika* **57** 97–109
[20] Kawasaki K 1972 *Phase Transitions and Critical Phenomena* vol 2 ed Domb C and Green M S (New York: Academic Press) p 443
[21] Glauber R J 1963 *J. Math. Phys.* **4** 294–307
[22] Ozeki Y and Ito N 2007 *J. Phys. A: Math. Theor.* **40** R149–R203
[23] Bittner E, Nußbaumer A, and Janke W 2008 *Phys. Rev. Lett.* **101** 130603
[24] Sasaki M and Hukushima K 2013 *J. Phys. Soc. Jpn.* **82** 094003
[25] Wang F and Landau D P 2001 *Phys. Rev. Lett.* **86** 2050–3
[26] Wang F and Landau D P 2001 *Phys. Rev.* E **64** 056101

Geometric allocation approaches in Markov chain Monte Carlo

S Todo[1] **and H Suwa**[2]

[1] Institute for Solid State Physics, University of Tokyo, 7-1-26-R501 Port Island South, Kobe 650-0047, Japan

[2] Department of Physics, Boston University, 590 Commonwealth Avenue, Boston, Massachusetts 02215, USA

E-mail: wistaria@issp.u-tokyo.ac.jp

Abstract. The Markov chain Monte Carlo method is a versatile tool in statistical physics to evaluate multi-dimensional integrals numerically. For the method to work effectively, we must consider the following key issues: the choice of ensemble, the selection of candidate states, the optimization of transition kernel, algorithm for choosing a configuration according to the transition probabilities. We show that the unconventional approaches based on the geometric allocation of probabilities or weights can improve the dynamics and scaling of the Monte Carlo simulation in several aspects. Particularly, the approach using the irreversible kernel can reduce or sometimes completely eliminate the rejection of trial move in the Markov chain. We also discuss how the space-time interchange technique together with Walker's method of aliases can reduce the computational time especially for the case where the number of candidates is large, such as models with long-range interactions.

1. Introduction

The Monte Carlo method for high-dimensional problems has a wide variety of applications as a versatile and interdisciplinary computational tool [1]. There are many interesting topics on strongly correlated systems in physics, e.g., the phase transitions and novel exotic phases, where the dimension of the state space for capturing the essential physics is often more than hundreds of thousand. The Markov chain Monte Carlo (MCMC) method that overcomes the curse of dimensionality has been effectively applied to many problems in high dimensions [2].

In principle, the MCMC method achieves a statistically exact importance sampling from any probability distribution function by introducing a special kind of random walk in the state space. The MCMC method keeps the sampling efficiency reasonably high even in high-dimensional problems. Instead of the curse of dimensionality, however, the MCMC method often suffers from the sample correlation. Since the next configuration is generated (updated) from the previous one, the samples are not independent of each other. Then the correlation gives rise to two problems; we have to wait for the distribution convergence (equilibration) before sampling, and the number of effective samples is decreased. The former convergence problem is quantified by the distance to the target distribution [3] or the spectral gap of the transition kernel. As an assessment for the latter problem, the decrease of the number of effective samples, the integrated autocorrelation time is defined as

$$\tau_{\text{int}} = \sum_{t=1}^{\infty} C(t) = \sum_{t=1}^{\infty} \frac{\langle O_{i+t} O_i \rangle - \langle O \rangle^2}{\langle O^2 \rangle - \langle O \rangle^2}, \tag{1}$$

where O_i is an observable at the i-th Monte Carlo step, and $C(t)$ is ideally independent of i after the distribution convergence. In terms of the autocorrelation time, the number of effective samples is roughly given by

$$M_{\text{eff}} \simeq \frac{M}{1 + 2\tau_{\text{int}}}, \tag{2}$$

where M is the total number of samples, i.e., the total Monte Carlo steps, in the simulation. Although an MCMC method satisfying the appropriate conditions, the balance condition and the ergodicity, guarantees asymptotically correct results in principle [4], variance reduction of relevant estimators is crucial for the method to work in practice. If the central limit theorem holds, the variance of expectations decreases as $v/M \simeq \text{var}(f)/M_{\text{eff}}$, where v is called the asymptotic variance that depends on the integrand function and the update method through the autocorrelation time.

In general, we should take the following four key points into account for achieving efficient updates in the MCMC method:

(i) Choice of ensemble and assignment (or modification) of the weights of configurations.

(ii) Generation of set of candidate configurations from the current configuration.

(iii) Construction of the transition kernel (probability), given a set of candidate configurations.

(iv) Algorithm for choosing a configuration among the candidates according to the transition probability.

As for (i), a series of extended ensemble methods, such as the multicanonical method [5], Wang-Landau method [6], simulated tempering [7], and exchange Monte Carlo method [8], have been proposed and successfully applied to the protein folding problems, the spin glasses, etc. The cluster algorithms, e.g., the Swendsen-Wang algorithm [9] and the loop algorithm [10], which overcome the critical slowing down near the critical point (the correlation-time growth in a power-law form of the system size) by taking advantage of mapping to graph configurations [11], are the representative improvements from the standpoint of (ii). The hybrid (Hamiltonian) Monte Carlo method [12] that performs a simultaneous move by generating the candidate state by a Newtonian dynamics with an artificial momentum is another successful example of (ii).

On the other hand, the key points (iii) and (iv) in the above list have *not* been studied in great depth so far. In this paper, we propose new approaches that are based on geometrical procedure on probabilities or weights. As we will show below, our unconventional approaches can improve the dynamics and scaling of the Monte Carlo simulation from the standpoints (iii) and (iv).

2. Markov chain Monte Carlo and balance condition

In the MCMC method, the *ergodicity* (irreducibility) of the Markov chain guarantees the consistency of estimators; the Monte Carlo average asymptotically converges in probability to a unique value irrespective of the initial configuration. The *total balance*, the invariance of the target distribution, is usually imposed though some interesting adaptive procedures have been proposed these days. Since the invention of the MCMC method in 1953 [13], the reversibility, namely the *detailed balance*, has been additionally imposed in most practical simulations as a sufficient condition for the total balance. Under the detailed balance, every elementary transition is forced to balance with a corresponding inverse process. Thanks to this condition, it becomes practically easy to find qualified transition probabilities in actual simulations. The standard update methods, such as the Metropolis (Metropolis-Hastings) algorithm [13, 14] and the heat bath algorithm (Gibbs sampler) [15], satisfy the reversibility. The performance of these seminal update methods has been analytically and numerically investigated in many papers.

There is a simple theorem about the reversible kernel as a guideline for the optimization of the transition matrix. Now we define an order of the matrices as $P_2 \geq P_1$ for any two transition matrices if each of the off-diagonal elements of P_2 is grater than or equal to the corresponding off-diagonal elements of P_1. The following statement is Theorem 2.1.1 of [16].

> **[Peskun (1973)]** Suppose each of the irreducible transition matrices P_1 and P_2 is reversible for a same invariant probability distribution π. If $P_2 \geq P_1$, then, for any f,
>
> $$v(f, P_1, \pi) \geq v(f, P_2, \pi), \tag{3}$$
>
> where
>
> $$v(f, P, \pi) = \lim_{M \to \infty} M \operatorname{var}(\hat{I}_M), \tag{4}$$
>
> and $\hat{I}_M = \sum_{i=1}^{M} f(x_i)/M$ is an estimator of $I = E_\pi(f)$ using M samples, $x_1, x_2, ..., x_M$, of the Markov chain generated by P.

According to this theorem, a modified Gibbs sampler, called "Metropolized Gibbs sampler," was proposed [17, 18]. By the usual Gibbs sampler, we choose the next state with forgetting the current state. By the Metropolized version, on the other hand, a candidate is chosen except the current state and it will be accepted/rejected by using the Metropolis scheme. This modified Gibbs sampler is reduced not to the usual Gibbs sampler but to the Metropolis algorithm in the case where the number of candidates is two. It is proved that the modified Gibbs sampler has a smaller asymptotic variance v than the original one [17, 18]. Furthermore, an iterative version of the Metropolized Gibbs sampler was proposed as well [17]. What we have learned from the Peskun's theorem is the guideline that the rejection rate (the diagonal elements of the transition matrix) should be minimized in general.

As we see, most of optimizations of the transition matrix have been proposed within the detailed balance. However, the reversibility is not a necessary condition for the invariance of the target distribution. The sequential update, where the state variables are swept in a fixed order, breaks the detailed balance, but can satisfy the total balance [19]. In the meanwhile, some modifications of reversible chain into an irreversible chain have been proposed so far. One example is a method of the duplication of the state space with an additional variable, such as a direction on an axis [20, 21, 22], and it has been shown that the autocorrelation time can be reduced drastically at least for specific models. The axis can be a combination of the state variables, the energy, or any quantity. The extended version with the multi-axes has been applied to some physical models [23]. Also the artificial momentum in the hybrid Monte Carlo method [12] performs partly as a direction in the state space. A similar idea with the addition of a direction has been proposed [24], where the next state in the Markov chain is generated depending on not only one step before but also the two (several) steps before. Then the resulting Markov chain can be irreversible because the history of the states has the direction. As other approaches, inserting a probability vortex in the state space was discussed [25], an asymmetric choice of the heading direction was applied in a hard-sphere system [26], and a global optimization of the transition matrix was discussed [27]. As seen above, the role of the net stochastic flow (irreversible drift) has caught the attention [28]. Note that the hybrid Monte Carlo method seems to break the detailed balance in the extended state space, but it is not essential because the additional update of the artificial momentum easily recovers the reversibility.

So far, most of the irreversible chains were based on the reversible update methods, such as the Metropolis algorithm. A more significant breaking of the reversibility can be achieved by applying the methods we will explain in this paper. In the following we introduce a new type of method breaking the detailed balance [29], which applies a geometric approach to solve the algebraic equations.

3. Geometric construction of irreversible kernel

In the MCMC method for a lattice system, the configuration is locally updated, and the huge transition kernel or matrix is implicitly constructed by the consecutive local updates. Following [29], let us consider a local update of a discrete variable as an elementary process. Now we have n next candidate configurations including the current one. The weight of each configuration is given by w_i ($i = 1, ..., n$), to which the target probability measure π_i is in proportion. We introduce a quantity $v_{ij} = w_i p_{i \to j}$ that corresponds to the amount of (raw) stochastic flow from state i to j. The law of probability conservation and the total balance are expressed as

$$w_i = \sum_{j=1}^{n} v_{ij} \qquad \forall i \tag{5}$$

$$w_j = \sum_{i=1}^{n} v_{ij} \qquad \forall j, \tag{6}$$

respectively. The average rejection rate is written as $\sum_i v_{ii} / \sum_i w_i$. It is easily confirmed that the Metropolis algorithm with the flat proposal distribution gives

$$v_{ij} = \frac{1}{n-1} \min[w_i, w_j] \qquad i \neq j, \tag{7}$$

and the heat bath algorithm (Gibbs sampler) does

$$v_{ij} = \frac{w_i w_j}{\sum_{k=1}^{n} w_k} \qquad \forall i, j. \tag{8}$$

The both satisfy the above conditions (5) and (6). The reversibility is manifested by the symmetry under the interchange of the indices:

$$v_{ij} = v_{ji} \qquad \forall i, j. \tag{9}$$

The aim here is to find a set $\{v_{ij}\}$ that minimizes the average rejection rate (the diagonal elements of the transition matrix) under the conditions (5) and (6). The procedure for the task can be understood visually as *weight allocation*, where we move (or allocate) weight (v_{ij}) from i-th state to j-th box with the entire shape of the weight boxes kept intact (figure 1). We propose the following algorithm [29]:

(i) Choose a configuration with maximum weight among the candidates. If two or more configurations have the same maximum weight, choose one of them. In the following, we assume w_1 is the maximum without loss of generality. The order of the remaining weights does not matter.

(ii) Allocate the maximum weight w_1 to the next box ($i = 2$). If the weight still remains after saturating the box, reallocate the remainder to the next ($i = 3$). Continue until the weight is all allocated.

(iii) Allocate the weight of the first filled box (w_2) to the last partially filled box in step (ii). Continue the allocation likewise.

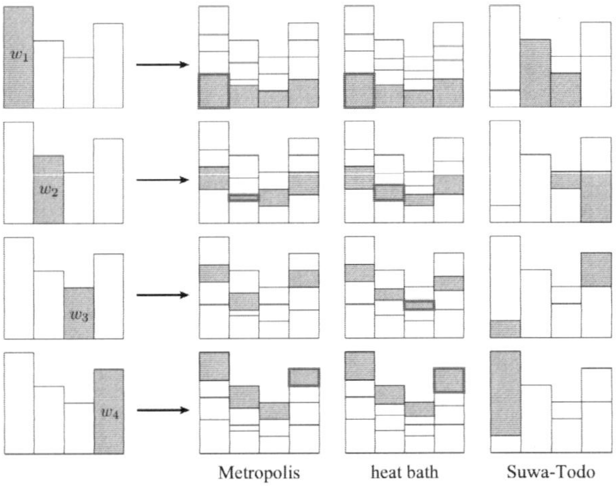

Figure 1. Example of weight landfill by the Metropolis, heat bath, and Suwa-Todo algorithms for $n = 4$. Our algorithm is rejection free, while there remain finite rejection rates in the conventional methods as indicated by the red thick frames (from [29]).

(iv) Repeat step (iii) for $w_3, w_4, \ldots w_n$. Once all the boxes with $i \geq 2$ are saturated, landfill the first box ($i = 1$) afterward.

By this procedure (the index 1 is such that w_1 has the maximum weight), $\{v_{i\to j}\}$ are determined as

$$v_{i\to j} = \max(0,\, \min(\Delta_{ij},\, w_i + w_j - \Delta_{ij},\, w_i,\, w_j)), \tag{10}$$

where

$$\begin{aligned} \Delta_{ij} &:= S_i - S_{j-1} + w_1 \qquad 1 \leq i,\, j \leq n & (11)\\ S_i &:= \sum_{k=1}^{i} w_k \qquad 1 \leq i \leq n & (12)\\ S_0 &:= S_n. & (13) \end{aligned}$$

It satisfies the conditions (5) and (6), but breaks the reversibility; for example, $v_{12} > 0$ but $v_{21} = 0$ in figure 1. The self-allocated weight that corresponds the rejection rate is expressed as

$$v_{ii} = \begin{cases} \max(0,\, w_1 - \sum_{i=2}^{n} w_i) & i = 1 \\ 0 & i \geq 2. \end{cases} \tag{14}$$

That is, a rejection-free solution is obtained if the condition

$$w_1 \leq \frac{S_n}{2} \equiv \frac{1}{2}\sum_{k=1}^{n} w_k \tag{15}$$

is satisfied. When it is not satisfied, the maximum weight has to be assigned to itself since it is larger than the sum of the rest. Thus, this solution is optimal in the sense that it minimizes the average rejection rate. Furthermore, the rejection rate expression (14) provides us a clear prospect; the rejection rate is certainly reduced as the number of candidates is increased. This idea was used to a quantum physical model and the rejection rate was indeed reduced to zero [29], and extended to the case with continuous variables [30].

In Figs. 2 and 3, the results of the benchmark test of our new method are presented. We investigate the convergence (equilibration) and the autocorrelations in the ferromagnetic q-state

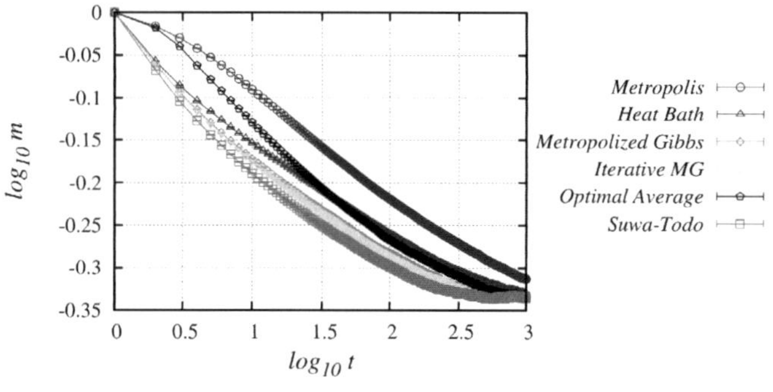

Figure 2. Convergence of magnetization squared in the ferromagnetic 4-state Potts model on the square lattice with $L = 32$ at the critical temperature. The horizontal axis is the Monte Carlo step. The simulation starts with the ordered (all "up") state.

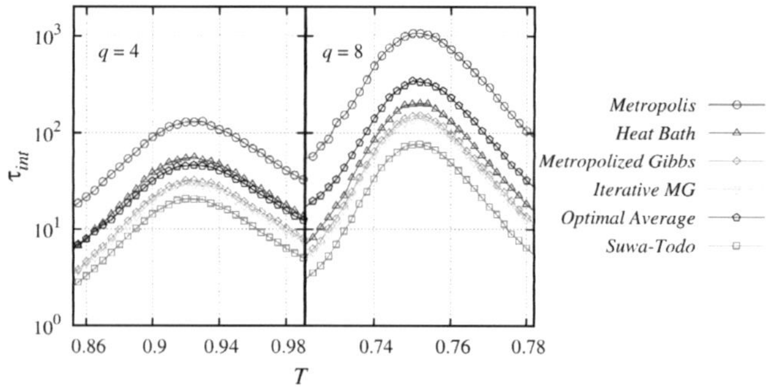

Figure 3. Autocorrelation time of order parameter near the transition temperature in the 4-state (left) and 8-state (right) Potts models. The system size is 16×16.

Potts model on the square lattice [31]; the local state at site k is expressed as σ_k that takes an integer ($1 \leq \sigma_k \leq q$) and the energy is expressed as $H = -\sum_{\langle i,j \rangle} \delta_{\sigma_i \sigma_j}$, where $\langle i, j \rangle$ is a pair of sites connected on the lattice. This system exhibits a continuous ($q \leq 4$) or first-order ($q > 4$) phase transition at temperature $T = 1/\ln(1 + \sqrt{q})$. We calculate the square of order parameter [32] (structure factor) for $q = 4$ and 8 by using the several algorithms.

The order parameter convergence is shown in figure 2, where the simulation starts with a fully-ordered (all "up") state and the local variables are sequentially updated. The square lattice with $L = 32$ on the periodic boundary conditions and the critical temperature $T = 1/\ln 3 = 0.9102392266\cdots$ are used. The Metropolis algorithm, the heat bath algorithm (Gibbs sampler), the Metropolized Gibbs sampler [18], the iterative Metropolized Gibbs sampler [17], the optimal average sampler [27], and the present algorithm (Suwa-Todo) [29] are compared. The validity of the all update methods are confirmed by comparing the asymptotic estimator convergence with each other (the Markov chain by the Gibbs sampler is ergodic). Our algorithm accomplishes the fastest convergence. This acceleration implies that the locally rejection-minimized algorithm reduces the second largest eigenvalue of the whole transition matrix in absolute value and increases the spectral gap of the Markov chain.

From figure 3, it is clearly seen that the present algorithm also reduces the autocorrelation time significantly in comparison with the conventional methods. The autocorrelation time $\tau_{\rm int}$ is estimated through the relation: $\sigma^2 \simeq (1 + 2\tau_{\rm int})\sigma_0^2$, where σ_0^2 and σ^2 are the variances of the estimator without considering autocorrelation and with calculating correlation from the binned data using a bin size much larger than the $\tau_{\rm int}$ [2], respectively. In the 4 (8)-state Potts model, the autocorrelation time becomes nearly 6.4 (14) times as short as that by the Metropolis algorithm, 2.7 (2.6) times as short as the heat bath algorithm, and even 1.4 (1.8) times as short as the iterative Metropolized Gibbs sampler. We investigated also the dynamical exponent of the autocorrelation time at the critical temperature. Unfortunately, the locally optimized method does not reduce the exponent. The factor over 6, however, is always gained against the Metropolis algorithm for all system sizes.

4. Walker's method of aliases and its generalization

As demonstrated in the previous section, increasing the number of candidate configurations can generally reduce the rejection rate. Then, we have to consider an efficient method to stochastically generate an event among a number of candidates. Such an event generation matters to various situations. In the Swendsen-Wang cluster algorithm for the long-range interacting spin models [33], for example, one has to choose $O(N)$ bonds among $O(N^2)$ candidates stochastically for making clusters, where N is the number of sites (spins). When the number of candidates increases, the computational cost for choosing configurations may dominate the total computing time. As seen below, a method based on geometric allocation of weights, called "Walker's method of aliases" [34, 35], and its generalization [36] give an efficient solution for the present problems.

Let us consider a random variable X that takes an integral value x according to probabilities $\{P(x)\}$ ($1 \le x \le M$ and $\sum_x P(x) = 1$). One of the simplest methods for generating such random numbers is the one based on the rejection. In this method the number of iterations required until a random number is obtained is approximately $M \times \max[P(x)]$ on average. It means that the method becomes inefficient quite rapidly as the variance of $P(x)$ increases. Instead, the binary search on the table of cumulative probabilities has been employed widely so far [33], where the number of operations can be reduced down to $O(\log M)$.

However, it is known that there exists a further effective method, Walker's method of aliases, which is rejection free and generates a random integer in a constant time. The Walker algorithm requires two tables of size M. One is the table of the modified probabilities $\{C(x)\}$ ($0 \le C(x) \le 1$ $\forall x$) and the other is of integral alias numbers $\{A(x)\}$ ($1 \le A(x) \le M$ $\forall x$). One has to set up these two tables in advance, so that

$$P(x) = \frac{1}{M}\Big\{C(x) + \sum_{k=1}^{M}[1 - C(k)]\,\delta_{A(k),x}\Big\} \qquad \forall x \tag{16}$$

is satisfied. Using these $\{C(x)\}$ and $\{A(x)\}$, a random integer is generated by the following procedure:

(i) Generate a uniform integral random variable r ($1 \le r \le M$).
(ii) Generate a uniform real random variable u ($0 < u \le 1$).
(iii) If $u < C(r)$, then $x = r$. Otherwise $x = A(r)$.

This procedure includes no iterations, and thus completes in a constant time.

The condition (16) for the modified probabilities and aliases can be illustrated graphically as demonstrated in figure 4. In figure 4(b), we have M bins of the same length. Each bin contains at most two colours, the original one and the colour pointed by its alias, and the total area of each colour is the same as the original probability (16). Note that the existence of Walker's tables for an arbitrary set of probabilities $\{P(x)\}$ is not trivial at all. However, a concrete and finite-step landfilling procedure of geometric allocation has been given explicitly [36], by which the existence of the solution is guaranteed, albeit the allocation is not unique and in general there exist many different solutions.

This Walker's technique can be generalized to the following problem: Let us consider M bits each of which takes either 0 (deactivated) or 1 (activated) independently according to the probability $P(x)$ ($1 \le x \le M$ and $0 \le P(x) \le 1$ $\forall x$). If one tries to determine the state of each bit one by one, it will naturally take the time proportional to M. However, in the case

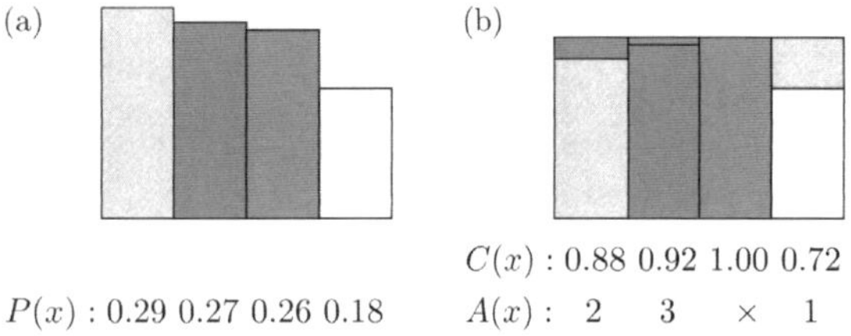

Figure 4. Example of (a) original probabilities $\{P(x)\}$, and (b) modified probabilities $\{C(x)\}$ and aliases $\{A(x)\}$ in Walker's method of aliases. The area of each colour is the same in (a) and (b).

where $\sum_x P(x)$ is much smaller than M, i.e. most of $P(x)$'s are very small or vanishing, one can adopt a different strategy by which the total computation time can be reduced from $O(M)$ to $O(\sum_x P(x))$ [36].

Our central idea is assigning a nonnegative integer to each bit instead of a binary, where zero (positive integers) corresponds to deactivated (activated) state. The integer to be assigned is generated according to the Poisson distribution:

$$f(\ell_x; \lambda_x) = \frac{e^{-\lambda_x} \lambda_x^{\ell_x}}{\ell_x!}, \tag{17}$$

where λ_x is the mean of the distribution. If one chooses λ_x as $-\log[1 - P(x)]$, then ℓ_x takes a positive integer with probability $P(x)$, since $\sum_{\ell_x \geq 1} f(\ell_x; \lambda_x) = 1 - e^{-\lambda_x}$. At first glance, one might think that the situation is getting worse by this modification, since a Poisson random number, instead of a binary, is needed for determining the state of each bit. At this point, however, we leverage an important property of the Poisson distribution; the Poisson process is that for random events and there is no statistical correlation between each two events. It allows us to interchange the space and time axes, i.e., we can realize the whole distribution by calculating just one Poisson random variable ℓ, the total number of events, from the distribution with mean $\lambda_{\text{tot}} = \sum_x \lambda_x$, and assigning each event to a bit afterward. The probability of choosing bit x at least once is given by

$$\begin{aligned} P_{\text{on}}(x) &= \sum_{\ell=1}^{\infty} f(\ell; \lambda_{\text{tot}}) \sum_{k=1}^{\ell} \frac{\ell!}{(\ell-k)!k!} \Big(\frac{\lambda_x}{\lambda_{\text{tot}}}\Big)^k \Big(1 - \frac{\lambda_x}{\lambda_{\text{tot}}}\Big)^{\ell-k} \\ &= 1 - \sum_{\ell=0}^{\infty} f(\ell; \lambda_{\text{tot}}) \Big(1 - \frac{\lambda_x}{\lambda_{\text{tot}}}\Big)^{\ell} = P(x). \end{aligned} \tag{18}$$

By employing Walker's algorithm, this procedure completes in $O(\lambda_{\text{tot}})$ on average, which can be much smaller than M, the number of bins, when most of $P(x)$'s are very small or vanishing [36].

As an example, let us consider the Swendsen-Wang cluster algorithm for the d-dimensional ferromagnetic Ising model of N sites with long-range interactions, $J_{i,j} \sim r_{i,j}^{-\sigma}$, where $r_{i,j}$ denotes the distance between two lattice sites, i and j [33]. The exponent σ should be larger than d, otherwise the ground state energy becomes non-extensive. At the cluster formation procedure in the Swendsen-Wang algorithm [9], each interacting bond is activated with probability $P_{i,j} = 1 - \exp(-2\beta J_{i,j})$, where β is the inverse temperature, and thus the total cost of this procedure is proportional to N^2 in a naive implementation. Since the interaction becomes weaker as the distance increases, however, one may reduce the computational cost by using the technique introduced above. Indeed, in the present specific case

$$\lambda_{\text{tot}} = -\sum_{i,j} \log(1 - P_{i,j}) = \sum_{i,j} 2\beta J_{i,j} \tag{19}$$

which is of $O(N)$ as long as $\sigma > d$. Thus, by using our space-time interchange trick explained above, the order of computational cost can be reduced greatly from $O(N^2)$ to $O(N)$ [36]

without any approximation, such as the introduction of a finite cutoff. One can extend this $O(N)$ technique in various ways; the measurement of energy and specific heat, Wang-Landau method [6], and exchange Monte Carlo [8], as well as the cluster update for long-range quantum spin models [36] and local-flip Monte Carlo update for long-range frustrated spin models [37].

5. Conclusion and discussions

In the present paper, we have shown that the unconventional approaches based on geometric allocation of probabilities or weights can improve the dynamics and scaling of the Monte Carlo simulation in several aspects. The approach using the irreversible kernel can reduce or sometimes completely avoid the rejection of trial move in the Markov chain. It can be applied to any MCMC sampling including the system with continuous variables and it is expected to improve the efficiency in general. (For recent application to biomolecule systems, see e.g., [38, 39].) Also, the space-time interchange technique together with Walker's method of aliases can reduce the computational time especially for the case where the number of candidates is large. We showed that by using this technique, the MCMC simulation for the models with long-range interactions can be executed quite efficiently in a rigorous way. This technique is also generic and expected to be applied to many MCMC methods. The algorithm for constructing an irreversible kernel for a generic set of weights has been implemented in C++ and released as an open-source library, BCL (Balance Condition Library) [40] together with an efficient implementation of Walker's method of aliases.

The most simulations in the present paper has been done by using the facility of the Supercomputer Center, Institute for Solid State Physics, University of Tokyo. The simulation code has been developed based on the ALPS library [41, 42]. The authors acknowledge the support by the Grant-in-Aid for Scientific Research Program (No. 23540438) from JSPS, HPCI Strategic Programs for Innovative Research (SPIRE) from MEXT, Japan, and the Computational Materials Science Initiative (CMSI). HS is supported by the JSPS Postdoctoral Fellow for Research Abroad.

References

[1] Robert C P and Casella G 2004 *Monte Carlo Statistical Methods* 2nd ed (New York: Springer)
[2] Landau D P and Binder K 2005 *A Guide to Monte Carlo Simulations in Statistical Physics* 2nd ed (Cambridge: Cambridge University Press)
[3] Tierney L 1994 *Ann. Stat.* **22** 1701
[4] Meyn S P and Tweedie R L 1993 *Markov Chains and Stochastic Stability* (New York: Springer)
[5] Berg B A and Neuhaus T 1992 *Phys. Rev. Lett.* **68** 9
[6] Wang F and Landau D P 2001 *Phys. Rev. Lett.* **86** 2050
[7] Marinari E and Parisi G 1992 *Europhys. Lett.* **19** 451
[8] Hukushima K and Nemoto K 1996 *J. Phys. Soc. Jpn.* **65** 1604
[9] Swendsen R H and Wang J S 1987 *Phys. Rev. Lett.* **58** 86
[10] Evertz H G, Lana G and Marcu M 1993 *Phys. Rev. Lett.* **70** 875
[11] Todo S 2013 *Strongly Correlated Systems: Numerical Methods (Springer Series in Solid-State Sciences)* ed Avella A and Mancini F (Berlin: Springer-Verlag) p 153
[12] Duane S, Kennedy A D, Pendleton B J and Roweth D 1987 *Phys. Lett. B* **195** 216
[13] Metropolis N, Rosenbluth A W, Rosenbluth M N, Teller A H and Teller E 1953 *J. Chem. Phys.* **21** 1087
[14] Hastings W K 1970 *Biometrika* **57** 97
[15] Geman S and Geman D 1984 *IEEE Trans. Pattn. Anal. Mach. Intel.* **6** 721
[16] Peskun P H 1973 *Biometrika* **60** 607
[17] Frigessi A, Hwang C R and Younes L 1992 *Ann. Appl. Probab.* **2** 610
[18] Liu J S 1996 *Biometrika* **83** 681
[19] Manousiouthakis V I and Deem M W 1999 *J. Chem. Phys.* **110** 2753
[20] Diaconis P, Holmes S and Neal R M 2000 *Ann. Appl. Probab.* **10** 726
[21] Turitsyn K S, Cherkov M and Vucelja M 2011 *Physica D* **240** 410
[22] Sakai Y and Hukushima K 2013 *J. Phys. Soc. Jpn.* **82** 064003
[23] Fernandes H C M and Weigel M 2011 *Comput. Phys. Commun.* **182** 1856

[24] Neal R M 2004 Improving asymptotic variance of MCMC estimators: non-reversible chains are better Tech. Rep. 0406 Department of Statistics, University of Toronto
[25] Sun Y, Gomez F J and Schmidhuber J 2010 *NIPS* p 2235
[26] Bernard E, Krauth W and Wilson D B 2009 *Phys. Rev. E* **80** 056704
[27] Hwang C R, Chen T L, Chen W K and Pai H M 2012 *Siam J. Control. Optim.* **50** 2743
[28] Hwang C R, Hwang-Ma S Y and Sheu S J 2005 *Ann. Appl. Probab.* **15** 1433
[29] Suwa H and Todo S 2010 *Phys. Rev. Lett.* **105** 120603
[30] Suwa H 2012 *Geometrically Constructed Markov Chain Monte Carlo Study of Quantum Spin-phonon Complex Systems* PhD thesis Department of Applied Physics, Graduate School of Engineering, The University of Tokyo
[31] Wu F Y 1982 *Rev. Mod. Phys.* **54** 235
[32] Zheng B 1998 *Int. J. Mod. Phys. B* **12** 1419
[33] Luijten E and Blöte H W J 1995 *Int. J. Mod. Phys. C* **6** 359
[34] Walker A J 1977 *ACM Trans. Math. Software* **3** 253
[35] Knuth D E 1997 *The Art of Computer Programming* vol 2, Seminumerical Algorithms (Reading: Addison Wesley) p 119 3rd ed
[36] Fukui K and Todo S 2009 *J. Comp. Phys.* **228** 2629
[37] Tomita Y 2009 *J. Phys. Soc. Jpn.* **78** 014002
[38] Itoh S G and Okumura H 2013 *J. Chem. Theory Comput.* **9** 570
[39] Kondo H X and Taiji M 2013 *J. Chem. Phys.* **138** 244113
[40] http://github.com/cmsi/bcl
[41] Bauer B, Carr L D, Evertz H G, Feiguin A, Freire J, Fuchs S, Gamper L, Gukelberger J, Gull E, Guertler S, Hehn A, Igarashi R, Isakov S V, Koop D, Ma P N, Mates P, Matsuo H, Parcollet O, Pawlowski G, Picon J D, Pollet L, Santos E, Scarola V W, Schollwöck U, Silva C, Surer B, Todo S, Trebst S, Troyer M, Wall M L, Werner P and Wessel S 2011 *Journal of Statistical Mechanics: Theory and Experiment* P05001
[42] http://alps.comp-phys.org/

Randomness in algorithm design

Shuji Kijima
Kyushu University, Fukuoka, 819-0395, Japan
E-mail: kijima@inf.kyushu-u.ac.jp

Abstract. What is the property that a randomized algorithm really requires for randomness? This article exhibits some topics related to randomization and derandomization in computation, the topics which the author has been concerned with motivated by the above question.

1. Introduction

Randomness is a powerful and indispensable tool for computation nowadays, supported by the probability theory standing on a long history. "What is the property that a randomized algorithm really requires for randomness?" Motivated by the question, the author has been investigating randomization and derandomization of algorithms. This article exhibits some topics related to this subject.

To demonstrate the power of randomness, we are concerned with a fundamental problem of finding frequent items in a data stream in Section 2. For the problem, we have developed a randomized approximation algorithm using memory of only $\mathrm{O}(\log\log n)$ bits in [49]. Our algorithm is based on an extension of *probabilistic counting* using memory of only $\mathrm{O}(\log\log n)$ bits, while it is known that any deterministic algorithm for counting requires $\Omega(\log n)$ bits, even for approximation. In Section 3, we briefly review the Markov chain Monte Carlo (MCMC) method, which is a sophisticated and useful technique for random sampling. The *coupling from the past*, devised by Propp and Wilson [50], is a wonderful algorithm for MCMC, which provides *perfect sampling*, meaning that a random sample *exactly* according to the limit distribution, by an ingenious simulation of Markov chain. We briefly review the coupling from the past algorithm in Section 4. In Section 5, we introduce the topic of *deterministic random* walks on finite graphs, which is an approach to replacing a random walk by a deterministic process, motivated by a question if it is possible to derandomize an MCMC.

2. Finding frequent items in a stream

2.1. Problem

Let us consider a problem of finding frequent items in a data stream (cf. [49]). Let Σ be a finite set[1], which we call "items", and let $\boldsymbol{x} = (x_1, x_2, \ldots) \in \Sigma^*$ denote a data stream of items. For instance, Σ denotes a set of IP addresses and $\boldsymbol{x}$ denotes an access-log of web page in a day. For another example, Σ denotes a set of goods in a supermarket and $\boldsymbol{x}$ denotes a POS data in a day. Suppose a threshold $\theta \in (0,1)$ is prescribed, and we want to know frequent items $s \in \Sigma$,

[1] For simplicity, we in this paper assume that Σ is finite where every item is identified with a constant number $\sigma := \lceil \lg(|\Sigma| + 1) \rceil$ of bits.

each of which appears at least $\theta\cdot n$ times in an input data stream $\boldsymbol{x} = (x_1, x_2, \ldots, x_n)$. How large memory space is required to find frequent items in $\boldsymbol{x}$, without knowing the value of n in advance?

Boyer and Moore [3, 4] gave an algorithm, known as *majority* algorithm, which solves the problem for $\theta = 1/2$ using $\mathrm{O}(\log n)$ bits, while Fischer and Salzburg [20] independently gave essentially the same algorithm. Karp, Shenker, and Papadimitriou [32] showed a lower bound that any online algorithm requires $\Omega(|\Sigma| \log(n/|\Sigma|))$ bits to find the frequent items *exactly*. Karp, Shenker, and Papadimitriou [32] also gave a simple and elegant deterministic approximation algorithm, which allows false positive outputs using memory of $\mathrm{O}(\theta^{-1} \log n)$ bits. Demaine, López-Ortiz, and Munro [9] gave an algorithm essentially the same as [32]. Manku and Motwani [44] proposed another deterministic algorithm, called *lossy counting*, which is based on approximate counting using $\mathrm{O}(\varepsilon^{-1} \log(\varepsilon^{-1} n))$ bits, that outputs items of frequency more than $(\theta - \varepsilon)$ for $\varepsilon \in (0, 1)$.

Many randomized algorithms are also known for the problem. Toivonen [54] proposed an algorithm using memory of $\mathrm{O}(\varepsilon^{-2} \log \delta^{-1} \log n)$ bits, which is based on uniform sampling of data in a stream, and outputs all frequent items with probability $1-\delta$. Manku and Motwani [44] gave another algorithm, called *sticky sampling*, which uses memory of $\mathrm{O}(\varepsilon^{-1} \log \delta^{-1} \log n)$ bits and outputs all frequent items with probability $1 - \delta$. Their algorithm samples items by adaptively changing the sampling rate according to the number of items that have arrived.

Each of deterministic or randomized approximation algorithms above mentioned uses memory of $\mathrm{O}(\log n)$ bits in terms of n, respectively. Thus, it can be a natural question: Is there an algorithm which uses memory of $\mathrm{o}(\log n)$ for the problem without knowing n in advance?

2.2. Probabilistic counting

Let us consider a simpler problem — just count the length n of an input stream $\boldsymbol{x} = (x_1, \ldots, x_n) \in \Sigma^n$ where Σ denotes the set of items. A trivial algorithm uses memory of $\lceil \lg(n+1) \rceil$ bits for counting *exact* n, and this is best possible because it demands $\lceil \lg(n+1) \rceil$ bits to represent exact n. It is thus natural to ask if it is possible to *approximate* n, e.g., to compute $\lg n$, using memory of $\mathrm{o}(\log n)$ bits. Unfortunately, any *deterministic* algorithm requires a working space of $\lg n$ bits even for an approximation of $\lg n$.

Morris [47] gave a fascinating randomized approximation algorithm, called *probabilistic counting*, for computing $\lg n$. The algorithm is as follows; set the initial value $h = 0$, and increment h by one with probability 2^{-h} for every input item. Then, the expectation of h is shown to be $\lg(n+2)$ (see also [21]). Morris also generalized it to treat with a sampling rate a^{-h} for any constants $a > 0$. Flajolet [21] gave detailed analyses on the variance of the probabilistic counting.

We can see that the probabilistic counting uses memory of only $\mathrm{O}(\log \log n)$ bits, which implies that a randomized algorithm reduces the space complexity of the problem, compared with a deterministic algorithm.

2.3. Finding frequent items in a stream using $\mathrm{O}(\log \log n)$ *bits*

Now, back to the problem of finding frequent items in a stream. Let Σ denote a finite set of items, and assume that every item is identified with a constant $\sigma := \lceil \lg |\Sigma| \rceil$ bits. For a sequence $\boldsymbol{x} = (x_1, \ldots, x_n)$ of items in Σ, let $f(s) \in \{0, \ldots, n\}$ for $s \in \Sigma$ denote the frequency of s in $\boldsymbol{x}$. Given a (constant) $\theta \in (0, 1)$, we say that an item $s \in \Sigma$ is *frequent* (in $\boldsymbol{x}$) if $f(s) \geq \theta n$.

In our approximation algorithm, suppose that parameters $\theta \in (0, 1)$, $\gamma \in (0, 1)$ and $\delta \in (0, 1)$ are prescribed by a user. Then with probability at least $1 - \delta$, the output of our algorithm contains all frequent items in $\{s \in \Sigma \mid f(s) \geq \theta n\}$ and does not contain any rare item in $\{s \in \Sigma \mid f(s) < (1 - \gamma)\theta n\}$. The following shows our algorithm for an input stream $\boldsymbol{x} = (x_1, \ldots, x_n)$, in which $b \in \mathbb{N}$ is a parameter for approximation. Let K denote a data structure which stores 2^b pairs of an item and a number $(s, K[s])$, where s uses σ bits, and $K[s]$ uses b bits.

Algorithm 2.1. ([49])

0. **Initialize** K. **Set** "exponent" $h := 0$.
1. **Read** an input x_i if it exists, otherwise **goto** 5. Suppose $x_i = s^\star$ for convenience.
2. "**Record**" $s^\star$ in K with probability 2^{-h}:
 add $s^\star$ in K unless $s^\star$ in K, and **increment** $K[s^\star]$ by one.
3. **Unless** $\sum_{s' \in K} K[s'] = 2^b$, **goto** 1.
4. "**Flush**":
 4-(i) **Increment** h by one.
 4-(ii) **For** each s in K,
  choose k with probability $\binom{K[s]}{k}/2^{K[s]}$, and
  set $K[s] := k$.
 4-(iii) **Delete** symbol $s \in K$ if $K[s] = 0$.
 Goto 3.
5. **Output** every s satisfying that $K[s] \geq (1-\gamma/2)\theta \sum_{s' \in K} K[s']$.

Let h^* denote the value of exponent when Algorithm 2.1 terminates. Then, we can show that Algorithm 2.1 uses $(b+\sigma)2^b + \lceil \lg h^* \rceil$ bits, and an extra working space of $\mathrm{O}(\lg h^*)$ bits for Steps 2 and 4. We also remark that the number of items output by Algorithm 2.1 is at most $1/(\theta(1-\gamma/2))$.

Theorem 2.1. ([49]) *Arbitrarily given* $\theta \in (0,1)$, $\gamma \in (0,1)$ *and* $\delta \in (0,1)$, *set*

$$b \geq \lceil \lg \left((\theta\gamma/2)^{-2} \ln(3((1-\gamma/2)\theta\delta)^{-1}) \right) \rceil + 3.$$

Then the output of Algorithm 2.1 contains all frequent items in $\{s \in \Sigma \mid f(s) \geq \theta n\}$ *and does not contain any rare item in* $\{s \in \Sigma \mid f(s) < (1-\gamma)\theta n\}$, *with probability at least* $1-\delta$, *for any input stream* $\boldsymbol{x} = (x_1, \ldots, x_n)$.

Theorem 2.1 indicates that our algorithm uses memory of

$$\begin{aligned} 2^b(b+\sigma) + \lg\lg n &\simeq \frac{32 \ln\left(\frac{3}{(1-\gamma/2)\theta\delta}\right)}{(\gamma\theta)^2} \lg \left(\frac{\ln\left(\frac{3}{(1-\gamma/2)\theta\delta}\right)}{(\gamma\theta)^2} \right) + \lg\lg n \\ &= \mathrm{O}\left(\frac{\log^2(\theta^{-1})}{\theta^2} + \log\log n \right) \end{aligned}$$

bits.

3. Markov chain Monte Carlo—for sampling combinatorial objects

Random sampling plays a fundamental role in randomized algorithms, especially from the viewpoint of computational efficiency. It is known, in the context of polynomial time computability, that there is a strong connection between sampling and approximate counting of combinatorial objects through the *self-reducibility* of the objects. For example, Dyer, Frieze, Kannan [19] gave a *fully polynomial time randomized approximation scheme* (*FPRAS*) for the volume computation of a convex body in a high dimensional space, which is a #P-hard problem, using random sampling in a convex body and self reducibility. As another example, Jerrum, Sinclair, Vigoda [30] gave an FPRAS for the computation of the permanent of a $n \times n$ non-negative matrix, which is also #P-hard. Both algorithms are based on the Markov chain Monte Carlo method.

The Markov chain Monte Carlo is a Monte Carlo method using a Markov chain for random sampling. The idea is simple; (1) design a Markov chain whose limit distribution is desired distribution, and (2) sample from the limit distribution simulating the chain. This section briefly reviews the MCMC method.

3.1. Designing a Markov chain with a desired distribution

To realize (1), there is a framework for designing a Markov chain with a desired limit distribution using *reversible Markov chain*, such as cerebrated Metropolis-Hastings, or Gibbs sampler.

Here, we are concerned with a finite Markov chain $\mathcal{M}$, defined by a transition matrix P on finite states Ω. A Markov chain is *irreducible* if $\exists t > 0$, $\Pr(X^t = y \mid X^0 = x) > 0$ holds for any pair $\{x, y\} \in \binom{\Omega}{2}$, where $X^t \in \Omega$ denotes the state of the Markov chain at time t. A Markov chain is *aperiodic* if $\forall x \in \Omega$, $\gcd(\{t \in \mathbb{Z}_{>0} \mid \Pr(X^t = x \mid X^0 = x) > 0\}) = 1$ holds, where $\gcd(A)$ denotes the greatest common deviser of $A \subseteq \mathbb{Z}_{>0}$. A Markov chain is ergodic if it is irreducible and aperiodic. It is well-known that an ergodic Markov chain has a unique stationary distribution π, which is a distribution satisfying $\pi P = \pi$, and the limit distribution is π.

The following Theorem provides a Markov chain with a desired stationary distribution.

Theorem 3.1. (cf. [26, 43]) *Suppose $\mathcal{M}$ is an ergodic Markov chain defined by a transition matrix P on a state space Ω. If a function $f\colon \Omega \to \mathbb{R}_{>0}$ satisfies the following* detailed balance equation

$$f(x)P(x, y) = f(y)P(y, x), \qquad \forall\{x, y\} \in \binom{\Omega}{2} \tag{1}$$

then the stationary distribution π of $\mathcal{M}$ satisfies $\pi(x) = C \cdot f(x)$ for any $x \in \Omega$ with a constant C, where $P(x, y)$ denotes the transition probability from x to y.

A Markov chain satisfying the detailed balance equation is said *reversible*. Many Markov chains appearing in MCMC are reversible.

3.2. Mixing time

On the other hand, estimation the speed of convergence into limit distribution, called *mixing time* of the Markov chain, for the purpose of (2) is not easy in general. Remark that "convergence" does not mean the convergence into a state of the Markov chain, and hence it is also difficult in practice to determine if the Markov chain converges to its limit distribution.

Given a pair of probability distributions ν_1 and ν_2 on the finite state space Ω, the *total variation distance* between ν_1 and ν_2 is defined by

$$\mathrm{d_{TV}}(\nu_1, \nu_2) \stackrel{\text{def.}}{=} \frac{1}{2} \sum_{x \in \Omega} |\nu_1(x) - \nu_2(x)|.$$

For any positive $\varepsilon < 1$, the *mixing time* $\tau(\varepsilon)$ of $\mathcal{M}$ with a state space Ω is defined by

$$\tau(\varepsilon) \stackrel{\text{def.}}{=} \max_{x \in \Omega}\{\min\{t \mid \forall s \geq t,\ \mathrm{d_{TV}}(\pi, P_x^s) \leq \varepsilon\}\}$$

where P_x^s denotes the distribution at time s of $\mathcal{M}$ starting from $x \in \Omega$. The definition of the mixing time implies that the total variation distance between the stationary distribution and the distribution of $\mathcal{M}$ after $\tau(\varepsilon)$ transitions is at most ε. We say a Markov chain is *rapidly mixing* if $\tau(\varepsilon) \leq \mathrm{poly}(\text{input size}, \varepsilon^{-1})$.

Mixing time is deeply related to the eigenvalue of the transition matrix. For instance, the following theorem is well known (see e.g., [51, 43]).

Theorem 3.2. *Suppose a Markov chain $\mathcal{M}$ defined by a transition matrix P is reversible and ergodic. Let λ_i ($i \in \{1, \ldots, |\Omega|\}$) denote the (multiple) eigenvalues of P, where without loss of generality we may assume that $\lambda_1 = 1$. Then, for any ε ($0 < \varepsilon < 1$),*

$$\frac{\lambda^* \ln((2\varepsilon)^{-1})}{1-\lambda^*} \le \tau(\varepsilon) \le \frac{\ln(\pi_{\min}^{-1}) + \ln(\varepsilon^{-1})}{1-\lambda^*}$$

holds where $\lambda^* \stackrel{\text{def.}}{=} \max\{|\lambda_i| \mid i \in \{2, \ldots, |\Omega|\}\}$ and $\pi_{\min} \stackrel{\text{def.}}{=} \min\{\pi(x) \mid x \in \Omega\}$ for the stationary distribution π of $\mathcal{M}$. Note that $\lambda^* < 1$ holds.

4. Perfect sampling

In stead of estimating the mixing time of a Markov chain, Propp and Wilson [50] devised an algorithm which enables *perfect sampling*, meaning that the algorithm provides a random sample according to *exactly* limit distribution. The algorithm is an ingenious simulation of a Markov chain, called *coupling from the past* (*CFTP*). Here we briefly review the CFTP.

4.1. Coupling from the past

Suppose that we have an ergodic Markov chain MC with finite state space Ω and transition matrix P. The transition rule of the Markov chain $X \mapsto X'$ can be described by a deterministic function ϕ: $\Omega \times [0,1) \to \Omega$, called an *update function*, as follows. Given a random number Λ uniformly distributed over $[0,1)$, an update function ϕ satisfies $\Pr(\phi(x,\Lambda) = y) = P(x,y)$ for any $x, y \in \Omega$. We can realize the Markov chain by setting $X' = \phi(X, \Lambda)$. Clearly, an update function corresponding to the given transition matrix P is not unique. The result of transitions of the chain from the time t_1 to t_2 ($t_1 < t_2$) with a sequence of random numbers $\boldsymbol{\lambda} = (\lambda[t_1], \lambda[t_1+1], \ldots, \lambda[t_2-1]) \in [0,1)^{t_2-t_1}$ is denoted by $\Phi_{t_1}^{t_2}(x, \boldsymbol{\lambda}) : \Omega \times [0,1)^{t_2-t_1} \to \Omega$ where $\Phi_{t_1}^{t_2}(x, \boldsymbol{\lambda}) \stackrel{\text{def.}}{=} \phi(\phi(\cdots(\phi(x, \lambda[t_1]), \ldots, \lambda[t_2-2]), \lambda[t_2-1])$. We say that a sequence $\boldsymbol{\lambda} \in [0,1)^{|T|}$ satisfies the *coalescence condition*, when $\exists y \in \Omega$, $\forall x \in \Omega$, $y = \Phi_T^0(x, \boldsymbol{\lambda})$.

With these preparations, standard Coupling From The Past algorithm is expressed as follows.

Algorithm 4.1. (Coupling from the past [50])

Step 1. Set the starting time period $T := -1$ to go back, and let $\boldsymbol{\lambda}$ be an empty sequence.

Step 2. Generate random real numbers $\lambda[T], \lambda[T+1], \ldots, \lambda[\lceil T/2 \rceil - 1] \in [0,1)$, and insert them to the head of $\boldsymbol{\lambda}$ in order, i.e., put $\boldsymbol{\lambda} := (\lambda[T], \lambda[T+1], \ldots, \lambda[-1])$.

Step 3. Start a chain from each element $x \in \Omega$ at time period T, and run each chain to time period 0 according to the update function ϕ with the sequence of numbers in $\boldsymbol{\lambda}$. (Here we note that every chain uses the common sequence $\boldsymbol{\lambda}$.)

Step 4. [Coalescence check] The state obtained at time period 0 can be denoted by $\Phi_T^0(x, \boldsymbol{\lambda})$.
(i)If $\exists y \in \Omega$, $\forall x \in \Omega$, $y = \Phi_T^0(x, \boldsymbol{\lambda})$, then return y and stop.
(ii)Else, update the starting time period $T := 2T$, and go to Step 2.

Theorem 4.1. ([50]) *Let MC be an ergodic finite Markov chain with state space Ω, defined by an update function $\phi : \Omega \times [0,1) \to \Omega$. If the CFTP algorithm (Algorithm 4.1) terminates with probability 1, then the obtained value is a realization of a random variable exactly distributed according to the stationary distribution.*

It is also well known that the time to couple is deeply related to the mixing time (see e.g., [50, 35]). We define the *coalescence time* T_* by

$$T^* \stackrel{\text{def.}}{=} \min\{t < 0 \mid \exists y \in \Omega,\ \forall x \in \Omega,\ y = \Phi_t^0(x, \boldsymbol{\lambda})\},$$

which denotes the minimum number of steps to obtain a sample by Algorithm 4.1.

Theorem 4.2. *For any* ε $(0 < \varepsilon < 1)$, $\tau(\varepsilon) \leq \mathrm{E}[T_*]\varepsilon^{-1}$ *holds.*

With some assumptions, we obtain tighter bounds.

4.2. Monotone CFTP

Theorem 4.1 gives a (probabilistically) finite time algorithm for infinite time simulation. However, it is often computationally infeasible to perform the simulations from all states executed in Step 3.

Suppose that there exists a partial order "$\succeq$" on the set of states Ω. A transition rule expressed by a deterministic update function ϕ is called *monotone* (with respect to "$\succeq$") if $\forall \lambda \in [0, 1)$, $\forall x, \forall y \in \Omega$, $x \succeq y \Rightarrow \phi(x, \lambda) \succeq \phi(y, \lambda)$. We also say that a chain is *monotone* if the chain has a *monotone* update function.

Theorem 4.3. (monotone CFTP [50]) *Suppose that a Markov chain defined by an update function* ϕ *is monotone with respect to a partially ordered set of states* $(\Omega, \succeq)$, *and* $\exists x_{\max}, \exists x_{\min} \in \Omega$, $\forall x \in \Omega$, $x_{\max} \succeq x \succeq x_{\min}$. *Then the CFTP algorithm (Algorithm 4.1) terminates with probability 1, and a sequence* $\boldsymbol{\lambda} \in [0, 1)^{|T|}$ *satisfies the coalescence condition, i.e.,* $\exists y \in \Omega$, $\forall x \in \Omega$, $y = \Phi_T^0(x, \boldsymbol{\lambda})$, *if and only if* $\Phi_T^0(x_{\max}, \boldsymbol{\lambda}) = \Phi_T^0(x_{\min}, \boldsymbol{\lambda})$.

When the given Markov chain satisfies the conditions of Theorem 4.3, we can modify Algorithm 4.1 by substituting Step 4 (a) by

Step 4. (a)′ If $\exists y \in \Omega$, $y = \Phi_T^0(x_{\max}, \boldsymbol{\lambda}) = \Phi_T^0(x_{\min}, \boldsymbol{\lambda})$, then return y and stop.

The algorithm obtained by the above modification is called a monotone CFTP algorithm. Propp and Wilson [50] gave a monotone CFTP for Ising model. Kijima and Matsui [35] gave a monotone Markov chain for sampling two rowed contingency tables (see also [56]).

5. Deterministic random walk

The *rotor-router model*, also known as the *Propp machine* [8], is a deterministic process analogous to a random walk on a graph. Instead of distributing tokens to randomly chosen neighbors, the Propp machine deterministically serves the neighbors in a fixed order by associating to each vertex a "rotor-router" pointing to one of its neighbors. The Propp machine sometimes appears under the name of *deterministic random walk*, meaning a "derandomized, hence *deterministic*, version of a *random walk*."

5.1. Rotor-router model

Let $\mathcal{G} = (V, \mathcal{E})$ be a multidigraph (i.e., directed graph with multiple edges) with vertex set V and edge set $\mathcal{E}$, where $n = |V|$ and $m = |\mathcal{E}|$. Here, we assume that $\mathcal{G}$ is strongly connected [8, 34, 31]. Let $N(v)$ denote the multiset of vertices consisting of destinations of outgoing edges from $v \in V$ in $\mathcal{G}$, and let $\delta(v) = |N(v)|$, i.e., the outdegree of $v \in V$ in $\mathcal{G}$. Note that v is a member of $N(v)$ if $\mathcal{E}$ includes a self-loop at v.

Let P be an $n \times n$ matrix representing a *standard random walk* on $\mathcal{G}$, meaning that

$$P(u, v) = \frac{|\{a \in \mathcal{E} \mid a = (u, v)\}|}{\delta(u)}.$$

Since $\mathcal{G}$ is strongly connected, P is irreducible. In this paper, we consider random walks of M tokens on $\mathcal{G}$, in which each token independently moves according to the transition matrix P. Let $\mu^{(0)} = (\mu_{v_1}^{(0)}, \ldots, \mu_{v_n}^{(0)}) \in \mathbb{Z}_{\geq 0}^n$ denote an initial configuration of tokens, and let $\mu^{(t)} = (\mu_{v_1}^{(t)}, \ldots, \mu_{v_n}^{(t)}) \in \mathbb{R}_{\geq 0}^n$ denote the *expected* configuration of tokens according to P at time $t \in \mathbb{Z}_{\geq 0}$, i.e., $\mu^{(t)} = \mu^{(0)} P^t$.

The *rotor-router model* is a deterministic process analogous to $\mu^{(t)}$. Let $\chi^{(0)}(= \mu^{(0)})$ be an initial configuration of tokens, and let $\chi^{(t)}$ denote a configuration of tokens in the rotor-router model at time t. In the rotor-router model, an ordering $\rho_v(0), \ldots, \rho_v(\delta(v)-1)$ of $N(v)$ is prescribed on each vertex $v \in V$. We define

$$\rho_v(i) \stackrel{\text{def.}}{=} \rho_v(i \bmod \delta(v)) \quad \left(\equiv \rho_v \left(i - \delta(v) \cdot \left\lfloor \frac{i}{\delta(v)} \right\rfloor \right) \right)$$

for all $i \in \mathbb{Z}_{\geq 0}$ by extension. This ordering ρ_v is called a *rotor-router* on v, which sequentially launches all $\chi_v^{(t)}$ tokens at each time $t \in \mathbb{Z}_{\geq 0}$. More precisely, since $\chi_v^{(0)}$ tokens are located on $v \in V$ at time $t = 0$, during the time interval $[0, 1)$, the rotor-router on v launches each token j ($j \in \{1, \ldots, \chi_v^{(0)}\}$) into vertex $\rho_v(j-1) \in N(v)$, where they arrive at the neighbors of v at time $t = 1$, Since at time $t \geq 1$, $\chi_v^{(t)}$ tokens are located on $v \in V$, and the rotor-router on v points to $\rho_v(\sum_{s=0}^{t-1} \chi_v^{(s)})$, the rotor-router on v launches each token j ($j \in \{1, \ldots, \chi_v^{(t)}\}$) into vertex $\rho_v(j - 1 + \sum_{s=0}^{t-1} \chi_v^{(s)}) \in N(v)$ during the time interval $[t, t+1)$, where they arrive at the neighbors of v at time $t+1$.

5.2. Discrepancy

Our goal is to estimate the discrepancy $\left|\chi_w^{(T)} - \mu_w^{(T)}\right|$ for $w \in V$ and $T \geq 0$. We have obtained the following results in [34, 31].

Theorem 5.1. ([34]) *If all the eigenvalues of P are nonnegative, then for any initial configuration of tokens and routers, we have*

$$\left|\chi_w^{(T)} - \mu_w^{(T)}\right| \;\leq\; (2m-n) \left(\max_{i \in \{1,\ldots,\kappa\}} n_i + n + 3 \right) \;\leq\; 4mn + \mathrm{O}(m)$$

for any $w \in V$ and $T \geq 0$, where n_i ($i \in \{1, \ldots, \kappa\}$) denotes the size of the i-th Jordan block of the Jordan normal form of P with κ the number of Jordan blocks.

Theorem 5.2. ([34]) *There exist a multidigraph and an initial configuration of tokens and routers such that $\left|\chi_w^{(T)} - \mu_w^{(T)}\right| = \Omega(m)$ holds for any node w and any positive integer T, even when the corresponding P is positive semidefinite, hence all the eigenvalues of P are nonnegative and P is diagonalizable.*

Theorem 5.3. ([31]) *Let P be the transition matrix of the standard random walk on a strongly connected multidigraph $\mathcal{G}$, and let $J = B^{-1}PB$ be a Jordan normal form of P where each column vector of B is normalized to norm one. Let $\lambda^* \stackrel{\text{def.}}{=} \max\{|\lambda_i| \mid \lambda_i$ is an eigenvalue of P, $|\lambda_i| \neq 1\}$, and let α^* be the maximum in the absolute values of elements of B^{-1}. Then, for any initial configuration of tokens and routers,*

$$\left|\chi_w^{(T)} - \mu_w^{(T)}\right| \;\leq\; \frac{2n(n-r)(m-n)\alpha^*}{1-\lambda^*}$$

holds for any $w \in V$ and $T \geq 0$, where r is the periodicity of P.

5.3. Deterministic random walk for irrational transition probabilities

Recently Shiraga et al. [52] gave a functional router model, which simulates Markov chains with irrational transition probabilities, in a similar fashion to the rotor-router model. They also gave a bound of the discrepancy for the functional router model as follows.

Theorem 5.4. ([52]) *Let $P \in \mathbb{R}_{\geq 0}^{n\times n}$ be a transition matrix which is irreducible, aperiodic, and reversible. Let $\pi \in \mathbb{R}_{\geq 0}^{n}$ be the stationary distribution of P, and let $\lambda^* \stackrel{\text{def.}}{=} \max\{|\lambda_i| \mid \lambda_i$ is an eigenvalue of P, $|\lambda_i| < 1\}$. Let $\chi^{(T)}$ denote the configuration of tokens at time $T \in \mathbb{Z}_{\geq 0}$ in a corresponding functional-router model with M tokens, then*

$$\left|\chi_w^{(T)} - \mu_w^{(T)}\right| \;\leq\; \sqrt{\frac{\pi(w)}{\pi_{\min}}} \cdot \frac{2m}{(1-\lambda^*)}$$

holds for any $w \in V$ and $T \geq 0$, where $\pi_{\min} \stackrel{\text{def.}}{=} \min_{u\in V} \pi(u)$.

It is well known that $1/(1-\lambda^*)$ provides (upper and lower bounds of) the mixing time of Markov chains (cf. [51]), and has been well investigated for several Markov chains, which are often reversible. Here we omit the detail of the functional router model (see [52]).

5.4. Related works on the rotor-router model

The Propp machine has been investigated intensively in recent years. Cooper and Spencer [8] dealt with a *parallel-walk* version of the Propp machine, and studied the discrepancy on a single vertex, meaning the difference between the number of tokens in the Propp machine and the expected number of tokens in a random walk. They showed that in the d-dimensional integer lattice $\mathbb{Z}^d$, the discrepancy on any node at any time is bounded by a constant c_d, which depends only on the dimension d but is independent of the number of tokens, for any initial configurations of tokens on 'even' vertices and for any configurations of rotor-routers. It was later shown that c_1 (for the infinite path) is approximately 2.29 due to Cooper, Doerr, Spencer, and Tardos [6] (cf [12]), and c_2 is approximately 7.83, if every rotor-router is in clockwise (or counterclockwise) order, and 7.29 otherwise due to Doerr and Friedrich [12]. For the infinite k-regular tree, Cooper, Doerr, Friedrich, and Spencer [5] showed that there exist initial configurations of tokens, which are not in the same parity, and rotor-routers such that the discrepancy on a node at time T is $\Omega(\sqrt{kT})$.

In contrast to those results dealing with parallel-walks on infinite graphs, Holroyd and Propp [27] analyzed a *single-walk* version of the rotor-router machine on *finite* graphs. In [27], they analyzed the hitting time of the rotor-router machine, and gave a bound that $|\nu^{(t)}(v)/t - \pi(v)| = \mathrm{O}(mn/t)$ on any node v at any time t, where $\nu^{(t)}(v)$ denotes the frequency of visits of the token on node v at time t in the machine, π denotes the stationary distribution of corresponding random walk, and n and m denote the number of vertices and edges of the graph, respectively. Recently, Friedrich and Sauerwald [23] studied the cover time of the rotor-router machine for several basic finite graphs such as tree, star, torus, hypercube and complete graph.

As another topic on the Propp machine, the aggregation model has been investigated [40, 37, 41, 42]. For a random walk, tokens in the Internal Diffusion-Limited Aggregation (IDLA) model on $\mathbb{Z}^d$ asymptotically converge to the Euclidean ball [39], and the difference between the largest distance of a token arriving from the origin and the smallest distance of a point where no token arrived is bounded by $\mathrm{O}(n^{1/3}\mathrm{poly}(\log n))$ after $\Theta(n^d)$ steps [38]. For the Propp machine, Levine and Peres [40, 41, 42] showed that tokens in the rotor-router aggregation model also form the Euclidean ball, and showed several bounds for the difference between the largest distance of a token arriving from the origin and the smallest distance of a point where no token arrived. Kleber [37] gave some computational results.

Doerr, Friedrich, and Sauerwald [14] showed that information spreading by the rotor-router machine is faster than the one by a random walk on some specific graphs, namely trees with the common depth and the common degree of inner nodes, and random graphs with restricted connectivity. Doerr, Friedrich, Künnemann, and Sauerwald [13] gave some computational results for this phenomena. There is much other research on information spreading by the rotor-router machine on some graphs [2, 11, 15, 16, 17, 28].

6. Concluding remarks

This article has exhibited some topics concerning randomization and derandomization of computations. Unfortunately, the author has not come to the answer for the question "What is the property that a randomized algorithm really requires for randomness?" yet. There exist several fascinating open questions, including P≠ZPP, related to the subject.

Acknowledgment

This work was/is supported in part by KAKENHI No. 23650007, 24106005 and 25700002.

References

[1] Aldous D 1983 *Lect. Notes Math.* **986** 243–297
[2] Angelopoulos S, Doerr B, Huber A and Panagiotou K 2009 *Electron. J. Comb.* **16** #R102
[3] Boyer R and Moore J 1981 A fast majority vote algorithm *Tech. Repo.* ICSCA-CMP-32
[4] Boyer R and Moore J 1991 MJRTY—a fast majority vote algorithm *Automated Reasoning: Essays in Honor of Woody Bledsoe* 105–117
[5] Cooper J, Doerr B, Friedrich T and Spencer J 2010 *Random Struct. Algor.* **37** 353–366
[6] Cooper J, Doerr B, Spencer J and Tardos G 2007 *Eur. J. of Combin.* **28** 2072–2090
[7] Cooper C, Ilcinkas D, Klasing R and Kosowski A 2011 *Distrib. Comput.* **24** 91–99
[8] Cooper J and Spencer J 2006 *Comb. Probab. Comput.* **15** 815–822
[9] Demaine E, López-Ortiz A, and Munro J 2002 *Proc. of ESA* 348–360
[10] Diaconis P 2011 *J. Stat. Phys.* **144** 445–458
[11] Doerr P 2009 Introducing quasirandomness to computer science *Efficient Algorithms: Essays Dedicated to Kurt Mehlhorn on the Occasion of His 60th Birthday* 99–111
[12] Doerr B and Friedrich T 2009 *Comb. Probab. and Comput.* **18** 123–144
[13] Doerr B, Friedrich T, Künnemann M, and Sauerwald T 2009 *Proc. of ALENEX* 145–153
[14] Doerr B, Friedrich T, and Sauerwald T 2008 *Proc. of SODA* 773–781
[15] Doerr B, Friedrich T and Sauerwald T 2009 *Proc. of ICALP* 366–377
[16] Doerr B, Friedrich T, and Sauerwald T 2009 *Electron. Notes Discrete Math.* **34** 243–247
[17] Doerr B, Huber A and Levavi A 2009 *Proc. of ISAAC* 812–821
[18] Dyer M 2003 *Proc. of STOC* 693–699
[19] Dyer M, Frieze A, and Kannan R 1991 *J. ACM* **38** 1–17
[20] Fischer M and Salzburg S 1982 *J. Algorithm* **3** 376–379
[21] Flajolet P 1985 *BIT* **25** 113–134
[22] Friedrich T, Gairing M, and Sauerwald T 2010 *Proc. of SODA* 1620–1629
[23] Friedrich T and Sauerwald T 2010 *Electron. J. Comb.* **17** R167
[24] Gopalan P, Klivans A, and Meka R 2010 Polynomial-time approximation schemes for knapsack and related counting problems using branching programs arXiv:1008.3187v1
[25] Gopalan P, Klivans A, Meka R, Stefankovic D, Vempala S and Vigoda E 2011 *Proc. of FOCS* 817–826
[26] Häggström 2002 *Finite Markov chains and algorithmic application* (Cambridge University Press)
[27] Holroyd A and Propp J 2010 Rotor walks and Markov chains, *Algorithmic Probability and Combinatorics* (The American Mathematical Society) 105–126
[28] Huber H and Fountoulakis N 2009 *Electron. Notes Discrete Math.* **34** 553–559
[29] Ikeda S, Kubo I and Yamashita M 2009 *Theor. Comput. Sci.* **410** 94–100
[30] Jerrum M, Sinclair A and Vigoda E 2004 *J. ACM* **51** 671–697
[31] Kajino H, Kijima S and Makino K 2012 Discrepancy analysis of deterministic random walks on finite irreducible graphs, *Discussion paper*
[32] Karp R, Shenker S and Papadimitriou C 2003 *ACM T. Database Syst.* **28** 51–55
[33] Kijima S 2011 *Proc. of HJ* 227–236
[34] Kijima S, Koga K and Makino K 2012 *Proc. of ANALCO* 16–25.
[35] Kijima S and Matsui T 2006 *Random Struct. Algor.* **29** 243–256
[36] Kijima S and Matsui T 2008 *SIAM J. Comput.* **38** 1484–1503
[37] Kleber M 2005 *Math. Intell.* **27** 55–63
[38] Lawler G 1995 *Ann. Probab.* **23** 71–86

[39] Lawler G, Bramson M and Griffeath D 1992 *Ann. Probab.* **20** 2117–2140
[40] Levine L and Peres Y 2005 *Math. Intell.* **27** 9–11
[41] Levine L and Peres Y 2008 *Indiana U. Math. J.* **57** 431–450
[42] Levine L and Peres Y 2009 *Potential Anal.* **30** 1–27
[43] Levin D, Peres Y and Wilmer E 2008 *Markov Chains and Mixing Times* (American Mathematical Society)
[44] Manku G and Motwani R 2002 *Proc. of VLDB* 346–357
[45] Matsumoto M and Nishimura T 1998 *ACM T. Model. Comput. S.* **8** 3–30
[46] Mitzenmacher M and Upfal E 2005 *Probability and Computing: Randomized Algorithms and Probabilistic Analysis* (Cambridge University Press)
[47] Morris R 1978 *Commun. ACM* **21** 840–842
[48] Morris B and Sinclair A 2004 *SIAM J. Comput.* **34** 195–226
[49] Ogata M, Yamauchi Y, Kijima S and Yamashita M 2011 *Lect. Notes Comput. Sc.* **7074** 514–523
[50] Propp J and Wilson D 1996 *Random Struct. Algor.* **9** 223–252
[51] Sinclair A 1993 *Algorithms for Random Generation & Counting, A Markov chain approach* (Birkhäuser)
[52] Shiraga T, Yamauchi Y, Kijima S and Yamashita M 2012 Deterministic random walks for irrational transition probabilities, *Discussion paper*
[53] Stefankovic D, Vempala S and Vigoda E 2012 *SIAM J. Comput.* **41** 356–366
[54] Toivonen H 1996 *Proc. of VLDP* 134–145
[55] Wilson D 2000 *Random Struct. Algor.* **16** 85–113
[56] http://tcslab.csce.kyushu-u.ac.jp/~kijima/applets/2nTable/

Sharp thresholds and the partition function

Amin Coja-Oghlan[1] **and Daniel Reichman**[2]

[1] Goethe University, Mathematics Institute, 10 Robert Mayer St, Frankfurt 60325, Germany
[2] The Weizmann Institute, Rehovot, Israel

E-mail: acoghlan@math.uni-frankfurt.de, daniel.reichman@gmail.com

Abstract. Let $\mathbf{\Phi}$ be a random k-CNF formula over n variables with $m = \lceil r \cdot n \rceil$ clauses and let $Z(\mathbf{\Phi})$ be the number of satisfying assignments. Assume that k is sufficiently large and that $r \leq (1 - o_k(1))2^k \ln(k)/k$, where $o_k(1)$ denotes a certain function that tends to 0 as k gets large. We prove that in this case, $\mathbf{\Phi}$ is satisfiable with probability $1 - O(1/n)$. Together with a recent result of Abbe and Montanari [arXiv:1006.3786], this implies that for such k, r the limit $\lim_{n\to\infty} \frac{1}{n}\mathrm{E}\left[\ln(1 + Z(\mathbf{\Phi}))\right]$ exists. The existence of this limit is related to the existence of a sharp threshold for satisfiability.

1. Introduction

Let $k \geq 3$, let $V = V_n = \{x_1, \ldots, x_n\}$ be a set of n propositional variables and let $\mathbf{\Phi} = \mathbf{\Phi}_k(n, m)$ be a random k-CNF with m clauses over the variables V, chosen uniformly from the set of all $(2n)^{km}$ such formulas. Unless specified otherwise, we assume that $m = \lceil rn \rceil$ for a number $r > 0$ that remains fixed as $n \to \infty$. We say that $\mathbf{\Phi}$ has some property $\mathcal{A}$ with high probability *("w.h.p.") if* $\lim_{n\to\infty} \mathrm{P}\left[\mathbf{\Phi} \in \mathcal{A}\right] = 1$.

One of the best-known open problems in the theory of random constraint satisfaction problems is the *satisfiability threshold conjecture* (e.g., [6, 20]). It states that for any $k \geq 3$ there exists a *satisfiability threshold* r_k^* such that for any density $r < r_k^*$ (that remains fixed as $n \to \infty$), the random formula $\mathbf{\Phi}$ is satisfiable w.h.p., while $\mathbf{\Phi}$ is unsatisfiable w.h.p. for any $r > r_k^*$.

Although this conjecture remains open, Friedgut [13] proved that for any $k \geq 3$ there exists a *sharp threshold sequence* $r_k(n)$ such that for any (fixed) $\varepsilon > 0$, the following is true:

- if $m/n < r_k(n) - \varepsilon$, then $\mathbf{\Phi}$ is satisfiable w.h.p.
- if $m/n > r_k(n) + \varepsilon$, then $\mathbf{\Phi}$ is unsatisfiable w.h.p.

The issue is that the sequence $r_k(n)$ is not known to converge. Thus, Friedgut's result allows for the possibility that the density at which the satisfiability probability drops from $1 - o(1)$ to $o(1)$ varies with n.

Over the past decade, properties of random k-CNF formulas have been studied via sophisticated, albeit non-rigorous methods from statistical physics [21, 23, 25, 26]. A key quantity of interest in these methods is the *partition function*, i.e., the number $Z(\mathbf{\Phi})$ of satisfying assignments. More precisely, because in the satisfiable regime $Z(\mathbf{\Phi})$ scales exponentially in n, it is natural to consider the "quenched average" $\frac{1}{n}\mathrm{E}[\ln(1 + Z(\mathbf{\Phi}))]$, where we add 1 inside the logarithm because $Z(\mathbf{\Phi})$ might be equal to zero. The limit

$$\phi_k(r) = \lim_{n\to\infty} \frac{1}{n}\mathrm{E}[\ln(1 + Z(\mathbf{\Phi}))] \qquad (1)$$

is referred to as the *free entropy density* [24].

A priori the limit (1) may not exist for all $r > 0$. However, Abbe and Montanari [1] proved (among other things) the following *sufficient* condition for the existence of $\phi_k(r)$.

Theorem 1.1 **([1])** *Suppose that $k \geq 3$ and $r > 0$ are such that for some number $\varepsilon = \varepsilon(k, r) > 0$ we have*

$$\mathrm{P}\left[\mathbf{\Phi} \textit{ is satisfiable}\right] \geq 1 - O(\ln^{-1-\varepsilon} n). \tag{2}$$

Then the limit $\phi_k(r)$ exists.

Abbe and Montanari also observed that the condition (2) is satisfied for $r < 1$ (for any $k \geq 3$). Our second result gives a significantly better bound for sufficiently large values of k.

The existence of the limit $\phi_k(r)$ is related to the convergence of the sharp threshold sequence.

Theorem 1.2 *Let $k \geq 3$. If the limit $\phi_k(r)$ exists for all $r > 0$, then the sequence $r_k(n)$ converges.*

Theorem 1.2 may actually be known to some experts in the area, but we are not aware of a proof of it appearing in the literature. We are going to give the (fairly easy) proof in Section 3.

There are two possible perspectives on Theorem 1.2 and its relationship to Theorem 1.1. On the one hand, one might argue that Theorem 1.2 just implies that proving the existence of $\phi_k(r)$ is a difficult problem. On the other hand, more optimistically, given Theorems 1.1 and 1.2, in order to prove the existence of a sharp threshold for satisfiability, one "just" has to establish the bound (2). Indeed, a natural line of attack might be to extract bounds on $\mathrm{P}[\mathbf{\Phi}$ is unsatisfiable$]$ for $m/n < r_k(n)$ from Friedgut's argument [13]. Unfortunately, meeting the condition (2) via this approach appears to be difficult. Yet there might be alternatives to Friedgut's proof that do give such a bound. In fact, our second (and main) result establishes a much stronger bound than (2) for a range of densities.

Theorem 1.3 *There is a sequence ε_k that tends to 0 in the limit of large k such that*

$$\mathrm{P}\left[\mathbf{\Phi} \textit{ is satisfiable}\right] \geq 1 - O(1/n) \qquad \textit{if } r < (1 - \varepsilon_k) \cdot 2^k \ln(k)/k. \tag{3}$$

We emphasize that there is a number $l = l(k) > 0$ such that for any $k \geq 3, r > 0$ we have

$$\mathrm{P}\left[\mathbf{\Phi} \text{ is satisfiable}\right] \leq 1 - \Omega(n^{-l}) \tag{4}$$

(where $l = l(k) \to \infty$ in the limit of large k). This is because bounded-sized unsatisfiable sub-formulas appear with probability $\Omega(n^{-l})$. It would be interesting to figure out the best possible exponent $l(k)$ for which (4) holds. For densities below the "pure literal threshold" (which is far below the density covered in (3)), the most likely unsatisfiable formulas can be quantified precisely [28].

The proof of Theorem 1.3 draws on ideas from algorithms for finding a satisfying assignment to a random k-SAT formula [8]. The possibility that such an approach is possible has been raised explicitly (but not carried out) in [1, following Remark 2]. However, rather than actually deriving the lower bound (3) directly by (going through the painstaking work of) tracing an algorithm, we pursue a more generic approach that may be applicable to other problems beyond random k-SAT. More precisely, we show that for any fixed $\eta > 0$, with probability $1 - \exp(-\Omega(n))$ the random formula $\mathbf{\Phi}$ admits an assignment σ under which only ηn clauses are unsatisfied. Furthermore, with probability $1 - O(1/n)$ such an "almost-satisfying" assignment σ can be turned into an actual satisfying assignment via a local-search type process. The latter is inspired by [8], but the fact that we only need to mend a fairly small number of violated clauses simplifies the analysis considerably (by comparison to [8]); see Section 4 for details.

Notation. For an integer z we let $[z] = \{1, 2, \dots, z\}$. Moreover, for $i \in [m]$ we denote by $\mathbf{\Phi}_i$ the ith clause of the formula $\mathbf{\Phi}$. Further, for $j \in [k]$ we let $\mathbf{\Phi}_{ij}$ be the jth literal of $\mathbf{\Phi}_i$. Hence,

$$\mathbf{\Phi} = \mathbf{\Phi}_1 \wedge \dots \wedge \mathbf{\Phi}_m, \qquad \mathbf{\Phi}_i = \mathbf{\Phi}_{i1} \vee \dots \vee \mathbf{\Phi}_{ik}.$$

For a literal $l \in \{x_1, \neg x_1, \dots, x_n, \neg x_n\}$ we let $|l|$ denote the underlying variable.

We use asymptotic notation both with respect to n and with respect to the clause length k. Asymptotics in n are denoted by $O(\cdot)$ and asymptotics in k by $O_k(\cdot)$. In particular, $o_k(1)$ signifies a term that tends to 0 for sufficiently large k.

2. Related work

Friedgut's result [13] proves the *existence* of the threshold sequence $r_k(n)$, but does not reveal its location. For general values of k, the best current (rigorous) bounds on $r_k(n)$ derive from the first and the second moment method [3, 4, 9]: we have $r_k(n) = 2^k \ln 2 - O_k(1)$. For $k = 3$, the best current lower bound is obtained by tracing a certain linear-time satisfiability algorithm [17, 18]. Upper bounds have been obtained via (tricky) first moment arguments [11, 19, 22].

For large k, the best current (rigorously analyzed) polynomial-time algorithm finds satisfying w.h.p. for $r < (1 - \varepsilon_k)2^k \ln(k)/k$, with $\varepsilon_k \to 0$ in the limit of large k [8]. It is no coincidence that this matches the bound in Theorem 1.3. More precisely, there are two reasons why the two bounds match. First, we are going to build upon ideas from [8]. Second, and more substantially, the point $r \sim 2^k \ln(k)/k$ marks the so-called *dynamical replica symmetry breaking* transition. This transition was predicted on the basis of physics arguments [21], and (to an extent) established rigorously [2]. Roughly speaking, the physics picture suggests that for $r > (1 + \varepsilon_k)2^k \ln(k)/k$ the geometry of the set of satisfying assignments renders local-search techniques ineffective. Since the proof of Theorem 1.3 builds upon local-search ideas, it is unsurprising that the argument stops working at $r \sim 2^k \ln(k)/k$.

The quantity $\phi_k(r)$ is closely related to the so-called *free energy density at inverse temperature* β, which is defined as follows. For a formula Φ and an assignment σ let $E(\Phi, \sigma)$ denote the number of clauses of Φ that are violated under σ. Let

$$Z_\beta(\Phi) = \sum_{\sigma \in \{0,1\}^n} \exp(-\beta \cdot E(\Phi, \sigma)) \tag{5}$$

be the *partition function*. Then the free energy density is

$$f_k(r, \beta) = -\lim_{n\to\infty} \frac{1}{\beta n} \mathrm{E}\left[\ln Z_\beta(\mathbf{\Phi})\right]. \tag{6}$$

Bayati, Gamarnik and Tetali [5] recently proved the existence of the limit (6) for any k and $\beta, r \in (0, \infty)$. The proof is based on the interpolation method [12, 16].

As β grows large, the partition function $Z_\beta(\mathbf{\Phi})$ is going to be concentrated on "good" assignments that only violate "relatively few" clauses. Indeed, in a sense $Z(\mathbf{\Phi})$ corresponds to the partition function $Z_\beta(\mathbf{\Phi})$ with $\beta = \infty$, and thus one might expect that for densities r in the satisfiable regime, $\mathrm{E}\left[\ln Z_\beta(\mathbf{\Phi})\right]$ provides a good approximation to $\mathrm{E}\left[\ln(1 + Z(\mathbf{\Phi}))\right]$. To this extent, [5] instills hope that the limit $\phi_k(r)$ might indeed exist for all $k \geq 2, r \in (0, \infty)$.

3. Proof of Theorem 1.2

Assuming that $\phi_k(r)$ exists for all $r > 0$, we are going to show that the sharp threshold sequence converges to $r^* = \inf\{r > 0 : \phi_k(r) = 0\}$. Observe that the function $\phi_k(r)$ is monotonically decreasing (because increasing r corresponds to adding more random clauses, which cannot increase Z). Hence,

$$\phi_k(r) > 0 \quad \text{if} \quad 0 < r < r^*, \text{ while } \quad \phi_k(r) = 0 \quad \text{if} \quad r > r^*. \tag{7}$$

The proof of Theorem 1.2 consists of two steps: showing that $\mathbf{\Phi}$ is satisfiable w.h.p. if $r < r^*$, and that it is unsatisfiable w.h.p. if $r > r^*$.

Lemma 3.1 *If $r < r^*$, then $\mathbf{\Phi}$ is satisfiable w.h.p.*

Proof. Equation (7) shows that $\phi_k(r) > 0$. As a first step, we claim that

$$\liminf_{n\to\infty} \mathrm{P}\left[\mathbf{\Phi} \text{ is satisfiable}\right] \geq \phi_k(r) > 0. \tag{8}$$

For assume that $\mathrm{P}\left[\mathbf{\Phi} \text{ is satisfiable}\right] < \phi_k(r)$ for infinitely many n. Then for infinitely many n,

$$\mathrm{E}\left[\ln(1 + Z(\mathbf{\Phi}))\right] \leq \phi_k(r)\ln 2 + o(1) < 0.9\phi_k(r), \tag{9}$$

because the total number of satisfying assignments cannot exceed 2^n. However, (9) contradicts the fact that $\phi_k(r) = \lim_{n\to\infty} \mathrm{E}\left[\ln(1 + Z(\mathbf{\Phi}))\right]$. Thus, we have established (8).

Now, (8) implies that the sharp threshold sequence satisfies $r_k(n) \geq r - o(1)$. Because this holds for any $r < r^*$, we see that $\liminf_{n\to\infty} r_k(n) \geq r^*$. Therefore, for any density $r < r^*$ we have $\mathrm{P}\left[\mathbf{\Phi} \text{ is satisfiable}\right] = 1 - o(1)$, as desired. □

Lemma 3.2 *If $r > r^*$, then $\mathbf{\Phi}$ is unsatisfiable w.h.p.*

Proof. Assume towards a contradiction that $r > r^*$ but $\mathrm{P}\left[\mathbf{\Phi} \text{ is satisfiable}\right] \geq \delta = \Omega(1)$ for infinitely many n. It is well-known that the random formula $\mathbf{\Phi}$ has αn isolated variables (i.e., variables that do not appear in any clause) w.h.p., where $\alpha = \alpha(k, r) > 0$ is independent of n. Furthermore, if σ is a satisfying assignment of $\mathbf{\Phi}$, then another satisfying assignment σ' can be be obtained by altering the truth values of any set of isolated variables. Consequently,

$$\mathrm{P}\left[Z(\mathbf{\Phi}) \geq 2^{\alpha n}\right] \;\geq\; \mathrm{P}\left[\mathbf{\Phi} \text{ is satisfiable}\right] - \mathrm{P}\left[\text{there are at most } \alpha n \text{ isolated variables}\right] = \delta - o(1).$$

Therefore, we have $\mathrm{E}\left[\ln(1 + Z(\mathbf{\Phi})\right] \geq \alpha(\delta - o(1))n$ for infinitely many n, while (7) yields $\phi_k(r) = \lim_{n\to\infty} \mathrm{E}\left[\ln(1 + Z(\mathbf{\Phi}))\right] = 0$. □

Finally, Theorem 1.2 is immediate from Lemmas 3.1 and 3.2.

4. Proof of Theorem 1.3

Unless otherwise specified, we assume that $k \geq k_0$ for a certain (large) constant k_0, and that $r \leq (1 - \varepsilon_k)2^k \ln(k)/k$ for a sequence $\varepsilon_k = o_k(1)$ that tends to 0 sufficiently slowly in the limit of large k. We also assume that n is sufficiently big for our various estimates to hold.

4.1. Proof strategy

To prove Theorem 1.3, we take a detour through the k-NAESAT ("not-all-equal satisfiability") problem. Recall that an assignment $\sigma : V \to \{0, 1\}$ is *NAE-satisfying* for a formula Φ if both σ and its complement $\bar{\sigma}$ (defined by changing the ith bit of σ for every i, σ_i to $1 - \sigma_i$) satisfy Φ. For a formula Φ and an assignment σ we let $E_{\mathrm{NAE}}(\Phi, \sigma)$ denote the number of clauses that are violated under either σ or $\bar{\sigma}$. Thus, σ is NAE-satisfying iff $E_{\mathrm{NAE}}(\Phi, \sigma) = 0$. For a number $\eta > 0$ we let $Y_\eta(\Phi)$ be the number of assignments σ such that $|E_{\mathrm{NAE}}(\Phi, \sigma)| \leq \eta n$. Using arguments developed in [2, 10], we are going to establish the following in Section 4.2.

Proposition 4.1 *For any $\delta > 0$ there is $\eta_0 > 0$ such that for any $0 < \eta < \eta_0$ we have*

$$\mathrm{P}\left[Y_\eta(\mathbf{\Phi}) < \mathrm{E}\left[Y_\eta(\mathbf{\Phi})\right] \cdot \exp(-\delta n)\right] = O(n^{-2}). \tag{10}$$

Moreover, $\mathrm{E}\left[Y_\eta(\mathbf{\Phi})\right] = 2^{(1-o_k(1))n}$.

The reason why we consider NAESAT satisfiability as opposed to the "usual" satisfiability problem is that the statement corresponding to Proposition 4.1 would have been false. This is due to the inherent asymmetry of the k-SAT problem, which also causes the difficulty of computing $\frac{1}{n}\mathrm{E}\ln(1+Z(\mathbf{\Phi}))$; we refer to [9] for a detailed discussion.

Proposition 4.1 shows not only that w.h.p. the random formula $\mathbf{\Phi}$ has plenty of assignments σ such that $|E_{\mathrm{NAE}}(\mathbf{\Phi},\sigma)| \leq \eta n$, but also allows us to study the typical properties of such assignments. Formally, let us define the *Gibbs distribution* denoted by G_η (we omit the obvious dependency of n and m in notation) as the distribution on pairs $(\mathbf{\Phi},\sigma)$ induced by the following experiment:

- Sample a random formula $\mathbf{\Phi}$.
- Then, sample an assignment $\sigma \in \{0,1\}^n$ such that $E_{\mathrm{NAE}}(\mathbf{\Phi},\sigma) \leq \eta n$ uniformly.

The Gibbs distribution is cumbersome to work with directly. But Proposition 4.1 allows us to approximate this distribution by another one that is easier to cope with. Namely, the so-called *planted distribution*, denoted P_η, is induced by the following experiment (cf. [2]):

- Sample a random assignment $\sigma \in \{0,1\}^n$.
- Choose a k-CNF $\mathbf{\Phi}(\sigma)$ uniformly out of all formulas with n variables and m clauses such that $E_{\mathrm{NAE}}(\mathbf{\Phi}(\sigma),\sigma) \leq \eta n$.

Combining Proposition 4.1 with a double-counting argument akin to the one used in [2], we obtain

Proposition 4.2 *Assume that δ,η are such that (10) is satisfied. Then for any event $\mathcal{B}$ we have*

$$\mathrm{P}_{G_\eta}[\mathcal{B}] \leq \exp(2\delta n)\cdot \mathrm{P}_{P_\eta}[\mathcal{B}] + O(n^{-2}).$$

Equipped with Proposition 4.2, we perform the key step of the proof in Section 4.3, where we prove the following. Let us say that a map $\tau: V \to \{0,1,*\}$ *satisfies* a clause $\mathbf{\Phi}_i$ if the clause contains a literal l such that $\tau(x)=1$ if $l=x\in V$ is a positive literal, and $\tau(x)=0$ if $l=\neg x$ is negative.

Proposition 4.3 *There exist numbers $\eta_0>0,\gamma>0$ such that for all $0<\eta<\eta_0$ in the distribution P_η with probability at least $1-\exp(-\gamma n)$ the following is true.*

There is a map $\tau: V\to\{0,1,*\}$ such that $\left|\tau^{-1}(*)\right| \leq 4^{-k}n$ and such that each clause has one of the following two properties.

T1. *either the clause is satisfied under τ,*

T2. *or there occur three variables in the clause that are set to $*$ under τ.*

Combining Propositions 4.2 and 4.3, we obtain

Corollary 4.4 *With probability $1-O(n^{-2})$ the random formula $\mathbf{\Phi}$ admits an assignment τ as in Proposition 4.3.*

Furthermore, a standard first moment argument yields the following.

Lemma 4.5 *With probability $1-O(1/n)$ the random formula $\mathbf{\Phi}$ has the following property.*

Let $S\subset V$ be a set of size $|S|\leq 4^{-k}n$. Let $I\subset[m]$ be the set of indices such that for all $i\in I$ there are three indices $1\leq j_1<j_2<j_3\leq k$ such that $|\mathbf{\Phi}_{ij_1}|,|\mathbf{\Phi}_{ij_2}|,|\mathbf{\Phi}_{ij_3}|\in S$. Then $|I|\leq|S|$. (11)

Proof. For a set $S \subset V$ and a set $I \subset [m]$ let $\mathcal{E}(S,I)$ be the event that each clause $\mathbf{\Phi}_i$, $i \in I$, features three occurrences of variables in S. Let $s \leq n/4^k$ be an integer and let $\mathcal{E}(s)$ be the event that that there exist sets $S \subset V$ and $I \subset [m]$ of size $|S| = |I| = s$ such that $\mathcal{E}(S,I)$ occurs. By the union bound and because the clauses of $\mathbf{\Phi}$ are chosen uniformly and independently, we have

$$\mathrm{P}\left[\mathcal{E}(s)\right] \leq \sum_{S,I:|S|=|I|=s} \mathrm{P}\left[\mathcal{E}(S,I)\right] \leq \binom{n}{s}\binom{m}{s}\left[\binom{k}{3}\left(\frac{s}{n}\right)^3\right]^s \leq (k^3 2^k s/n)^s.$$

Hence, again by the union bound $\mathrm{P}\left[\bigcup_{1 \leq s \leq n/4^k} \mathcal{E}(s)\right] \leq \sum_{1 \leq s \leq n/4^k} (k^3 2^k s/n)^s = O(1/n)$. □

Proof of Theorem 1.3. Suppose that $\mathbf{\Phi}$ admits an assignment τ as in Proposition 4.3. Let $\mathcal{M}$ be the set of indices $i \in [m]$ such that τ fails to satisfy $\mathbf{\Phi}_i$. Moreover, let $S = \tau^{-1}(*)$. Set up a bipartite graph B on $\mathcal{M} \cup S$ in which $i \in \mathcal{M}$ is adjacent to all variables $x \in S$ that occur in $\mathbf{\Phi}_i$. If (11) holds, then by Hall's theorem, B contains a matching M that covers all of $\mathcal{M}$. Thus, we can construct a satisfying assignment σ by setting $\sigma(x) = \tau(x)$ if $\tau(x) \neq *$, and by setting $\sigma(x)$ so that the clause that x is matched to under M is satisfied if x is covered by M. (If $\tau(x) = *$ but x is not covered by M, we can set $\sigma(x)$ to either value.) Finally, Theorem 1.3 follows because Corollary 4.4 shows that there is an assignment τ as above with probability $1 - O(n^{-2})$, and because (11) holds with probability $1 - O(1/n)$ by Lemma 4.5. □

4.2. Proof of Proposition 4.1

The proof is based on studying the number $Z_{\mathrm{NAE}}(\mathbf{\Phi})$ of NAE-satisfying assignments of $\mathbf{\Phi}$ along with the partition function

$$Z_{\beta,\mathrm{NAE}}(\mathbf{\Phi}) = \sum_{\sigma \in \{0,1\}^V} \exp(-\beta \cdot E_{\mathrm{NAE}}(\mathbf{\Phi},\sigma)) \qquad \text{for } \beta \in (0,\infty). \tag{12}$$

Notice that, in contrast to (5), in (12) we work with the number $E_{\mathrm{NAE}}(\mathbf{\Phi},\sigma)$ of clauses that are NAE-violated. Clearly, for any $\beta > 0$ we have $Z_{\beta,\mathrm{NAE}} \geq Z_{\mathrm{NAE}}$. Hence, to obtain a lower bound on $Z_{\beta,\mathrm{NAE}}$, we are going to establish one for Z_{NAE}. The starting point is the following bound on the second moment of Z_{NAE}.

Lemma 4.6 ([3]) *For any $k \geq 3$ there is a number $C(k) > 0$ such that for $r < 2^{k-1}\ln 2 - 2$ we have $\mathrm{E}\left[Z_{\mathrm{NAE}}(\mathbf{\Phi})^2\right] \leq C(k) \cdot \mathrm{E}\left[Z_{\mathrm{NAE}}(\mathbf{\Phi})\right]^2$. Moreover, $\mathrm{E}\left[Z_{\mathrm{NAE}}(\mathbf{\Phi})\right] = \exp(\Omega(n))$.*

To turn Lemma 4.6 into a lower bound on $Z_{\mathrm{NAE}}(\mathbf{\Phi})$, we remember the *Paley-Zygmund inequality*: if $Y \geq 0$ is a random variable with $0 < \mathrm{E}\left[Y^2\right] < \infty$, then

$$\mathrm{P}\left[Y \geq t \cdot \mathrm{E}\left[Y\right]\right] \geq (1-t)^2 \mathrm{E}\left[Y\right]^2 / \mathrm{E}\left[Y^2\right] \qquad \text{for any } t \in (0,1).$$

Hence, Lemma 4.6 yields the following result (similar bounds have been used in [2, 10]).

Corollary 4.7 *Suppose that $k \geq 3$ and $r < 2^{k-1}\ln 2 - 2$. There exists a number $\alpha = \alpha(k) > 0$ such that $\mathrm{P}\left[Z_{\mathrm{NAE}} \geq \alpha \cdot \mathrm{E}\left[Z_{\mathrm{NAE}}\right]\right] \geq \alpha$.*

We proceed by showing (by a standard argument, cf. [5]) that Z_{NAE} is tightly concentrated.

Lemma 4.8 *Let $k \geq 3$ and let $\beta, r > 1$ be arbitrary but independent of n. Then*

$$\mathrm{P}\left[\left|\ln Z_{\beta,\mathrm{NAE}}(\mathbf{\Phi}) - \mathrm{E}\left[\ln Z_{\beta,\mathrm{NAE}}(\mathbf{\Phi})\right]\right| > 100\beta\sqrt{m}\ln n\right] \leq n^{-100}.$$

Proof. Let $\mathbf{\Phi}[t] = \mathbf{\Phi}_1 \wedge \cdots \wedge \mathbf{\Phi}_t$ be the formula comprising the first t clauses of $\mathbf{\Phi}$ $(0 \le t \le m)$. Fix an assignment σ. Then

$$\exp(-\beta E_{\mathrm{NAE}}(\mathbf{\Phi}[t+1], \sigma)) = \begin{cases} \exp(-\beta E_{\mathrm{NAE}}(\mathbf{\Phi}[t], \sigma)) & \text{if } \sigma \text{ NAE-satisfies } \mathbf{\Phi}_{t+1}, \\ \exp(-\beta E_{\mathrm{NAE}}(\mathbf{\Phi}[t], \sigma) - \beta) & \text{otherwise.} \end{cases}$$

Hence, for any $0 \le t < m$ we have $|\ln Z_{\beta,\mathrm{NAE}}(\mathbf{\Phi}[t+1]) - \ln Z_{\beta,\mathrm{NAE}}(\mathbf{\Phi}[t])| \le \beta$, i.e., the random variables $(\ln Z_{\beta,\mathrm{NAE}}(\mathbf{\Phi}[t]))_{0 \le t \le m}$ satisfy a Lipschitz condition. Because the clauses are independent, it follows from Azuma's inequality that

$$\mathrm{P}\left[|\ln Z_{\beta,\mathrm{NAE}}(\mathbf{\Phi}) - \mathrm{E}\left[\ln Z_{\beta,\mathrm{NAE}}(\mathbf{\Phi})\right]| > \lambda\right] \le 2 \exp\left[-\frac{\lambda^2}{2\beta^2 m}\right] \qquad (\lambda > 0).$$

Setting $\lambda = 100\beta\sqrt{m}\ln n$ completes the proof. □

Corollary 4.9 *Let $k \ge 3$ and let $\beta, r > 1$ be arbitrary but independent of n. Then*

$$\mathrm{E}\left[\ln Z_{\beta,\mathrm{NAE}}(\mathbf{\Phi})\right] \ge \ln \mathrm{E}\left[Z_{\mathrm{NAE}}(\mathbf{\Phi})\right] - n^{0.51}.$$

Proof. Once more we use a standard concentration argument (reminiscent of Frieze's argument for proving the concentration of the chromatic number of random graphs). Assume towards a contradiction that $\mathrm{E}\left[\ln Z_{\beta,\mathrm{NAE}}(\mathbf{\Phi})\right] < \ln \mathrm{E}\left[Z_{\mathrm{NAE}}(\mathbf{\Phi})\right] - n^{0.51}$. Let $\alpha > 0$ be as in Corollary 4.7. Since $Z_{\beta,\mathrm{NAE}} \ge Z_{\mathrm{NAE}}$, Lemma 4.8 implies

$$\begin{aligned} \mathrm{P}\left[Z_{\mathrm{NAE}}(\mathbf{\Phi}) \ge \alpha \mathrm{E}\left[Z_{\mathrm{NAE}}(\mathbf{\Phi})\right]\right] &\le \mathrm{P}\left[\ln Z_{\beta,\mathrm{NAE}}(\mathbf{\Phi}) \ge \ln \mathrm{E}\left[Z_{\mathrm{NAE}}(\mathbf{\Phi})\right] - O(1)\right] \\ &\le \mathrm{P}\left[\ln Z_{\beta,\mathrm{NAE}}(\mathbf{\Phi}) \ge \mathrm{E}[\ln\left[Z_{\beta,\mathrm{NAE}}(\mathbf{\Phi})\right]] + n^{0.51} - O(1)\right] \\ &\le n^{-100}, \end{aligned}$$

in contradiction to Corollary 4.7. □

Proof of Proposition 4.1. Let $\eta > 0$ be sufficiently small and choose $\beta = \beta(\eta) > 0$ big enough so that $2^n \exp(-\beta\eta n) < 1$. Lemma 4.8 and Corollary 4.9 imply that with probability $1 - O(n^{-100})$ we have $Z_{\beta,\mathrm{NAE}} \ge \mathrm{E}\left[Z_{\mathrm{NAE}}\right] \exp(-n^{2/3}) = \exp(\Omega(n))$ (by Lemma 4.6). In this case, we have

$$\begin{aligned} \mathrm{E}\left[Z_{\mathrm{NAE}}(\mathbf{\Phi})\right] \exp(-n^{2/3}) &\le Z_{\beta,\mathrm{NAE}} \le Y_\eta(\mathbf{\Phi}) + \sum_{\sigma: E_{\mathrm{NAE}}(\sigma, \mathbf{\Phi}) > \eta n} \exp(-\beta E_{\mathrm{NAE}}(\sigma, \mathbf{\Phi})) \\ &\le Y_\eta(\mathbf{\Phi}) + 1 \qquad [\text{because } 2^n \exp(-\beta\eta n) < 1]. \end{aligned} \tag{13}$$

Furthermore, because a given assignment NAE-satisfies a random clause with probability $1 - 2^{1-k}$, and because the m clauses of $\mathbf{\Phi}$ are chosen independently, we have

$$\mathrm{E}\left[Y_\eta(\mathbf{\Phi})\right] \ge \mathrm{E}\left[Z_{\mathrm{NAE}}(\mathbf{\Phi})\right] = 2^n(1 - 2^{1-k})^m = 2^{(1-o_k(1))n} \quad [\text{as } r \le 2^k \ln(k)/k]. \tag{14}$$

Conversely,

$$\begin{aligned} \mathrm{E}\left[Y_\eta(\mathbf{\Phi})\right] &= 2^n \cdot \mathrm{P}\left[\mathrm{Bin}(m, 2^{1-k}) \le \eta n\right] \sim 2^n \binom{m}{\eta n} 2^{(1-k)\eta n}(1 - 2^{1-k})^{m-\eta n} \\ &\le 2^n(1 - 2^{1-k})^m \cdot \exp\left[\eta n\left(1 - \ln\eta + \ln r + (1-k)\ln 2 - \ln(1 - 2^{1-k})\right)\right] \\ &= \mathrm{E}\left[Z_{\mathrm{NAE}}\right] \cdot \exp\left[\eta n\left(1 - \ln\eta + \ln r + (1-k)\ln 2 - \ln(1 - 2^{1-k})\right)\right]. \end{aligned} \tag{15}$$

Since $\lim_{\eta \to 0} \eta\left(1 - \ln\eta + \ln r + (1-k)\ln 2 - \ln(1 - 2^{1-k})\right) = 0$, (15) shows that for any given $\alpha > 0$ we can choose η small enough so that $\mathrm{E}\left[Y_\eta\right] \le \exp(\alpha n)\mathrm{E}\left[Z_{\mathrm{NAE}}\right]$. Thus, the assertion follows from (13) and (14). □

4.3. Proof of Proposition 4.3

Throughout this section we assume that $0 < \eta < \eta_0$ for a small enough number $\eta_0 = \eta_0(k) > 0$. We start from a pair $(\mathbf{\Phi}, \sigma)$ chosen from P_η; without loss of generality, we may assume that $\sigma = \mathbf{1}$ is the all-true assignment. We are going to construct the desired map τ via the following process.

(i) Initially, let $\tau = \sigma$.

(ii) While there is a clause $\mathbf{\Phi}_i$ that is unsatisfied under τ and that does not contain at least three variables that take the value $*$ under τ, pick the least such index i.

(iii) Let us say that a variable x *supports* a clause $\mathbf{\Phi}_l$ under τ if
- $\tau(x) \in \{0, 1\}$,
- $\mathbf{\Phi}_l$ has at most two positive literals, and
- assigning x the value $1 - \tau(x)$ while keeping the values of all other variables deprives $\mathbf{\Phi}_l$ of satisfied literals (where literals whose underlying variables are set to $*$ do not count as satisfied).

If there are at least three indices $j \in [k-3]$ such that the variable $|\mathbf{\Phi}_{ij}|$ does not support a clause under τ, then pick any three such indices j and set them to $*$ under τ.

(iv) If such three indices do not exist, set $\tau(|\mathbf{\Phi}_{ij}|) = *$ for $j = k-2, k-1, k$.

The map τ that results from this process clearly satisfies conditions **T1** and **T2**. However, it is non-trivial to upper bound the number of variables assigned $*$ in the course of the above process. To accomplish this, we need a few basic properties of the random pair $(\mathbf{\Phi}, \sigma)$. Let us call a clause $\mathbf{\Phi}_i$ *critical* if there is a variable x that supports $\mathbf{\Phi}_i$; hence, $\mathbf{\Phi}_i$ has no more than two positive literals, and flipping a single variable may render $\mathbf{\Phi}_i$ unsatisfied.

Lemma 4.10 *With probability* $\geq 1 - \exp(-\Omega(n))$ *the following two statements hold.*

(i) The total number of clauses that are critical under σ *is at most* $n(1 - \varepsilon_k + 1/k)\ln k$.

(ii) There are at least $\frac{1}{2}k^{\varepsilon_k - 1}n$ *variables that do not support a clause under* τ_0.

Lemma 4.10 follows from our assumption that $r \leq (1-\varepsilon_k)2^k \ln(k)/k$ and standard concentration bounds.

Lemma 4.11 *Let* $\Omega(1) < \alpha < k^{-4}$. *With probability* $1 - \exp(-\Omega(n))$ *the following is true.*

> *Let* $S \subset V$ *be a set of size* $|S| = \alpha n$. *The total number of clauses that have at most 2 positive literals, one of which belongs to* S, *is bounded by* $-k^3\alpha\ln(\alpha)\cdot n$. (16)

Proof. Fix a set S of size αn. Then the number X_S of clauses that have at most 2 positive literals, one of which belongs to S, has a binomial distribution with mean $\leq \alpha k^2 n$. Hence, by the Chernoff bound $\mathrm{P}\left[X_S > -k^3\alpha\ln(\alpha)\, n\right] \leq \exp(k^3\alpha\ln(\alpha)\, n)$. Thus, by the union bound

$$\begin{aligned}\mathrm{P}\left[\exists S : X_S > -k^3\alpha\ln(\alpha)\right] &\leq \binom{n}{\alpha n}\exp(k^3\alpha\ln(\alpha)\, n)\\ &\leq \exp\left[\alpha n\left(1 - \ln\alpha + k^3\ln\alpha\right)\right] \leq \exp(-\Omega(n)),\end{aligned}$$

as claimed. □

Corollary 4.12 *With probability* $1 - \exp(-\Omega(n))$ *the following is true.*

> *Let* S *be a set of size* $|S| \leq n/k^6$. *The number of clauses that are not critical under* σ *but that become so once we set the variables in* S *to* $*$ *is bounded by* n/k^3. (17)

After these preliminaries we come to the actual analysis of the above process that yields τ. We characterize its state at time t by maps $\tau_t : V \to \{0, 1, *\}$ and

$$\pi_t : [m] \times [k] \to \{\pm 1, x_1, \bar{x}_1, \dots, x_n, \bar{x}_n\}.$$

Let $\tau_0 = \sigma$. Moreover, we define

$$\pi_0(i,j) = \begin{cases} \mathbf{\Phi}_{ij} & \text{if } |\mathbf{\Phi}_{ij}| \text{ supports clause } \mathbf{\Phi}_i, \\ \mathrm{sign}(\mathbf{\Phi}_{ij}) & \text{otherwise.} \end{cases}$$

Each time step of the process corresponds to mending one clause $\mathbf{\Phi}_i$ (i.e., making sure that $\mathbf{\Phi}_i$ receives three $*$-variables). As time progresses, π_t will summarise the information about $\mathbf{\Phi}$ that has been revealed thus far.

More precisely, suppose we have already defined $\pi_0, \dots, \pi_{t-1}$. Then π_t is obtained as follows. Suppose at time t we mend clause $\mathbf{\Phi}_{i_t}$. Let τ_t be the assignment obtained after steps (iii) and (iv) above, i.e., τ_t is obtained from τ_{t-1} by setting up to three of the variables in $\mathbf{\Phi}_{i_t}$ to $*$. We define

$$\pi_t(i,j) = \begin{cases} \mathbf{\Phi}_{ij} & \text{if either } i = i_t, \ |\mathbf{\Phi}_{ij}| \text{ supports clause } \mathbf{\Phi}_i, \text{ or } \tau_t(|\mathbf{\Phi}_{ij}|) = *, \\ \pi_{t-1}(i,j) & \text{otherwise.} \end{cases}$$

Let $\mathcal{A}_t = \tau_t^{-1}(*)$. Moreover, let T be the stopping time of the process, i.e., the first time when all clauses are either satisfied under τ_t or contain at least three variables x with $\tau_t(x) = *$. In addition, let $(\mathcal{F}_t)_{t \ge 0}$ be the filtration generated by π_t, τ_t. The following observation ("principle of deferred decisions") is immediate from the construction.

Fact 4.13 *For any $t \ge 0$, given $\mathcal{F}_t$ the variables $|\mathbf{\Phi}_{ij}|$ with $\pi_t(i,j) = \pm 1$ are independently uniformly distributed over $V \setminus \mathcal{A}_t$.*

Let $\theta = 2\eta n$. We aim to show that $T \le \theta$ with probability $1 - \exp(-\Omega(n))$. To prove this, we need to bound the number of times that step (iv) is executed (observe that new unsatisfied clauses not containing three $*$ can only generated in step (iv), hence it suffice to consider the number of iterations of step (iv)). Thus, let us call a variable x *t-free* if $\tau_{t-1}(x) \ne *$ and if x does not support a clause under τ_{t-1}. Moreover, let us call step t a *failure* if there are fewer than three t-free variables in $\{|\mathbf{\Phi}_{i_t j}| : j \in [k-3]\}$.

Lemma 4.14 *Suppose we run the process for $t \le \theta$ steps. Then with probability $1 - \exp(-\Omega(n))$ there are at most θ / k^9 failures.*

Proof. By Corollary 4.12 we may assume that (17) holds. In addition, we may assume that 1. and 2. in Lemma 4.10 are satisfied. We claim that at any time $t \le \theta$,

$$\mathrm{P}\left[\text{step } t \text{ is a failure} | \mathcal{F}_{t-1}\right] \le k^{-10}. \tag{18}$$

Indeed, by the definition of the process, the clause $\mathbf{\Phi}_{i_t}$ has at least $k-6$ literals $j \in [k-3]$ such that $\pi_{t-1}(i_t, j) \in \{\pm 1\}$. Fact 4.13 shows that for all such j, the variable $|\mathbf{\Phi}_{i_t j}|$ is uniformly distributed over $V \setminus \mathcal{A}_{t-1}$, independently of the others. Furthermore, we claim that the total number of variables that are t-free is at least $\frac{1}{4} k^{\varepsilon_k - 1} n$. Indeed, there are at least $\frac{1}{2} k^{\varepsilon_k - 1} n$ variables that do not support a clause under $\tau_0 = \sigma$ (by item 2. in Lemma 4.10). Moreover, by (17) (assuming η is small enough) the number of such variables that fail to be t-free is bounded by n/k^3. Hence, for each variable $|\mathbf{\Phi}_{i_t j}|$ with $\pi_{t-1}(i_t, j) \in \{\pm 1\}$ and $j \in [k-3]$ the probability of being t-free is at least $\frac{1}{4} k^{\varepsilon_k - 1}$. Thus, the probability that at most two of them are t-free is bounded by

$$\xi = k^2 (1 - k^{\varepsilon_k - 1}/4)^{k-5} \le k^2 \exp(-(k-5) k^{\varepsilon_k - 1}/4) \le k^{-10}.$$

Finally, because (18) provides a uniform bound for all $t \le \theta$, the total number of failures is stochastically dominated by a binomial variable $\mathrm{Bin}(\theta, k^{-10})$. Thus, the assertion follows from the Chernoff bound. □

Corollary 4.15 *With probability* $1 - \exp(-\Omega(n))$ *the stopping time of the process is* $T \le \theta$.

Proof. By Lemma 4.11 we may assume that (16) holds. In addition, we may assume that 1. and 2. in Lemma 4.10 are satisfied, and that the total number of failures $t \le \theta$ is bounded by θ/k^9 (by Lemma 4.14). Let F be the set of all $t \le \theta$ that are failures and let $\mathcal{B} = \bigcup_{t \in F} \mathcal{A}_t \setminus \mathcal{A}_{t-1}$ contain all the variables that were added to $\mathcal{A}_T$ at steps $t \in F$. Let us say that clause $\mathbf{\Phi}_i$ is *t-unhappy* if there is a time $s \le t$ such that $\pi_s(i,j) \ne 1$ for all $j \in [k]$ and $|\{j \in [k] : \pi_s(i,j) = -1\}| \ge k-2$. In words, at some time $s \le t$ clause $\mathbf{\Phi}_i$ did neither contain three $*$-variables nor at least one positive literal.

Because the pair $(\mathbf{\Phi}, \sigma)$ was chosen from the distribution P_η, no more than ηn clauses are 0-unhappy. Hence, if $T > \theta$, then there must be at least $\theta/2$ clauses that are θ-unhappy but not 0-unhappy. Furthermore, all of these clauses have at most two positive literals, at least one of which belongs to $\mathcal{B}$. This contradicts (16), because $|\mathcal{B}| \le 3|F| \le 3\theta/k^9$. □

Proof of Proposition 4.3. Corollary 4.15 yields $\mathrm{P}[T \le \theta] \ge 1 - \exp(-\Omega(n))$. Since in each step of the process at most three variables are set to $*$, we conclude that $\mathrm{P}[|\mathcal{A}_T| \le 3\theta] \ge 1 - \exp(-\Omega(n))$. Finally, for sufficiently small η we have $3\theta/n \le 4^{-k}$. □

Acknowledgments

The first author thanks David Gamarnik, Will Perkins and Prasad Tetali for helpful discussions. The second author has been supported in part by The Israel Science Foundation (grant No. 621/12) and by the Citi Foundation. The research leading to these results has received funding from the European Research Council under the European Union's Seventh Framework Programme (FP/2007-2013) / ERC Grant Agreement n. 278857–PTCC.

References

[1] Abbe E and Montanari A 2010 On the concentration of the number of solutions of random satisfiability formulas *Preprint* CoRR/1006.3786
[2] Achlioptas D and Coja-Oghlan A 2008 Algorithmic barriers from phase transitions *Proc. 49th FOCS* pp 793–802
[3] Achlioptas D and Moore C 2006 Random k-SAT: two moments suffice to cross a sharp threshold *SIAM J. Comput.* **36** 740–62
[4] Achlioptas D and Peres Y 2004 The threshold for random k-SAT is $2^k \ln 2 - O(k)$ *J. Amer. Math. Soc.* **17** 947–73
[5] Bayati M, Gamarnik D and Tetali P 2010 Combinatorial approach to the interpolation method and scaling limits in sparse random graphs *Proc. 42nd STOC* pp 105–14
[6] Cheeseman P, Kanefsky B and Taylor W 1991 Where the *really* hard problems are *Proc. IJCAI* pp 331–7
[7] Chvátal V and Reed B 1992 Mick gets some (the odds are on his side) *Proc. 33th FOCS* pp 620–7
[8] Coja-Oghlan A 2010 A better algorithm for random k-SAT *SIAM J. Comput.* **39** 2823–64
[9] Coja-Oghlan A and Panagiotou K 2013 Going after the k-SAT threshold *Proc. 45th STOC* pp 705–14
[10] Coja-Oghlan A and Zdeborová L 2012 The condensation transition in random hypergraph 2-coloring *Proc. 23rd SODA* pp 241–50
[11] Dubois O, Boufkhad Y and Mandler J 2000 Typical random 3-SAT formulae and the satisfiability threshold *Proc. 11th SODA* pp 126–7
[12] Franz S and Leone M 2003 Replica bounds for optimization problems and diluted spin systems *J. Statist. Phys.* **111** 535–64
[13] Friedgut E 1999 Sharp thresholds of graph properties, and the k-SAT problem *J. Amer. Math. Soc.* **12** 1017–54
[14] Frieze A and Suen S 1996 Analysis of two simple heuristics on a random instance of k-SAT *J. Algorithms* **20** 312–55

[15] Frieze A and Wormald N 2005 Random k-Sat: a tight threshold for moderately growing k *Combinatorica* **25** 297–305
[16] Guerra F 2003 Broken replica symmetry bounds in the mean field spin glass model *Comm. Math. Phys.* **233** 1–12
[17] Hajiaghayi M and Sorkin G 2003 The satisfiability threshold of random 3-SAT is at least 3.52 IBM Research Report RC22942
[18] Kaporis A, Kirousis L and Lalas E 2006 The probabilistic analysis of a greedy satisfiability algorithm *Random Structures Algorithms* **28** 444–80
[19] Kirousis L, Kranakis E, Krizanc D and Stamatiou Y 1998 Approximating the unsatisfiability threshold of random formulas *Random Structures Algorithms* **12** 253–69
[20] Kirkpatrick S and Selman B 1994 Critical behavior in the satisfiability of random boolean expressions *Science* **264** 1297–301
[21] Krzakala F, Montanari A, Ricci-Tersenghi F, Semerjian G and Zdeborová L 2007 Gibbs states and the set of solutions of random constraint satisfaction problems *Proc. National Academy of Sciences* **104** pp 10318–23
[22] Maneva E and Sinclair A 2008 On the satisfiability threshold and clustering of solutions of random 3-SAT formulas *Theoret. Comput. Sci.* **407** 359–69
[23] Mertens S, Mézard M and Zecchina R 2006 Threshold values of random K-SAT from the cavity method *Random Structures Algorithms* **28** 340–73
[24] Mézard M and Montanari A 2009 *Information, Physics and Computation* (Oxford: Oxford University Press)
[25] Mézard M, Parisi G and Zecchina R 2002 Analytic and algorithmic solution of random satisfiability problems *Science* **297** 812–5
[26] Monasson R and Zecchina R 1996 Entropy of the K-satisfiability problem *Phys. Rev. Lett.* **76** 3881–3885
[27] Panchenko D and Talagrand M 2004 Bounds for diluted mean-fields spin glass models *Probab. Theory Related Fields* **130** 319–36
[28] Scott A and Sorkin G 2010 Structure of random r-SAT below the pure literal threshold *Preprint* CoRR/1008.1260

A survey on pattern formation of autonomous mobile robots: asynchrony, obliviousness and visibility

Yukiko Yamauchi
Department of Informatics, Graduate School of ISEE, Kyushu University,
744 Motooka, Nishi-ku, Fukuoka, Japan.

E-mail: yamauchi@inf.kyushu-u.ac.jp

Abstract. A robot system consists of autonomous mobile robots each of which repeats Look-Compute-Move cycles, where the robot observes the positions of other robots (Look phase), computes the track to the next location (Compute phase), and moves along the track (Move phase). In this survey, we focus on self-organization of mobile robots, especially their power of forming patterns. The formation power of a robot system is the class of patterns that the robots can form, and existing results show that the robot system's formation power is determined by their asynchrony, obliviousness, and visibility. We briefly survey existing results, with impossibilities and pattern formation algorithms. Finally, we present several open problems related to the pattern formation problem of mobile robots.

1. Introduction

Motion planning of mobile robots has been studies in many areas such as sensing, exploration, tracking, boundary patrolling, vehicle routing, and so on. As mobile robots become inexpensive and large-scale robot systems become widely used, distributed control of robot systems has attracted increasing attention. In this paper, we consider a robot system consisting of autonomous mobile robots with very weak capabilities, that is, each robot observes the locations of other robots, computes a track to the next location, and move along the track. There is no centralized control, the robots cannot explicitly exchange messages, and they may stop on the track before reaching the destinations.

One of the most important properties of such robot system is its *autonomy*. Because initial deployment of robots can be arbitrary, states of robots may not be initialized, and the robots may not be activated simultaneously, an initial configuration can be arbitrary. It is expected that the robots autonomously adapt to a given initial configuration and start their task. A system is *self-stabilizing* if it autonomously recovers correct behavior starting from any arbitrary initial configuration [3]. Self-stabilization also guarantees fault-tolerance for any finite number of transient faults, because the configuration after the last fault is considered as an arbitrary initial configuration.

The *pattern formation problem* is to form a given pattern from a given initial configuration. See Figure 1, which is an example of circle formation of 24 robots. A robot system is *self-organizing* if the robots autonomously form a given pattern. Since Yamashita and Suzuki [11] shed light to the theoretical aspect of distributed control of robot systems, many papers have revealed self-organization and self-stabilization of robot systems. In this survey, we focus on the pattern formation problem of mobile robot systems.

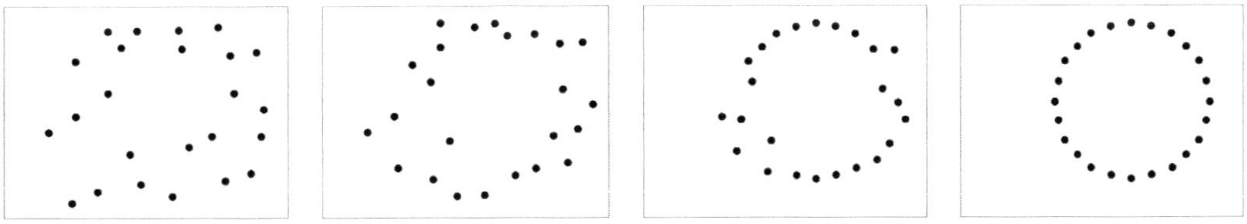

(a) Initial configuration (b) (c) (d) Final configuration

Figure 1. Circle formation

We model a robot by a point in a two dimensional Euclidean space. Each robot repeats "Look-Compute-Move" cycle, where it observes the locations of other robots (Look phase), computes a track to the next location using a given algorithm (Compute phase), and move along the track (Move phase). The robots are *anonymous*, that is, they have no identifiers and they cannot recognize each other. They are *oblivious*, that is, they do not remember the past cycles and they compute with the observation of the current cycle. They observe and move with its *local coordinate* system. They neither know the global coordinate system nor agree on the local coordinate system. Finally, their Look-Compute-Move cycles are neither instantaneous nor synchronized. This model is called *asynchronous* (ASYNC). Note that any look phase is instantaneous, in the sense that it returns the positions of robots at a time. Two stronger models are the *semi-synchronous* (SSYNC) model and the *fully-synchronous* (FSYNC) model. In the SSYNC model, Look-Compute-Move cycles are instantaneous, i.e., the time to execute a cycle is negligible, and in the FSYNC model, all robots execute the i-th cycle simultaneously. The essential difference between the ASYNC and the SSYNC models is that a robot may be observed while moving in the ASYNC model, while such a situation never occurs in the SSYNC model.

Suzuki and Yamashita first revealed the pattern formation power of non-oblivious SSYNC and FSYNC robots [11]. They characterized pattern F formable from a given initial configuration I: I is formable from F if the geometric symmetricity of I divides that of F. Here, geometric symmetricity is the number of rotations centered at the smallest enclosing circle of I (F, respectively) that produces I (F, respectively) itself. Later, Yamashita and Suzuki showed that oblivious robots in the SSYNC model have the same formation power as non-oblivious robots in the FSYNC model except the gathering of two robots [12]. The *gathering problem* is to gather all robots at some point, which is a special case of the pattern formation problem. They also pointed out that any oblivious pattern formation algorithm is self-stabilizing which motivates many researches on oblivious robots. Flocchini, Prencipe, Santoro, and Widmayer introduce the ASYNC model [5]. The main difficulty in the ASYNC model is that the robots may observe other robots while they are moving, and never recognize it. Fujinaga, Ono, Kijima, and Yamashita introduced the *embedded pattern formation problem* where robots can "see" the target pattern as if is drawn on the ground [7]. They proposed an algorithm in ASYNC model that outputs an optimal matching between the initial configuration and the embedded target pattern in the sense that the sum of Euclid distance of the matching is minimized[1]. Hence, even when the robots move asynchronously, the matching never changes. Based on the embedded pattern formation, Fujinaga, Yamauchi, Kijima, and Yamashita showed that oblivious robots in the ASYNC model have the same formation power as non-oblivious robots in the FSYNC model [8]. They showed that robots can agree on the embedding and keep it during the formation even in the ASYNC model.

[1] Their algorithm considers target patterns without multiplicity, that is, no two robots occupy the same point in the target pattern. The algorithms in [11, 12] allows multiplicities in the target pattern.

All these existing works assume unlimited visibility of robots, that is, each robot can observe the positions of all other robots. A few papers work on the pattern formation problem of robots with limited visibility. The *convergence problem* is to make robots move arbitrarily close to some point. Ando, Oasa, Suzuki, and Yamashita provided a convergence algorithm for oblivious robots with limited visibility in the SSYNC and FSYNC models [1]. Flocchini, Prencipe, Santoro, and Widmayer provided a gathering algorithm for oblivious robots with limited visibility in the ASYNC model [4]. Recently, Yamauchi and Yamashita showed that oblivious robots with limited visibility have essentially weak pattern formation power compared with robots with unlimited visibility. This is because robots with limited visibility increase the symmetricity without knowing the positions of other robots outside their visibility. However, they also provide positive results for non-oblivious robots by showing that they can regain the initial symmetricity if the system is FSYNC model or robots' movements are rigid, that is, robots reaches destinations in every cycle.

The sequence of papers on the pattern formation problem has shown that the pattern formation power of robot system is determined by the asynchrony, obliviousness, and visibility. This paper provides over view of the existing impossibility results and distributed algorithm mainly from [1, 7, 4, 8, 11, 12, 13]. See also the book by Flocchini, Prencipe, and Santoro [6].

Organization: This survey is organized as follows. In Section 2, we define the robot system and the pattern formation problem. We start with the simplest case with the gathering problem and the convergence problem in Section 3, which shows essential difficulties in distributed settings. Then, we proceed to the pattern formation problem in Section 4. We conclude and show future directions in Section 5.

2. Robot Model and Problems

Robot System: Let $R = \{r_1, r_2, \ldots, r_n\}$ $(n \geq 3)$ be a set of anonymous robots in a two-dimensional Euclidean space. Each robot r_i does not have an identifier, and we use r_i just for description.

We consider discrete time $1, 2, \ldots$. Let $p_i(t)$ (in the global coordinate system Z_0) be the position of r_i at time t ($r_i \in R$). $P(t) = \{p_1(t), p_2(t), \ldots, p_n(t)\}$ is a configuration of robots at time t. The robots initially occupy distinct locations, i. e, $|P(0)| = n$.

The robots do not agree on the coordinate system, and each robot r_i has its *local coordinate system* $Z_i(t)$ such that the origin of $Z_i(t)$ is the position of r_i at time t (i.e., $\mathbf{0} = Z_i(t)[p_i(t)]$)[2]. However, we assume all local coordinate systems are right-handed.

We denote the distance (in Z_0) between two points p, q by $dist(p, q)$. *Unlimited visibility* provides each robot r_i with the positions of all robots irrespective of their distance from r_i while *limited visibility* provides r_i with the positions of robots in distance V (in Z_0) from r_i. Each robot does not know the value of V. Let $R_i(t)$ be the set of robots visible from r_i at time t, and $P_i(t) = \{p_j | r_j \in R_i(t)\}$. In a look phase, r_i obtains $S_i(t) = Z_i(t)[P_i(t)]$ at some time t in the look phase. We call $S_i(t)$ the *local view* of r_i at time t. Because the visibility range V is common to all robots, $r_i \in S_j(t)$ means $r_j \in S_i(t)$. We define an undirected graph $G(t) = (R, E(t))$ for $P(t)$, called *mutual visibility graph* at time t where $(r_i, r_j) \in E(t)$ if and only if $r_i \in S_j(t)$. We assume that $G(0)$ is connected. In the following, we use $P(t)$ and $G(t)$ ($S_i(t)$ and the subgraph of $G(t)$ induced by $R_i(t)$, respectively) interchangeably.

Multiplicity detection allows robots to determine whether a point is occupied by more than two robots. A robot without multiplicity detection only recognizes a point as occupied, but does not know the number of robots on the point. In the following, we assume robots without multiplicity detection, unless explained explicitly.

[2] We assume that during a move phase, the origin of $Z_i(t)$ is fixed to the point of r_i in the look phase, and does not change.

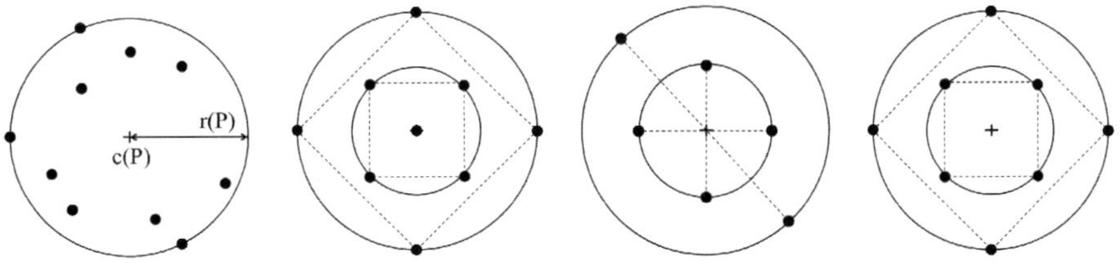

(a) $r(P)$ and $l(P)$ (b) $\rho(P) = 1$ (c) $\rho(P) = 2$ (d) $\rho(P) = 4$

Figure 2. Symmetricity and decomposition: Black dots represent the points in P.

An algorithm is a function, say ψ, that returns a curve to the next location in the Euclidean space, and each robot moves along the curve. A robot is *oblivious* if the Compute phase only depends on the Look phase of the cycle, otherwise *non-oblivious*. A move phase is *rigid* if r_i moves along ψ and to the endpoint of ψ. On the other hand, a *non-rigid* move phase may finish while r_i is still on the way to the next location. However, we assume that each robot moves at least δ (in Z_0), or reaches the next location if the length of the curve is shorter than δ. We use non-rigid moves without any explicit explanation when it is clear from the context.

An execution is a sequence of configurations, $P(0), P(1), P(2), \ldots$. The execution is not uniquely determined even when it starts from a fixed initial configuration I. Rather, there are many possible executions depending on the distance that the robots move, the activation schedule of robots in SSYNC and ASYNC model, and the length of the move phase in ASYNC model.

Pattern Formation: A target pattern F is given to every robot r_i as a set of points $Z_0(F) = \{Z_0(p) : p \in F\}$. (Remember that r_i does not have access to the global coordinate system Z_0.) We assume that $|Z_0| = n$ and $|F| \leq n$. In the following, as long as it is clear from the context, we identify p with $Z_0[p]$, and write, e.g., "F is given to r_i", instead of "$Z_0[F]$ is given to r_i".

Let $\mathbb{T}$ be the set of all coordinate systems, which can be identified with the set of all transformations consisting of transformations, rotations, and uniform scalings. Let $\mathcal{P}_n$ be the set of all patterns of n points. For any $P, P' \in \mathcal{P}_n$, P is *similar* to P', if there exists $Z \in \mathbb{T}$ such that $Z[P] = P'$, denoted by $P \simeq P'$.

We say that algorithm ψ forms pattern $F \in \mathcal{P}_n$ from initial configuration I, if for any execution $P(0)(= I), P(1), \ldots$, there exists a time instant t such that $P(t') \simeq F$ for all $t' > t$.

For any $P \in \mathcal{P}_n$, let $C(P)$ be the smallest enclosing circle of P, and $c(P)$ be the center of $C(P)$. Formally, the *symmetricity* $\rho(P)$ of P is defined by

$$\rho(P) = \begin{cases} 1 & \text{if } c(P) \in P, \\ |\{Z \in \mathbb{T} : P = Z[P]\}| & \text{otherwise.} \end{cases}$$

We can also define $\rho(P)$ in the following way [11]: P can be divided into regular k-gons with co-center $c(P)$, and $\rho(P)$ is the maximum of such k. Here, any point is a regular 1-gon with an arbitrary center, and any pair of points $\{p, q\}$ is a regular 2-gon with its center $(p + q/2)$.

A point on the circumference of $C(P)$ is said to be "on circle $C(P)$" and "the interior of $C(P)$" does not include the circumference. We denote the interior of $C(P)$ by $Interior(P)$, and the radius of $C(P)$ by $r(P)$.

Journal of Physics: Conference Series **473** (2013) 012016 doi:10.1088/1742-6596/473/1/012016

3. Gathering and convergence

We start with a special case with $|F| = 1$, which is called the *gathering problem*, that is, the problem is to make the robots form a point. It is also called *point formation* problem. One weaker version of the gathering problem is the *convergence problem* that requires the robots to move arbitrarily close to some point. We say a robot system solves the gathering problem (the convergence problem, respectively) if the robots can form a point (converge to a point, respectively) in any execution from a given initial configuration.

3.1. Asynchrony and obliviousness

Clearly, gathering is easy in the FSYNC model because the robots are activated simultaneously and can agree the gathering point based on the common observation (in their own coordinate system). For example, Algorithm 3.1 gathers the robots at the center of gravity of the current positions of robots.

Algorithm 3.1 Gathering algorithm for oblivious robots in the FSYNC model [12]

For robots r_i,
Move to the center of gravity of the current position of robots.

However, Algorithm 3.1 cannot guarantee gathering in the SSYNC model, because semi-synchrony do not always activate all robots. Now, we consider a simplest case of two robots ($n = 2$). Consider an approach that makes a robot to move the position of the other robot. This approach does not work because in the SSYNC model, the two robots may forever swaps their positions. Though we need more theoretical discussion, this is the intuition for the impossibility of gathering of $n = 2$ robots in the SSYNC and ASYNC models. We note that Algorithm 3.1 guarantees convergence in the ASYNC and SSYNC models. Now, we have the following three theorems.

Theorem 1 *[12] For the gathering problem and the convergence problem of oblivious robots, we have the following:*

(i) The gathering problem is solvable in the FSYNC model.
(ii) The convergence problem is solvable in the SSYNC model.
(iii) When $n = 2$, the gathering problem is unsolvable in the SSYNC model.

Even when $n = 2$, *non-oblivious* robots with fixed origins of local coordinate systems can solve the gathering problem in the SSYNC model. Suzuki and Yamashita proposed a solution (Algorithm 3.2) that makes the two robots show their initial positions and gather at the midpoint of their initial positions [12].

Algorithm 3.2 Gathering for $n = 2$ non-oblivious robots [12]

For robot r_i,

(i) At the first activation, rotate Z_i so that the other robot r_j is on the positive y-axis, and move slightly along the positive x axis. (See Figure 3 (*i*).)
(ii) After the first activation, move to the same direction as the first activation until r_j's position changes twice.
(iii) Then, r_i and r_j can recognize their initial positions, and move to the midpoint.

Depending on the timing of the first observations of the two robots, we have the following two cases:

(*a*) x-y coordinate at r_i (*b*) x-y coordinates at r_i and r_j (Case 2)

Figure 3. Movement of the two robots to draw x-y coordinate

Case 1: The two robots observe their positions simultaneously. They move in parallel and each robot can recognize the initial position of the other robot by using its initial position.

Case 2: The two robots observe their positions at different time. Let r_i observe first, and r_j observe r_i's position after r_i moves (See Figure 3 (ii)). In this case, r_j's trajectory intersects r_i's negative x-axis. Thus, both robots can recognize the case when they observed each other. Additionally, r_i observed initial positions of r_i and r_j, and r_j can recognize r_i's initial position by the intersection of its own initial position and r_i's positive y-axis.

Consequently, the two robots can agree on the rendezvous point.

When $n > 2$ gathering is solvable for ASYNC (thus, SSYNC) robots if the robots have the multiplicity detection or an agreement on the coordinate system. Assume, for contradiction, that there exists an algorithm $\mathcal{A}$ that solves the gathering problem for oblivious SSYNC robots without multiplicity detection and any agreement on the coordinate system. Let R be divided into $R_1 = r_i$ and $R_2 = R \setminus \{r_i\}$, where all robots in R_2 has the same coordinate system, which is different from r_i's coordinate system. Consider an execution of $\mathcal{A}$, say E, where gathering of robots R terminates at time t and the robots gather at point p. Thus, at time $t-1$, the robots in R occupy different positions. We have the following two cases:

Case 1: When r_i is not on p. Then, consider an execution E' that is same as E from time 0 to $t-1$, and at time t, only r_i is not activated. Thus, at time $t+1$, the robots in R_2 occupy p, and R_1 is not on p. We denote this configuration by C_1. It's the same configuration as the gathering for $n = 2$ SSYNC robots. From Theorem 1, algorithm $\mathcal{A}$ cannot solve the gathering problem.

Case 2: Otherwise. Thus, there exists some r_j in R_2 that is not on p at time $t-1$. Then, consider an execution E'' that is same as E from time 0 to $t-1$, and at time t, only $R'_1 = \{r_j\}$ is not activated. Thus, at time $t+1$, the robots in $R'_2 = R_1 \setminus \{r_j\} \cup \{r_i\}$ occupies p and R'_1 is not at p. We denote this configuration as C_2. It is easily shown that from configuration C_2, there exists an execution that either reaches C_1 configuration or keeps forming C_2 configuration, which means gathering is impossible.

Consequently, we have the following theorem.

Theorem 2 *[10] Oblivious robots without multiplicity detection and any agreement on the coordinate system cannot solve the gathering problem in the SSYNC model.*

Cielievak, Flocchini, Prencipe, and Santoro proposed a gathering algorithm for $n > 2$ oblivious SSYNC robots with multiplicity detection [2]. The strategy is as follows; first the robots make a single "dense" point with multiple robots, and all the other robots gathers at the point. However, during the gathering, the robots should avoid collisions that make dense points.

Theorem 3 *[2] When $n > 2$, oblivious robots with multiplicity detection can solve the gathering problem in the ASYNC (thus, SSYNC) model.*

Clearly, if the robots have a common coordinate system, they can gather at, for example, the position of the lowest rightmost robot. Motivated by this fact Izumi et al. [9] introduced unreliable compasses and characterized the solvable cases in terms of the types of compasses and their maximum deviation angles.

3.2. Visibility

When the robots have only limited visibility, gathering and convergence can be used as a very important first step for the pattern formation problem. Once they get closer enough so that each robot can observe all the other robots, they can execute existing pattern formation algorithms for unlimited visibility robots[3].

Ando, Oasa, Suzuki, and Yamashita proposed a point convergence algorithm for oblivious ASYNC robots [1]. Their algorithm is based on the following simple idea: (i) robots in a connected visibility graph "get closer" and (ii) robots that are mutually visible at time t remain so for all $t' > t$. Intuitively, (i) is achieved by making each robot r_i move toward the center of the smallest enclosing circle of its local view. To predict the moving distance of other robots, their algorithm checks a common point of circles with radius $V/2$ and centered at the midpoint between r_i and r_j for all r_j in its local view. Then, (ii) is achieved by stopping r_i at the furthest common point on the way to the center of the smallest enclosing circle. By the above two properties, the algorithm ensures that the convex hull of the positions of robots shrinks and eventually robots converge to a point.

Flocchini, Prencipe, Santoro, and Widmayer investigated the gathering problem in ASYNC model with common compass [4]. In their algorithm, each robot moves to the bottom and right most robot in its local view. Eventually, the robots gather at the unique bottom and right most robot.

4. Pattern formation

In this section, we briefly overview existing pattern formation algorithms for $n \geq 3$ [11, 12, 7, 13]. Suzuki and Yamashita [11] first showed that the symmetricity of a given initial configuration I and the symmetricity of a target pattern F determines whether F is formable from I. Specifically, F is formable from I if and only if $\rho(I)$ divides $\rho(F)$. Even though oblivious robots cannot remember the symmetricity of the initial configuration, they showed that oblivious robots have the same pattern formation power as non-oblivious robots in the SSYNC and FSYNC models. In other words, during the pattern formation, robots encodes $\rho(I)$ to their global positions. Later, the result is extended to the ASYNC model [12] by using the embedded pattern formation algorithm by Fujinaga, Ono, Kijima, and Yamashita [7]. All these results assume unlimited visibility of robots. When the visibility of the robots are limited, the robots can neither recognize $\rho(I)$ nor encode $\rho(I)$ to their global positions. Yamauchi and Yamashita showed this impossibility for oblivious robots with limited visibility [13], and they showed non-oblivious robots still have the same formation power in the SSYNC and FSYNC models.

4.1. Obliviousness

Suzuki and Yamashita [11] first characterized the class of patterns formable by non-oblivious robots in the SSYNC and FSYNC models. They pointed out that geometric symmetricities of initial configuration $I = \{p_1(0), P_2(0), \ldots, p_n(0)\}$ and target pattern $F = \{f_1, f_2, \ldots, f_n\}$ determine whether F is formable from I. From definition, I (F, respectively) can be partitioned

[3] For example, the pattern formation algorithm in [8] does not move any robot to the outside of the smallest enclosing circle of the initial configuration during execution

into regular $\rho(I)$-gons ($\rho(F)$-gons, respectively) all with center $c(I)$ ($c(F)$, respectively). Let $n/\rho(I) = k_I$ and $n/\rho(F) = k_F$. An important property is that all robots observe I even in its local coordinate system can agree on an order of I_i's, such that the distance of the points in I_i from $c(I)$ is no greater than the distance of the points in I_i, and that each robots is conscious of the group I_i it belongs to. The same property holds for F. Hence, when $\rho(I)$ divides $\rho(F)$, the robots can recognize the correspondence between $I_1, I_2, \ldots, I_{k_I}$ and $F_1, F_2, \ldots, F_{k_F}$ such that in the order of $I_0, I_1, \ldots$, robots occupy the empty F_j with the smallest j ($1 \leq j \leq k_F$). Then, the robots in I_0 is considered as (part of) F_0, and other robots move in the order of $I_1, I_2, \ldots$ to occupy the nearest position in the corresponding $\rho(F)$-gon. This is an overview of the pattern formation algorithm for non-oblivious robots in the SSYNC and FSYNC models in [11].

The authors also pointed out that when $\rho(I)$ does not divide $\rho(F)$, it is impossible to agree this ordering [11]. This is the source of impossibility of pattern formation. For example, consider pattern formation of F from I where $\rho(I)$ does not divide $\rho(F)$ and $|I_1| > |F_1| = \rho(F)$. The robots in I_1 first check the matching between F_1 by selecting $\rho(F)$ robots from I_1. Remember that the robots are anonymous. The robots in I_1 can use the activation schedule, and observation to agree the selection of $\rho(F)$ robots. However, in the worst case, they are activated at the same time and obtain the same observation by the symmetric local coordinate system. Hence, robots can never agree on the selection, and it is impossible to break the symmetricity.

Theorem 4 *[11] Let $I, F \in \mathcal{P}_n$ for any $n \geq 3$. Then F is formable by non-oblivious robots from an initial configuration I in the FSYNC and SSYNC model if and only if $\rho(I)$ divides $\rho(F)$.*

Later, Yamashita and Suzuki investigated the formation power of oblivious robots, and showed that oblivious robots have the same formation power except the gathering problem for two robots [12]. Now, we present the sketch of the algorithm in [12]. The robots in I_{k_I} first move slightly outside $C(I)$ if $I_{k_I - 1}$ is on $C(I)$. This first step is necessary to keep the symmetricity smaller than $\rho(F)$ during an execution, because when $\rho(I) = \rho(F)$, and $\rho(F)$ contains a regular k-gon ($k > \rho(F)$) in its interior, the symmetricity becomes larger than $\rho(F)$. Oblivious robots have no way to reduce the symmtericity thereafter. After this perturbation, the robots in $I_{k_I - 1}$ stays there until other robots occupies their target positions in F so that the smallest enclosing circle remains unchanged.

Then, each robot r_i moves to the closest position of the corresponding $\rho(F)$-gon according to the above order by using the following coordinate system Z_i^*: the origin is $c(I)$, the positive x-axis is the half line $c(P)p_i(0)$, and the unit distance is $r(P)$. When r_i and r_j are in the same I_k, then their track is symmetric without intersection. Finally, the robots in I_{k_I} move to their destinations with carefully keeping the smallest enclosing circle. During the execution, the symmetricity may decrease, however, the robots recognize the new symmetricity and move according to the new ordering.

Theorem 5 *[12] Let $I, F \in \mathcal{P}_n$ for any $n \geq 3$. Then F is formable by oblivious robots from an initial configuration I in the FSYNC and SSYNC model if and only if $\rho(I)$ divides $\rho(F)$.*

4.2. Asynchrony

The crucial difference between the ASYNC model and the SSYNC model is that the robots may observe other robots while some of them are moving. Hence, we cannot use the algorithm for the SSYNC model.

Fujinaga, Ono, Kijima and Yamashita proposed *the embedded pattern formation problem*, that is, the target pattern is embedded on the ground, and the robots autonomously move to form the pattern [7]. The robots should avoid forming a dense point because once two or more robots occupy a position, they may move completely the same thereafter. Hence, the embedded pattern formation problem is to obtain a matching between the initial configuration I and the embedded

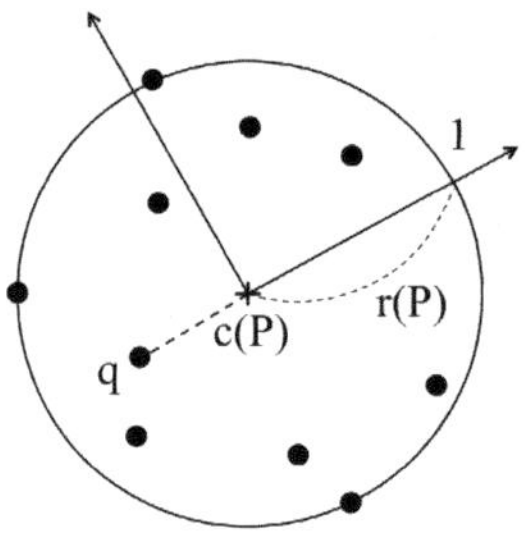

Figure 4. Coordinate system $Z(P, q)$

target pattern F. In [7], they defined and showed the existence of a *clockwise matching* (CWM) $f_{CWM} : I \to F$ that is (i) a perfect matching between I and F, (ii) a optimum matching in the sense that the sum of the Euclidean distance between $p_i(0)$ and $f_{CWM}(p_i(0))$ for all r_i is minimized. The optimality guarantees that any straight movement of r_i to $f_{CWM}(p_i(0))$.

Theorem 6 *[7] Let $I, F \in \mathcal{P}_n$ for any $n \geq 3$ without multiplicity. Then an embedded pattern F is formable from I in the ASYNC (thus, SSYNC and FSYNC) model.*

Fujinaga, Yamauchi, Kijima, and Yamashita proposed a pattern formation algorithm based on the embedded pattern formation [8]. Their algorithm first forms an agreement on a coordinate system and embeds the target pattern F, and moves the robots according to the CWM algorithm in [7]. Given a pattern P, let $Z(P, q)$ be the coordinate system defined by the smallest enclosing circle and position $q \in P$ as shown in Figure 4. An important property is that, when $\rho(I)$ divides $\rho(F)$, $Z(I, r_i)[F] = Z(I, r_j)[F]$ for any $r_i, r_j \in I_0$. In [8], the proposed a pattern formation algorithm that embeds F by using $Z(I, r_i)$ ($r_i \in I_0$). The difficulty is that during the formation, no robot cannot go inside $C(I_0)$. The authors used a mapping from the two dimensional Euclidean space to a cylinder defined by the distance from $c(I)$ and the radius from the x-axis. The CWM is computed on the cylinder, and the corresponding track in the two dimensional Euclidean space is a curve in the $Interior(I) \setminus \{C(I_0) \cup Interior(I_0)\}$. Hence, the robots moving along the curve never go insider $C(I_0)$.

Theorem 7 *Let $I, F \in \mathcal{P}_n$ for any $n \geq 3$ without multiplicity. Then F is formable from an initial configuration I by oblivious robots in the ASYNC model if and only if $\rho(I)$ divides $\rho(F)$.*

4.3. Visibility

Robots with limited visibility can observe other robots within distance V, and they are not aware of the global configuration, such as their smallest enclosing circle, and the symmtriciry of their positions.

A necessary and sufficient condition for the pattern formation with unlimited visibility is based on the "global" symmetricity [11]. The pattern formation algorithms [11, 12, 8] also use the positions of robots as the indicator of the formation process while robots are oblivious.

Yamauchi and Yamashita showed that robots with limited visibility cannot form target pattern F from initial configuration I even when $\rho(I)$ divides $\rho(F)$ in the FSYNC model [13]. The counter example is a visibility graph with low connectivity, where local observation leads the robots to increase the global symmetricity. They further showed that non-oblivious robots can regain the symmetricity of the initial configuration by recording all local observations and all outputs of compute phases.

The algorithm first executes the convergence algorithm [1], and after robots get closer enough, executes existing a pattern formation algorithm for robots with unlimited visibility. Because algorithm [1] guarantees that the output track is a straight line, the robots can compute the rotation angle between any two consecutive outputs. Hence, it is clear that when all moves are rigid, each robot can compute its initial position. The authors showed that the robots can draw its initial position after they get closer enough so that the robots obtains initial positions all other robots. When the moves are non-rigid, they further showed that the robots can regain $\rho(I)$ in the FSYNC model because the local history of observations and output traces break the symmetricity.

Theorem 8 *[13] Let $I, F \in \mathcal{P}_n$ for any $n \geq 3$ without multiplicity.*

(i) There exist infinitely many patterns F that is not formable from I by non-oblivious robots with limited visibility in the FSYNC model, even when $\rho(I)$ divides $\rho(F)$.

(ii) F is formable by non-oblivious robots with limited visibility from an initial configuration I in the FSYNC model if and only if $\rho(I)$ divides $\rho(F)$.

(iii) F is formable by non-oblivious robots with limited visibility and rigid moves from an initial configuration I in the SSYNC model if and only if $\rho(I)$ divides $\rho(F)$.

5. Concluding Remark and Future Directions

We surveyed existing results on pattern formation problems by mobile robots. The pattern formation power of mobile robots is determined by their asynchrony, obliviousness, and visibility. One of the future directions is to reveal formation power of robots with limited visibility. For example, the pattern formation problem for robots with limited visibility in the ASYNC model has not been solved. Another direction is to develop a randomized solution for the problems that have been shown to be deterministically unsolvable. Such a randomized and localized mobile robot system model would be related to wireless sensor networks, vehicular networks, chemical reactions, and so on. Recently, mobile robot system model founds new applications in molecular robotics where system size (i.e., the number of robots) is in the Avogadro number scale and robots with very small capabilities cooperate by interacting with each other randomly. We are interested in whether and how the size of the system has effect on their computational power.

Acknowledgment

This work is supported in part by JSPS Grant-in-Aid for Scientific Research on Innovative Areas (No.24104519), and JSPS Grant-in-Aid for Scientific Research (No. 23700019).

References

[1] Ando H, Oasa Y, Suzuki I and Yamashita M 1999 *IEEE T. Robotic. Autom.* **15** 818–28
[2] Cielibak M, Flocchini P, Principe G and Santoro N 2012 *SIAM J. Comput.* **41** 829–79
[3] Dijkstra E 1974 *Commun. ACM* **17** 643–4
[4] Flocchini P, Prencipe G, Santoro N and Widmayer P 2005 *Theoret. Comput. Sci.* **337** 147–68
[5] Flocchini P, Prencipe G, Santoro N and Widmayer P 2008 *Theoret. Comput. Sci.* **407** 412–47
[6] Flocchini P, Prencipe G and Santoro N, 2012 Distributed computing by oblivious mobile robots *Synthesis lectures on distributed computing theory* (Morgan and Claypool Publishers)
[7] Fujinaga N, Ono H, Kijima S and Yamashita M 2010 *Proc. of OPDIS 2010* 1–15
[8] Fujinaga N, Yamauchi Y, Kijima S and Yamashita M 2012 *Proc. of DISC 2012*, 312–25
[9] Izumi T, Soussi S, Katayama Y, Inuzuka N, Défago X, Wada K and Yamashita M 2012 *SIAM J. Comput.* **41** 26–46
[10] Principe G, 2007 *Theoret. Comput. Sci.* **384** 222–31
[11] Suzuki I and Yamashita M 1999 *SIAM J. Comput.* **28** 1347–63
[12] Yamashita M and Suzuki I 2010 *Theoret. Comput. Sci.* **411** 2433–53
[13] Yamauchi Y and Yamashita M 2013 *Proc. of SIROCCO 2013* to appear

The physics of distributed information systems

Erik Aurell

Aalto University School of Science, Helsinki, Finland, ACCESS Linnaeus Centre, KTH, Stockholm Sweden and Dept. Computational Biology, AlbaNova University Centre, 106 91 Stockholm, Sweden

E-mail: eaurell@kth.se

Abstract. This paper aims to introduce Distributed Systems as a field where the ideas and methods of physics can potentially be applied, and to provide entry points to a wide literature. The contributions of Leslie Lamport, inspired by Relativity Theory, and of Edsger Dijkstra, which has the flavor of a growth process, are discussed at some length. The intent of the author is primarily to stimulate interest in the statistical physics community, and the discussions are therefore framed in a non-technical language; the author apologizes in advance to readers from the computer science side for the unavoidable impreciseness and ambiguities.

1. Distributed Systems

This paper aims to present distributed systems as a new (interesting) area of applications of statistical physics, and to make the the case that statistical physics can be quite useful in understanding such systems. Most working physicists, as members of the informed general public, are most likely aware of the practical and societal importance of distributed systems. But relatively few physicists are familiar with the problems and main results of the the branch of Computer Science called Distributed Systems. To mitigate this state of affairs, to some extent, is the main motivation of this paper [1].

Our presentation of Distributed Systems is frankly biased and instrumental. For the interested reader, who would like to go further, I recommend the monograph of Tannenbaum and van Steen [1], or the more introductory text of Tel [2]. For personal perspectives I also recommend selected chapters in the collection edited by Edsger W. Dijkstra [3] and the *catalogue raisonné* of Leslie Lamport [4]. With these caveats, a suitable point of departure is a famous quote, from an e-mail sent by Leslie Lamport in May 1987 [5], which states

> A distributed system is one in which the failure of a computer you didn't even know existed can render your own computer unusable.

To a physicist, the phrase "a computer you did not even know existed" suggests a division of the world into events inside and outside a light-cone of possible causes. Indeed, the physical light-cone sets the absolute upper limit of information transfer; in a real distributed system however, information transformation is limited also by processing at the sender and at the receiver, as well by the finite speed over wires or wireless connections. It is not hard to imagine a geometry of delays between a set of computers, such that there is a minimum delay τ_{AB} from computer

[1] In this paper I refer to the academic discipline as Distributed Systems, and to the objects under study as distributed systems.

A to B, and that events in computer A at time t can hence only influence events at computer B at time $t + \tau_{AB}$, or later. In the time interval $t - \tau_{BA} < t' < t + \tau_{AB}$, events at B at t' cannot be influenced by events at A at t, and neither can events at B at t' influence events at A at t. Such events are called *concurrent* in Distributed Systems, and the problems facing a designer or manager of a system with concurrent events is called *concurrency* [2]. By analogy with Relativity, in Physics we would call such events simultaneous, with respect to the particular geometry of delays. The space-time picture has been a major influence in Distributed Systems, as we will see below in Section 3.

The first and last notions in Lamport's definition, "one in which the failure" and "render [...] unusable", stress that things may fail, and that failures have unexpected, and negative, consequences. Failures and how to handle them is a classic problem in Distributed Systems. In section 4 I will describe how a very large numbers of simultaneous failures can be handled by algorithms called "Byzantine" after [6], and which certainly appear to the neophyte to be of Byzantine complexity. I will also describe other classic algorithms that handle (non-Byzantine) failures, some of which are at least conceptually simple, and even have a physical flavor. From today's perspective, in a very large system concurrent changes are unavoidable, and a failure is just a name for a change which we do not like. The process by which a large system is continuously subject to change is called *churn*. A system under churn differs from a system with changes isolated in space-time in that under churn the system, as a whole, does not have time to recover from a change (somewhere) before another change occurs (somewhere else). How to cope with churn is a main topic of Distributed Systems over the last decade or so, and I will give a few suggestions for further reading in that direction.

The outline of the paper is as follows. In Section 2 I introduce in a non-technical manner the concept of state machines, and in Sections 3 and 4 I describe work by Leslie Lamport (mainly) and others (partly) in this direction, still in a non-technical manner. In Section 5 I introduce the distributed self-stabilizing algorithm of Edsger Dijkstra and the concept of "self-star", and in Section 6 I give some further pointers to the literature.

2. State machines, and the replicated state machine approach

I base the discussion in this Section on the concept of a state machine. I distinguish between a big state machine (one for each system) and small state machines (one for each element in the system) [3]. At this point, the physically inclined reader can think of a big state is a microstate, a big state machine is a complete description of the time evolution of a microstate, and a small state machine is the local dynamics of local state variables.

Algorithmic complexity theory lies outside the scope of this review, but as a motivational device I invoke the Turing machine, which most readers will know as a (the) universal model of computation. A Turing machine reads and writes on a tape. The state of a Turing machine is specified by the content of the tape, an internal state, and the position of the read/write head on the tape. Most (and perhaps all) computational devices can similarly be described by a set $\mathcal{S}$ of states, a set of initial states $\mathcal{I}$, and a next-state relation $\mathcal{N}$, such that $\mathcal{I} \subset \mathcal{S}$ and $\mathcal{N} \subset \mathcal{S} \times \mathcal{S}$. Such an abstract device is called a *state machine*, and the computations it can perform are sequences of states $s_1 \to s_2 \to s_3 \to \cdots$, or *state behaviors* [7]. In a deterministic state machine, *e.g.* a Turing machine, the next-state relation is a function.

A distributed system can be seen as one big state machine, where $\mathcal{S}$ is the set of internal states of all its elements (including, if so desired, the state of the communication medium), $\mathcal{I}$ is the set of initial states and $\mathcal{N}$ specifies the combined behavior of all the elements. The reader can imagine $\mathcal{S}$ as the set of spins in a lattice and $\mathcal{N}$ as some lattice dynamics. This is a

[2] I here base concurrency on which messages could have been sent. The alternative definition, based on messages actually sent, is taken up in Section 3

[3] May I be excused of this notational innovation.

deterministic state machine if the spins are always updated in the same order, and if the local update rule is deterministic; otherwise the state machine is non-deterministic. A distributed system can also be seen as many small state machines, one for each element, where $\mathcal{S}_i$ is the set of internal states of element i. The reader can in this case think of a single spin in a lattice. In contrast to the big state machine picture, the many small state machine picture is incomplete, since in all cases of interest the next-state relation of element i depends on the states of at least some other elements. The reader may think of of the update rule of a single spin which depends on the neighboring spins.

Distributed systems are typically open systems, which interact with an external world (communication media, hardware states, system users, other distributed systems *etc.*). While it would be possible to include the relevant states of the external world in the state description, as in Physics it is often more convenient to separate the two. To the set of states is then added a set $\mathcal{A}$ of actions, such that the next state relation is a subset of $\mathcal{S} \times \mathcal{A} \times \mathcal{S}$, and a *state-action behavior* is a sequence $s_1 \rightarrow_{\alpha_1} s_2 \rightarrow_{\alpha_2} s_3 \rightarrow_{\alpha_3} \cdots$ [7]. If the state-action machine is deterministic, then the state machine will change its state in one way only under a given action, and the state machine then obeys the actions as externally issued commands. If the state-action machine is not deterministic, then the action sequence does not determine the state behavior, but only which state behaviors are possible.

I separate big states and small states by the convention that in a big state actions are only generated by events deemed external to the system. The absence of any external events is a null event, which also generates an action, and the big state can therefore change, even if there are no external events (but it does not have to do so). In a small state, actions can be generated both by external events relevant to this element, and by other elements in the system changing their states. The set of all the small state-action machines can obviously be as a complete description (if done correctly) of the system as the big state-action machine. When describing a real distributed system it is largely a matter of art to chose what set of states to include as small states in the big state picture, and which states of the world will be deemed external to the system.

As touched upon above, distributed systems have several natural failure modes which distinguish their study from other computational devices. One is that some elements may fail entirely. This is similar to some spins in a lattice for some reason using the wrong (or random) update rule. Another mode is that the elements and/or the connections between them may work at different rates, and so get unsynchronized. This is analogous to the spins in a lattice updating in random order, and some spins updating twice or many more times before some other spins have even updated once, which, if one is not careful, could lead to incorrect thermodynamics. It is also largely a matter of art whether a particular deviation from a normative behavior should be described by an action, or as a failure mode of the system.

The reader may at this point think of a system which is large, but not in the thermodynamic limit, and on which are placed strict conditions that some bad behavior should never occur. More concretely, the reader may think of the servers running the on-line banking system of a very large bank, and the unwanted behavior that some client manages to withdraw more money than he or she has in his or her account. Such a system can consist of hundreds or thousands of servers in many countries. At least sometimes some of these servers may be down (temporary failures), most likely the servers are not identical (running at different speeds), and almost surely there are unpredictable delays in the inter-server communications (which translates to different effective speeds, and perhaps also to effective failures). Showing that the conditions are fulfilled can then be translated as a Distributed Systems task of the type "Prove that distributed system X, if it has property S at time t_0, will have property A for all times $t > t_0$".

As Physics problems, such tasks have, if anything, the flavor of Mechanics. One way to proceed (in a physical setting) would be to show that the time evolution conserves some invariant.

For example, if a spin update rule preserves an energy function, and if energy is positive at time t_0, then energy will be positive for all time. Another way to proceed would be to show that although the system is not forbidden by invariants to reach a certain state, in fact, if one follows the time evolution, one can see that it can never do so. This is obviously the more difficult assignment, and which depends more on system details. Arguably, Mechanics is rather far from the on-line banking example, however, both ways to proceed do have parallels in Distributed Systems, where reasoning about invariants is called an *assertional proof*, and reasoning about the time evolution a *behavioral proof*, and similarly to Mechanics, an assertional proof is much preferred, when it is available. In actual practice, both assertional and behavioral proofs in Distributed Systems can be quite fantastically complicated, even for problems which at first glance look fairly straightforward, and such proofs often cannot be done manually. Physics has few examples to offer of such reasoning, and will hence have little to say to such problems.

On the other hand, if satisfying a condition on no bad behavior is the most important property we want from a system, then Distributed System has the option to change the system under study until the conditions are met. Alternatively, the system can be modified to a kind of normal form, and then the task is to show that some kind of generic condition is satisfied by systems in normal form. The main approach in this direction is called *state machine replication*. This means that one and the same state machine is faithfully replicated either in all the elements of a distributed system, or in a qualified majority of them, or in some other suitable number, depending on the scenarios in which the conditions are to be satisfied. At this point, I leave that number vague; examples will follow. Suppose for illustration that the state machine to be replicated is simple. For instance, to borrow an example from [8], suppose the state is how much money a client has in his or her account and that the client issues withdrawal requests. Leaving out of the description how the client is informed if his or her request is granted, and how he or she actually gets the money, then the state is just given by one number $c \geq 0$, and the request is an action represented by another number $r > 0$. The good behavior is described by a deterministic state machine where $c \rightarrow_r (c - r)$ if $r \leq c$, and $c \rightarrow_r c$ if $r > c$. That the state machine is replicated across a number of computers in a big bank means that they all (or a qualified majority, or some other suitable set) hold the same number c, and respond to the same action r. If so, the computers will agree between themselves how much money the client has in the account, and the client will not be able to withdraw more than that amount.

The question in normal form is hence whether a state machine can be suitably replicated when the elements of this system run with different speeds and with failures. The answer to this question is a "yes" – provided that the elements properly wait for one another, and/or that there are enough back-up copies to handle failures. I will describe what it means for elements to properly wait for one another, and then how many back-up copies are needed to handle different kinds of failures. But let us also remark that while this approach is general and versatile it also limits what the full system system can do to what one of its components could also do on its own. The main interest is therefore either to build extremely fault-tolerant systems, since the system as whole can be much more robust to failures than one of its elements, and to critical aspects of a system, where the elements must agree – even if they disagree about other things.

3. The space-time view of concurrency

Leslie Lamport, known to most physicists as the designer of LaTeX, is also the author of a paper entitled "Time, Clocks, and the Ordering of Events in a Distributed System" [9],which, although obviously inspired by Special Relativity, is surprisingly little known among physicists [4]. Lamport solves the problem of replicating any state machine by a distributed system where elements run at different speeds but without failures. The method works by loading timestamps on all commands

[4] In fact, by an informal survey, the author has not found a single colleague physicist who has read this paper.

and all messages issued by and sent between the elements. Quite strikingly, Lamport begins by pointing out that a set of elements which send messages between themselves can be illustrated by a space-time diagram as in Figure 1. The wide use of space-time diagrams to describe and

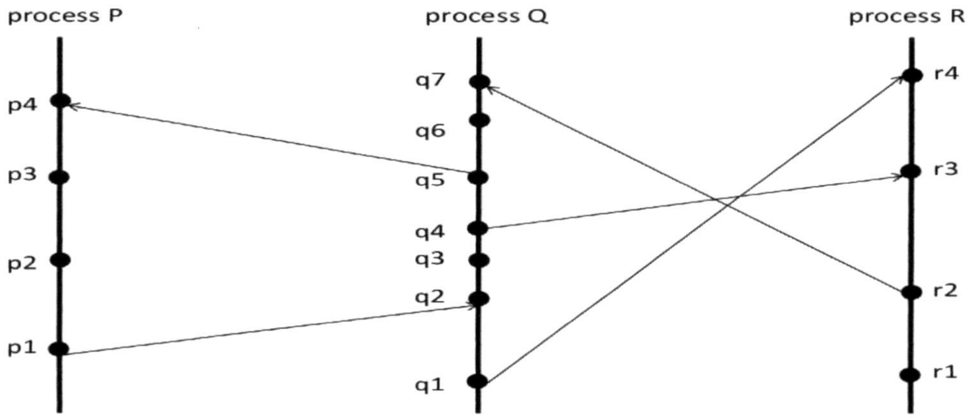

Figure 1. This figure reproduces the left panel of Fig 1 of the paper "Time, clocks and the ordering of events in a distributed system" by L. Lamport published in Communications of the ACM 21, 7 (July 1978), 558-565 [9]. Vertical lines are time lines of processes (or processors, or the entities of any distributed system under study) where time is running upwards. Nodes on the time lines are events in the respective processes, event count in each process increasing with time. Diagonal lines are time-and-space traces of messages sent between processes. The line event $p1$ (in process P) to event $q2$ (in process Q) is thus a message sent from process P to process Q. Messages may travel at different speeds between process and may arrive out of order as illustrated by the pair $q1 \rightarrow r4$ and $q4 \rightarrow r3$.

reason about communication protocols originates, as far as this author understands, in this paper from 1978 [9, 10]. An event a can causally affect event b if there is a path from a to b in the space-time diagram, always going forwards along world-lines and messages. As in Relativity, potential causality imparts a partial ordering of events. If neither a can causally affect b, nor b causally affect a, then a and b will be called concurrent events. Note that this notion of concurrency depends on the messages actually sent, *i.e.* on the space-time diagram.

A key insight of Lamport was that even without proper relativistic effects, there is no invariant total ordering of events in a distributed system. If two elements actually know absolute Newtonian time they will agree on event order. However, time must be measured by physical clocks, one in each element, and such clocks run at different rates. Unless the clocks be synchronized and all relevant events sufficiently well separated in time, two elements in the system can disagree on the ordering of some pair of events. Furthermore, physical clocks are expensive, and more expensive the better they keep time. Therefore, it is of interest to consider as bad clocks as possible, which Lamport calls "Logical Clocks". Assume that all events in a single element are totally ordered, *e.g.* by an advancing event counter [5]. A set of clocks C_i with readings $C_i(a)$, where i is an element in a distributed system and a is an event which happens in that element, running at otherwise arbitrary speeds, respect the partial ordering of events when they obey the following two conditions:

C1 if a and b are events in the same element i, and a comes before b, then $C_i(a) < C_i(b)$;

C2 if a is the sending of a message from element i and b is the receipt of the message by element j, then $C_i(a) < C_j(b)$.

Respecting the partial ordering of events means that if a can casually affect b, then $C_i(a) < C_j(b)$. Condition **C1** can be met by advancing the clocks C_i at least as quickly as the event counters, but it will be appreciated that a set of event counters will not satisfy condition **C2**. However, by introducing a timestamp T_m of every message m, conditions **C1** and **C2** can be met by the following two further implementation rules:

IR2(a) if a is the sending of a message m from element i, then m contains a timestamp $T_m = C_i(a)$;

IR2(b) upon receiving message m, element j sets C_j to be greater or equal to its present value, and greater than T_m.

Lamport extends the ordering of events to a total ordering by using logical clocks and any ordering between the elements in the system. Event a in element i is then before event b in element j if *either* $C_i(a) < C_j(b)$ *or* $C_i(a) = C_j(b)$ *and* element i is before element j. This ordering of events is not invariant, and does not necessarily agree with the order of occurrence of events in Newtonian time. However, it allows any state machine to be simulated by requiring that an element can only execute a command timestamped T when it has learned of all commands issued by all the other elements with timestamps less than or equal to T. It is therefore a kind of general "waiting for the others" scheme.

In the toy example of above, at time T any of the bank's computers i could receive a request of withdrawal for the amount r (a deposit could be a withdrawal of $r < 0$). The computer has its own state c and if r is greater than c it could just refuse the request and let the client know this. Hence, assume that $c \geq r$, such that the request looks like a valid one to the computer receiving it. The computer then timestamps this request with T, and puts it on a queue. The action to be taken in response to the client's request is, when this action is taken, to withdraw money from the account (and send the money to the client) or to refuse the request (and inform the client). The message computer i needs to send to the others when it grants such a request is therefore "The client has withdrawn r from the account from me. New balance is c.". That message is timestamped with T_c, the time at which this computer takes this action. In response to such a message, the other computers will also update their value of c and send the messages "Following the action of i, I have set the balance to c.", timestamped by T_a, the time at which this acknowledgment was sent. The receiving computer puts both kinds of messages together with the client requests on its queue. At any one time, the computer's queue hence contains a number of pending commands and acknowledgments. That list is ordered such that if two items have different timestamps, the one with earlier timestamp comes first, and if two items have the same timestamp, then a client request received by computer i takes the order of computer i itself, while a message from computer j takes the order of computer j. Computer i removes any number of leading acknowledgments from the list, and looks at the first command and its timestamp T_f. By above, it cannot execute this command until it has learned of all the other commands in all the other computers with timestamps less than or equal to T_f. This means that there is at least one message on its list timestamped later than T_f from all computers ranked

[5] This is formalized as implementation rule IR1 in [9].

before, and at least one message timestamped T_f or later from all computers ranked after. If now computer i can execute this first command and it is a message from another computer, then it updates the account to the new value c, removes the first command from the list, and sends the acknowledgment message to all the other computers. If, on the other hand, the first command is a client request put on the list by computer i itself, then computer i takes the appropriate action (based on r and its current value c), sends a message to the other computers if the request is granted, and removes that command from the list. One can reason about the correctness of this protocol by following the fate of a request. First, a computer never grants a request unless it thinks itself that this is OK. Second, it never grants a request until it has learned of all the withdrawals granted by the other computers and which are timestamped earlier, and of all the withdrawals with the same timestamp granted by computers which come before. Third, from the opposite direction, until the computer has dealt with the request, no other computer will grant a request which is timestamped later, or which has the same timestamp but arrives in a computer which comes later. This is so because when the computer deals with the request it sends a acknowledgment message with this timestamp, and all the other computers have to wait for a message with such a timestamp. It has been assumed that no computers fail, no messages are lost, and no messages are received out of order.

The paragraph is an example of a behavioral proof written informally. The subtlety of Lamport's procedure is perhaps better appreciated by the concrete example he presents, which is that of a fixed collection of elements sharing a single resource (a problem which will be taken up in another version and from another angle in Section 5). Suppose that the resource ought to be granted to the elements in a manner which satisfies the following three conditions:

I An element which has been granted the resource must release it before it can be granted to another element;

II Different requests for the resource must be granted in the order in which they were made;

III If every element which is granted the resources eventually releases it, then every request is eventually granted.

The solution is built on timestamped *request* and *release* commands and messages. Note that as in the toy example, the order of requests in condition **II** above is then that of the total ordering. But why such complexity? For one thing, a central scheduling process, which grants requests in the order in which they are received, will not satisfy condition **II**. The interested reader is encouraged to look at [9] for the (short) proof that this is impossible, as well as the (slightly longer) description of the distributed algorithm which does work.

What about physical clocks? This question is also treated in [9] by asking how small ϵ in a synchronicity condition $|C_i(t) - C_j(t)| < \epsilon$ can be, if it is to hold for all clock pairs $C_i(t)$ and $C_j(t)$ and for all time t, and if single clocks, running without resetting, are accurate to a fraction κ. Note that the clock speeds are assumed to vary unpredictably within the bounds set by κ as otherwise the clocks readings could be re-calibrated. Note also that a naive algorithm, with a master clock which the other clocks adjust to periodically after receiving messages transmitted with maximum delay ξ every τ seconds will achieve $\epsilon \approx 2\kappa\tau + \xi$ (the master sends his time to element i which receives it ξ seconds later, and while waiting for the next time, element i and the master can drift an additional $2\kappa\tau$ seconds apart). Such an algorithm will however not satisfy clock condition **C1** (if clock $C_i(t)$ has outrun the master clock and needs to be set back), and requires that all clocks communicate directly with the master. I refer to to [9] for a precise statement of the theorem Lamport proves, and just mention that if messages can be sent on every link every τ seconds, and under other suitable assumptions, then $\epsilon \approx d(2\kappa\tau + \xi)$, where d is the diameter of the graph. In a complete graph (d equal to one), a fully distributed algorithm which obeys both clock conditions **C1** and **C2**, and where all elements send the same number of messages as the master in the naive algorithm, hence achieves the same synchronicity.

4. Coping with failures

One way to cope with failure is to have a back-up, and, in some sense, this must be the only way. However, things can fail in different ways, and they can fail singly or fail together. For instance, trivially, if some elements fail together, then there must be at least one back-up left. This Section will describe methods invented to deal with single, multiple, and more or less malicious failures in distributed systems which replicate state machines.

The main distinction is between an arbitrary failure and limited failure mode. An arbitrary failure is called Byzantine after the parable of the Byzantine Generals told in [11], and other failures are accordingly called non-Byzantine. A collection of nodes which fails in a Byzantine manner can be imagined to have been taken over by an adversary who is malicious, *i.e.* has the objective to disrupt the function of the state machine. The adversary must not all-powerful, *e.g.*, as in the method described below, be computationally limited in the style of cryptographic analysis, but is at least able to send any message he wants at any time to any of the other nodes. As told by Lamport in [12] the term "Byzantine" was inspired by the "Chinese Generals Problem" and the problem was originally called "The Albanian Generals Problem" but modified to "Byzantine" lest no Albanian would be offended. As an aside, the dictionary term "Byzantine" can refer to an inhabitant or to the art of Byzantium, the Eastern half of the Roman empire, which survived until 1453 A.D., but also has, in most European languages, the meaning of a procedure of great and even dizzying but largely or completely unnecessary complexity. The use of Byzantine in [11] was certainly apt also from this perspective. Starting with the Oral Message (OM) algorithm of [11] (which will not be described here), these methods are complex in themselves, and later improvements (of which I will describe one) were intended for use in real implementations. The original publications can therefore describe what are in effect system designs (even benchmarked prototypes), and typically then contain several protocol variants and incremental additions. A presentation in full detail is out of scope of this paper; the reader should however be aware of that there is more (and sometimes much more) in the sources we follow, mainly [13].

Let us start by considering a distributed system of N small state-action machines, and ask how many machines f with Byzantine failures can the system tolerate. The external actions address one, some or all small state machines simultaneously and identically, and are called *requests*. The system functions correctly if sufficiently many of the small state-action machines change state correctly in response to an action in a finite amount of time. These properties are called *safety* and *liveness*. State change is witnessed by a *response*. Requests and responses can be thought of as messages exchanged between the system and an external client, and can be specified as in [13]. The property to have failed is externally given. Mechanisms for a failed node to recover are not included in the description. The question could therefore perhaps be more properly phrased as what is the smallest number $N_c(f)$ of correctly functioning small state-action machines necessary that the system can cope with f faulty machines. That is, all the $N_c(f)$ correctly functioning machines stay correct in the time interval between a request and a response.

Contrary to what one would perhaps naively expect, $N_c(f) = 2f + 1$ (not $f + 1$). This result, first shown in [6], follows from the following argument in [13]. If the client gets $N_c(f)$ messages as responses, then this must be enough for the client to be sure that the system has executed the state transition correctly. This is a liveness requirement, because the failed machines may fail in the way of not sending any responses at all. However, any of the failed machines may also fail in a way where it does send a response (Byzantine failure), and the client has no way of knowing which of the $N_c(f)$ messages it has received which are from the failed ones. Hence, the responses from the correctly functioning machines must always outnumber the failed ones, within a total number of $N_c(f)$ responses. This means $N_c(f) > 2f$, which is a safety requirement. The result is usually quoted as $N = 3f + 1$.

To argue the correctness of more complex algorithms of this nature is not easy, even if many more details are given, and I have to refer the interested reader to [13] for an example where there are three rounds of messages exchanged among the machines themselves, a number of messages exchanged growing quadratically with the number of machines, and with three checkpoints, two among the machines, and one at the client when it receives the responses. Nevertheless, this algorithm has been reported to work well as a basis for a distributed file server (BFS, a Byzantine-fault-tolerant NFS service) with only a quite small overhead compared to the standard NFS system [13].

5. The self-star approach

Edsger Dijkstra more than a generation ago invented an algorithm to ensure cooperation between processors without central control [14]. Dijkstra takes cooperation in general to mean that the system is is some set of legitimate states, defined such that one out of $N+1$ processor arranged around a ring is "privileged", and that the privilege rotates around the ring. Dijkstra's algorithm therefore also solves a version of the problem of fixed collection of elements sharing a single resource discussed above in Section 3. The algorithm is "self-stabilizing", meaning that regardless of initial state, the system eventually arrives in a legitimate state, and is therefore fault-tolerant, even if the number of steps needed to correct for a failure increases somewhat quickly with the number of processes.

Dijkstra's algorithm is extremely well known, and presented in many places, but for completeness I will also give it here. The small state of each processor is a counter x_i, a number modulo K, where K has to be greater or equal than N for the algorithm to work. The big state is hence a vector $(x_0, x_1, \ldots, x_N)$. The actions are for one processor i to perform an update according to the rule

$$\text{For } i = 1, 2, \ldots, N: \quad \text{if } x_i \neq x_{i-1} \quad \text{then } x_i = x_{i-1} \tag{1}$$

$$\text{For } i = 0: \quad \text{if } x_0 = x_N \quad \text{then } x_0 = (x_0 + 1) \text{mod} K \tag{2}$$

The most important property of Dijkstra's algorithm is that even if all the processors (except 0) look at the counter of their left-side neighbor twice per update, it does not matter if in the meantime the left-side neighbor changes the value of its counter. If the test $x_i = x_{i-1}$ evaluates to true this is so because then x_i will not be changed. That has the same effect as if the test had not been performed at all. If one the other hand the test $x_i = x_{i-1}$ evaluates to false and afterward x_{i-1} is updated, then either the new value of x_{i-1} is x_i, in which case again x_i will not be changed, or the new value of x_{i-1} is not x_i, in which case the test would have evaluated to false also.

It has probably not escaped the reader that Dijkstra's algorithm can be used implement the toy example of a bank which tolerates processors working at different speeds. Suppose that the processors on the ring are initiated in a legitimate state with processor 0 having the privilege. Each processor opens a ledger and initializes a current account variable c_i. Processor 0 is told what the client has in his her account, sets c_0 to this value, and then gives the privilege to processor 1. The bank is now open for business, and the client can contact any processor with a request to withdraw an amount r ($r < 0$ could a deposit). When processor i receives such a request it puts the new request on its ledger, and then checks whether its counter x_i is equal to x_{i-1} (processor 0 checks with processor N). If the condition is true, then processor i knows it does not have the priority, and answers the client with something like "Request for r received, transaction pending". If, on the other hand, x_i is not equal to x_{i-1}, then the processor knows it has priority. First it then sets its current account variable c_i equal to c_{i-1} (processor 0 sets c_0 to c_N). If it has n requests on its ledger it forms $c' = c_i - r_1 - r_2 - \cdots - r_n$. If

c' is positive it sets c_i equal to c', erases the requests from the ledger, and sends something like "Requests $r_1, r_2, \ldots, r_n$ accepted, new balance is c'". Otherwise the processor looks for the largest k such that the partial sum $c' = c_i - r_1 - r_2 - \cdots - r_k$ is positive, sets c_i to this c', erases requests $r_1, r_2, \ldots, r_k$, and sends something like "Requests $r_1, r_2, \ldots, r_k$ accepted, new balance is c'. Requests $r_{k+1}, r_{k+2}, \ldots, r_n$ still pending, please put more money in the account". Only after doing all this the does processor i set x_i equal to x_{i-1} (processor 0 sets x_0 equal to $(x_0 + 1)\text{mod}K$. If the processor i receives another request while it is has the privilege it is put on a buffer, and dealt with after the processor has given up the privilege. If the state is legitimate it cannot happen that the left-hand neighbor of i changes its current account value after processor i has evaluated the condition $x_i = x_{i-1}$ to false, but before processor i has given up the privilege. When the bank closes for the day, the processor with the privilege holds the current account. It can be objected that the requests are not necessarily processed in the same order as the client feels it has issued them. But this was not so in the approach using state machine replication either. It can also be objected that this bank based on Dijkstra's algorithm does not behave in exactly the same way as the other bank based on Lamport's algorithm. But both implementations respect in some sense the condition that the bank does not pay out money the client does not have in his or her account, and unless we have given more precise specifications, it is not clear which implementation is better. The real advantage of the Lamport bank is that it will function correctly if some processors fail, while the Dijkstra bank – at least this toy one — will not.

Dijkstra's paper eventually gave rise to the whole field of self-stabilization, and this idea of "self-" has more recently resulted in "self-management", "self-configuration", "self-healing" and even many more "self-", such as "self-organization". Hence the present term "self-star" or "self-$\star$" for any system property which is deemed "self-". I are not aware of any definition of "self-" which is both well established in the Distributed Systems literature, and simple enough for the purposes of this paper. So I offer one:

> A distributed system is self-P for some property P if it achieves P without replicating a state machine which has P.

Dijkstra's algorithm is of course "self-stabilizing", as well as (at least) "self-configuring", since privilege is not even defined as a property of a small state only.

Physics has quite a few examples which can be thought of as "self-$\star$". In Statics, a system of weights, springs and spokes is at rest if the forces on each weight balance. This is self-$\star$ because the forces depend on the states of the other weights (*e.g.* their heights), and typically are not the same on each. A hydrodynamic velocity is self-$\star$, because the single particles move with very different velocities, and hydrodynamic stability and the whirls and jiggles of fluid flow are therefore both self-$\star$. In fact, practically any property in Condensed Matter Physics is self-$\star$, because it is defined by the interactions of the constituents. For instance, to move beyond the classical domain, the lattice mass of a conducting electron in a metal is self-$\star$ (as well as the whole band structure) because the single electrons are particles with another mass.

All the physical properties evoked above are however self-$\star$ because they are macroscopic. A microscopic system of weight, strings and spokes, such as a molecular motor, will not be at rest, because it is driven by thermal forces. Likewise, in a small enough sample of a liquid or a gas, average velocity fluctuates too much to be a useful concept, and a small enough collection of metal atoms will not form a crystal and will not have a clear band structure. Therefore, self-$\star$ properties of physical systems are more or less the same as *emergence*, and the difference between the microscopic and the mesoscopic or the macroscopic. This then suggests the following:

> Physics is useful to analyze self-$\star$ properties of some distributed system which hold only almost surely if the system is large enough.

The work outlined below in Section 6 was done in this spirit. Let us note that Dijkstra's algorithm as a physical device would also only be "thermodynamically self-⋆". Since it does not obey detailed balance it must be a driven system, therefore cannot be at thermodynamic equilibrium at zero temperature, and therefore must have fluctuations. In a microscopic device these fluctuations would translate into occasional mistakes, and such a device would therefore at times leave the set of legitimate states. This suggests then that the effects of failures in a (large) distributed system can be analyzed by the methods used in Statistical Physics to go between the microscopic and the macroscopic, and in particular the mesoscopic. I have found this a fruitful approach.

6. A guide to further reading

The discussion above has attempted to convey the impression that it is dealing with concurrency and failures together with makes Distributed Systems a complex science. I have also sketched that while methods to cope with any number of failures exist, those methods are expensive, and their cost grows fairly quickly with system size. One reason that Distributed Systems is a complicated science is that it has aimed to solve very difficult problems, that is to provide performance guarantees under very general perturbations to the system. In a very large system concurrent changes are unavoidable, and a failure is just a name for a change which we do not like. The process by which a large system is continuously subject to change is called *churn*. In Physics we are accustomed to much weaker adversaries than in distributed systems (certainly not Byzantine Generals), but on the other hand have techniques to handle very large systems. One does not expect to be able to follow every event on the atomic scale in a liquid or a gas, and even if one could, one would not know how to use that information. But we can hope to predict average properties, which will then almost surely hold in a single, but large enough, system. To do so we must describe the statistically stationary state where changes, of different kinds, and on the average, balance. For a distributed system to function under churn the stationary state must furthermore be one of acceptable expected functionality. Contrary to the possibility evoked by Lamport in the opening quote defining a distributed system, your own computer should *not* be rendered unusable by a single failure in another computer, or even by many such failures, except at most very rarely. Such a property could loosely be described as *robustness*.

In the late 1990'ies the field of Distributed Computing became intertwined with Peer-to-Peer computing which was then widely believed to be the future of computing technology and services [15]. Two key theoretical developments were the development of consistent hashing by Karger and collaborators in [16], allowing for distributed storage that functions under churn, and the Plaxton, Rajaraman and Richa scheme [17] to find and access copies in a distributed storage. A storage architecture developed slightly later and incorporating these ideas was "Chord" where processors and data items have logical addresses in a (large) ring, and ring nodes know of each other in a manner which today would be called "scale-free" (though ordered, not random) [18]. Such systems are robust in that they can provide stable services, between most pairs of nodes in the system, even if other nodes in the meantime arrive or leave. On the other hand, some effort has to be expended by the nodes in the system to keep it functioning, such as correcting incorrect links or look-up tables from information received or by actively polling neighbors a node can talk to directly.

Comparing the performance of different Peer-to-Peer architectures (node conductivities) and network maintenance schemes to the efforts needed to keep the network functioning was of interest, and was quite early evaluated by simulations or by probabilistic techniques [19, 20]. However, they can also, and usually more accurately, be evaluated by statistical physics techniques. Chord under churn can for instance be modeled as a generalized dynamic spin model where the spins indicate the states of different nodes in the system, and the change in system state can be modeled by a master equations [21]. This allows for rather precise quantitative understanding of the performance-cost trade-off, and can be used to compare different network maintenance strategies [22]. Similar techniques can also be used to evaluate the performance of certain network aggregation systems, with excellent results [23].

Acknowledgments

This paper was started as a section in a longer work with Mikko Alava and Supriya Krishnamurthy which never materialized; I thank them for interesting discussions. I also thank Seif Haridi and Sameh El-ansary for discussions on Distributed Systems and Seif Haridi for valuable suggestions. This work was supported by the Academy of Finland through its Finland Distinguished Program, project 129024/Aurell, and through its Centre of Excellence COIN.

References

[1] Tanenbaum A S and van Steen M 2002 *Distributed Systems: Principles and Paradigms* (Prentice Hall Inc)
[2] Tel G 2000 *Introduction to Distributed Algorithms* (Cambridge University Press)
[3] Dijkstra E W 1982 *Selected Writings on Computing: A Personal Perspective* (Springer Verlag)
[4] Lamport L My writings last modified 20 March 2013
[5] Lamport L 1987 Distribution In 'My Writings', item 75, Email message sent to a DEC SRC bulletin board at 12:23:29 PDT on 28 May 87
[6] Pease M, Shostak R and Lamport L 1980 *Journal of the Association for Computing Machinery* **27** 2
[7] Lamport L 2008 Computation and state machines In 'My Writings', item 165
[8] Lamport L 2005 Generalized consensus and paxos Tech. rep. Microsoft Research
[9] Lamport L 1978 *Communications of the ACM* **21** 558
[10] Lamport L 1978 Time, clocks and the ordering of events in a distributed system In 'My Writings', item 27
[11] Lamport L, Pease M and Shostak R 1982 *ACM Transactions on Programming Languages and Systems* **4** 382–401
[12] Lamport L 1982 The byzantine generals problem (with marshall pease and robert shostak) In 'My Writings', item 46
[13] Castro M and Liskov B 1999 *Proc. Third Symposium on Operating Systems Design and Implementations*
[14] Dijkstra E W 1974 *Comm. ACM* **17** 643
[15] Oram A 2001 *Peer-to-Peer: Harnessing the Power of Disruptive Technologies* (O'Reilly)
[16] Karger D, Lehman E, Leighton T, Levine M and Panigrahy R 1997 *ACM Symposium on Theory of Computing* 654
[17] Plaxton C G, Rajaraman R and Richa A W 1997 *ACM Symposium on Theory of Computing* 311
[18] Stoica I, Morris R, Karger D, Kaashoek M and Balakrishnan H 2001 *ACM SIGCOMM* 149
[19] Li J, Stribling J, Morris R, Kaashoek M F and Gil T M 2005 *Proc. 24th Infocom* pp 225–235
[20] Liben-Nowell D, Balakrishnan H and Karger D 2002 *ACM Principles of Distributed Computing (PODC)* 233
[21] Krishnamurthy S, El-Ansary S, Aurell E and Haridi S 2005 *Proc IPTPS, Lecture Notes in Computer Science* **3650** 93
[22] Krishnamurthy S, El-Ansary S, Aurell E and Haridi S 2008 *IEEE/ACM Transactions on Networking* **16** 814
[23] Krishnamurthy S, Ardelius J, Aurell E, Dam M, Stadler R and Wuhib F 2010 *PODC '10 Proceedings of the 29th ACM SIGACT-SIGOPS symposium on Principles of distributed computing* (ACM New York, NY, USA) pp 291–292 ISBN 978-1-60558-888-9

ELC International Meeting on Inference, Computation, and Spin Glasses (ICSG2013) IOP Publishing
Journal of Physics: Conference Series **473** (2013) 012018 doi:10.1088/1742-6596/473/1/012018

Renormalization group for quantum walks

S Boettcher[1], **S Falkner**[1] **and R Portugal**[2]

[1]Department of Physics, Emory University, Atlanta, GA 30322; USA
[2]Laboratorio Nacional de Computacao Cientıfica, Petropolis, RJ 25651-075; Brazil

E-mail: `http://www.physics.emory.edu/faculty/boettcher`

Abstract. We present a detailed introduction to the discrete-time quantum walk problem, in close analogy with the classical ordinary and persistent random walk. This approach facilitates a uniform application of the renormalization group that highlights similarities and differences between the classical and the quantum walk problem. Specifically, we discuss the renormalization group treatment for the mean-square displacement of a walker starting from a single site on the $1d$-line for ordinary and persistent random walks and the quantum walk. We outline the significance of universality for quantum walks and the control this might provide for quantum algorithms. We use our RG method to verify that all 2-state quantum walks on the $1d$-line are in the same universality class.

1. Introduction

A renormalization group (RG) treatment [1] of quantum walks (QW) holds significant promise for a better understanding of search algorithms for quantum computing [4, 5, 6, 7] or of quantum transport phenomena [3]. Much numerical and analytical work has been done to extract properties of QW in specific networks and lattice, typically with specific settings for the quantum coin, for a recent overview see [2]. Surprisingly little attention has been paid to categorizing asymptotic properties of QW in their relation with inherent symmetries of the quantum problem. For example, from the existing work it may stand to reason that a QW starting from a single site of a translational-invariant lattice may have a mean-square displacement (MSD) of

$$\left\langle r^2 \right\rangle \sim t^{\frac{2}{d_w}} \tag{1}$$

with an exponent $d_w = 1$, describing ballistic spread in all directions, but a formal proof of such a statement has not yet been attempted. While it seems clear that quantum interference effects are the cause of this rapid spread, the strength and limits of such a causal relation have not been tested. Unlike for a stochastic classical random walk, the unitary description of a QW immediately entails additional internal degrees of freedom (the "coin" space) whose dimensionality depends on the neighborhood degrees in the lattice, which may vary for different lattices (square, triangular, or hexagonal, for instance) even within the same spatial dimension. This raises the question as to why there would be such a uniform result for d_w. More generally: What are the relevant parameters that determine universality classes [1] in QW and other quantum transport problems? Much of the work on QW has been driven by a computer science perspective that have lead to very strong results, for instance, regarding QW as a universal computing framework [7], similar to well-known results for random walks in classical computation, or on optimizing search performance. We hope to contribute to the basic understanding of QW by connecting asymptotic scaling properties to symmetries and dynamics of the process, the ultimate tenants of control physicists have historically derived from RG.

In the following, we first review many of the technical details of the RG formalism as applied to RW. Then, we observe that the classical subject of persistent random walks (PRW) can be formulated (as a 2nd-order Markov process) in a manner identical to a QW on the same geometry. We explore the RG formalism for these PRW at length so that our QW discussion significantly shortens. Finally, we demonstrate how this RG approach carries over to QW and highlight some of the important new aspects that inevitably arise when extending the RG into the complex plane. We then reproduce with the RG the familiar exponent $d_w = 1$ for QW on the $1d$-line.

2. General Formulation of the Walk Problem

The generic master-equation for a discrete-time walk with a coin, whether classical or quantum, is

$$|\Psi(t+1)\rangle = \mathcal{U}\,|\Psi(t)\rangle\,, \qquad \mathcal{U} = \mathcal{S}\,(\mathcal{C}\otimes\mathcal{I})\,, \tag{2}$$

where the time-evolution operator (or propagator) $\mathcal{U}$ is written in terms of the "shift" operator $\mathcal{S}$ and the "coin" $\mathcal{C}$ [2]. In the d-dimensional site-basis $|\vec{n}\rangle$, we can describe the state of the system in terms of the site amplitudes $\psi_{\vec{n},t} = \langle\vec{n}|\Psi(t)\rangle$, simply the probability density to be at that site for a classical walk, but representing in the quantum walk a vector in coin-space with each component holding the amplitude for transitioning out of site $\vec{n}$ along one of its edges. Application of the coin $\mathcal{C}$ entangles these components, with subsequent redistribution of the walk to neighboring sites by the shift operator $\mathcal{S}$, based on those amplitudes. The general shift operator for a homogeneous nearest-neighbor walk on the line is

$$\mathcal{S} = \sum_{\vec{n}}\sum_{\mu=1}^{d}\left\{P\otimes|\vec{n}-\hat{e}_\mu\rangle\langle\vec{n}| + Q\otimes|\vec{n}+\hat{e}_\mu\rangle\langle\vec{n}| + R\otimes|\vec{n}\rangle\langle\vec{n}|\right\} \tag{3}$$

with the shift matrices P, Q, and R for moving left or right, or not at all. From (2), we get

$$\mathcal{U} \;=\; \sum_{\vec{n}}\sum_{\mu=1}^{d}\left\{A\otimes|\vec{n}-\hat{e}_\mu\rangle\langle\vec{n}| + B\otimes|\vec{n}+\hat{e}_\mu\rangle\langle\vec{n}| + M\otimes|\vec{n}\rangle\langle\vec{n}|\right\} \tag{4}$$

with $A = P\mathcal{C}$, $B = Q\mathcal{C}$, and $M = R\mathcal{C}$.

In a quantum walk, the "hopping" operators A, B, and M are constrained by the requirement of unitary propagation, $\mathcal{I} = \mathcal{U}^\dagger\mathcal{U}$, which results in the conditions in coin-space,

$$\mathcal{I}_d \;=\; A^\dagger A + B^\dagger B + M^\dagger M, \quad 0 = A^\dagger M + M^\dagger B = A^\dagger B. \tag{5}$$

As $\mathcal{C}$ is unitary, these conditions equally apply to P, Q, and R. They can not be satisfies by scalars (except for trivial cases). The algebra in (5) requires at least two-dimensional matrices, and the common interpretation is that the dimension c of the coin space has to match the degree of each site, i.e., each component of the site amplitudes $\psi_{\vec{n},t}$ refers to the transmission along a specific edge [2]. Thus, on a d-dimensional hypercubic lattice the coin-space dimension is $c = 2d$ (or $c = 2d+1$ to allow for a component for remaining at the same site).

In a typical discussion of QW on the *1d*-line, most authors simply choose trivial shift matrices that satisfy (5), such as $P = \begin{pmatrix} 1 & 0 \\ 0 & 0 \end{pmatrix}$, $Q = \begin{pmatrix} 0 & 0 \\ 0 & 1 \end{pmatrix}$, and $R = \begin{pmatrix} 0 & 0 \\ 0 & 0 \end{pmatrix}$ and explore some version(s) of the most general unitary coin, in this case,

$$\mathcal{C} = \begin{pmatrix} \sqrt{\rho} & \sqrt{1-\rho}\,e^{i\phi} \\ \sqrt{1-\rho}\,e^{i\theta} & -\sqrt{\rho}\,e^{i(\phi+\theta)} \end{pmatrix}. \tag{6}$$

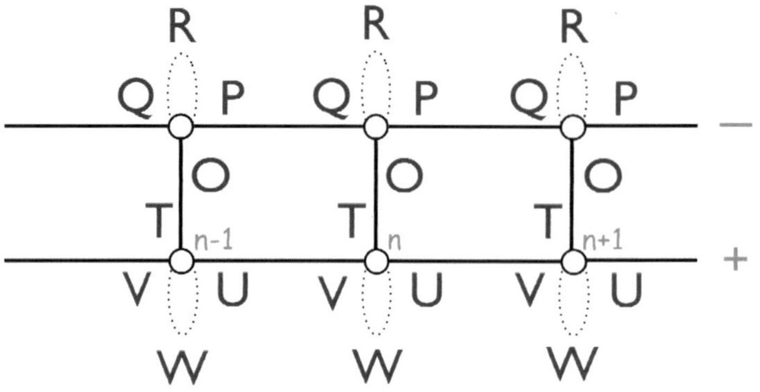

Figure 1. Depiction of the hopping operators for a quantum walk on a 1d-ladder. Each "rung" of the ladder is labeled by a consecutive site-index n while the upper and lower rail is labeled "-" and "+", respectively. Note that each vertex has degree 3, while the dashed loops are meant to indicate a hopping that remains at a site. Thus, the wavefunction on each site has either three or four components.

In general, the algebra in (5) is not limited to its lowest-ranked representatives (here, $c = 2$). It is not obvious that any higher-dimensional matrices obeying (5) would yield new universality classes, of course.

Classical RW are also represented by the propagator $\mathcal{U}$ in (4). In that case, the condition in (5) does not apply and we merely require $\mathcal{U}$ to be *stochastic*, i.e.,

$$A + B + M = \mathcal{C}, \tag{7}$$

where the coin $\mathcal{C}$ must be a stochastic matrix. In fact, (7) is satisfiable with scalars, say, $A = p(1-q)$, $B = (1-p)(1-q)$, and $M = q$, in which case $S = 1$. We note that the stochasticity constraint on $\mathcal{U}$ in (7) can be satisfied also by higher-dimensional matrices. For example, using the stochastic coin on the $1d$-line,

$$\mathcal{C} = \begin{pmatrix} \rho & 1-\sigma \\ 1-\rho & \sigma \end{pmatrix}, \qquad 0 \le \rho, \sigma \le 1, \tag{8}$$

together with the basic shift matrices P, Q, and R (as above) reproduces the well-studied *persistent* RW [8, 9]. For this 2nd-order Markov process it is known that the freedom in the parameters ρ and σ does not provide a distinct universality from the ordinary RW, i.e., it behaves diffusive or ballistic for $\rho = \sigma$ or $\rho \neq \sigma$, respectively.

To illustrate the questions regarding universality in QW, consider the walk on a $1d$-ladder, as shown in figure 1: We have in the site-basis $|n, \pm\rangle$ for the shift operator,

$$\begin{aligned} \mathcal{S} &= \sum_n \{O\,|n,+\rangle\langle n,-| + P\,|n+1,-\rangle\langle n,-| + Q\,|n-1,-\rangle\langle n,-| + R\,|n,-\rangle\langle n,-| \\ &\qquad + T\,|n,-\rangle\langle n,+| + U\,|n+1,+\rangle\langle n,+| + V\,|n-1,+\rangle\langle n,+| + W\,|n,+\rangle\langle n,+|\}, \\ &= \sum_n \left\{ \begin{bmatrix} R & O \\ T & W \end{bmatrix} |n\rangle\langle n| + \begin{bmatrix} P & 0 \\ 0 & U \end{bmatrix} |n+1\rangle\langle n| + \begin{bmatrix} Q & 0 \\ 0 & V \end{bmatrix} |n-1\rangle\langle n| \right\}, \end{aligned} \tag{9}$$

where the first row of the matrix correspond to "$|-\rangle$". Allowing even for two distinct coins for the upper and lower rail-sites,

$$\mathcal{C} = \begin{bmatrix} C_- & 0 \\ 0 & C_+ \end{bmatrix}, \tag{10}$$

we obtain the propagator $\mathcal{U}$ in (4) with

$$M = \begin{bmatrix} RC_- & OC_+ \\ TC_- & WC_+ \end{bmatrix}, \quad A = \begin{bmatrix} PC_- & 0 \\ 0 & UC_+ \end{bmatrix}, \quad B = \begin{bmatrix} QC_- & 0 \\ 0 & VC_+ \end{bmatrix}. \tag{11}$$

For $\mathcal{U}$ to be unitary, A, B, and M again should satisfy (5). It appears from the degree-3 vertices of the $1d$-ladder that the hopping operators in $\mathcal{S}$ should be at least 3×3 matrices (or 4×4, in case that $R\neq 0$ and $W\neq 0$ in figure 1 and an amplitude is needed for remaining at a site), which would make A, B, and M each 6×6 (or 8×8) matrices. Does a QW on this ladder always renormalize into the same universality class of the ordinary 2-state QW on a line? Can coins $C_\pm$ be devised such that the loops along the ladder alter the asymptotic scaling?

The fundamental quantity of interest for any walk is the amplitude $\psi_{\vec{n},t} = \langle \vec{n}|\Psi(t)\rangle$ to be at site $\vec{n}$ at time t [10]. Starting from some initial condition (IC) $\psi_{\vec{n},t=0}$, the time-evolution of the walk is governed by the master equation in (2), which now reads

$$\psi_{\vec{n},t+1} = \sum_{\vec{m}} U_{\vec{n},\vec{m}}\psi_{\vec{m},t} \tag{12}$$

with $U_{\vec{n},\vec{m}} = \langle \vec{n}|\mathcal{U}|\vec{m}\rangle$, for discrete "hops" from connected sites during a single time-step. To eliminate the aspect of time from this dynamic process in preparation for an RG treatment, we introduce generating functions (i.e., discrete Laplace transforms), here given by

$$\tilde{\psi}_{\vec{n}}(z) = \sum_{t=0}^{\infty} \psi_{\vec{n},t} z^t. \tag{13}$$

We typically suppress the argument z unless it appears explicitly. For simplicity, we merely consider IC localized at the origin, $\psi_{\vec{n},0} = \delta_{\vec{n},0}$, where $\delta_{\vec{n},\vec{m}}$ is the Kronecker symbol. Applying (13) to the master equation in (12) obtains

$$\tilde{\psi}_{\vec{n}} = \delta_{\vec{n},0} + z\sum_{\vec{m}} U_{\vec{n},\vec{m}}\tilde{\psi}_{\vec{m}}. \tag{14}$$

3. RG for a Classical Random Walk

For RW on a line, (12) with (4) reduces to

$$\tilde{\psi}_n = \delta_{n,0} + zq\tilde{\psi}_n + z(1-q)\,p\tilde{\psi}_{n-1} + z(1-q)(1-p)\,\tilde{\psi}_{n+1}, \tag{15}$$

where we have allowed for a probability q to remain at a given site and a bias p for jumps to the right, i.e., $1-p$ for jumps to the left. The RG start from (15) [11, 10], written for any even $n\neq 0$:

$$\begin{aligned}
\tilde{\psi}_{n-1} &= a_k\tilde{\psi}_{n-2} + m_k\tilde{\psi}_{n-1} + b_k\tilde{\psi}_n = \frac{a_k}{1-m_k}\tilde{\psi}_{n-2} + \frac{b_k}{1-m_k}\tilde{\psi}_n, \\
\tilde{\psi}_n &= a_k\tilde{\psi}_{n-1} + m_k\tilde{\psi}_n + b_k\tilde{\psi}_{n+1}, \\
\tilde{\psi}_{n+1} &= a_k\tilde{\psi}_n + m_k\tilde{\psi}_{n+1} + b_k\tilde{\psi}_{n+2} = \frac{a_k}{1-m_k}\tilde{\psi}_n + \frac{b_k}{1-m_k}\tilde{\psi}_{n+2},
\end{aligned} \tag{16}$$

defining $a_0 = z(1-q)p$, $b_0 = z(1-q)(1-p)$, and $m_0 = zq$ as the "raw" hopping parameters. Eliminating algebraically all odd-site amplitudes (for $\ldots, n-3, n-1, n+1, n+3\ldots$) and identifying in the remaining relations the renormalized hopping parameters,

$$\tilde{\psi}_n = a_{k+1}\tilde{\psi}_{n-2} + m_{k+1}\tilde{\psi}_n + b_{k+1}\tilde{\psi}_{n+2}, \tag{17}$$

we obtain the RG-flow recursions

$$a_{k+1} = \frac{a_k^2}{1-m_k}, \qquad b_{k+1} = \frac{b_k^2}{1-m_k}, \qquad m_{k+1} = m_k + \frac{2a_k b_k}{1-m_k}. \tag{18}$$

The fixed points (FP) arising from this RG-flow for $k \sim k+1 \to \infty$ are $(a^*, b^*, m^*) = (0, 0, m^*)$, $(1-m^*, 0, m^*)$, or $(0, 1-m^*, m^*)$ for any value of m^* (presumably on the unit interval). The 2nd and 3rd FP yield the ballistic solutions for $z \to 1$, with $d_w = 1$ easily obtained from the eigenvalues of the Jacobian

$$J = \left.\frac{\partial(a_{k+1}, b_{k+1}, m_{k+1})}{\partial(a_k, b_k, m_k)}\right|_{k\to\infty} = \begin{pmatrix} \frac{2a^*}{1-m^*}, & 0, & \frac{2b^*}{1-m^*} \\ 0, & \frac{2b^*}{1-m^*}, & \frac{2a^*}{1-m^*} \\ \frac{(a^*)^2}{(1-m^*)^2}, & \frac{(b^*)^2}{(1-m^*)^2}, & 1 + \frac{2a^*b^*}{(1-m^*)^2} \end{pmatrix}, \tag{19}$$

(From the exact solution one finds that in this case the FP value m^* depends in the initial values for the RG-flow, $1 - m^* = |a_0 - b_0|_{z=1}$, which can not be obtained from the local analysis near FP.) The indeterminedness of m^* in the 1st FP is peculiar. In fact, only two possible values for m^* can be obtained. For any $|z| < 1$, only the trivial solution $m^* = 0$ can be reached. For $z = 1$, $a_k + b_k + m_k = 1$ for all k and starting from symmetric initial values $a_0 = b_0$, both remain identical and vanish together, $a_k \equiv b_k \to 0$, and $m_k \to m^* = 1$. Since both numerators and denominators in the Jacobian vanish, a correlated solution has to be constructed that "peals off" the leading behaviour to glance into the boundary layer. Using $a_k \equiv b_k \sim A_k\alpha^k$ and $m_k \sim 1 - M_k\alpha^k$ assuming large k and $|\alpha| < 1$, results in

$$A_{k+1} = \frac{A_k^2}{\alpha M_k}, \qquad M_{k+1} = \frac{1}{\alpha}M_k - \frac{2A_k^2}{\alpha M_k} \tag{20}$$

with a single FP that self-consistently determines $\frac{A^*}{M^*} = \alpha = \frac{1}{2}$. The Jacobian of these recursions at its FP gives $\lambda = 4$ as the largest eigenvalue, i.e., $d_w = 2$ for the diffusive solution. In this formulation, even if we start with vanishing self-term, $q = 0$, initially, the self-term ultimately dominates, reflecting the fact that in diffusion the renormalized domain of size $L = 2^k$ outgrows the walk such that almost all hops remain within that domain.

We can conclude that the RG projects the salient, asymptotic properties of the walk into two universality classes (for any $0 \leq q < 1$): ballistic motion, either to the left ($p < \frac{1}{2}$) or right ($p > \frac{1}{2}$), or diffusion ($p = \frac{1}{2}$), each characterized by a distinct exponent d_w. Each class reflects a fundamental (a)symmetry of the process.

4. Persistent Random Walks

The persistent random walk (PRW) [9, 8] on a $1d$-line is a well-studied process with many applications. On the line, persistence (or, anti-persistence) have a clear, intuitive meaning that can be described with a single parameter. PRW is a very useful precursor for the study of QW. In fact, we find that PRW is isomorph to QW up to the point when observables are considered. Then, of course, the difference between a dissipative stochastic process with direct probabilistic interpretation and one evolving a wave-function unitarily becomes very noticeable. But both are discrete processes in time and space described by a master equation that evolves a state variable with internal degrees of freedom. In fact, as a 2nd-order Markov process [9], a PRW requires two separate pieces of initial conditions, for location and preferred direction, similar in structure to a Schroedinger wave-function in QW. The quantum coin-matrix in the QW has its equivalent here in a stochastic matrix driving the PRW. But the requirements for a stochastic matrix leads to very different physics than for a unitary coin. The results for the PRW on $1d$-lines provide hardly any new results compared to the ordinary RW studied above. On other lattices, such as the Hanoi networks [12], quite interesting and novel results can be achieved for PRW, though.

We will label the state variable describing a PRW also as $\psi_{n,t}$ with the understanding that it now represents a 2-component vector. Say, the upper component, $\psi^{+}_{n,t}$, refers to a walker with the preference to step to the right in the next time-step, and the lower component, $\psi^{-}_{n,t}$, indicates a left-hop preference. The value of each component describes the probability of finding a walker at that site n and time t in state "$\pm$", and the total probability of finding a walker there, irrespective of preference, is simply the sum of the two, $p_{n,t} = \psi^{+}_{n,t} + \psi^{-}_{n,t}$. Otherwise, the formalism above leading to the master equation (14) for the generating function applies unchanged, providing us again with the basis of our RG-analysis. However, due to the internal structure of our state variable, the renormalizable quantities describing the effective hopping probabilities between domains now possess matrix form, and we will have to reconcile with the fact that our RG-flow equations will take on the form of non-linear matrix recursions.

Ignoring a possible left-right bias here, we consider a walker coming from the left (right) to have a probability ρ to continue to move right (left), and a probability $1-\rho$ to reverse direction in the next step. For $\rho > \frac{1}{2}$ ($\rho < \frac{1}{2}$) the walker exhibits persistence (anti-persistence), and for $\rho = \frac{1}{2}$ becomes again an ordinary unbiased RW without memory. The master equations then read:

$$\begin{aligned} \tilde{\psi}^{+}_{n} &= z\rho\tilde{\psi}^{+}_{n-1} + z\left(1-\rho\right)\tilde{\psi}^{-}_{n-1} + \delta_{n,0}\tilde{\psi}^{+}_{IC}, \\ \tilde{\psi}^{-}_{n} &= z\left(1-\rho\right)\tilde{\psi}^{+}_{n+1} + z\rho\tilde{\psi}^{-}_{n+1} + \delta_{n,0}\tilde{\psi}^{-}_{IC}, \end{aligned} \tag{21}$$

where $\tilde{\psi}_{IC}$ (with $\tilde{\psi}^{+}_{IC} + \tilde{\psi}^{-}_{IC} = 1$) represents the IC of the PRW, which we place again at the origin. Initially, before RG, $\tilde{\psi}^{+}_{n}$ ($\tilde{\psi}^{-}_{n}$) only depends on hops from its left (right) neighbor; it is that inflow which induces the "+" ("−") state. (21) is conveniently rewritten in matrix notation as

$$\tilde{\psi}_{n} = M\tilde{\psi}_{n} + A\tilde{\psi}_{n-1} + B\tilde{\psi}_{n+1} + \delta_{n,0}\tilde{\psi}_{IC} \tag{22}$$

with "raw" hopping matrices

$$A_0 = z\begin{pmatrix} \rho & 1-\rho \\ 0 & 0 \end{pmatrix}, \qquad B_0 = z\begin{pmatrix} 0 & 0 \\ 1-\rho & \rho \end{pmatrix}, \tag{23}$$

and $M_0 = 0$ before renormalization, $k = 0$. Notice that A and B are different but do not express a bias as in (15). It is more revealing to decompose these matrices further and write

$$A_k = P_k\mathcal{C}, \qquad B_k = Q_k\mathcal{C}, \qquad M_k = R_k\mathcal{C} \tag{24}$$

with initially

$$P_0 = z\begin{pmatrix} 1 & 0 \\ 0 & 0 \end{pmatrix}, \qquad Q_0 = z\begin{pmatrix} 0 & 0 \\ 0 & 1 \end{pmatrix}, \qquad R_0 = z\begin{pmatrix} 0 & 0 \\ 0 & 0 \end{pmatrix} \tag{25}$$

and the *stochastic* "coin" matrix $\mathcal{C}$ in (8) (choosing $\sigma = \rho$), where each column sums to unity. Separating the generalized hopping matrices P_k, Q_k, R_k as the renormalizable quantities from the coin $\mathcal{C}$ will prove convenient to determine scalar recursion relations describing the RG-flow. The same procedure will apply also for the coin operator in QW below. Clearly, different coins may lead to different RG-flows, which is exactly the purpose of the RG: discriminating universality classes amongst the coins. For PRW on the $1d$-line, the RG confirms that there are no new universality classes beyond those of an ordinary RW. Note that the most general stochastic coin in $1d$ has to be a 2×2 real matrix with two constraints on the rows, leaving room for at most two free parameters. By fixing $\sigma = \rho$, we already eliminated the directional asymmetry that leads to ballistic motion discussed for RW in section 3 for $p \neq \frac{1}{2}$.

As in section 3, to apply RG we write (22) as

$$\begin{aligned} \tilde{\psi}_{n-1} &= M_k \tilde{\psi}_{n-1} + A_k \tilde{\psi}_{n-2} + B_k \tilde{\psi}_n, \\ \tilde{\psi}_n &= M_k \tilde{\psi}_n + A_k \tilde{\psi}_{n-1} + B_k \tilde{\psi}_{n+1}, \\ \tilde{\psi}_{n+1} &= M_k \tilde{\psi}_{n+1} + A_k \tilde{\psi}_n + B_k \tilde{\psi}_{n+2}. \end{aligned} \tag{26}$$

Again, solving for the central site yields

$$\tilde{\psi}_n = M_{k+1} \tilde{\psi}_n + A_{k+1} \tilde{\psi}_{n-2} + B_{k+1} \tilde{\psi}_{n+2} \tag{27}$$

with the RG-flow

$$\begin{aligned} A_{k+1} &= A_k \left(I - M_k\right)^{-1} A_k, \\ B_{k+1} &= B_k \left(I - M_k\right)^{-1} B_k, \\ M_{k+1} &= M_k + A_k \left(I - M_k\right)^{-1} B_k + B_k \left(I - M_k\right)^{-1} A_k, \end{aligned} \tag{28}$$

applying $M_0 = 0$ and (23). The key realization here is not to explore the FP of these matrix recursions directly. Instead, trial-and-error suggests that the RG-flow can be parameterized as

$$P_k = \begin{pmatrix} a_k & 0 \\ 0 & 0 \end{pmatrix}, \qquad Q_k = \begin{pmatrix} 0 & 0 \\ 0 & a_k \end{pmatrix}, \qquad R_k = \begin{pmatrix} 0 & b_k \\ b_k & 0 \end{pmatrix}. \tag{29}$$

Then, one iteration of the RG-flow in (28) provides a closed set of exact recursions:

$$\begin{aligned} a_{k+1} &= \frac{\rho a_k^2}{(1-b_k)\left[1-(1-2\rho)b_k\right]}, \\ b_{k+1} &= b_k + \frac{a_k^2\left[1-\rho-(1-2\rho)\,b_k\right]}{(1-b_k)\left[1-(1-2\rho)b_k\right]}. \end{aligned} \tag{30}$$

These recursions have only two FP, $(a^*, b^*) \in \left\{(0, b^*), \left(\frac{\rho}{2\rho-1}, \frac{1-\rho}{1-2\rho}\right)\right\}$. The 2nd FP is physical only for $\rho = 0$ or $\rho = 1$ but not for any other value of ρ in between. It is important that we set $\rho = 0$ or $\rho = 1$ both, in the FP and the Jacobian, before we evaluate the Jacobian at the respective FP, since this FP is not physical on any open set of its analytic continuation. For $\rho = 1$ the Jacobian at this FP provides a degenerate eigenvalue, $\lambda = 2$, because the PRW starts with a given direction from the IC and can never reverse direction, i.e., it moves in a trivial, lock-step manner ballistically with $d_w = 1$. On the other hand, for $\rho = 0$, the totally anti-persistent walk should remain forever vacillating in a confined domain around its IC. Here, it is essential to evaluate the Jacobian for $\rho = 0$ first before we apply the FP; only then do we get the eigenvalues $\lambda = 1, 0$, suggesting that the walk maintains its initial position.

The first FP contains the trivial case $z < 1$ for $b^* = 0$. The general indeterminacy of b^* hints at the possibility of a scaling Ansatz again, as in section 3. With $a_k \sim \alpha^k a'_k$ and $b_k \sim 1 - \alpha^k b'_k$ for $\alpha < 1$ and $k \to \infty$ we get $a'_{k+1} = \left(a'_k\right)^2 / \left(2\alpha b'_k\right)$ and $b'_{k+1} = b'_k - a'_{k+1}$, which has an FP at $a'^* = b'^*$ and $\alpha = \frac{1}{2}$, *independent* of ρ, and its Jacobian $J' = \partial\left(a'_{k+1}, b'_{k+1}\right) / \partial\left(a'_k, b'_k\right)$ has the desired diffusive eigenvalues, $\lambda = 4, 1$. Thus, we arrive at the well-known result that a PRW has the same universality classes as the ordinary RW on a line.

5. Quantum Walks on a Line

Quantum walks (QW) are isomorph to PRW, but fundamentally differ in two respects: (1) Instead of a stochastic matrix, the evolution in a QW is driven by a unitary coin; (2) while asymptotic large-time behaviour of observables in ordinary walks are obtained in the limit of real $z \to 1$ from the generating functions, in QW observables typically result from integrals of the modulo-square of generating functions over the unit circle in the complex-z plane. While the first point provides conceptual challenges – and a much richer phenomenology – for QW, the integral in the second point is the main source of technical problems in the RG-analysis of QW: Instead of one FP, it seems that we have to study a wide-ranging set of asymptotic behaviours for the RG-flow anywhere near that circle. Those asymptotic behaviours may be grouped into a finite set of classes, yet, in some of those the RG-flow does not converge but remains forever oscillatory in a seemingly chaotic fashion. Interestingly, all of these features are already present for QW on a $1d$-line.

In close correspondence to the discussion of the PRW in section 4, we have the master equations for the QW in terms of generating functions

$$\tilde{\psi}_n = M\tilde{\psi}_n + A\tilde{\psi}_{n-1} + B\tilde{\psi}_{n+1} + \delta_{n,0}\psi_{IC} \tag{31}$$

with

$$A_k = P_k\mathcal{C}, \qquad B_k = Q_k\mathcal{C}, \qquad M_k = R_k\mathcal{C}, \tag{32}$$

where initially

$$P_0 = z\begin{pmatrix} 1 & 0 \\ 0 & 0 \end{pmatrix}, \qquad Q_0 = z\begin{pmatrix} 0 & 0 \\ 0 & 1 \end{pmatrix}, \qquad R_0 = z\begin{pmatrix} 0 & 0 \\ 0 & 0 \end{pmatrix} \tag{33}$$

with the most general quantum "coin" given in (6).

The RG-analysis is formally *identical* to that already discussed in section 4. Starting from the same master equations in (26), we arrive at the same RG-flow equations in (28). Evolving those recursions from the initial conditions numerically for a few iterations (initially, for a conveniently chosen coin, such as Hadamard: $\rho = \frac{1}{2}$, $\phi = \theta = 0$), a recursive pattern emerges that suggest a similar Ansatz to that in (29):

$$P_k = \begin{pmatrix} a_k & 0 \\ 0 & 0 \end{pmatrix}, \qquad Q_k = \begin{pmatrix} 0 & 0 \\ 0 & -a_k \end{pmatrix}, \qquad R_k = \begin{pmatrix} 0 & b_k \\ b_k & 0 \end{pmatrix}, \tag{34}$$

the minute difference being due to having a unitary instead of a stochastic coin. We attain the initial values $a_1 = \sqrt{\rho}z^2, b_1 = \sqrt{1-\rho}z^2$ only after the first RG-step, the general behaviour ensuing thereafter. The resulting recursions are quite similar in appearance to the corresponding (30) for PRW and read

$$\begin{aligned} a_{k+1} &= \frac{\sqrt{\rho}a_k^2}{1 - 2\sqrt{1-\rho}b_k + b_k^2}, \\ b_{k+1} &= b_k + \frac{\left(b_k - \sqrt{1-\rho}\right)a_k^2}{1 - 2\sqrt{1-\rho}b_k + b_k^2} \end{aligned} \tag{35}$$

for general ρ and $\phi = \theta = 0$. In any case, however, a_k and b_k in general are complex variables that need to satisfy additional unitarity constraint which derive from (5).

In figure 2 we plot the modulo-square of the (1,1)-element of A_k (essentially, $|a_k|^2$), evolving the RG-flow recursions in (35) up to $k = 4$, starting with $z = e^{i\alpha}$. It demonstrates the complicated, oscillatory dependence on z of the renormalized quantities. In figure 3 we show the corresponding poles of that expression in the complex-z plane. A similar plot for PRW has

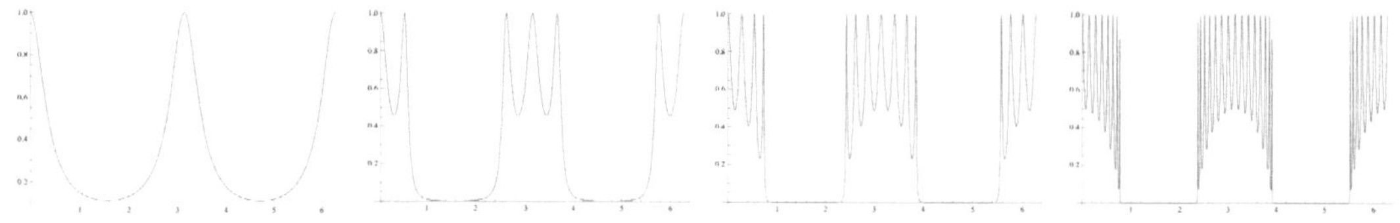

Figure 2. Plot of $\left|A_{1,1}^{(k)}\left(e^{i\alpha}\right)\right|^2$ as a function of $\alpha = \arg(z)$, $0 \leq \alpha \leq 2\pi$, for $k = 1, 2, 3$, and 4. Note that the function is periodic with period π and has significant support only between $-\frac{\pi}{4} < \alpha < \frac{\pi}{4}$ and its periodic iteration.

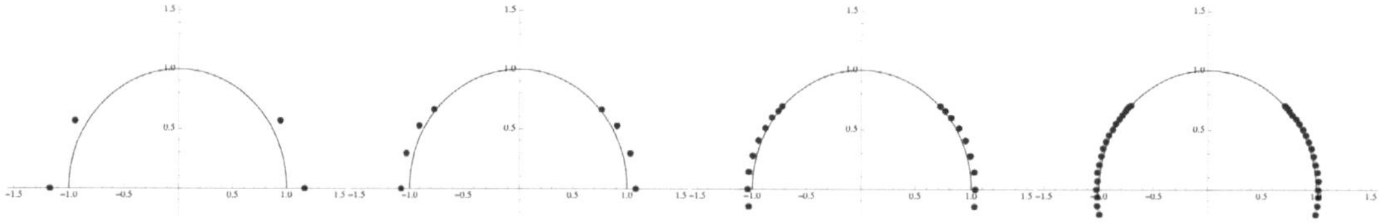

Figure 3. Plot of the poles of $A_{1,1}^{(k)}(z)$ in the complex-z plane for $k = 1, 2, 3$, and 4. Note that the poles cluster between $-\frac{\pi}{4} < \arg z < \frac{\pi}{4}$ and its periodic iteration, just outside and ever closer to the unit circle. Since $A^*(z) = A\left(\frac{1}{z}\right)$, the modulus has also a corresponding set of mirror-poles on the inside of the circle.

a set of entirely *real* poles that accumulate on the positive real axis just above, and ever closer to, $z = 1$ [10]. Thus, while in the classical case a fixed-point analysis near $z \to 1$ suffices, in QW we have to study the RG-flow on extensive parts of the unit circle in the complex-z plane.

We now focus on the analysis of the recursions in (35) for unimodular z and asymptotically for large k. As figures 2 and 3 reveal, a FP-analysis for some fixed limiting value, such as $z \to 1$ for the ordinary RW, appears to be insufficient. Rather, we have to confront the behaviour of the RG-flow over the entire circle of $\alpha = \arg(z)$. Since (35) do not explicitly contain z, it is straightforward to apply a conventional FP analysis, which result in the FP $(a^*, b^*) = \left\{(0, b^*), \left(\sqrt{\rho}, \sqrt{1-\rho}\right)\right\}$. The 2$^{\rm nd}$ FP can be reached only for one single value of z, namely that at $\alpha = 0$. As for ordinary RW above, the first FP requires a more extensive scaling Ansatz.

For $\alpha = 0$ (or $z \to 1$) we are apparently conducting the analysis typical for an ordinary RW. Yet, this turns out to be a "sweet-spot" at the center of the quantum domain, where the phases of the renormalized quantities evolves the least with the RG index k; in fact, the phase remains zero. Although the FP itself possesses an interesting explicit ρ-dependence, the Jacobian matrix and its eigenvalues at this FP are independent of ρ. The eigenvalue $\lambda = 2$ is degenerate but otherwise resembles the known result for QW on a $1d$-line, $d_w = 1$. However, for $\alpha \neq 0$ such a traditional fixed-point analysis is useless, and we have developed a more sophisticated approach that confirms that $d_w = 1$ over the *entire* oscillatory regime.

Exactly at the edges of the oscillatory regime, i.e., at $\alpha = (2j+1)\frac{\pi}{4}$ for $\rho = \frac{1}{2}$, we touch on the first FP in the same singular manner we also encountered for RW: The scaling Ansatz $a_k \sim i\epsilon^k a'_k$ and $b_k \sim \frac{1+i}{\sqrt{2}} - \epsilon^k b'_k$, near the pole of the denominator, leads to the new system, $\epsilon a'_{k+1} = \left(a'_k\right)^2 / \left(2b'_k\right)$ and $\epsilon b'_{k+1} = b'_k - a'_{k+1}$. Its FP provides $\epsilon = \frac{1}{2}$ and $a'^* = b'^*$, and a Jacobian with eigenvalues of $\lambda = 4, 1$. This isolated incidence of diffusive behaviour does not seem to affect the known true scaling. The entire discussion for this special case looks very much similar to the typical analysis for ordinary RW but here represents only a vanishingly small set of all options for the QW.

For $\frac{\pi}{4} < \alpha < \frac{3\pi}{4}$ at $\rho = \frac{1}{2}$, and the other rapidly decaying regimes, no such scaling Ansatz will balance the vanishing numerator with a cancellation from the denominator of the a_k-recursion. Instead, the denominator simply approaches a complex constant and both a_k and b_k phase-lock with a phase difference of α, apparently. The Ansatz $a_k \sim \delta_k e^{i\mu}$ with vanishing δ_k for large k leads to $\delta_{k+1} \sim \delta_k^2$ while $b_k \sim -e^{i\mu}$ remains a constant (with μ undetermined, but see below). The exponentially decaying solution, $\delta_k \sim e^{-c2^k} / \left(2 + 2\sqrt{2}\cos\mu\right)$, provides no eigenvalue (or $\lambda = 0$) while the constant b_k-solution has $\lambda = 1$, i.e., there is no large-scale quantum transport in this entire sector of α. Any other domain for α simply reiterates these results due to the periodicity in $\alpha \to \alpha + \pi$.

Finally, it should be pointed out that the generic picture for the $1d$-behaviour remains unaltered for different choices of the quantum coin in (6). While figure 2 was generated with the Hadamard coin, $\rho = \frac{1}{2}$, $\phi = \theta = 0$, it is easy to show that any variation in either phase, ϕ or θ, merely rotates the unit circle (i.e., offsets α), whereas a variation in ρ alters the sizes of either type of domain without changing the behaviours within those. In all cases, our analysis reproduces the well-know fact [2] that $d_w = 1$ for the MSD in the QW is universal, unaffected by any variation in the coin parameters. Notice, that for either $\rho \to 0$ or $\rho \to 1$, the domains degenerate and either cover the entire circle or squeeze into a point, respectively, and special behaviour may arise.

Our presentation here has been largely focused on method and no new results have been produced, as we have only demonstrated our method to the well-known cases of RW, PRW, and QW on the $1d$-line. However, we are currently extending our analysis to more complicated cases, such a Sierpinski gaskets, where the exact RG-flow can be found and interesting new phenomena may arise due to the lack of translational invariance, which is bound to affect the universality class.

Acknowledgments

SB thanks LNCC for its support and hospitality. RP thanks Emory University for its hospitality. This work has been supported also by grant 1207431 from the Division of Materials Research at the National Science Foundation.

References

[1] Plischke M and Bergersen B 1994 *Equilibrium Statistical Physics, 2nd edition* (World Scientifc, Singapore)
[2] Portugal R 2013 *Quantum Walks and Search Algorithms* (Springer, Berlin)
[3] Boettcher S, Varghese C and Novotny M A 2011 *Phys. Rev. E* **83** 041106
[4] Ambainis A, Bach E, Nayak A, Vishwanath A and Watrous J 2001 *Ann. ACM Symp. Th. Comp.* **33** 37–49
[5] Bach E, Coppersmith S, Goldschen M P, Joynt R and Watrous J 2004 *J. Comp. and Sys. Sci.* **69** 562–592
[6] Childs A M and Goldstone J 2004 *Phys. Rev. A* **70** 022314
[7] Childs A M 2009 *Phys. Rev. Lett.* **102** 180501
[8] Weiss G H 1994 *Aspects and Applications of the Random Walk* (North-Holland, Amsterdam)
[9] Chen A and Renshaw E 1994 *J. Appl. Prob.* **31** 869–884
[10] Redner S 2001 *A Guide to First-Passage Processes* (Cambridge: Cambridge University Press)
[11] Boettcher S, Goncalves B and Azaret J 2008 *J. Phys. A* **41** 335003
[12] Boettcher S and Goncalves B 2008 *EuroPhys. Lett.* **84** 30002

ELC International Meeting on Inference, Computation, and Spin Glasses (ICSG2013) IOP Publishing
Journal of Physics: Conference Series **473** (2013) 012019 doi:10.1088/1742-6596/473/1/012019

Stability of topologically-protected quantum computing proposals as seen through spin glasses

H G Katzgraber
Department of Physics & Astronomy, Texas A&M University, College Station, Texas 77843
Materials Science & Engineering, Texas A&M University, College Station, Texas 77843
E-mail: hgk@tamu.edu

R S Andrist
Theoretische Physik, ETH Zurich, CH-8093 Zurich, Switzerland
E-mail: andrist@phys.ethz.ch

Abstract.
Sensitivity to noise makes most of the current quantum computing schemes prone to error and nonscalable, allowing only for small proof-of-principle devices. Topologically-protected quantum computing aims at solving this problem by encoding quantum bits and gates in topological properties of the hardware medium that are immune to noise that does not impact the entire system at once. There are different approaches to achieve topological stability or active error correction, ranging from quasiparticle braidings to spin models and topological colour codes. The stability of these proposals against noise can be quantified by their error threshold. This figure of merit can be computed by mapping the problem onto complex statistical-mechanical spin-glass models with local disorder on nontrival lattices that can have many-body interactions and are sometimes described by lattice gauge theories. The error threshold for a given source of error then represents the point in the temperature-disorder phase diagram where a stable symmetry-broken phase vanishes. An overview of the techniques used to estimate the error thresholds is given, as well as a summary of recent results on the stability of different topologically-protected quantum computing schemes to different error sources.

1. Introduction

Topological error-correction codes represent an appealing alternative to traditional [1, 2, 3] quantum error-correction approaches. Interaction with the environment is unavoidable in quantum systems and, as such, efficient approaches that are robust to errors represent the holy grail of this field of research. Traditional approaches to error correction require, in general, many additional quantum bits, thus potentially making the system more prone to failures. However, topologically-protected quantum computation presents a robust and scalable approach to quantum error correction: The quantum bits are encoded in delocalized, topological properties of a system, i.e., logical qubits are encoded using physical qubits on a nontrivial surface [4]. Topological quantum error-correcting codes [5, 6, 7, 8, 9, 10] are thus instances of stabilizer codes [11, 12], in which errors are diagnosed by measuring check operators (stabilizers). In topological codes these check operators are local, thus keeping things simple. The ultimate goal is not only to achieve good quantum memories, but also to reliably perform computations.

The first topological quantum error-correction code was the Kitaev toric code [5]. Other proposals followed, such as colour codes [6, 7, 8], as well as stabilizer subsystem codes [13, 14]. Interestingly, topological quantum error correction has a beautiful and deep connection to classical spin-glass models [15] and lattice gauge theories [16, 4]: When computing the error stability of the quantum code to different error sources (e.g., qubit flips, measurement errors, depolarization, etc.) the problem maps onto disordered statistical-mechanical Ising spin models on nontrivial topologies with N-body interactions. Furthermore, for some specific error sources, the problem maps onto novel lattice gauge theories with disorder.

This paper outlines the close relationship between several realizations of these error-correction strategies based on topology to classical statistical-mechanical spin models. The involved mapping associates faulty physical qubits with "defective" spin-spin interactions, as well as imperfections in the error-correction process with flipped domains. Thus a disordered spin state—characterized by system-spanning domain walls—can be identified with the proliferation of minor errors in the quantum memory. As a result, the mapping can be used to calculate the error threshold of the original quantum proposal [4, 17, 18, 19, 20, 21, 22, 23]: If the spin system remains ordered, we know that error correction is feasible for a given error source and an underlying quantum setup. Because the quantum problem maps onto a disordered Ising spin-glass-like Hamiltonian [15], no analytical solutions exist. As such, the computation of the error thresholds strongly depends on numerical approaches.

Different methods to compute the error thresholds exist, ranging from zero-temperature approaches that use exact matching algorithms (see, for example, [21]), to duality methods [24, 25, 26, 27]. Unfortunately, the former only delivers an upper bound, while the latter is restricted to problems defined on planar graphs. A generic, albeit numerically-intensive approach that allows one to compute the error threshold for any error source (i.e., for any type of N-body Ising spin glass on any topology) is given via Monte Carlo simulations [28, 29].

In section 2 we outline the quantum–to–statistical mapping for the case of the toric code [4, 5]. In this particular case, computing the error tolerance of quantum error correction due to qubit flips maps onto a two-dimensional random-bond Ising model [15] with additional requirements imposed on the random couplings. The Monte Carlo methods used are described in section 3. Beyond the toric code, an equivalent mapping is also possible for more involved error-correction schemes, more realistic error sources, as well as under the assumption of an imperfect quantum measurement apparatus. Section 4 summarizes our results for different topologically-protected quantum computing proposals to different error sources and a summary is presented in section 5.

Besides providing new and interesting classical statistical-mechanical models to study, the results accentuate the feasibility of topological error correction and raise hopes in the endeavor towards efficient and reliable quantum computation.

2. Mapping topological qubits onto spin glasses: Example of the toric code

During the error-correction process, different errors can have the same error syndrome [5, 4] and we cannot determine which error occurred. The best way to proceed is by classifying errors into classes with the same effect on the system, i.e., errors that share the same error-correction procedure. Once the classification is complete, we correct for the most probable error class. Successful error correction then amounts to the probability of identifying the correct error class.

In topological error-correction codes, this is achieved by measuring local stabilizer operators. These are projective quantum measurements acting on multiple neighboring qubits in order to determine, for example in the case of qubit flip errors, their flip-parity. The actual quantum operators are chosen carefully to allow for the detection of a flipped qubit in a group without measuring (and thus affecting) the encoded quantum information. Due to this limitation, the stabilizer measurements can only provide some information about the location of errors, which is then used to determine the most probable error class.

Journal of Physics: Conference Series **473** (2013) 012019 doi:10.1088/1742-6596/473/1/012019

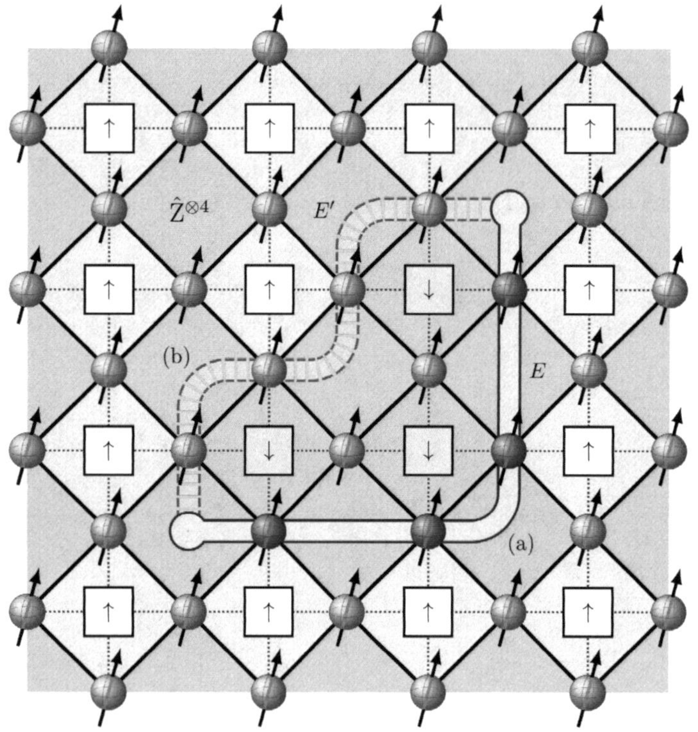

Figure 1. For the toric code, qubits are arranged on a square lattice, with stabilizer operators acting on plaquettes of four qubits. The figure shows the spin placement to sample chains from an error class $\bar{E}$: (a) A reference error chain E defines the error class; the interactions are anti-ferromagnetic along this chain. (b) A chain which differs from the reference chain by a product of spin-plaquettes: Together with the reference chain it forms the boundary $E + E'$ of a flipped domain of spins. The terms along E' are excited, while the ones along E are now satisfied. Thus the Boltzmann weight for this configuration is proportional to $(p/(1-p))^{|E'|-|E'|}$.

The Kitaev proposal for the toric code arranges qubits on a square lattice with stabilizer operators of the form $\hat{Z}^{\otimes 4}$ ($\hat{Z}$ a Pauli operator) and $\hat{X}^{\otimes 4}$ ($\hat{X}$ a Pauli operator) acting on the four qubits around each plaquette. When only qubit-flip errors are considered, it is sufficient to treat only stabilizers of type $\hat{Z}^{\otimes 4}$ which are placed on the dark tiles of the checkerboard decomposition, see figure 1. The measurement outcome of each stabilizer applied to its four surrounding qubits is ± 1, depending on the parity of the number of flipped qubits. These parity counts are not sufficient to locate the exact position of the qubit errors, but for sufficiently low error rates it is still possible to recover using this partial information (see [5] for details). This can be achieved by interpreting sets of neighboring errors as chains and classifying them into error classes with the same effect on the encoded information. During the error-correction process, all stabilizer operators are measured and the resulting error syndrome represents the end points of error chains. We refer to these sets of errors as chains, because two adjacent errors cause a stabilizer to signal even flip-parity – only the end points of the chain are actually detected. This information is still highly ambiguous in terms of the actual error chain E, where the errors occurred. Fortunately, we do not need to know where exactly the error occurred: Successful error correction amounts to applying the error-correction procedure for an error from the correct error class, i.e., such that no system spanning loop is introduced. The question of whether error recovery is feasible therefore is determined by the probability of identifying the correct error class. This likelihood is what can be calculated through the mapping to classical spin glasses.

For a constant qubit error rate p, the probability $\mathcal{P}$ for a specific error chain E is determined by the number of faulty qubits $|E|$:

$$\mathcal{P}(E) = (1-p)^{N-|E|}(p)^{|E|}, \tag{1}$$

where N is the number of qubits in the setup. Equivalently, we can describe this error chain with Boolean values $n_\ell^E \in \{0,1\}$ for each qubit ℓ, describing whether an error occurred. The probability in (1) can then be written as

$$\mathcal{P}(E) = \prod_\ell (1-p)^{N-n_\ell^E}(p)^{n_\ell^E} = (1-p)^N \prod_\ell \left(\frac{p}{1-p}\right)^{n_\ell^E}, \tag{2}$$

where the product is over all qubits ℓ. Because the stabilizer measurements only yield the boundary of the error chain, there are many other error chains E' that are compatible with the same error syndrome. If two chains E and E' share the same boundary, then they can only differ by a set of cycles C, which have no boundary. The relative probability of $E' = E + C$, compared to E, depends on the relative number of faulty qubits, which increases for every qubit in $C \setminus E$ but decreases for qubits in $C \cap E$. Therefore, using analogous Boolean descriptors for the chain, n_ℓ^C, we can write the relative probability $\mathcal{P}(E')/\mathcal{P}(E)$ as:

$$\frac{\mathcal{P}(E')}{\mathcal{P}(E)} = \prod_\ell \left(\frac{p}{1-p}\right)^{n_\ell^C(1-2n_\ell^E)} \propto \prod_\ell \exp[\beta J \tau_\ell \underbrace{(1-2n_\ell^C)}_{u_\ell}]. \tag{3}$$

The newly-defined variable $u_\ell \in \{\pm 1\}$ is negative for all links in C and we have introduced carefully-chosen coupling constants $\tau_\ell \in \{\pm 1\}$ and a factor βJ such that

$$e^{-2\beta J \tau_\ell} = [p/(1-p)]^{1-2n_\ell^E}. \tag{4}$$

Note that the sign of τ_ℓ is dictated by the presence of an error in the reference chain E, and J is related to the error probability via the Nishimori condition [30]:

$$-2\beta J = \ln\left(\frac{p}{1-p}\right). \tag{5}$$

The constraint for C to be cyclic (no boundary) imposes the additional requirement that the number of adjacent faulty qubits $\ell \in C$ must be even for every plaquette. One way to satisfy this condition is to introduce Ising variables $s_i \in \{\pm 1\}$ for each plaquette of the *opposite* colour. That is, each spin represents an elementary cycle around a plaquette and larger cycles are formed by combining several of these elementary loops. For any choice of the spin variables s_i, the variables $u_\ell = s_i s_j$, with ℓ the edge between plaquettes i and j, describes such a cyclic set C (see figure 1).

We have therefore found that the spin configurations enumerate all error chains E' that differ from the reference chain E by a tileable set of cycles. With the Nishimori relation (5) their Boltzmann weight is also proportional to the probability of the respective error chain. Therefore, it is possible to sample the fluctuations of error chains within the same error class by sampling configuration from the classical statistical model described by the partition function

$$\mathcal{Z}_{\{\tau_{ij}\}} = \sum_{\{s\}} \exp\big(-\beta J \sum_{\langle i,j\rangle} \tau_{ij} s_i s_j\big). \tag{6}$$

Here J is dictated by the Nishimori condition (5) and τ_{ij} are quenched, disordered interactions which are negative if the associated qubit is faulty in the reference chain. Because the mapping identifies error chains with domain walls and their difference with a flipped patch of spins, we can identify the ordered state with the scenario where error chain fluctuations are small and correct error classification is feasible. And while this sampling does not implicitly consider homologically nontrivial cycles, we can interpret percolating domain walls as error fluctuations which are too strong to reliably distinguish cycles of different homology.

Because βJ and p can only be related on the Nishimori line for the mapping between the quantum problem and the statistical-mechanical counterpart to work, we need to compute the point in the disorder p and critical temperature $T_c(p)$ plane where the Nishimori line (5) intersects the phase boundary between a paramagnetic and a ferromagnetic phase, see figure 3. This point, p_c then corresponds to the error threshold of the underlying topologically-protected quantum computing proposal. For the case of the toric code with qubit flip errors, the problem maps onto a two-dimensional random-bond Ising model described by the Hamiltonian:

$$\mathcal{H} = -J \sum_{\langle i,j \rangle} \tau_{ij} s_i s_j \, , \tag{7}$$

where J is a global coupling constant chosen according to the Nishimori condition (5), $s_i \in \{\pm 1\}$, the sum is over nearest neighbor pairs, and the mapping requires τ_{ij} to be quenched bimodal random interactions, distributed according to the error rate p:

$$\mathcal{P}(\tau_{ij}) = \begin{cases} +1\,; & 1-p \\ -1\,; & p\,. \end{cases} \tag{8}$$

For the toric code with qubit flip errors $p_c \approx 10.9\%$ [31, 32, 25, 33], i.e., as long as the fraction of faulty physical qubits does not exceed 10.9%, errors can be corrected. In the next section we outline the procedure used to estimate different error thresholds using Monte Carlo methods.

3. Estimating error thresholds using Monte Carlo methods

The partition function found in the mapping for Kitaev's toric code can be interpreted as a classical spin model where the different spin configurations are weighted proportional to the likelihood of the error chain they represent. The existence of such a relationship is instrumental in understanding the fluctuations of error chains, because it allows for the computation of the thermodynamic value of the error threshold using tools and methods from the study of statistical physics of disordered systems.

3.1. Algorithms

The simulations are done using the parallel tempering Monte Carlo method [34, 35, 36, 37, 38]. The method has proven to be a versatile "workhorse" in many fields [39]. Similar to replica Monte Carlo [34], simulated tempering [40], or extended ensemble methods [41], the algorithm aims to overcome free-energy barriers in the free energy landscape by simulating several copies of a given Hamiltonian at different temperatures. The system can thus escape metastable states when wandering to higher temperatures and relax to lower temperatures again in time scales several orders of magnitude smaller than for a simple Monte Carlo simulation at one fixed temperature. A delicate choosing of the individual temperatures is key to ensure that the method works efficiently. See, for example, [38].

The classical Hamiltonians obtained when computing error thresholds in topological quantum computing proposals all have quenched bond disorder, i.e., a complex energy landscape with many metastable states. As such, parallel tempering is the algorithm of choice when simulating these systems, especially when temperatures are low and disorder high (for example, close to the error threshold) where thermalization is difficult.

Equilibration is typically tested in the following way: We study how the results for different observables vary when the simulation time is successively increased by factors of 2 (logarithmic binning). We require that the last three results for all observables agree within error bars.

3.2. Determination of the phase boundary

To determine the error threshold we have to first determine the phase boundary between an ordered and disordered phase. In most cases this phase boundary can be determined by studying dimensionless functions of local order parameters like the magnetization. Note, however, that some models map onto lattice gauge theories where averages of all local order parameters are zero. In these specific cases we use other approaches outlined below.

As an example, for the case of the toric code, we study a transition between a paramagnetic and a ferromagnetic phase. The transition is determined by a finite-size scaling of the dimensionless two-point finite-size correlation length divided by the system size [42, 43, 44, 45]. We start by determining the wave-vector-dependent susceptibility

$$\chi(k) = \frac{1}{N}\sum_{ij}^{N}\langle S_i S_j\rangle_T \; e^{i\mathbf{k}\cdot(\mathbf{R}_i-\mathbf{R}_j)} \, . \tag{9}$$

Here, $\langle\cdots\rangle_T$ represents a thermal (Monte Carlo time) average and $\mathbf{R}_i$ is the spatial location of the N spins. The correlation length is then given by

$$\xi_L = \frac{1}{2\sin(k_{\rm min}/2)}\sqrt{\frac{[\chi(k=0)]_{\rm av}}{[\chi(k_{\rm min})]_{\rm av}} - 1}, \tag{10}$$

where $k_{\rm min} = (2\pi/L, 0)$ is the smallest nonzero wave vector and $[\cdots]_{\rm av}$ represents an average over the different error configurations (bond disorder). The finite-size correlation length divided by the system size has the following simple finite-size scaling form:

$$\xi_L/L \sim \widetilde{X}(L^{1/\nu}[T - T_{\rm c}]), \tag{11}$$

where ν is a critical exponent and $T_{\rm c}$ represents the transition temperature we need to construct the phase boundary. Numerically, finite systems of linear size L are studied. In that case the function ξ_L/L is independent of L whenever $T = T_{\rm c}$ because then the argument of the function $\tilde{X}$ is zero. In other words, if a transition is present, the data cross at one point (up to corrections to scaling). This is illustrated in figure 2 for the case of the toric code and 4% faulty qubits: Data for different system sizes cross at $T_{\rm c} \approx 1.960(2)$ signaling a transition. Because finite-size scaling corrections are typically small, one can use the estimate of $T_{\rm c}$ obtained as a very good approximation to the thermodynamic limit value.

The simulations must now be repeated for different fractions p of faulty qubits, i.e., for different fractions of ferromagnetic-to-antiferromagnetic bonds between the classical Ising spins. This then allows one to build a temperature–disorder phase diagram as shown in figure 3 for the case of the toric code. The error threshold then corresponds to the point where the Nishimori line (dashed line in figure 3) intersects the phase boundary (solid line in figure 3). In this particular case this occurs for $p_{\rm c} \approx 10.9\%$, i.e., as long as there are less than 10.9% physical qubit flips, errors can be corrected in the quantum code.

4. Results

We now summarize our results for different topological codes, as well as different sources of error. Note that the mappings onto the statistical-mechanical models are often complex and, for the sake of brevity, we refer the reader to the individual manuscripts cited.

4.1. Toric code with qubit-flip errors

The toric code with qubit-flip errors has already been described in detail in section 2. As stated before, the error-correction process maps onto a two-dimensional random-bond Ising model:

$$\mathcal{H} = -J\sum_{i,j}\tau_{\langle ij\rangle}s_i s_j \, , \qquad \mathcal{P}(\tau_{ij}) = \left\{ \begin{array}{ll} +1 \, ; & 1-p \\ -1 \, ; & p \end{array}\right. \tag{12}$$

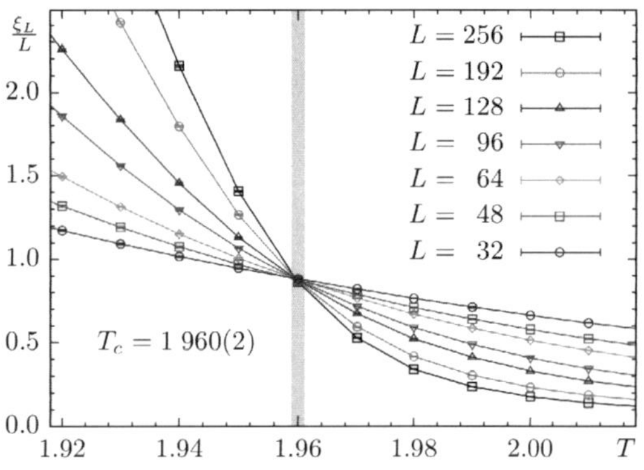

Figure 2. Two-point finite-size correlation length ξ_L/L as a function of temperature for different system sizes L for the two-dimensional random-bond Ising model with a fraction of $p = 0.04$ antiferromagnetic bonds. This corresponds to a toric code with 4% faulty qubits (bit flip errors). The data cross at $T_c \approx 1.960(2)$, signaling a transition (shaded area; the width represents the statistical error bar). The data cross cleanly and show only small finite-size corrections (these become stronger close to p_c).

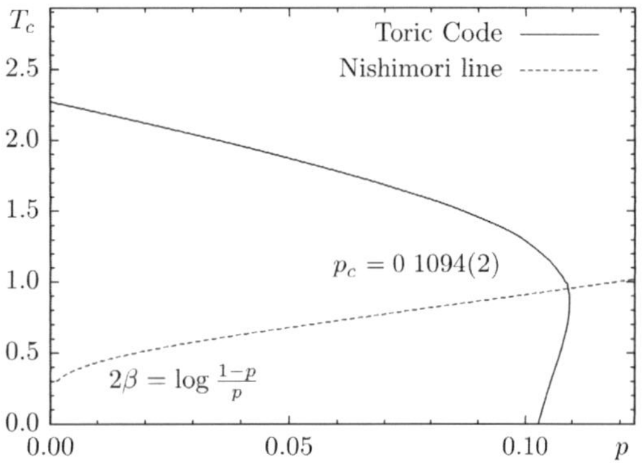

Figure 3. Phase diagram for the two-dimensional random-bond Ising model. p represents the fraction of antiferromagnetic bonds (fraction of flipped qubits) and T_c the transition temperature. The dashed line is the Nishimori line. The point where it intersects the phase boundary (solid line) represents the error threshold which, in this case, is $p_c \approx 0.1094(2)$ [31, 46, 32, 25, 33]. This means that errors can be corrected as long as no more than 10.9% physical qubits have flipped. Phase boundary data taken from [47].

where J is a global coupling constant chosen according to the Nishimori condition (5), $s_i \in \{\pm 1\}$, and the sum is over nearest neighbors. The error threshold for the toric code was computed by Dennis *et al.* in [4], $p_c = 0.1094(2)$, with a lower bound given by Wang *et al.* in [46] (the phase diagram is reentrant [46, 33, 47]). Note that more detailed estimates of p_c followed later [31, 32, 25, 33]. The associated phase boundary is shown in figure 3.

4.2. Colour codes with qubit-flip errors

In 2006 Bombín and Martín-Delgado showed that the concept of topologically-protected quantum bits could also be realized on trivalent lattices with three-colourable faces. The *topological colour codes* they introduced [6] share similar properties to the toric code: stabilizer generators are intrinsically local and the encoding is in a topologically-protected code subspace. However, colour codes are able to encode more qubits than the toric code and, for some specific lattice structures, even gain additional computational capabilities.

In colour codes, qubits are arranged on a trivalent lattice (hexagonal or square octagonal), such that each qubit contributes a term of the form $\beta J \tau_{ijk} s_i s_j s_k$ in the mapping. In the case of hexagonal lattices, the partition function takes the form [18]

$$\mathcal{Z}_{\{\tau_{ijk}\}} = \sum_{\{s\}} \exp\Big(- \beta J \sum_{i,j,k \in \{\triangle\}} \tau_{ijk} s_i s_j s_k \Big) . \tag{13}$$

Equation (13) describes a disordered statistical system with three-spin interaction for each plaquette. Note that the spins s_i defined for the mapping are located on the triangular lattice

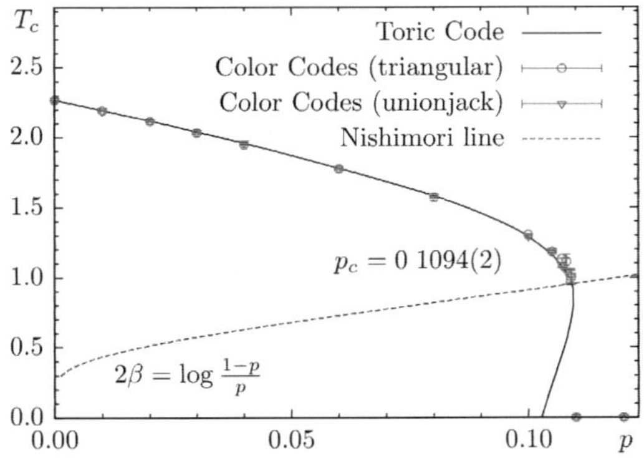

Figure 4. Comparison of the p–T_c phase diagram for the toric code (solid black line) and the results for random three-body Ising model on a triangular lattice (red), as well as on a Union Jack lattice (green). The error threshold is indicated by the intersection of the phase boundary with the Nishimori line (dashed blue) For $p > p_c \approx 0.109$ the ferromagnetic order is lost. Note the agreement with the toric code. Data taken from Refs. [18], [50], and [47].

which is *dual* to the original hexagonal arrangement. Every qubit corresponds to a triangle in the new lattice and dictates the sign of the associated plaquette interaction via τ_{ijk}. Thus, the statistical-mechanical Hamiltonian for the system related to colour codes [as described by (13)] is given by

$$\mathcal{H} = -J \sum_{i,j,k \in \{\triangle\}} \tau_{ijk} s_i s_j s_k \,, \tag{14}$$

where J is a global coupling constant chosen according to the Nishimori condition, $s_i \in \{\pm 1\}$, and the mapping requires τ_{ijk} to satisfy

$$\mathcal{P}(\tau_{ijk}) = \begin{cases} +1\,; & 1-p \\ -1\,; & p \end{cases} \,. \tag{15}$$

Note that the disordered three-body Ising model on the triangular lattice with $p = 0.5$ is NP-hard and therefore numerically difficult to study [48]. We would like to emphasize that colour codes on square-octagonal lattices are of particular interest, because, contrary to both toric codes and colour codes on honeycomb lattices, they allow for the transversal implementation of the whole Clifford group of quantum gates.

Figure 4 shows the p–T_c phase diagram for colour codes on hexagonal (maps onto a triangular lattice; empty circles), as well as square octagonal lattices (maps onto a Union Jack lattice; empty triangles). In addition, the solid (black) line is the phase boundary for the toric code. Surprisingly, the phase boundaries for all three models agree within statistical error bars, suggesting that colour as well as toric codes share similar error thresholds p_c. This is surprising, because the underlying statistical models have very different symmetries and are in different universality classes. For example, in the absence of randomness, the three-body Ising model on a triangular lattice is in a different universality class than the two-dimensional Ising model. Whereas for three-body Ising model on a two-dimensional triangular lattice $\nu = \alpha = 2/3$ [49], for the two-dimensional Ising model $\nu = 1$ and $\alpha = 0$. The disordered three-body Ising model on a triangular lattice has not been studied before, therefore highlighting again the fruitful relationship between quantum information theory and statistical physics.

Finally, our results also show that the enhanced computing capabilities of colour codes on the square-octagonal lattice do not come at the expense of an increased susceptibility to noise.

4.3. Depolarizing channel

The effects of single-qubit operations can be decomposed into qubit flips and phase flips, as well as a combination thereof; represented by the three Pauli matrices $\hat{\mathrm{X}}$, $\hat{\mathrm{Z}}$, and $\hat{\mathrm{Y}}$. When describing decoherence effects as a noisy channel, depolarizing noise is characterized by equal probability

Figure 5. When computing the stability of the toric code to depolarization, the problem maps onto a classical statistical Ising model on two stacked square lattices with both two-body and four-body interactions.

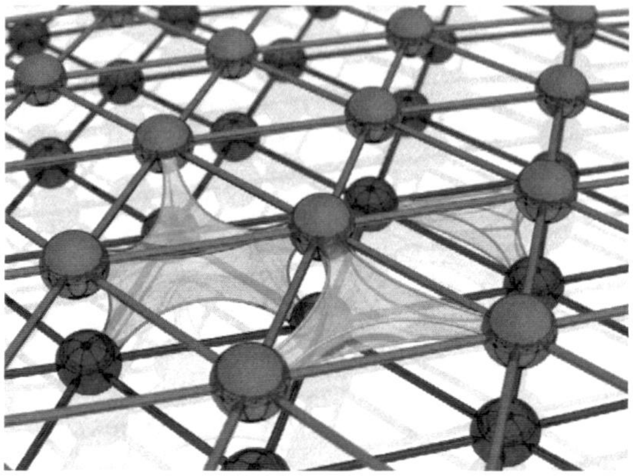

Figure 6. For colour codes, the spins are arranged on triangular lattices. In addition to the three-body interactions found for qubit-flip errors, both layers are connected via six-body interactions (light green).

for each type of error to occur, i.e., $p_x = p_y = p_z := p/3$. Note that the depolarizing channel is more general than the bit-flip channel, because it allows for the unified, correlated effect of all three basic types of errors [8, 23].

As in the previous mappings, in order to express the probability of an error class in terms of a classical Boltzmann weight, we need to associate with each elementary error loop a classical Ising spin. However, because of the error correlations, we cannot treat errors of different types independently any more. Instead, the resulting model contains spins of different types, according to the types of stabilizers used in the code. In fact, the mapping can be carried out in a very general way that requires no assumptions on the individual error rates or the actual quantum setup, see [23]. However, here we merely provide a brief explanation of the resulting Hamiltonian for the Toric code within the depolarizing channel.

In addition to the stabilizers of type $\hat{\mathrm{Z}}^{\otimes 4}$ (see section 2), the toric code also places stabilizers of type $\hat{\mathrm{X}}^{\otimes 4}$ on the remaining squares in the checkerboard decomposition. These allow for the concurrent detection of possible phase errors on the physical qubits. As a result, whenever a qubit flips, this is signaled by adjacent $\hat{\mathrm{Z}}$-stabilizers, whereas a qubit attaining a phase error is identified by $\hat{\mathrm{X}}$-stabilizers. Additionally, a combined qubit flip and phase error affects both the neighboring $\hat{\mathrm{Z}}$ and $\hat{\mathrm{X}}$-stabilizers. Therefore, the resulting Hamiltonian contains three terms per qubit; one describing each of the aforementioned scenarios:

$$\mathcal{H} = -J \sum_{ijk\ell} (\tau_{ij}^x s_i^z s_j^z + \tau_{k\ell}^z s_k^x s_\ell^x + \tau_{ijk\ell}^y s_i^z s_j^z s_k^x s_\ell^x) \,, \tag{16}$$

where the sum is over all qubits, the indices i, j, k and ℓ denote the four affected elementary equivalences and the sign of τ^w is dictated by whether the qubit has suffered an error of type $w \in \{x, y, x\}$. This Hamiltonian describes a classical Ising model that can be interpreted as two stacked square lattices which are shifted by half a lattice spacing, see figure 5. In addition to the standard two-body interactions for the top and bottom layers, the Hamiltonian also includes four-body terms (light green in the figure) that introduce correlations between the layers. Interestingly, toric codes under the depolarizing channel are related to the eight-vertex model introduced by Sutherland [51], as well as Fan and Wu [52], and later solved by Baxter [53, 54, 55].

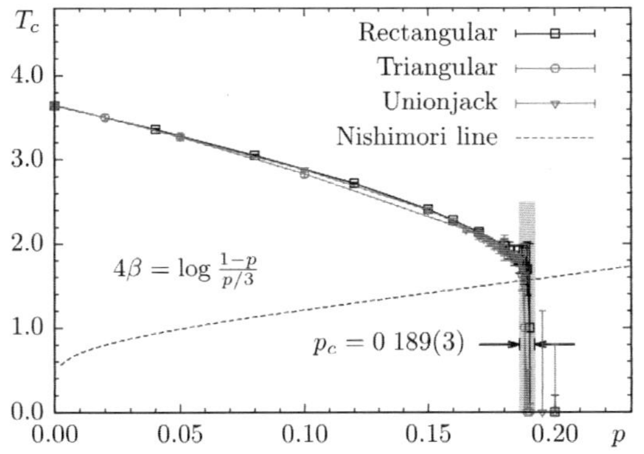

Figure 7. Estimated phase boundary for the models related to the depolarizing channel. The individual data sets represent the toric code (black), as well as colour codes on triangular (red) and Union Jack (green) lattices. The error threshold $p_c = 0.189(3)$ corresponds to the point where the Nishimori line (dashed, blue) intersects the phase boundary. Remarkably, the phase boundaries for all three codes agree within error bars. Data taken from [23].

For topological colour codes, qubits are arranged on trivalent lattices (hexagonal or square-octagonal) and the problem then maps onto either a triangular or Union Jack lattice (see section 4.2). For the depolarizing channel, an analogous mapping to the previous one relates this quantum setup to a Hamiltonian of the form:

$$\mathcal{H} = -J\sum_{ijk}(\tau^x_{ijk}s^z_i s^z_j s^z_k + \tau^z_{ijk}s^x_i s^x_j s^x_k + \tau^y_{ijk}s^z_i s^z_j s^z_k s^x_i s^x_j s^x_k)\,. \tag{17}$$

The details of this mapping are also contained in [23]. In addition to the in-plane three-body plaquette interactions that appear already when studying individual qubit-flips (see section 4.2), here additional six-body interactions add the necessary correlations between the planes (see figure 6). The resulting model maps to an eight-vertex model on a Kagomé lattice [56].

We perform simulations using the sublattice magnetization [23] to compute the susceptibility, and construct again the $p-T_c$ phase diagram for the toric code, as well as colour codes on both hexagonal and square-octagonal lattices. Note also, that the Nishimori condition changes slightly in this particular context (see the equation in figure 7 and [23]). Interestingly, the phase boundaries for all three error-correction models agree and we estimate conservatively an error threshold of $p_c = 0.189(3)$ for all three models. A similar study based on duality considerations [23, 24] yields results that agree within error bars. It is remarkable that the error threshold to depolarization errors for different classes of topological codes studied is larger than the threshold for expected uncorrelated errors, $p'_c = (3/2)p_{c,\text{flip}} \approx 16.4\%$. This is encouraging and shows that topological codes are more resilient to depolarization than previously thought. It also suggests that a detailed knowledge of the correlations in the error source can allow for a more efficient, custom-tailored code to be designed.

4.4. Subsystem Codes

All topological codes share the advantage that the quantum operators involved in the error-correction procedure are local, thus rendering actual physical realizations more feasible. However, in practice the decoherence of quantum states is not the only source of errors: Both syndrome measurement and qubit manipulations are delicate tasks and should be kept as simple as possible to minimize errors. For the toric code and topological colour codes, the check operators, while local, still involve the combined measurement of 4, 6, or even 8 qubits at a time.

By using concepts from subsystem stabilizer codes, Bombin was able to introduce a class of *topological subsystem codes* [8] that only requires pairs of neighboring qubits to be measured for syndrome retrieval. This is achieved by designating some of the logical qubits as "gauge qubits"

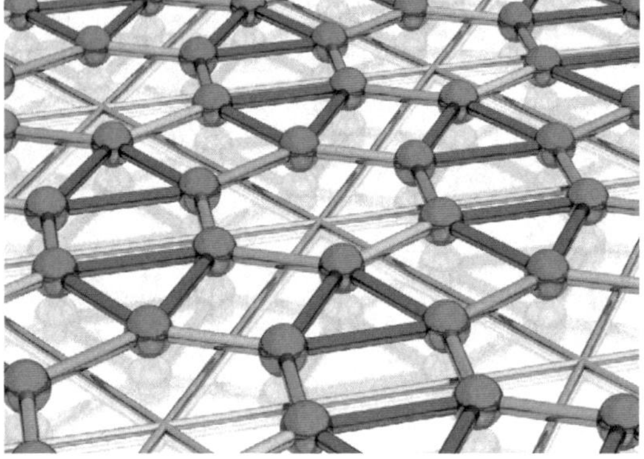

Figure 8. Topological subsystem codes combine aspects of both topological error-correction codes and subsystem codes. Starting from a triangular lattice, the construction places three qubits in each triangle with stabilizers acting on pairs of them. Here the red spheres represent individual qubits, interconnected with stabilizers of type $\hat{X}$ (yellow), $\hat{Y}$ (green) and $\hat{Z}$ (blue). The original triangular lattice for the construction is shown in grey. Despite only relying on two-qubit stabilizers, this setup is able to preserve quantum information up to a threshold of 5.5%.

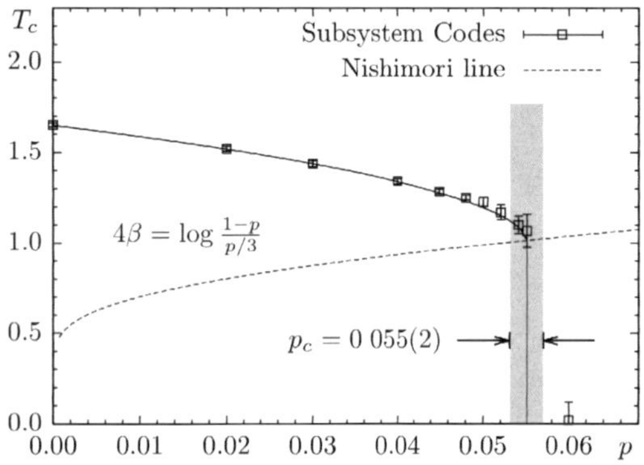

Figure 9. Disorder–temperature phase diagram for subsystem codes under depolarizing noise. The error threshold is given by the intersection of the phase boundary (solid line) with the Nishimori line (dashed). Our (conservative) numerical estimate is $p_c = 0.055(2)$ (orange vertical bar). Despite the low threshold compared to other proposals, this code is very promising due to the simpler stabilizers involved in the error-correction process.

where no information is stored. Due to the resulting simplicity of the error-correction procedure, these are very promising candidates for physical implementations.

The generation and mapping of a quantum setup that incorporates all these desired concepts is rather involved; we refer the reader to the relevant papers in Refs. [8] and [57]. One possible arrangement is shown in figure 8, consisting of qubits arranged in triangles, squares and hexagons with stabilizer operators of different types connecting neighboring pairs. In the mapping to a classical system, the setup then corresponds to a set of Ising spins (one for each stabilizer) with interactions dictated by how each stabilizer is affected by errors on adjacent qubits. This gives rise to a Hamiltonian of the general form

$$\mathcal{H} = -J \sum_{\ell} \sum_{w=x,y,z} \tau_{\ell}^{w} \prod_{i} s_i^{g_{i\ell}^{w}}, \tag{18}$$

where ℓ enumerates all qubit sites, w the three error types and i iterates over all Ising spins, respectively. The exponent $g_{i\ell}^{w} \in \{0, 1\}$ determines whether the stabilizer i is affected by an error of type w on qubit ℓ. Thus, for every qubit ℓ and error type w, the Hamiltonian contains a term of the form $-J\tau_{\ell}^{w} s_i s_j s_k$, where J is a constant, τ_{ℓ}^{w} is a quenched random variable (representing a possible qubit error) and the product contains all Ising spins corresponding to stabilizers affected by such an error.

Using Monte Carlo simulations, we compute the temperature–disorder phase diagram for the aforementioned statistical-mechanical model (see figure 9) and estimate an error threshold of $p_c = 0.055(2)$ [57], which is remarkable given the simplicity of the error-correction procedure.

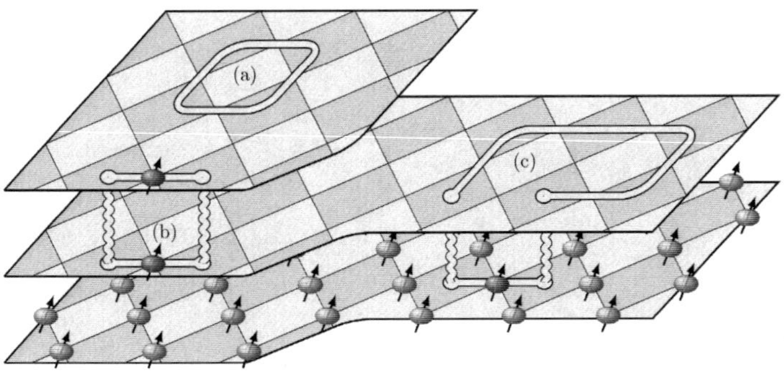

Figure 10. Toric code with measurement errors. (a) Elementary loops are error chains as seen before (figure 1). (b) Vertical links represent stabilizer measurements. This minimal error history consists of two consecutive qubit flips that remain unnoticed due to faulty measurements. (c) A more complex error history.

Note that this critical error rate is (numerically) smaller than the threshold calculated for the toric code, as well as topological colour codes. This is a consequence of a compromise for a much simpler syndrome-measurement and error-correction procedure: with a streamlined syndrome measurement process, the physical qubits are given less time to decohere and the error rate between rounds of error correction in an actual physical realization will be smaller.

4.5. Topological codes with measurement errors

So far we have only considered different types of errors that might occur on the physical qubits (at a rate p), while the process of syndrome measurement was assumed to be flawless. However, if additional measurement errors occur (at a rate q), we need to devise a scheme that can preserve quantum information over time even if intermediate measurements are faulty. This leads to the notion of so-called "fault-tolerant" codes: in this case, our best option is actually to continuously measure and correct errors as they are detected. Note that this introduces errors whenever the syndrome is faulty and these pseudo errors can only be detected and corrected in a later round of error correction.

This process of alternating measurement and correction phases can be modeled by considering vertically stacked copies of the original quantum setup, each representing one round of error correction. In this simplified scenario, errors only occur at discrete time steps, and it is instructive to think of the additional vertical dimension as a *time*. In particular, the measurements are represented by vertical connections between the plaquettes where the corresponding stabilizer resides. Then, the state of each layer is related to the layer immediately before it by the effect of the error channel, followed by one round of syndrome measurement and error correction. For no measurement errors (i.e., $q = 0$) all errors are detected perfectly and are corrected within one step. Consequently, there is no inter-relation between the errors found in consecutive layers. If $q > 0$, however, some errors can remain and new ones might be introduced due to the faulty syndrome measured. In analogy to the error chains seen earlier, we refer to these errors persevering over time as "error histories."

Mapping the qubit-flip and measurement problem in the toric code to a statistical-mechanical Ising model to compute the error threshold [4, 46, 58] yields a Hamiltonian of them form

$$\mathcal{H} = -J \sum_{\square} \tau_{\square}^{\mathrm{s}} s_i^{\mathrm{s}} s_j^{\mathrm{s}} s_k^{\mathrm{t}} s_\ell^{\mathrm{t}} - K \sum_{\square} \tau_{\square}^{\mathrm{t}} s_i^{\mathrm{t}} s_j^{\mathrm{t}} s_k^{\mathrm{t}} s_\ell^{\mathrm{t}} , \qquad (19)$$

where the first [second] sum is over all qubits [vertical links] in the lattice. Furthermore, we have introduced positive interaction constants J and K to be chosen according to the (adapted) Nishimori conditions [30] $\exp(-2\beta J) = p/(1-p)$ and $\exp(-2\beta K) = q/(1-q)$. Note that each of the spatial spins s^{s} represents an in-plane (i.e., horizontal) elementary loop that consists

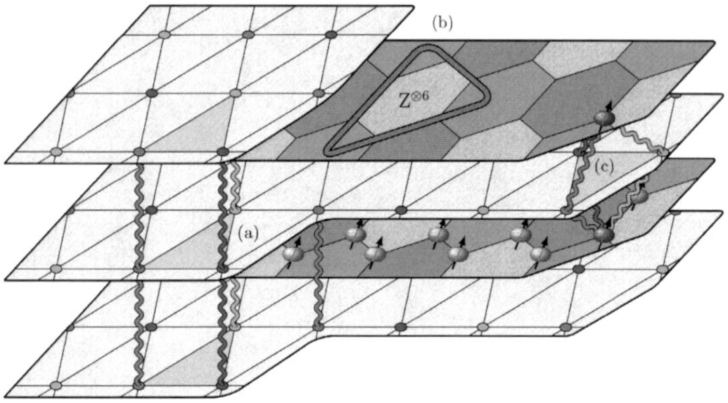

Figure 11. Lattice structure for colour codes that consists of stacked triangular and hexagonal layers. (a) Vertical connections represent the measurement history, while the qubits reside on hexagonal lattice sites. The lattice gauge theory has two elementary equivalences: (b) Coloured loops in the hexagonal planes. (c) Temporal equivalences consisting of two consecutive qubit errors and three faulty measurements.

purely of qubit flip errors, while the time-like spins s^{t} represent minimal error histories (vertical loops) that consist of two qubit flip errors and two faulty measurements (see figure 10). As for the toric code without measurement errors, these loops are used to tile the difference between two error chains. Thus different spin configurations represent error chains of the same class (sharing the same end points), albeit with different qubit and measurement errors. And, given the Nishimori condition, the Hamiltonian ensures that the Boltzmann weight corresponds to the relative probability of each scenario.

Equation (19) describes a disordered Ising lattice gauge theory with multi-spin interactions and four parameters, βJ, βK, p and q. The mapping is valid along the two-dimensional Nishimori sheet. We have treated spatial and time-like equivalences separately to allow for different qubit and measurement error rates. Interestingly, for the special case $p = q$, the resulting Hamiltonian is isotropic, i.e.,

$$\mathcal{H} = \sum_{\Box} \tau_{\Box} s_i s_j s_k s_l \, , \tag{20}$$

where $\Box$ represents a sum over *all plaquettes* in the lattice (both vertical and horizontal). Because $p = q$ also implies $J = K$ via the Nishimori condition, the model to investigate for this special case has only two parameters, namely βJ and p.

The model is a $\mathbb{Z}_2$ lattice gauge theory [46, 58], so we cannot use a local order parameter to determine the phase transition in our numerical simulations. Instead we consider the peak in the specific heat and the distribution of Wilson loop values to identify ordering in the system [59]. The latter observable is interesting because the first-order transition present in this system causes a double-peak structure near the transition. Even though the effect is smeared out when disorder is introduced, we can still reliably detect this shift of weight from one peak to the other by performing a finite-size scaling analysis of the skewness (third-order moment of the distribution). The temperature where the skewness is zero represents the point where the distribution of Wilson loops is double-peaked and symmetric, i.e., the phase transition. Comparisons to more traditional methods, such as a Maxwell construction, show perfect agreement. Note that this approach is generic and can be applied to any Hamiltonian that has a first-order transition.

An analogous mapping and analysis is also possible for topological colour codes with measurement errors. In this case, one considers a three-dimensional lattice consisting of stacked triangular layers, each representing one round of error syndrome measurement. The qubits reside on intermediate hexagonal layers and are connected to their respective check operators via vertical links as indicated in figure 11. In this case, the mapping to compute the error

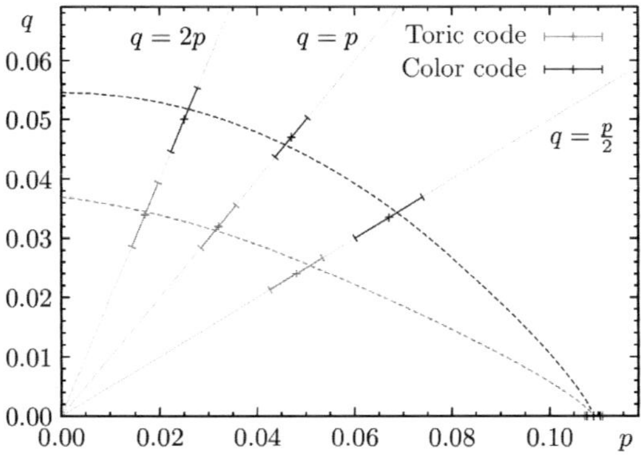

Figure 12. Preliminary results: Error threshold for bit flip errors (probability p) with different probabilities of measurement errors q for the toric code (red) and colour codes (black). For perfect error syndrome measurement, i.e., $q = 0$ (horizontal axis), the error thresholds of both codes agree, $[p_c = 0.109(2)]$. However, for a non-vanishing measurement error rate ($q > 0$), the error thresholds differ. The dashed lines are guides to the eye. Numerical values listed in Table 1.

threshold for faulty measurements and qubit flips produces a $\mathbb{Z}_2 \times \mathbb{Z}_2$ lattice gauge theory given by the Hamiltonian

$$\mathcal{H} = -J \sum_{\triangle} \tau_{\triangle} s_i^{\mathrm{s}} s_j^{\mathrm{s}} s_k^{\mathrm{s}} s_l^{\mathrm{t}} s_m^{\mathrm{t}} - K \sum_{\circ} \tau_{\circ} s_i^{\mathrm{t}} s_j^{\mathrm{t}} s_k^{\mathrm{t}} s_l^{\mathrm{t}} s_m^{\mathrm{t}} s_n^{\mathrm{t}} , \qquad (21)$$

where the first [second] sum is, again, over all qubits [measurements], represented by the product of the five [six] equivalences affected by the corresponding errors. The choice of the constants J and K is done as for the toric code following the modified Nishimori conditions. This Hamiltonian also describes an Ising lattice gauge theory, but even for the choice of $p = q$ it is *not* isotropic as was the case for the toric code with measurement errors.

Both models for topological codes with measurement errors are studied numerically using Monte Carlo simulations [59]. Interestingly, the thresholds calculated for the toric code and topological colour codes do *not* agree when possible measurement errors are taken into account. While the toric code can only correct up to 3.3% errors when $p = q$, colour codes remain stable up to 4.8%. This remarkable discrepancy is also seen in numerical studies where we allow p and q to be different (as long as $q > 0$), see figure 12.

5. Summary and conclusions

In Table 1 we summarize our results for different combinations of topological codes and error sources. As can be seen, the different proposals for topologically-protected quantum computation are very resilient to the different error sources.

Note that the results for different error channels in Table 1 cannot be compared directly: For the qubit-flip channel, p_c only refers to the maximum amount of flip errors that can be sustained, while for the depolarizing channel p_c is the sum of all three basic error types. Furthermore, the lower threshold for subsystem codes is the result of a compromise for a simpler error-correction procedure. Likewise, a lower value of p_c is to be expected in fault-tolerant schemes due to the additional presence of measurement errors. Remarkably, the error stability of the toric code and topological colour codes appears to be different only in the fault-tolerant regime where the mapping to a statistical-mechanical model produces a lattice gauge theory, despite finding perfect agreement in all other error channels.

We have outlined the mapping and subsequent analysis of several topological error-correction codes to classical statistical-mechanical Ising spin models with disorder. Because error chains correspond to domain walls under this mapping, an ordered state in the classical model can be identified with the scenario of feasible error correction, while the proliferation of errors in the quantum setup is associated with a disordered state in the classical model. After numerically calculating the disorder–temperature phase diagram of the (classical) statistical spin models, the error threshold can be identified with the intersection point of the phase boundary with the Nishimori line. This critical error threshold represents the maximum amount of perturbation each setup can sustain and does not include the effects of realistic device implementations. However, the fact that these "theoretical" best-case thresholds are so high is rather promising.

We conclude by highlighting again the beautiful synergy between quantum error correction and novel disordered spin models in statistical physics. We hope that these results will encourage scientists specialized in analytical studies of disordered systems to tackle these simple yet sophisticated Hamiltonians.

Error Source	Topological Error Code	Threshold p_c
Qubit-Flip	Toric code	0.109(2)
Qubit-Flip	Colour code (honeycomb lattice)	0.109(2)
Qubit-Flip	Colour code (square-octagonal lattice)	0.109(2)
Depolarization	Toric code	0.189(2)
Depolarization	Colour code (honeycomb lattice)	0.189(2)
Depolarization	Subsystem code	0.055(2)
Qubit-Flip & Measurement	Toric code	0.033(5)
Qubit-Flip & Measurement	Toric code ($q = 2p$)	0.017(3)
Qubit-Flip & Measurement	Toric code ($q = p/2$)	0.047(6)
Qubit-Flip & Measurement	Colour code ($p = q$)	0.048(3)
Qubit-Flip & Measurement	Colour code ($q = 2p$)	0.025(4)
Qubit-Flip & Measurement	Colour code ($q = p/2$)	0.066(7)

Table 1: Summary of error thresholds calculated numerically. Note that our estimate for the toric code with qubit-flip and measurement errors agrees with the results of Wang *et al.* [46] and Ohno *et al.* [58]. The most precise estimate for the toric code with qubit-flip errors is $p_c = 0.10919(7)$, see [33].

Acknowledgments
Part of the figures shown in this work are from the PhD Thesis of Ruben S. Andrist (Eidgenössische Technische Hochschule Diss. No. 20588). Work done in collaboration with H. Bombin (Perimeter Institute), M. A. Martin-Delgado (Universidad Complutense de Madrid), and C. K. Thomas (Texas A&M University). H.G.K. acknowledges support from the SNF (Grant No. PP002-114713) and the NSF (Grant No. DMR-1151387) and would like to thank ETH Zurich for CPU time on the Brutus cluster and Texas A&M University for CPU time on the Eos, as well as Lonestsar clusters.

References

[1] Shor P W 1995 *Phys. Rev. A* **52** R2493
[2] Steane A M 1996 *Phys. Rev. Lett.* **77** 793
[3] Knill E and Laflamme R 1997 *Phys. Rev. A* **55** 900
[4] Dennis E, Kitaev A, Landahl A and Preskill J 2002 *J. Math. Phys.* **43** 4452
[5] Kitaev A Y 2003 *Ann. Phys.* **303** 2
[6] Bombin H and Martin-Delgado M A 2006 *Phys. Rev. Lett.* **97** 180501
[7] Bombin H and Martin-Delgado M A 2007 *Phys. Rev. B* **75** 075103
[8] Bombin H 2010 *Phys. Rev. A* **81** 032301
[9] Bravyi S, Terhal B M and Leemhuis B 2010 *New J. Phys.* **12** 083039
[10] Haah J 2011 *Phys. Rev. A* **83** 042330

[11] Gottesman D 1996 *Phys. Rev. A* **54** 1862
[12] Calderbank A R and Shor P W 1996 *Phys. Rev. A* **54** 1098
[13] Poulin D 2005 *Phys. Rev. Lett.* **95** 230504
[14] Bacon D 2006 *Phys. Rev. A* **73** 012340
[15] Binder K and Young A P 1986 *Rev. Mod. Phys.* **58** 801
[16] Nishimori H 2001 *Statistical Physics of Spin Glasses and Information Processing: An Introduction* (New York: Oxford University Press)
[17] Raussendorf R and Harrington J 2007 *Phys. Rev. Lett.* **98** 190504
[18] Katzgraber H G, Bombin H and Martin-Delgado M A 2009 *Phys. Rev. Lett.* **103** 090501
[19] Barrett S D and Stace T M 2010 *Phys. Rev. Lett.* **105** 200502
[20] Duclos-Cianci G and Poulin D 2010 *Phys. Rev. Lett.* **104** 050504
[21] Wang D S, Fowler A G and Hollenberg L C L 2011 *Phys. Rev. A* **83** 020302
[22] Landahl A, Anderson J T and Rice P 2011 *(arXiv:quant-phys/1108.5738)*
[23] Bombin H, Andrist R S, Ohzeki M, Katzgraber H G and Martin-Delgado M A 2012 *Phys. Rev. X* **2** 021004
[24] Ohzeki M, Nishimori H and Berker A N 2008 *Phys. Rev. E* **77** 061116
[25] Ohzeki M 2009 *Phys. Rev. E* **79** 021129
[26] Ohzeki M 2009 *Phys. Rev. E* **80** 011141
[27] Ohzeki M, Thomas C K, Katzgraber H G, Bombin H and Martin-Delgado M A 2011 *J. Stat. Mech.* P02004
[28] Landau D P and Binder K 2000 *A Guide to Monte Carlo Simulations in Statistical Physics* (Cambridge University Press)
[29] Katzgraber H G 2009 Introduction to Monte Carlo Methods (arXiv:0905.1629)
[30] Nishimori H 1981 *Prog. Theor. Phys.* **66** 1169
[31] Honecker A, Picco M and Pujol P 2001 *Phys. Rev. Lett.* **87** 047201
[32] Merz F and Chalker J T 2002 *Phys. Rev. B* **65** 054425
[33] Parisen Toldin F, Pelissetto A and Vicari E 2009 *J. Stat. Phys.* **135** 1039
[34] Swendsen R H and Wang J 1986 *Phys. Rev. Lett.* **57** 2607
[35] Geyer C 1991 *23rd Symposium on the Interface* ed Keramidas E M (Fairfax Station, VA: Interface Foundation) p 156
[36] Hukushima K and Nemoto K 1996 *J. Phys. Soc. Jpn.* **65** 1604
[37] Marinari E, Parisi G, Ruiz-Lorenzo J and Ritort F 1996 *Phys. Rev. Lett.* **76** 843
[38] Katzgraber H G, Trebst S, Huse D A and Troyer M 2006 *J. Stat. Mech.* P03018
[39] Earl D J and Deem M W 2005 *Phys. Chem. Chem. Phys.* **7** 3910
[40] Marinari E and Parisi G 1992 *Europhys. Lett.* **19** 451
[41] Lyubartsev A P, Martsinovski A A, Shevkunov S V and Vorontsov-Velyaminov P N 1992 *J. Chem. Phys.* **96** 1776
[42] Cooper F, Freedman B and Preston D 1982 *Nucl. Phys. B* **210** 210
[43] Palassini M and Caracciolo S 1999 *Phys. Rev. Lett.* **82** 5128
[44] Ballesteros H G, Cruz A, Fernandez L A, Martin-Mayor V, Pech J, Ruiz-Lorenzo J J, Tarancon A, Tellez P, Ullod C L and Ungil C 2000 *Phys. Rev. B* **62** 14237
[45] Martín-Mayor V, Pelissetto A and Vicari E 2002 *Phys. Rev. E* **66** 026112
[46] Wang C, Harrington J and Preskill J 2003 *Ann. Phys.* **303** 31
[47] Thomas C K and Katzgraber H G 2011 *Phys. Rev. E* **84** 040101(R)
[48] Thomas C K and Katzgraber H G 2011 *Phys. Rev. E* **83** 046709
[49] Baxter R J and Wu F Y 1973 *Phys. Rev. Lett.* **31** 1294
[50] Katzgraber H G, Bombin H, Andrist R S and Martin-Delgado M A 2010 *Phys. Rev. A* **81** 012319
[51] Sutherland B 1970 *J. Math. Phys.* **11** 3183
[52] Fan C and Wu F Y 1970 *Phys. Rev. B* **2** 723
[53] Baxter R J 1971 *Phys. Rev. Lett.* **26** 832
[54] Baxter R J 1972 *Ann. Phys.* **70** 193
[55] Baxter R 1982 *Exactly Solved Models in Statistical Mechanics* (London: Academic Press)
[56] Baxter R J 1978 *Phil. Trans. Roy. Soc.* **289** 315
[57] Andrist R S, Bombin H, Katzgraber H G and Martin-Delgado M A 2012 *Phys. Rev. A* **85** 050302
[58] Ohno T, Arakawa G, Ichinose I and Matsui T 2004 *Nucl. Phys. B* **697** 462
[59] Andrist R S, Katzgraber H G, Bombin H and Martin-Delgado M A 2011 *New J. Phys.* **13** 083006

Journal of Physics: Conference Series **473** (2013) 012020 doi:10.1088/1742-6596/473/1/012020

Packing hard spheres with short-range attraction in infinite dimension: phase structure and algorithmic implications

M Sellitto

DIII, Seconda Università di Napoli, Real Casa dell'Annunziata, I-81031 Aversa (CE), Italy

F Zamponi

LPT, École Normale Supérieure, UMR 8549 CNRS, 24 Rue Lhomond, 75005 Paris, France

Abstract. We study, via the replica method of disordered systems, the packing problem of hard-spheres with a square-well attractive potential when the space dimensionality, d, becomes infinitely large. The phase diagram of the system exhibits reentrancy of the liquid-glass transition line, two distinct glass states and a glass-to-glass transition, much similar to what has been previously obtained by Mode-Coupling Theory, numerical simulations and experiments. The presence of the phase reentrance implies that for a suitable choice of the intensity and attraction range, high-density sphere packings more compact than the one corresponding to pure hard-spheres can be constructed in polynomial time in the number of particles (at fixed, large d) for packing fractions $\varphi \leq 6.5\, d\, 2^{-d}$. Although our derivation is not a formal mathematical proof, we believe it meets the standards of rigor of theoretical physics, and at this level of rigor it provides a small improvement of the lower bound on the sphere packing problem.

1. Introduction

Packing problems [1] are ubiquitous in science and engineering and arise in a variety of contexts ranging from biology (e.g. in the crowded cellular environment where physiological processes conspire significantly with excluded-volume effects [2]) to communication technology (e.g. in connection with the design of error correcting codes for signal transmission over noisy channels [3]). From a mathematic [4, 5] and algorithmic [1, 6, 7, 8] point of view the problem of packing sphere efficiently is most challenging when the dimension of the physical space d becomes increasingly large. This is mainly due to several unusual features which are at the origin of the so-called "curse of dimensionality". It is well known, for example, that hypersphere packings cannot be very dense because the volume of the empty spaces left unoccupied by the spheres tends to become dominant in high-dimensional spaces thus giving a vanishingly small sphere packing density. Lattice packings with simple unit cells and simple symmetries are particularly inefficient in filling space when d is large [4], hence it is reasonable to guess that dense packing with complex unit cells might be the densest ones in large d. However, such packings have complex symmetries and they are hard to construct explicitly in generic dimensions [4]. At the same time, numerical simulations require a minimal number of particles that grow exponentially with d [6, 7, 8, 9] and become extremely hard for large d. In this situation it is by no means obvious whether the dense packings should have a periodic or rather an irregular structure and how to devise an algorithmic procedure to construct explicitly them [7].

In statistical physics, hard-sphere systems have been used for a long time to describe the gas-liquid transition (e.g. by van der Waals [10]) and the geometric structure of dense liquids (starting from Bernal [11]). More recently, they have been much studied as models of colloids (with steric and electrostatic stabilization). The interest has been especially motivated by the possibility of introducing a (depletion-induced) attraction between colloidal particles by adding a suitable amount of non-adsorbing polymers to the colloidal solution. The intensity and the range of the attraction can be tuned by changing the polymer concentration and the polymer coil radius, respectively. By doing so, one can thus explore a wide range of static and dynamic behaviours which are not directly accessible in a simple liquid system. Some of the fascinating properties that have been identified in these systems include the reversible melting-by-cooling of the colloidal glass state and structurally distinct types of dynamically arrested states [12].

The above colloid-polymer solutions can be modeled by an assembly of particles interacting through a potential made by a repulsive hard-core plus a very short-range attractive part. In the following, we shall be focused on the phase behaviour of such systems when the space dimensionality becomes infinitely large. This is interesting for several reasons. First, in this limit the approach we use is arguably exact [13] and this allows for a comparison with the results obtained by alternative methods. Moreover, our study provides evidence for the possibility of improving slightly the lower bound on the sphere packing problem in large dimensions.

A short account of our work was presented in [14]. In the present contribution we report extensively on the method we have used and discuss thoroughly the results of our calculations. In the remaining part of this introduction we explain further the motivations behind our work, including their possible algorithmic relevance. In section 2, we briefly outline the Franz-Parisi effective potential method [15] (the related replica calculation are reported in Appendix A). In section 3 we present the derivation of the phase diagram in an exemplary case and in the sticky limit as well. We then conclude by a qualitative description of the dynamical behaviour implied by our results, and by discussing further perspectives of our work.

1.1. Short-range attractive colloids: a re-entrant liquid-glass transition and two glasses

Colloidal systems with short-range attraction display a very rich phase diagram. When the range of the attraction is not too short, the phase diagram is characterized by a re-entrant liquid-glass transition line. This means that for a suitable intensity of the attraction (or, equivalently, temperature), the liquid-glass transition occurs at density *higher* than for the pure hard-sphere system without attraction. Moreover, when the range of the attraction is short enough, a glass-glass transition appears at high packing density. This transition separates a "repulsive glass" phase dominated by the hard core repulsion from an "attractive glass" phase dominated by the short range attraction [12]. The attractive glass is particularly interesting because it forms also at quite small densities where the slowing down is induced by a "gelation mechanism": filamentary chains of bonded particles form and percolate the system.

This interesting phase diagram has been first obtained by MCT [16, 17], and later confirmed by numerical simulations [18, 19, 20], experiments [21, 22], and within the heterogeneous facilitation picture [23, 24]. MCT is often thought to be part of a more general Random First Order Transition (RFOT) scenario for the glass transition. This idea is based on the analogy between the glass problem and a class of spin glass models [25, 26, 27]: indeed, several studies found glass-glass transitions in these spin glass (or lattice glass) models [28, 29, 30, 31, 32] and gave further insight on the relaxation dynamics around this transition [28]. According to the general RFOT scenario [27], the same phase diagram should be obtained using the replica method [33, 27]. The advantage of using the replica method is that one can also access the glass phase (e.g. compute the equation of state of the glass). Moreover, critical properties are easier to compute in the replica formalism [34].

So far only one attempt to use the replica method for this problem has been reported [35]. It is based on the replicated Hypernetted Chain (HNC) formalism, which is known to give a correct qualitative phase diagram, even if its quantitative accuracy is rather poor. Within this formalism the re-entrant glass transition was found, however the glass-glass transition and in particular the attractive glass phase were not found. This is probably due to the fact that the replicated HNC scheme is known to fail badly when the "cage size" is very small [36], which is the case in the attractive glass [12]. It is therefore important to carry out this computation by using the "small cage expansion" scheme [33, 36], which is more appropriate for this situation and normally gives much better results from the quantitative point of view. In this paper we report such a computation and we show that its results are consistent with the one obtained from MCT, which is nicely consistent with the general RFOT picture.

1.2. Large space dimension: algorithmic implications for the sphere packing problem

In the following, we will stick to the $d \to \infty$ limit, for two reasons. First of all, computations are simplified, the theory can be shown to be exact at the level of rigor of theoretical physics [13], and one is able to access all the transitions that characterize the RFOT scenario (the dynamical transition, the Kauzmann transition, and the jamming/glass close packing point) [36]. A second and more important motivation is that the presence of a re-entrant glass transition has an algorithmic interest in $d \to \infty$, which we now explain.

Rigorous results on high dimensional sphere packings– The problem of packing spheres is very hard when the space dimensionality becomes larger than $d \sim 20$ [4, 5]. From an analytic point of view, the recursive strategy of constructing optimal packings by slicing d-dimensional packings in $d+1$ dimensions turns out to be very inefficient [4, 5] and the few good packings which are known for dimension smaller than $d \sim 200$ have been mostly "handcrafted". Moreover, because in some dimensions there are special symmetries that lead to the existence of extremely dense packings, it is very hard to extrapolate the trend of the maximum packing density to larger dimensions. From the algorithmic point of view, one could try to solve the problem by adapting the "simulated annealing" procedure, which in this case consists in constructing good packings by slowly compressing low density configurations of spheres. This procedure is known as the Lubachevsky-Stillinger (LS) algorithm [37, 38]. Unfortunately, as soon as $d > 5$, crystallization becomes exceedingly rare [38, 39, 9]: the system remains stuck "forever" in an amorphous phase, which at high enough density is a glass. It seems that the time needed to crystallize increases at least exponentially with dimension. Other smart algorithms have been used to construct dense lattice packings [6, 7, 8] but unfortunately their running time increases fast with dimension and for the moment the use of these algorithms is limited to dimensions smaller than $d \sim 40$. It is also important to stress that, at least to investigate amorphous packings, the minimal number of particles that need to be studied increases exponentially with d [9], which is an important limitation to study large spatial dimensions.

Because finding optimal packings is so difficult for large d, only *non-constructive* upper and lower bounds to the best packing fraction (the fraction of volume covered by the spheres) φ are known. Roughly speaking (see [4, 5, 1] for more precise formulations), the known best lower bound is $\varphi \geq (6d/e)2^{-d}$ [40] (note that it took 20 years to gain a factor of $3/e$ with respect to the previous best lower bound $\varphi \geq 2d2^{-d}$ [41]), while the known best upper bound is $\varphi \leq 2^{-0.5990d}$ [42]. Therefore, the upper and lower bounds are exponentially divergent with d, leaving a huge gap open where the densest packing might be located.

Non-rigorous results on disordered packings– It has been proposed, based on an analysis of consistency conditions for the pair correlation functions, that sphere packings could exist up to a density $\varphi \sim 2^{-0.77865d}$, i.e. somewhere in between the lower and upper bounds but exponentially larger than the best lower bound [1]. The fact that the pair correlation function used in the analysis has no structure suggests that these packings, if they exist, could be lattice packings with a very complex fundamental cell or even disordered packings. Unfortunately, no way to construct such packings, or to transform the conjecture in a more rigorous bound, has been obtained.

The study of disordered packings in large space dimension is therefore extremely interesting to obtain further insight into the problem. The analytical study of a system of amorphous hard spheres in large dimensions within the RFOT scenario and using the replica method [43, 36] gives the following predictions.

- An ergodic liquid phase exists up to a density φ_d that scales asymptotically as $\varphi_d = 4.8\,d\,2^{-d}$. At φ_d, the liquid phase fragments into a large number of non-ergodic components. Under very general conditions, one can prove rigorously [44] that if such an ergodicity breaking transition exists, the equilibration time of any local dynamics is polynomial in the number of particles N for $\varphi < \varphi_d$, while it is exponential in N for $\varphi > \varphi_d$. Hence at φ_d, a dynamical arrest of the MCT type towards a glass phase happens. Note however that this statement is not rigorous for hard spheres because the very existence of φ_d is not rigorously proven in that case.
- An exponential number of *amorphous* packings exist up to a density φ_{GCP} scaling asymptotically as $\varphi_{\mathrm{GCP}} = d\log(d)\,2^{-d}$. Yet, based on the statement above, the time needed to explore these packings using standard local dynamics is exponentially large in the number of particles.

This approach can be shown to be exact at the level of rigor of theoretical physics [13], and these predictions are consistent with numerical simulations in dimensions d ranging from 3 to 13 [9, 45]. We are therefore led to conclude that *exponentially many disordered sphere packings exist up to* $\varphi_{\mathrm{GCP}} = d\log(d)\,2^{-d}$, *but sampling them using local hard sphere dynamics is exponentially hard in the number of particles* N *when the density is larger than* $\varphi_d = 4.8\,d\,2^{-d}$. Note that this situation resembles closely the one that is encountered in many random optimization problems for which the RFOT scenario is exact, for example the coloring of random graphs in the limit of a large number of colours [46]. Note also that sampling the packings uniformly is exponentially hard in N for $\varphi > \varphi_d$, but one can still construct packings in polynomial time in N above φ_d by adiabatic compression of the packings at φ_d [9] (again, in close analogy with the coloring problem [47]). This allows to construct packings with density slightly larger than φ_d, but probably still proportional to $d\,2^{-d}$ (the computation of the proportionality constant remains however an open problem, and should be done by extending the replica method following [48]).

An interesting conclusion of the above discussion is that at the level of rigor of theoretical physics, we can state that *disordered sphere packings exist and can be sampled with simulated annealing in polynomial time in* N *at fixed* d *for density at least equal to* $\varphi_d = 4.8\,d\,2^{-d}$, which provides a non-rigorous (but constructive) small improvement of the lower bound of Vance $\varphi \geq (6d/e)2^{-d}$ [40] mentioned above (the reader should however keep in mind that, unfortunately, the minimal number of particles N must scale exponentially with d [9], which in practice renders the numerical simulations prohibitive for current computers above $d \sim 13$). In this paper we will show that when a short range attraction is added to the hard sphere potential, the glass transition line φ_d moves to even higher densities. The presence of this re-entrance of the line φ_d implies that one is able to equilibrate the system in polynomial time in N up to a higher threshold. We will show that unfortunately the improvement is only in the prefactor, which is brought to at most $\varphi_{d,max} \sim 6.5\,d\,2^{-d}$ for a suitable choice of the range and intensity of the attraction.

2. The Franz-Parisi potential and the replica method

We assume that the potential has a soft repulsive core of diameter $D = 1$. The most general potential we want to consider in this paper has the form

$$v(r) = \epsilon(1-r)^2\theta(r<1) - U_0\theta(1<r<1+\sigma) \tag{1}$$

However, in the following we restrict for simplicity to the case $\epsilon \to \infty$ which corresponds to hard spheres with a square-well attraction. We scale the attraction width in such a way that $\widehat{\sigma} = \sigma\, d$ is finite. Moreover, the natural scale of packing fraction is the scaled $\widehat{\varphi} = 2^d\varphi/d$ where φ is the packing fraction of the repulsive core.

The basic idea of the replica approach to the glass transition is that the kinetic slowing down on approaching the glass phase is due to the sudden appearance of a bunch of long-lived metastable states [49, 50, 51]. Under this assumption, the glass transition can be detected by looking at the free energy of a system constrained to be at a fixed distance from a reference equilibrium configuration. This is known as the *Franz-Parisi potential* [52, 53, 54], see also [51, 55] for an alternative but very related approach. In the present context, the best way of making this construction explicit is the following. One considers an equilibrium configuration $X = \{x_i\}_{i=1\cdots N}$ of the liquid at a given state point, and a second configuration Y that is constrained to be closed to the first one, in such a way that the mean square displacement

$$\frac{d}{N}\left\langle (x_i - y_i)^2 \right\rangle \le 2\widehat{A}\ , \tag{2}$$

where $\widehat{A}$ is a fixed constant. One then computes the free energy of the system Y for fixed X, and then averages it over the equilibrium distribution of X. The result is the average free energy $V_{\rm FP}(\widehat{A})$ of a system constrained to be at distance $\widehat{A}$ from a typical liquid configuration. In the liquid phase, it is always possible for particles to diffuse away from any reference configuration, and correspondingly $V_{\rm FP}(\widehat{A})$ has a unique minimum at $\widehat{A} = \infty$. On the contrary, in the glass phase, there is a metastable phase – corresponding to a local minimum of $V_{\rm FP}(\widehat{A})$ at finite $\widehat{A}$ – in which the second system remains spontaneously close to the first one, signaling the presence of a caging effect. The secondary minimum appears discontinuously at a finite $\widehat{A}$ below the *dynamical transition* line $T_d(\varphi)$. At the mean-field level, this secondary minimum has an infinite life time. Very remarkably, such a construction can be done explicitly for realistic models of glass formers, both analytically (using standard liquid theory approximations) or numerically [15, 54, 56, 57]. From an analytic point of view, the computation of the Franz-Parisi potential requires the use of the replica method. This procedure has been developed in [58, 54, 59, 36]. For the specific case of the potential in (1), the best known approximation has been discussed (for $U_0 = 0$) in [60]. Moreover, it has been shown (at a theoretical physics level, i.e. not rigorously) in [13] that this approximation becomes exact in the limit $d \to \infty$ where space dimensionality becomes very large. The extension of these results to $U_0 \neq 0$ is straightforward and it is discussed in Appendix A.

In summary, we show in Appendix A that

$$V_{\rm FP}(\widehat{A}) = \text{const.} + \frac{d}{2}\mathcal{V}_{\rm FP}(\widehat{A})\ , \tag{3}$$

and $\mathcal{V}_{\rm FP}$ can be computed by means of (A.13). Moreover,

$$\mathcal{V}'_{\rm FP}(\widehat{A}) = -\frac{1}{\widehat{A}}\left[1 - \widehat{\varphi}\mathcal{F}_1(\widehat{A})\right]\ . \tag{4}$$

Therefore the stationary points (maxima and minima) of the Franz-Parisi potential are located at values of $\widehat{A}$ which are the solutions of

$$\frac{1}{\widehat{\varphi}} = \mathcal{F}_1(\widehat{A})\ , \tag{5}$$

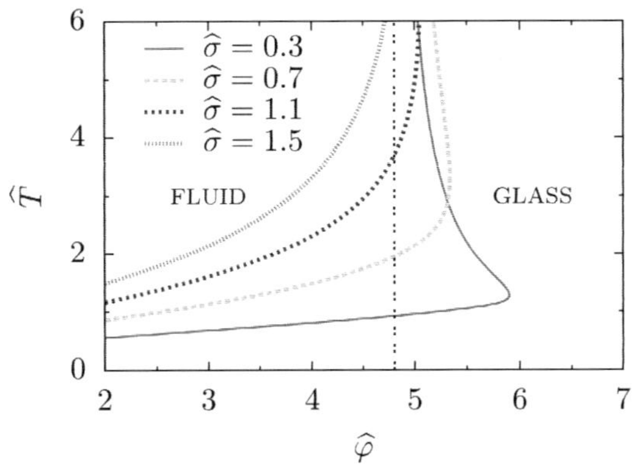

Figure 1. Phase diagram in the rescaled variables, temperature $\widehat{T} = k_B T/U_0$ vs packing fraction $\widehat{\varphi}$, for moderately large square-well attraction width $\widehat{\sigma} \geq 0.3$. The vertical line is the packing fraction at the dynamic glass transition for the purely hard-sphere potential. In this case there is only one glass phase but the fluid-glass transition line is reentrant: in a range of the control parameters the amorphous solid melts upon cooling (and the liquid freezes upon heating).

with the function $\mathcal{F}_1(\widehat{A})$ defined in (A.12). The values of $\widehat{A}$ corresponding to the local minima of $V_{\rm FP}(\widehat{A})$ correspond to the long-time mean square displacement (i.e. the Debye-Waller factor) in the glass phase. In this context the dynamical transition corresponds to the disappearance of all the local minima of $V_{\rm FP}(\widehat{A})$, which happens when $\widehat{\varphi}^{-1} > \max_{\widehat{A}} \mathcal{F}_1(\widehat{A})$. Hence, the equation for the dynamical transition is

$$\frac{1}{\widehat{\varphi}_d} = \max_{\widehat{A}} \mathcal{F}_1(\widehat{A}) \ . \tag{6}$$

3. The static phase diagram of the square-well potential

Let us now discuss the *static* phase diagram, which is derived by studying the behaviour of the functions $\mathcal{V}_{\rm FP}$ and $\mathcal{F}_1$. The connection with dynamics will be briefly discussed in the conclusions.

3.1. Medium-range attraction: a single glass phase

When the attraction width is large enough (above $\widehat{\sigma} \simeq 0.19$), the function $\mathcal{F}_1$ has a single maximum for all densities and temperatures, and the resulting phase structure is easily determined: for each temperature, a single glass phase exists for densities larger than $\widehat{\varphi}_d$ defined by (6). In figure 1 we show the phase diagram in rescaled variables: reduced temperature, $\widehat{T} = k_B T/U_0$, and reduced packing fraction, $\widehat{\varphi}$. Interestingly, there is a range of $\widehat{\varphi}$ above ≈ 5 in which the glass melts upon cooling and the resulting fluid freezes when temperature is further lowered. This reentrance effect is driven by the width of the square-well potential. The smaller the attraction width the deeper the fluid phase enters into the glass region. This effect was already found in [35] by using the replica method in a different approximation scheme.

3.2. Short-range attraction: two glass phases

When the attraction width is below $\widehat{\sigma} \simeq 0.19$, the function $\mathcal{F}_1$ can have two maxima in some range of temperatures and densities, and the determination of the phase diagram requires some care in this case because of the appearance of multiple glass phases. To discuss in detail the several interesting features we find for the square-well potential we focus on a representative case with $\widehat{\sigma} = 0.06$.

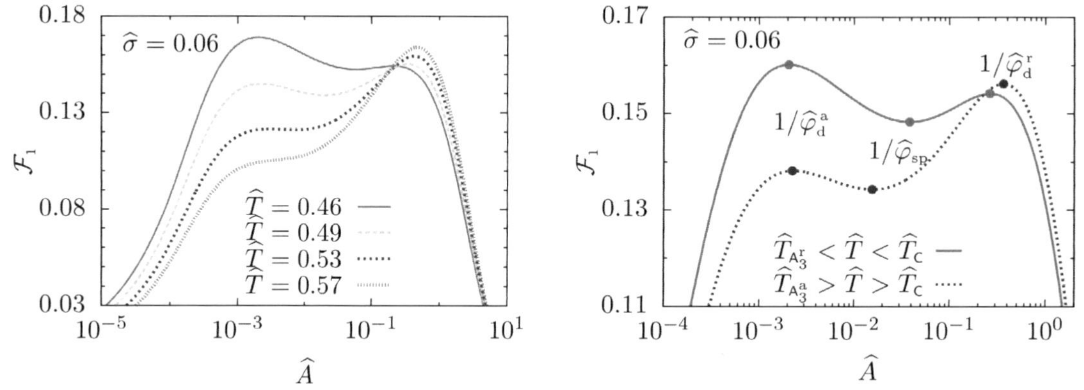

Figure 2. (Left) Shape of the function $\mathcal{F}_1$ for $\widehat{\sigma} = 0.06$ at some reduced temperatures $\widehat{T} = k_{\rm B}T/U_0$. (Right) The definitions of (7) are illustrated at temperatures $\widehat{T} = 0.5$ and $\widehat{T} = 0.47$. In the appropriate range of density $\max\{\widehat{\varphi}_{\rm d}^{\rm r}, \widehat{\varphi}_{\rm d}^{\rm a}\} < \widehat{\varphi} < \widehat{\varphi}_{\rm sp}$, (5) has four solutions, two of them corresponding to the glasses and the two others to local maxima of $\mathcal{V}_{\rm FP}$.

The function $\mathcal{F}_1(\widehat{A})$ is reported in figure 2 for $\widehat{\sigma} = 0.06$ and two selected values of temperature. A crucial observation is that $\mathcal{F}_1$ does not depend on density. It is seen that in these cases, $\mathcal{F}_1$ has two local maxima separated by a local minimum. Let us give the following definitions, illustrated in figure 2:

$$\mathcal{F}_1 = \begin{cases} 1/\widehat{\varphi}_{\rm d}^{\rm a} & \text{at the maximum at smaller } \widehat{A}, \\ 1/\widehat{\varphi}_{\rm d}^{\rm r} & \text{at the maximum at larger } \widehat{A}, \\ 1/\widehat{\varphi}_{\rm sp} & \text{at the minimum;} \end{cases} \tag{7}$$

we also consider the analytic continuations of these densities when they exist. Therefore in some range of densities (5) has four solutions: two of them (the first and the third, upon increasing $\widehat{A}$) correspond to local minima and the two others correspond to local maxima of the Franz-Parisi potential. This is illustrated in figure 3, and remember that while $\mathcal{F}_1$ does not depend on density, $\mathcal{V}_{\rm FP}$ does. The two local minima correspond to two different metastable glass states, and we call them the *attractive glass* (the one with smaller $\widehat{A}$) and *repulsive glass* (with larger $\widehat{A}$). Therefore, in this region of density we have coexistence between the two glasses.

To obtain the phase diagram we must analyze the behaviour of $\mathcal{F}_1(\widehat{A})$ systematically as a function of temperature. From figure 2 we deduce the existence of several different temperature ranges:

- For $\widehat{T} < \widehat{T}_{{\rm A}_3^{\rm r}}$, $\mathcal{F}_1$ has a unique maximum at small $\widehat{A}$, corresponding to $\widehat{\varphi}_{\rm d}^{\rm a}$. The other two densities in (7) do not exist, only the attractive glass exists for these temperatures, in the region $\widehat{\varphi} \in [\widehat{\varphi}_{\rm d}^{\rm a}, \infty)$.
- For $\widehat{T} = \widehat{T}_{{\rm A}_3^{\rm r}}$, a pair of stationary points of $\mathcal{F}_1$ (a local minimum and a local maximum at large $\widehat{A}$) appear (figure 4): this leads to $\widehat{\varphi}_{\rm sp} = \widehat{\varphi}_{\rm d}^{\rm r}$. At this temperature, the repulsive glass solution exists only marginally: it corresponds to a very flat maximum of the Franz-Parisi potential, see figure 5(top).
- For $\widehat{T}_{{\rm A}_3^{\rm r}} < \widehat{T} < \widehat{T}_{\rm C}$, there are two local maxima separated by a local minimum, and the maximum at small $\widehat{A}$ is higher. In other words, $\widehat{\varphi}_{\rm d}^{\rm a} < \widehat{\varphi}_{\rm d}^{\rm r}$. In this region the attractive glass

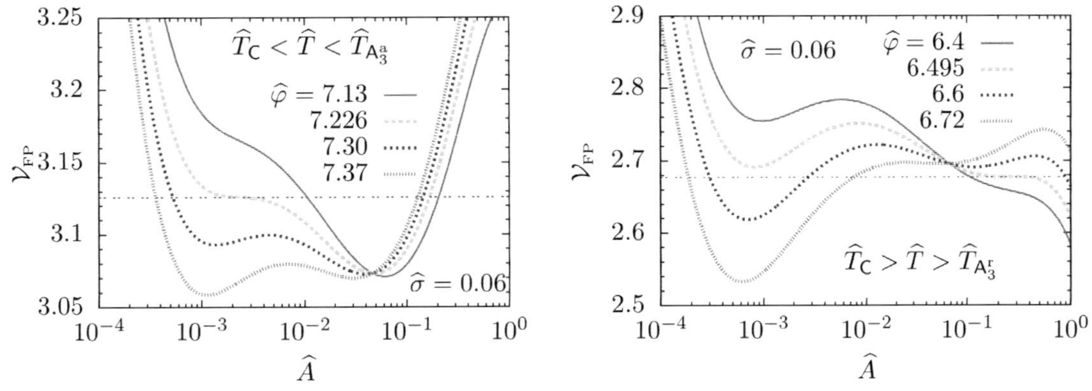

Figure 3. The Franz-Parisi potential $\mathcal{V}_{\mathrm{FP}}(\widehat{A})$ for two temperatures, $\widehat{T} = 0.5$ (left) and $\widehat{T} = 0.47$ (right), at several values of the density. The corresponding functions $\mathcal{F}_1$ are reported in figure 2.

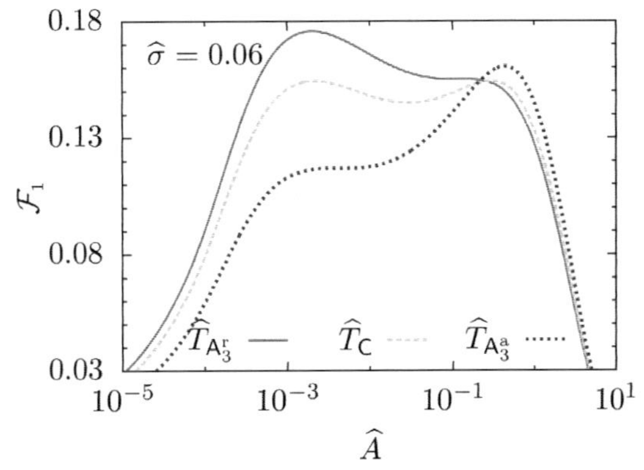

Figure 4. Shape of the function $\mathcal{F}_1$ for $\widehat{\sigma} = 0.06$ for the special temperatures $\widehat{T}_{\mathsf{A}_3^{\mathrm{r}}}$, $\widehat{T}_{\mathsf{C}}$, $\widehat{T}_{\mathsf{A}_3^{\mathrm{a}}}$ defined in the text.

exists for $\widehat{\varphi} \in [\widehat{\varphi}_{\mathrm{d}}^{\mathrm{a}}, \infty)$, while the repulsive glass exists for $\widehat{\varphi} \in [\widehat{\varphi}_{\mathrm{d}}^{\mathrm{r}}, \widehat{\varphi}_{\mathrm{sp}}]$. The Franz-Parisi potential is reported in figure 3(right).

- At $\widehat{T}_{\mathsf{C}}$ the two maxima of $\mathcal{F}_1$ have the same height (figure 4), hence the two lines cross $\widehat{\varphi}_{\mathrm{d}}^{\mathrm{a}} = \widehat{\varphi}_{\mathrm{d}}^{\mathrm{r}}$. The Franz-Parisi potential is in figure 5(middle).
- For $\widehat{T}_{\mathsf{C}} < \widehat{T} < \widehat{T}_{\mathsf{A}_3^{\mathrm{a}}}$ the situation is reversed, $\widehat{\varphi}_{\mathrm{d}}^{\mathrm{a}} > \widehat{\varphi}_{\mathrm{d}}^{\mathrm{r}}$, but for the rest nothing changes. The Franz-Parisi potential is in figure 3(left).
- When $\widehat{T} = \widehat{T}_{\mathsf{A}_3^{\mathrm{a}}}$, the local minimum and the small-$\widehat{A}$ maximum of $\mathcal{F}_1$ coalesce (figure 4). Here $\widehat{\varphi}_{\mathrm{sp}} = \widehat{\varphi}_{\mathrm{d}}^{\mathrm{a}}$ and the attractive glass becomes a marginally stable point. The Franz-Parisi potential is in figure 5(bottom).
- Finally, for $\widehat{T} > \widehat{T}_{\mathsf{A}_3^{\mathrm{a}}}$ there is a unique maximum at large $\widehat{A}$ corresponding to the attractive glass, which is the only state in this region and exists for $\widehat{\varphi} \in [\widehat{\varphi}_{\mathrm{d}}^{\mathrm{r}}, \infty)$.

In figure 6 we show the transition lines in the phase diagram that delimit the regions defined above. The two lines $\widehat{\varphi}_{\mathrm{d}}^{\mathrm{a}}(\widehat{T})$ and $\widehat{\varphi}_{\mathrm{d}}^{\mathrm{r}}(\widehat{T})$ cross each other at point C. Below $\widehat{T}_{\mathsf{C}}$, the liquid-glass transition is given by $\widehat{\varphi}_{\mathrm{d}}^{\mathrm{a}}(\widehat{T})$ and the liquid transforms into the attractive glass, while above $\widehat{T}_{\mathsf{C}}$,

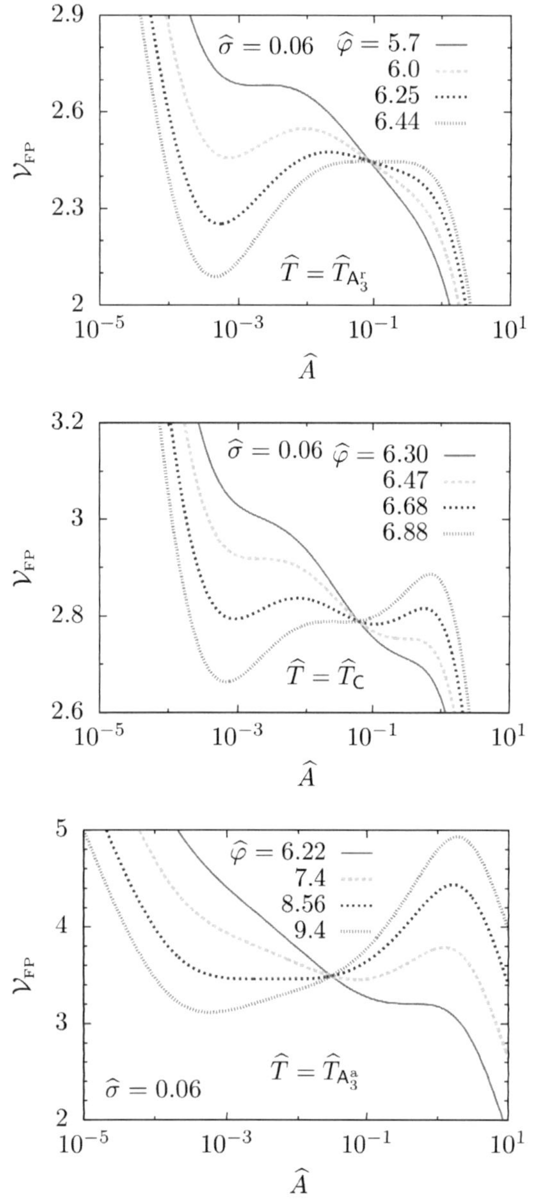

Figure 5. Shape of the Franz-Parisi potential $\mathcal{V}_{FP}$ for $\widehat{\sigma} = 0.06$ for the special temperatures $\widehat{T}_{A_3^r}$ (top), $\widehat{T}_C$ (middle), $\widehat{T}_{A_3^a}$ (bottom) defined in the text, for several densities at each temperature. The corresponding functions $\mathcal{F}_1$ are reported in figure 4.

the liquid-glass transition is given by $\widehat{\varphi}_d^r(\widehat{T})$ and the liquid transforms into the repulsive glass. Hence

$$\widehat{\varphi}_d(\widehat{T}) = \min\{\widehat{\varphi}_d^r(\widehat{T}), \widehat{\varphi}_d^a(\widehat{T})\} \ . \tag{8}$$

and the liquid is ergodic for $\widehat{\varphi} < \widehat{\varphi}_d(\widehat{T})$.

For $\widehat{\varphi} > \widehat{\varphi}_d(\widehat{T})$, the system is arrested but the two different glass phases can exist. The line $\widehat{\varphi}_d^a(\widehat{T})$ can be continued from point C to point A_3^a where it merges with the line $\widehat{\varphi}_{sp}$. The point A_3^a is a critical endpoint representing a cusp singularity that corresponds to a very singular point of the Franz-Parisi potential. Similarly, the line $\widehat{\varphi}_d^r(\widehat{T})$ can be continued from point C to

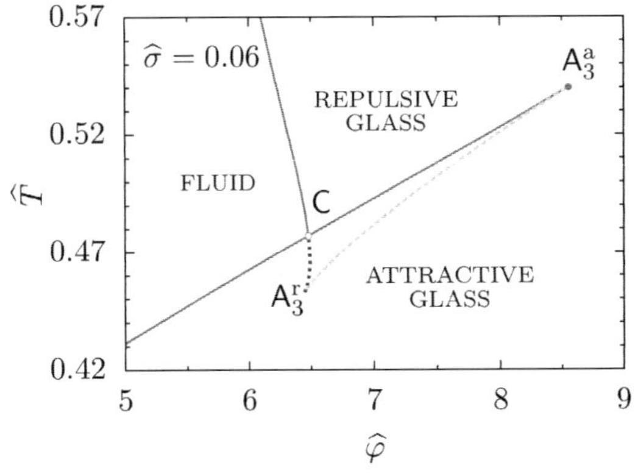

Figure 6. Phase diagram for the representative case $\widehat{\sigma} = 0.06$. There are two distint types of glass phases and fluid-glass transitions plus a glass-glass transition line. The two fluid-glass transition lines meet at a crossing point C along with the glass-glass transition line. The line C-A_3^a corresponds to a spinodal for the attractive glass, above which this phase does not exist anymore. Similarly, the lines C-A_3^r and A_3^r-A_3^a are spinodals of the repulsive glass. The glass-glass coexistence region is delimited by these three lines. The two special points A_3^a and A_3^r are cusp singularities; they correspond to points where the two glasses coalesce giving rise to a higher order singularity in the Franz-Parisi potential.

point A_3^r where it merges with the line $\widehat{\varphi}_{sp}$. Together, these three lines delimit the region of temperature and density where the two glasses coexist, while outside this region only one glass is present. This scenario is the standard one associated with a first order phase transition between the two glasses, as described by the Franz-Parisi potential that exhibits two stable minima in the coexistence region.

3.3. The sticky limit

It is interesting to consider a sticky sphere limit where $\widehat{\sigma} \to 0$, $U_0 \to \infty$ while $\mu = -\beta U_0 - \log \widehat{\sigma}$ is kept constant. In this limit the control parameters are therefore $\widehat{\varphi}$ and μ. One can show (Appendix A) that in this limit the small-$\widehat{A}$ maximum of $\mathcal{F}_1(\widehat{A})$, corresponding to the attractive glass, moves to $\widehat{A} = 0$. Because of the sticky attraction, particles do not move at all in the attractive glass phase. The height of this maximum can be computed explicitly and gives $\widehat{\varphi}_d^a = 2e^{\mu}$, see (A.18). Hence in this case the spinodal point of the attractive glass can be computed analytically and reaches arbitrarily large densities, therefore the point A_3^a moves to infinite density. Still, the other two lines $\widehat{\varphi}_d^r$ and $\widehat{\varphi}_{sp}$ –corresponding respectively to the large-$\widehat{A}$ local maximum and the local minimum– have to be computed numerically. The resulting phase diagram is reported in figure 7.

3.4. The evolution of the phase diagram with $\widehat{\sigma}$: A_3 and A_4 singularities

We now summarize the evolution of the phase diagram when the range of the attraction $\widehat{\sigma}$ is changed. At large $\widehat{\sigma}$, the phase diagram displays a single re-entrant liquid-glass transition line, see figure 1. At very small $\widehat{\sigma}$, a phase coexistence region between two glasses is observed, and is delimited by the triangle A_3^a-C-A_3^r in figure 6. This triangle is quite large at very small $\widehat{\sigma}$ (its size diverges in the sticky limit, see figure 7). Its size decreases with increasing $\widehat{\sigma}$, until it disappears at $\widehat{\sigma} \approx 0.19$ where the three points A_3^a-C-A_3^r merge into a further higher-order singularity of type A_4, also known as swallowtail bifurcation.

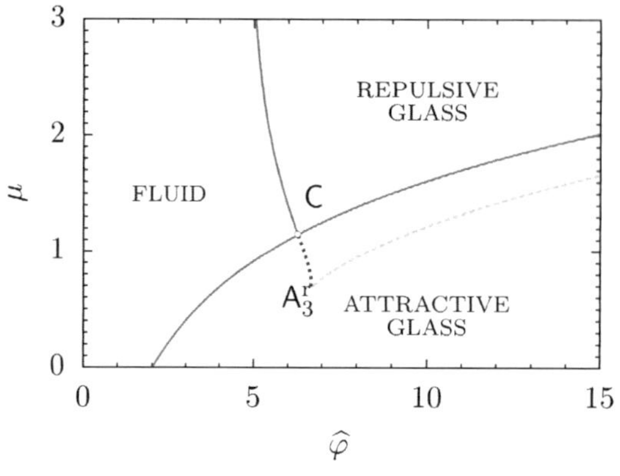

Figure 7. Phase diagram in the sticky limit $\widehat{\sigma} \to 0$, $U_0 \to \infty$ while keeping $\mu = -\beta U_0 - \log \widehat{\sigma}$ constant.

The higher-order singularities, A_3 and A_4, are special singular points of the Franz-Parisi potential. From (4) it is easy to show that if (5) is satisfied (i.e. on stationary points of the potential), the vanishing of the n-th derivative of $\mathcal{F}_1(\widehat{A})$ leads to the vanishing of the $(n+1)$-th derivative of $\mathcal{V}_{\rm FP}(\widehat{A})$. On the points A_3, the first two derivatives of $\mathcal{F}_1$ vanish (see figure 4) and therefore the first three derivatives of the Franz-Parisi potential vanish (see figure 5). On the point A_4, the first three derivatives of $\mathcal{F}_1$ vanish and therefore the first four derivatives of the Franz-Parisi potential vanish (not shown). This higher-order singularities lead to peculiar properties of the relaxation dynamics, that have been explored in great detail in the Mode-Coupling theory framework [61, 17, 20, 28, 62] and numerically [63, 64]. However, note that point $\mathsf{A}_3^{\rm r}$ is a *maximum* of the Franz-Parisi potential and therefore it is unstable, see figure 5(top). This instability has also been discussed within Mode-Coupling Theory [62].

3.5. The Kauzmann transition

Finally, it is important to stress that the complexity contains a term proportional to $\log(d)$, which makes it positive for all finite values of $\widehat{\varphi}$. Hence the Kauzmann transition happens at $\widehat{\varphi} \to \infty$ and $\widehat{A} \to 0$ [36] and it is therefore out of the range of the phase diagrams discussed above. In particular, this implies that the Kauzmann transition always happens in a region of the phase diagram where there is no coexistence between different glasses. The complexity associated to both glasses is therefore positive and for this reason we did not discuss the Kauzmann transition.

4. Conclusion

The main result of this paper is that a static replica picture based on the Franz-Parisi potential allows to re-derive many of the results that have been previously obtained using Mode-Coupling theory for attractive colloids [20, 12, 62], namely the re-entrance of the glass transition line and the coexistence of two glass phases for very short range attractions. Here we limited ourselves to the $d \to \infty$ limit where computations are easier and the replica theory is exact [13], but the computations can in principle be generalized to any finite dimension [36]. Moreover, we found that for suitable attraction range $\widehat{\sigma} \approx 0.06$ and temperature $\widehat{T} \approx 0.48$, the fluid phase extends to density $\widehat{\varphi} \approx 6.5$. This shows that sphere packings exist up to packing fraction $\varphi \approx 6.5\, d\, 2^{-d}$ and can be produced by slow compression of the fluid phase of attractive systems with the above parameters.

Let us conclude by some speculations on the connection between the static replica picture presented in this paper and the dynamic MCT picture of [20, 12, 62], and by presenting some perspective for future work.

4.1. Dynamics

It is very important to discuss the consequences that the static phase diagram obtained above has for dynamics. This is crucial to compare theory with numerical simulations and experiments, that in the glass phase are necessarily out of equilibrium. In simple spin glass models this connection is perfectly understood (both at equilibrium and out of equilibrium) within the general RFOT framework [49, 50, 51, 65, 66, 27]. Unfortunately, when one tries to translate this to approximate theories of particle systems based on MCT and replicas, the situation is much more complicated and controversial results have been obtained in the past [49, 67, 68, 69, 36, 9]. This is due to the fact that, although the general RFOT picture is perfectly consistent at least for large enough dimensions [70], its practical implementation using liquid theory approximate closure can lead to inconsistencies (a fact which is well known in the theory of simple liquids [71]).

Moreover, the discussion of out of equilibrium dynamics is not obvious because in general this dynamics depends a lot on the preparation history and the protocol that is used to explore the glass phase. This is particularly true in the present situation where two distinct glass phases can coexist. In principle, if one considers a slow annealing in the glass phase, one can obtain precise information through the so-called *state-following* method [48]. However, such a computation goes beyond the scope of this work.

Given these important remarks, we can discuss qualitatively the evolution of the cage radius in the glass phase based on our results for the Franz-Parisi potential. When $\widehat{\varphi} < \widehat{\varphi}_{\rm d}(\widehat{T})$, the system is in a liquid phase and it is ergodic. Particles can diffuse and the long time limit of connected density correlations is zero. As usual, when $\widehat{\varphi} \to \widehat{\varphi}_{\rm d}(\widehat{T})$ from below, dynamics slows down until at $\widehat{\varphi}_{\rm d}(\widehat{T})$ diffusion is arrested and the long time limit of the mean square displacement becomes equal to a constant according to (2). For densities very close to $\widehat{\varphi}_{\rm d}(\widehat{T})$, the mean square displacement plotted as a function of time exhibits a long plateau before crossing over to the diffusive regime, and the value of this plateau corresponds roughly to the long time mean square displacement in the glass phase.

We therefore have two different behaviour depending on temperature. When $\widehat{T} < \widehat{T}_{\rm C}$, the liquid arrests into the attractive glass and the value of mean square displacement corresponds to the small $\widehat{A}$ solution. Instead, when $\widehat{T} > \widehat{T}_{\rm C}$ the liquid arrests into the repulsive glass with larger $\widehat{A}$.

Suppose that we now continue to compress the system slowly and isothermally, out of equilibrium into the glass phase. We have several possibilities:

- At low $\widehat{T} < \widehat{T}_{\rm C}$, the system jumps in the attractive glass at $\widehat{\varphi}_{\rm d}$. Then it remains in this phase upon compression. This is because, even if the repulsive glass solution can appear in some range of density, dynamics is arrested in the attractive glass, which does not have a spinodal that could make it unstable. Therefore the blue and green lines in figure 6 cannot be observed under isothermal compression below $\widehat{T}_{\rm C}$.
- At intermediate temperature $\widehat{T}_{\rm C} < \widehat{T} < \widehat{T}_{\rm A_3^a}$ the system is stuck in the repulsive glass at $\widehat{\varphi}_{\rm d}$. However upon further increasing density, the repulsive glass solution disappears at large enough density (corresponding to $\widehat{\varphi} = \widehat{\varphi}_{\rm sp}(\widehat{T})$, the green line in figure 6). Around this point, the system undergoes a transition to the attractive glass, which then remains stable upon further compression. Note however that strong history dependent out of equilibrium effects are surely present, therefore the transition can happen everywhere between the red and green lines in figure 6.

- At $\widehat{T} > \widehat{T}_{\mathsf{A}_3^{\mathsf{a}}}$, there is no attractive glass. The system jumps into the repulsive glass at $\widehat{\varphi}_d$ and remains there at any other density because this is the only solution.

In general, it is reasonable to expect that the repulsive glass phase will be very hard to observe below $\widehat{T}_{\mathsf{C}}$. Although this phase exists in a strictly mean-field picture as a metastable minimum, one can easily see from figure 3(bottom) that the barriers separating this phase from the liquid and the attractive glass are very small. It is therefore most probable that in finite dimensions these barriers are easily crossed in this region, leading to a very short lifetime of the repulsive glass. The most relevant line in the triangle in figure 6 seems therefore the line $\mathsf{A}_3^{\mathsf{a}}$-$\mathsf{C}$, consistently with MCT results [20, 62].

In particular, the $\mathsf{A}_3^{\mathsf{r}}$ singularity corresponds to the point where the repulsive glass disappears leading to a *local quartic maximum* of the Franz-Parisi potential, as shown by the curve for $\widehat{\varphi} = 6.44$ in figure 5(top): it is therefore dynamically unobservable as the repulsive glass will always be unstable around this singularity. This is consistent with results from MCT [62]. On the contrary, the $\mathsf{A}_3^{\mathsf{a}}$ singularity corresponds to a *local quartic minimum* of the Franz-Parisi potential, see the curve for $\widehat{\varphi} = 8.56$ in figure 5(bottom). This leads to a unique glass phase with peculiar dynamical properties and a very slow logarithmic relaxation, as it has been shown within MCT [62].

4.2. Perspectives

The discussion of dynamics reported above is very qualitative and preliminary. Exploring further the connection between static and dynamical pictures will surely lead to a better understanding of the rich out of equilibrium dynamics in the glass-glass coexistence phase, where one could naturally expect hysteresis effects. We believe that such a study is a very interesting subject for future work. Performing numerical simulations in high dimensions, following [9], could be very useful to remove undesired metastability effects that complicate the picture. Of course, the hope is to put back these effects in a controlled way once the mean field picture is fully understood.

Our work opens the way to several other studies. First of all one should understand better the nature of the glass phase close to the glass-glass transition. It is to be expected that the complexity function will have two distinct branches, corresponding to the two different sets of glassy states in coexistence. The solution that maximizes the complexity should then correspond to the "typical" phase, in which equilibrium configurations will be found with probability 1 in the thermodynamic limit. This corresponds as usual to choosing the solution that minimizes the Franz-Parisi potential and would provide an "equilibrium" definition of the glass-glass transition line in the coexistence region. How activated barrier processes change the picture within RFOT theory remains an open problem that would be very interesting to investigate. Another question that would be important to address is the role of the Kauzmann transition in finite dimension, which we did not discuss here because in the limit $d \to \infty$ this transition moves at infinite $\widehat{\varphi}$, and is therefore outside the range of relevant densities for the glass-glass coexistence region. Furthermore, one could compute the equation of state of the two glasses, the jump of specific heat at the glass transition, and so on.

For soft matter applications, one would like of course to repeat this calculation in $d = 3$. We expect that the qualitative phase diagram will remain the same in low dimensions. Unfortunately, for the moment the small cage expansion scheme does not give good quantitative results for the dynamical transition in low dimensions, but there is hope to improve it.

For algorithmic applications, it would be very important to perform a "state following" calculation [48], to study how much the states at φ_d can be compressed adiabatically: this would give the true threshold beyond which packings cannot be constructed in polynomial time by simulated annealing.

Acknowledgments – We warmly thank P. Charbonneau and E. Zaccarelli for stimulating discussions.

Appendix A. Gaussian replica equations for a generic potential

We collect here all the equations that are needed to obtain the results presented in this work. Because these are based on extensions of previous work [36, 60], we do not provide the general derivation but rather highlight the extensions that are needed with respect to the previous works.

We assume that the potential has a soft repulsive core of diameter $D = 1$. The most general potential we consider has the form

$$v(r) = \epsilon(1-r)^2\theta(r<1) - U_0\theta(1<r<1+\sigma) \tag{A.1}$$

Here T is temperature and $\beta = 1/(k_B T)$. We define $\widehat{\sigma} = \sigma\, d$ and $\widehat{\epsilon} = \beta\epsilon/d^2$ and $\widehat{\varphi} = 2^d\varphi/d$ where φ is the packing fraction of the repulsive core. We also define $\beta U_0 = \widehat{U}_0 = 1/\widehat{T}$.

Appendix A.1. Finite dimensions

The approximation scheme used here holds for $\beta\epsilon$ large enough and is based on [60, (22) and (23)], which give the replicated free entropy separated between the harmonic and the liquid contributions

$$\begin{aligned} \mathcal{S}(m,A) &= S_h(m,A) + \mathcal{S}_{\rm liq}(T/m,\varphi) + 2^{d-1}\varphi y_{\rm liq}^{\rm HS}(\varphi) G_m(A) \ , \\ G_m(A) &= d\int_0^\infty dr\, r^{d-1}\left[q(A;r)^m - e^{-\beta m v(r)}\right] \ , \\ S_h(m,A) &= \frac{d}{2}(m-1)\log(2\pi A) - \frac{d}{2}(1-m-\log m) \ , \end{aligned} \tag{A.2}$$

for m replicas at temperature T, in a Gaussian cage of variance $2A$. The function $q(A;r) = \int d^d r' \gamma_{2A}(\vec{r}')e^{-\beta v(|\vec{r}-\vec{r}'|)}$ is defined in [60, just after (16)] where γ_{2A} is a normalized and centered Gaussian of variance $2A$, and $y_{\rm liq}^{\rm HS}$ is the contact value of the hard-spheres cavity function. Introducing bipolar coordinates, as in [36, Appendix C.2.a], we obtain the generalization of [36, (C16)] to a generic potential $v(r)$

$$q(A;r) = \int_0^\infty du\, e^{-\beta v(u)} \left(\frac{u}{r}\right)^{\frac{d-1}{2}} \frac{e^{-\frac{(r-u)^2}{4A}}}{\sqrt{4\pi A}} \left[e^{-\frac{ru}{2A}}\sqrt{\pi\frac{ru}{A}} I_{\frac{d-2}{2}}\left(\frac{ru}{2A}\right)\right] . \tag{A.3}$$

From the replicated entropy we can obtain the Franz-Parisi potential

$$\begin{aligned} \beta V_{\rm FP}(A) &= -\left.\frac{d(\mathcal{S}/m)}{dm}\right|_{m=1} = \mathcal{S}_{\rm liq}(T) - d - \frac{d}{2}\log(2\pi A) - 2^{d-1}\varphi y_{\rm liq}^{\rm HS}(\varphi) H_1(A) \ , \\ H_m(A) &= m\frac{\partial G_m(A)}{\partial m} \ . \end{aligned} \tag{A.4}$$

Note that the Franz-Parisi potential corresponds to the complexity at $m = 1$ as a function of A.

The local minima of $V_{\rm FP}(A)$ correspond to stable glass phases. Because $\mathcal{S}(m,A)$ is independent of A at $m = 1$, one can show that the stationary points in A of $V_{\rm FP}(A)$ are the same as those of $\mathcal{S}(m,A)$ for $m \to 1$. In general, the stationary points of $\mathcal{S}(m,A)$ are given by the condition

$$\frac{d}{2^d\varphi y_{\rm liq}^{\rm HS}(\varphi)} = \frac{A}{1-m}\frac{\partial G_m(A)}{\partial A} \equiv F_m(A) \ . \tag{A.5}$$

Appendix A.2. Infinite dimension

We follow [36]: in the limit of infinite dimension, we are interested in the scaling $A = \widehat{A}/d^2$. One can show that

$$\lim_{d\to\infty} e^{-d^2 z}\sqrt{2\pi d^2 z}\, I_{\frac{d-2}{2}}(d^2 z) = e^{-\frac{1}{8z}} \,. \tag{A.6}$$

Using this result in (A.3), we have

$$q(A;r) = \int_0^\infty du\, e^{-\beta v(u)} \left(\frac{u}{r}\right)^{\frac{d-1}{2}} \frac{e^{-\frac{(r-u)^2}{4A}}}{\sqrt{4\pi A}} e^{-\frac{\widehat{A}}{4ru}} \tag{A.7}$$

Changing variables again to $t = \frac{r-1}{\sqrt{4A}}$ and $s = \frac{u-1}{\sqrt{4A}}$ and using

$$\left(\frac{u}{r}\right)^{\frac{d-1}{2}} = \left(\frac{1+\frac{s\sqrt{4\widehat{A}}}{d}}{1+\frac{t\sqrt{4\widehat{A}}}{d}}\right)^{\frac{d-1}{2}} \sim e^{(s-t)\sqrt{\widehat{A}}} \,, \tag{A.8}$$

we get

$$\begin{aligned} q(\widehat{A};t) &\sim e^{-\frac{\widehat{A}}{4}} \frac{1}{\sqrt{\pi}} \int_{-\infty}^{\infty} ds\, e^{-\beta v\left(1+s\sqrt{4\widehat{A}}/d\right)} e^{(s-t)\sqrt{\widehat{A}}-(t-s)^2} \\ &= e^{-\frac{\widehat{A}}{4}} \frac{1}{\sqrt{\pi}} \int_{-\infty}^{0} ds\, e^{-4\widehat{A}\widehat{\varepsilon}s^2+(s-t)\sqrt{\widehat{A}}-(t-s)^2} + e^{-\frac{\widehat{A}}{4}} \frac{1}{\sqrt{\pi}} \int_0^{\widehat{\sigma}/\sqrt{4\widehat{A}}} ds\, e^{\widehat{U}_0+(s-t)\sqrt{\widehat{A}}-(t-s)^2} \\ &+ e^{-\frac{\widehat{A}}{4}} \frac{1}{\sqrt{\pi}} \int_{\widehat{\sigma}/\sqrt{4\widehat{A}}}^{\infty} ds\, e^{(s-t)\sqrt{\widehat{A}}-(t-s)^2} \,. \end{aligned} \tag{A.9}$$

We can use this result to compute $\mathcal{G}_m(\widehat{A}) = \lim_{d\to\infty} G_m(Ad^2)$:

$$\begin{aligned} \mathcal{G}_m(\widehat{A}) &= \lim_{d\to\infty} \sqrt{4\widehat{A}} \int_{-d/\sqrt{4\widehat{A}}}^{\infty} dt \left(1+t\frac{\sqrt{4\widehat{A}}}{d}\right)^{d-1} \left[q(\widehat{A};t)^m - e^{-\beta m v\left(1+t\sqrt{4\widehat{A}}/d\right)}\right] \\ &= \int_{-\infty}^{\infty} dy\, e^y \left[q(\widehat{A};y)^m - e^{-\beta m v(1+y/d)}\right] \\ &= \int_{-\infty}^{0} dy\, e^y \left[q(\widehat{A};y)^m - e^{-m\widehat{\varepsilon}y^2}\right] + \int_0^{\widehat{\sigma}} dy\, e^y \left[q(\widehat{A};y)^m - e^{m\widehat{U}_0}\right] \\ &+ \int_{\widehat{\sigma}}^{\infty} dy\, e^y \left[q(\widehat{A};y)^m - 1\right] \end{aligned} \tag{A.10}$$

where $q(\widehat{A};y) = q(\widehat{A};t/\sqrt{4\widehat{A}})$.

The equation for $\widehat{A}$ is

$$\frac{1}{\widehat{\varphi}} = \frac{\widehat{A}}{1-m}\frac{d\mathcal{G}_m(\widehat{A})}{d\widehat{A}} = \mathcal{F}_m(\widehat{A}) = \frac{m\widehat{A}}{1-m}\int_{-\infty}^{\infty} dy\, e^y q(\widehat{A};y)^{m-1}\frac{\partial q(\widehat{A};y)}{\partial \widehat{A}} \tag{A.11}$$

At $m = 1$ the equation is

$$\frac{1}{\widehat{\varphi}} = \mathcal{F}_1(\widehat{A}) = -\widehat{A}\int_{-\infty}^{\infty} dy\, e^y \log[q(\widehat{A};y)]\frac{\partial q(\widehat{A};y)}{\partial \widehat{A}} \tag{A.12}$$

hence the dynamical transition is reached when $1/\widehat{\varphi}_d = \max_{\widehat{A}} \mathcal{F}_1(\widehat{A})$, as stated in (6). The non-trivial part of the Franz-Parisi potential, for $d \to \infty$, is proportional to

$$\mathcal{V}_{\rm FP}(\widehat{A}) = -\log(\widehat{A}) - \widehat{\varphi}\mathcal{H}_1(\widehat{A}) \ ,$$
$$\mathcal{H}_1(\widehat{A}) = \left[m \frac{\partial \mathcal{G}_m(\widehat{A})}{\partial m} \right]_{m=1} = \int_{-\infty}^{0} dy\, e^y \left[q(\widehat{A};y) \log q(\widehat{A};y) + \widehat{\epsilon} y^2 e^{-\widehat{\epsilon} y^2} \right]$$
$$+ \int_0^{\widehat{\sigma}} dy\, e^y \left[q(\widehat{A};y) \log q(\widehat{A};y) - \widehat{U}_0 e^{\widehat{U}_0} \right] + \int_{\widehat{\sigma}}^{\infty} dy\, e^y \left[q(\widehat{A};y) \log q(\widehat{A};y) \right] \qquad (A.13)$$

Appendix A.3. Square-well potential
We consider here the case $\widehat{\epsilon} = \infty$ corresponding to a square well potential. In this case the function $q(\widehat{A};y)$ is

$$q(\widehat{A};y) = (1 - e^{\widehat{U}_0}) \Theta\left(\frac{y + \widehat{A} - \widehat{\sigma}}{2\sqrt{\widehat{A}}} \right) + e^{\widehat{U}_0} \Theta\left(\frac{y + \widehat{A}}{2\sqrt{\widehat{A}}} \right) ,$$
$$\frac{\partial q(\widehat{A};y)}{\partial \widehat{A}} = (1 - e^{\widehat{U}_0}) \frac{\widehat{A} - y + \widehat{\sigma}}{4\widehat{A}^{3/2}\sqrt{\pi}} e^{-\frac{(\widehat{A}+y-\widehat{\sigma})^2}{4\widehat{A}}} + e^{\widehat{U}_0} \frac{\widehat{A} - y}{4\widehat{A}^{3/2}\sqrt{\pi}} e^{-\frac{(\widehat{A}+y)^2}{4\widehat{A}}} , \qquad (A.14)$$

which allow one to compute easily $\mathcal{F}_1(\widehat{A})$ and solve the equation for $\widehat{A}$.

Appendix A.4. Sticky spheres
A special case is the limit $\widehat{\sigma} \to 0$ and $\widehat{U}_0 \to \infty$ with $\widehat{\sigma} e^{\widehat{U}_0} = e^{-\mu}$ or $\mu = -\widehat{U}_0 - \log \widehat{\sigma}$. In this case

$$q(\widehat{A};y) = \Theta\left(\frac{y + \widehat{A}}{2\sqrt{\widehat{A}}} \right) + e^{-\mu} \frac{e^{-\frac{(\widehat{A}+y)^2}{4\widehat{A}}}}{\sqrt{4\pi\widehat{A}}} ,$$
$$\frac{\partial q(\widehat{A};y)}{\partial \widehat{A}} = \frac{\widehat{A} - y}{4\widehat{A}^{3/2}\sqrt{\pi}} e^{-\frac{(\widehat{A}+y)^2}{4\widehat{A}}} - e^{-\mu} \frac{2\widehat{A} + \widehat{A}^2 - y^2}{8\widehat{A}^{5/2}\sqrt{\pi}} e^{-\frac{(\widehat{A}+y)^2}{4\widehat{A}}} , \qquad (A.15)$$

In this limit the attractive glass has $\widehat{A} = 0$, hence it is useful to compute the limit $\widehat{A} \to 0$. We have

$$\widehat{A} \frac{\partial q(\widehat{A};y)}{\partial \widehat{A}} \to e^{-\mu} \frac{y^2 - 2\widehat{A}}{4\widehat{A}} e^{-y/2} \frac{e^{-\frac{y^2}{4\widehat{A}}}}{\sqrt{4\pi\widehat{A}}} ,$$
$$q(\widehat{A}; z\sqrt{\widehat{A}}) \to e^{-\mu} \frac{e^{-\frac{z^2}{4}}}{\sqrt{4\pi\widehat{A}}} , \qquad (A.16)$$

hence we can write, with $y = \sqrt{\widehat{A}} z$:

$$\mathcal{F}_1(\widehat{A} \to 0) = -\int_{-\infty}^{\infty} dz\, \log[q(\widehat{A}; z\sqrt{\widehat{A}})] e^{-\mu} \frac{z^2 - 2}{4} \frac{e^{-\frac{z^2}{4}}}{\sqrt{4\pi}}$$
$$= e^{-\mu} \int_{-\infty}^{\infty} dz\, \frac{z^2}{4} \frac{z^2 - 2}{4} \frac{e^{-\frac{z^2}{4}}}{\sqrt{4\pi}} = \frac{1}{2} e^{-\mu} \qquad (A.17)$$

The branch of the dynamical transition that corresponds to the attractive glass therefore reads

$$\frac{1}{\widehat{\varphi}_{\rm d}^{\rm a}} = \frac{1}{2} e^{-\mu} \ . \qquad (A.18)$$

References

[1] Torquato S and Stillinger F H 2010 *Rev. Mod. Phys.* **82** 2633–2672
[2] Rivas G, Ferrone F and Herzfeld J 2004 *EMBO Rep.* **5** 23–27
[3] Hamming R W 1997 *The Art of Doing Science and Engineering* (Gordon and Breach)
[4] Conway J H and Sloane N J A 1993 *Sphere Packings, Lattices and Groups* (New York: Spriger-Verlag)
[5] Cohn H 2010 *Proceedings of the International Congress of Mathematicians* **IV** 2416–2443 (*Preprint* arXiv:1003.3053)
[6] Torquato S and Jiao Y 2010 *Physical Review E* **82** 061302
[7] Andreanov A and Scardicchio A 2012 *Physical Review E* **86** 041117
[8] Marcotte E and Torquato S 2013 arXiv.org:1304.5003
[9] Charbonneau P, Ikeda A, Parisi G and Zamponi F 2011 *Phys. Rev. Lett.* **107**(18) 185702
[10] Van der Waals J D 1873 *Doctoral thesis*
[11] Bernal J and Mason J 1960 *Nature* **188** 910–911
[12] Sciortino F 2002 *Nature materials* **1** 145
[13] Kurchan J, Parisi G and Zamponi F 2012 *Journal of Statistical Mechanics: Theory and Experiment* **2012** P10012
[14] Sellitto M and Zamponi F 2013 arXiv.org:1306.2912, *to appear on Europhysics Letters*
[15] Franz S and Parisi G 1998 *Physica A: Statistical Mechanics and its Applications* **261** 317–339
[16] Frisch H L and Percus J K 1999 *Phys. Rev. E* **60** 2942–2948
[17] Fabbian L, Götze W, Sciortino F, Tartaglia P and Thiery F 1999 *Phys. Rev. E* **59**(2) R1347–R1350
[18] Zaccarelli E and Poon W C 2009 *Proceedings of the National Academy of Sciences* **106** 15203–15208
[19] Foffi G, Sciortino F, Zaccarelli E and Tartaglia P 2004 *Journal of Physics: Condensed Matter* **16** S3791
[20] Dawson K, Foffi G, Fuchs M, Götze W, Sciortino F, Sperl M, Tartaglia P, Voigtmann T and Zaccarelli E 2000 *Phys. Rev. E* **63**(1) 011401
[21] Eckert T and Bartsch E 2002 *Phys. Rev. Lett.* **89**(12) 125701
[22] Pham K N, Puertas A M, Bergenholtz J, Egelhaaf S U, Moussaıd A, Pusey P N, Schofield A B, Cates M E, Fuchs M and Poon W C 2002 *Science* **296** 104–106
[23] Sellitto M, De Martino D, Caccioli F and Arenzon J J 2010 *Phys. Rev. Lett.* **105** 265704
[24] Sellitto M 2012 *Phys. Rev. E* **86** 030502
[25] Kirkpatrick T R and Thirumalai D 1987 *Phys. Rev. Lett.* **58** 2091–2094
[26] Kirkpatrick T R and Wolynes P G 1987 *Phys. Rev. B* **36** 8552–8564
[27] Wolynes P and Lubchenko V (eds) 2012 *Structural Glasses and Supercooled Liquids: Theory, Experiment, and Applications* (Wiley)
[28] Caiazzo A, Coniglio A and Nicodemi M 2004 *Phys. Rev. Lett.* **93**(21) 215701
[29] Crisanti A and Leuzzi L 2006 *Phys. Rev. B* **73**(1) 014412
[30] Krakoviack V 2007 *Phys. Rev. B* **76**(13) 136401
[31] Crisanti A and Leuzzi L 2007 *Phys. Rev. B* **76**(13) 136402
[32] Krzakala F, Tarzia M and Zdeborová L 2008 *Physical Review Letters* **101** 165702 (pages 4)
[33] Mézard M and Parisi G 1999 *The Journal of Chemical Physics* **111** 1076–1095
[34] Franz S, Parisi G, Ricci-Tersenghi F and Rizzo T 2011 *The European Physical Journal E* **34** 1–17
[35] Velenich A, Parola A and Reatto L 2006 *Physical Review E* **74** 021410 (pages 7)
[36] Parisi G and Zamponi F 2010 *Rev. Mod. Phys.* **82** 789–845
[37] Lubachevsky B D and Stillinger F H 1990 *J. Stat. Phys.* **60** 561–583
[38] Skoge M, Donev A, Stillinger F H and Torquato S 2006 *Physical Review E* **74** 041127 (pages 11)
[39] van Meel J A, Charbonneau B, Fortini A and Charbonneau P 2009 *Phys. Rev. E* **80** 061110
[40] Vance S 2011 *Advances in Mathematics* **227** 2144 – 2156
[41] Ball K 1992 *International Math. Research Notices* **1992** 217–221
[42] Kabatiansky G A and Levensthein V I 1978 *Problems on Information Transmission* **14** 1–17
[43] Parisi G and Zamponi F 2006 *Journal of Statistical Mechanics: Theory and Experiment* **2006** P03017
[44] Montanari A and Semerjian G 2006 *J.Stat.Phys.* **125** 23–54
[45] Charbonneau P, Corwin E I, Parisi G and Zamponi F 2012 *Phys. Rev. Lett.* **109**(20) 205501
[46] Zdeborová L and Krzakala F 2007 *Physical Review E* **76** 031131 (pages 29)
[47] Krzakala F and Kurchan J 2007 *Physical Review E* **76** 021122 (pages 13)
[48] Krzakala F and Zdeborová L 2010 *EPL (Europhysics Letters)* **90** 66002
[49] Kirkpatrick T R and Wolynes P G 1987 *Phys. Rev. A* **35** 3072–3080
[50] Kirkpatrick T R and Thirumalai D 1988 *Phys. Rev. A* **37**(11) 4439–4448
[51] Kirkpatrick T R and Thirumalai D 1989 *Journal of Physics A: Mathematical and General* **22** L149
[52] Franz S and Parisi G 1995 *Journal de Physique I* **5** 1401–1415
[53] Franz S and Parisi G 1997 *Phys. Rev. Lett.* **79** 2486–2489

[54] Cardenas M, Franz S and Parisi G 1998 *Journal of Physics A: Mathematical and General* **31** L163–L169
[55] Monasson R 1995 *Phys. Rev. Lett.* **75** 2847–2850
[56] Dawson K A, Franz S and Sellitto M 2003 *Europhysics Letters* **64** 302–308
[57] Cammarota C, Cavagna A, Giardina I, Gradenigo G, Grigera T S, Parisi G and Verrocchio P 2010 *Phys. Rev. Lett.* **105**(5) 055703
[58] Mézard M and Parisi G 1996 *Journal of Physics A: Mathematical and General* **29** 6515
[59] Mezard M and Parisi G 2012 *Structural Glasses and Supercooled Liquids: Theory, Experiment and Applications* ed PGWolynes and VLubchenko (Wiley & Sons) (*Preprint* `arXiv:0910.2838`)
[60] Berthier L, Jacquin H and Zamponi F 2011 *Phys. Rev. E* **84**(5) 051103
[61] Bergenholtz J and Fuchs M 1999 *Phys. Rev. E* **59**(5) 5706–5715
[62] Götze W 2009 *Complex dynamics of glass-forming liquids: A mode-coupling theory* (Oxford University Press, USA)
[63] Sciortino F, Tartaglia P and Zaccarelli E 2003 *Phys. Rev. Lett.* **91**(26) 268301
[64] Charbonneau P and Reichman D R 2007 *Phys. Rev. Lett.* **99**(13) 135701
[65] Cugliandolo L F and Kurchan J 1993 *Phys. Rev. Lett.* **71** 173–176
[66] Castellani T and Cavagna A 2005 *Journal of Statistical Mechanics: Theory and Experiment* **2005** P05012
[67] Ikeda A and Miyazaki K 2010 *Phys. Rev. Lett.* **104** 255704
[68] Schmid B and Schilling R 2010 *Phys. Rev. E* **81** 041502
[69] Szamel G 2010 *Europhysics Letters* **91** 56004
[70] Charbonneau P, Ikeda A, Parisi G and Zamponi F 2012 *Proceedings of the National Academy of Sciences* **109** 13939–13943
[71] Hansen J P and McDonald I R 1986 *Theory of simple liquids* (London: Academic Press)

ELC International Meeting on Inference, Computation, and Spin Glasses (ICSG2013) IOP Publishing
Journal of Physics: Conference Series **473** (2013) 012021 doi:10.1088/1742-6596/473/1/012021

The hard-core model on random graphs revisited

Jean Barbier[†], **Florent Krzakala**[†], **Lenka Zdeborová**[⋆] **and Pan Zhang**[†,‡]
† ESPCI and CNRS UMR 7083, 10 rue Vauquelin Paris 75005 France
Laboratoire de Physique Statistique, CNRS UMR 8550, Université P. et M. Curie, Ecole Normale Supérieure, Paris, France
⋆ Institut de Physique Théorique, IPhT, CEA Saclay and URA 2306, CNRS, 91191 Gif-sur-Yvette, France.
‡ Santa Fe Institute, 1399 Hyde Park Road, Santa Fe, NM, USA

E-mail: jean.barbier@espci.fr, florent.krzakala@ens.fr, lenka.zdeborova@cea.fr, july.lzu@gmail.com

Abstract. We revisit the classical hard-core model, also known as independent set and dual to vertex cover problem, where one puts particles with a first-neighbor hard-core repulsion on the vertices of a random graph. Although the case of random graphs with small and very large average degrees respectively are quite well understood, they yield qualitatively different results and our aim here is to reconciliate these two cases. We revisit results that can be obtained using the (heuristic) cavity method and show that it provides a closed-form conjecture for the exact density of the densest packing on random regular graphs with degree $K \geq 20$, and that for $K > 16$ the nature of the phase transition is the same as for large K. This also shows that the hard-code model is the simplest mean-field lattice model for structural glasses and jamming.

Given a graph, how to put a large number of particles on the vertices avoiding any first-neighbor contact? This is the task one has to solve in the hard-core model (HC), also known as independent set (IS), or vertex cover (VC) when the meaning of covered/empty is switched. It is a NP-hard combinatorial optimization problem that has many applications such as scheduling problems [1], inference of phylogenetic trees [2] or in the communications [3] where one seeks for the minimal set of placed sensor devices in a service area so that the entire area is accessible.

The hard-core model is defined as follows: We consider a graph $G = (V, E)$ of size $N = |V|$ and associate an occupation number $\sigma_i \in \{0, 1\}$ to every vertex $i \in V$, where 0 stands for free and 1 for occupied. The Gibbs measure that corresponds to the hard-core model reads

$$P(\{\sigma_i\}_{i=1,\dots,N}) = \frac{1}{Z} e^{\mu \sum_{i\in V} \sigma_i} \prod_{(ij)\in E} (1 - \sigma_i \sigma_j) , \quad (1)$$

where μ is a constant called the chemical potential and Z is a normalization called the partition function. The average packing fraction $\rho = 1/N \sum_{j=1}^{N} \sigma_j P(\{\sigma_i\}_{i=1,\dots,N})$ increases with μ, the limit of $\mu \to \infty$ corresponding to the densest packing. The hard-core problem (1) is equivalent to the independent set problem where every edge is allowed at most one occupied end-node. This is dual (in the sense that $\sigma_i \to \neg\sigma_i$ and $\rho \to 1 - \rho$) to the vertex cover problem, where every edge must have at least one occupied neighbor.

In statistical physics the hard-core model is seen as a very natural lattice model for the behavior of hard-spheres. The peculiar glassy properties of packed hard spheres are widely studied and their understanding has important physical consequences, for a review see e.g. [4]. The hard-core model is well studied in the statistical physics literature [5, 6, 7, 8, 9, 10, 11]. It was shown that as the density of covered sites increases the hard-core model on random graphs with small average degree undergoes a continuous phase transition (towards a so-called full replica symmetry breaking phase) [12, 13, 14, 15, 16]. For this reason it is often disregarded as a good model for the behavior of hard spheres, because one desires a discontinuous phase transition (towards the so called dynamical one step replica symmetry breaking phase) as the density of particles is increased in such a model. More complicated lattice glass models were hence introduced and studied [13, 14]. In the class of models with a discontinuous phase transition there are in fact two such transitions, a dynamical one (also referred to as the mode coupling transition in the literature about glasses) and a condensation one (also called Kauzmann transition in the same literature) [17].

In mathematics and computer science the hard-core model is also extensively studied. On large Erdös-Rényi or regular random graphs with constant average degree K it has been shown that for sufficiently large K there exists an independent set of density (fraction of covered nodes) $\rho \approx 2\log K/K$ [18, 19, 20]. While it is simple to design a polynomial algorithm that construct independent sets (packings) [21] up to density $\rho = (1+o(1))\log K/K$, about half the maximum size, no algorithm is known (either provably, or heuristically) to produce in polynomial time a packing of density $\rho = (1+\epsilon)\log K/K$, with finite ϵ. Hence a question of central importance in theory of algorithms is how to construct larger independent sets or what property of the problem makes it algorithmically so hard [22]. Recently the work of [23] shed light on this question by showing that for sufficiently large K the space of all independent sets of density $\rho < \log K/K$ is simple (connected), for $\rho > \log K/K$ the space of independent sets splits into small cluster that are far away from each other. An exponential lower bound for a Metropolis process for sampling independent sets testifies that some onset of algorithmic difficulty happens at $\rho \approx \log K/K$ [23]. A remarkable recent work [24] showed, using the properties of clustering, that local algorithms cannot find dense configurations of the hard-core model. The existence of a discontinuous phase transition in the hard core model is also suggested by other mathematical literature. Indeed, the results of [25] on the reconstruction threshold for the hard-core model indicates that the dynamical transition appears at densities $\approx \log K/K$, whereas the results of [20] indicate that both the condensation transition and the densest packing appear at $\approx 2\log K/K$.

The main novel contribution of this paper is to settle down this apparent discrepancy between the physics and mathematical works concerning the nature of the phase transition in the hard-core problem on random graphs. The result is that whereas the current statistical physics picture with continuous phase transition is correct for graphs of small average degree ($K < 16$ for random regular graphs), the model has a discontinuous phase transition for degree $K > 16$ and the large K expansion of the corresponding equations is in agreement with all the mathematical results. This is showing that for large dimensional hard spheres (corresponding to large degree K) this simple hard-core model actually is a very good lattice model.

Another contribution of this paper is to point out that on the K-regular random graphs the one-step replica symmetry broken result for the size of largest independent set, or minimal vertex cover, is stable towards more steps of replica symmetry breaking for $K \geq 20$, and hence for $K \geq 20$ we provide a closed form conjecture about the *exact* values of this density.

We structure the paper as an overview of the results obtained for the hard-core model after a careful investigation of the cavity solution of the model on random regular graphs of degree K. We used the heuristic cavity method that is well known in statistical physics and has been derived and explained in details elsewhere, see e.g. [26] for an overview. In order to keep this paper accesible also to readers not familiar with the cavity method we give only the resulting expressions and mostly only in cases when they are simple to state and lead to interesting conjectures about the structure of dense packings or the density of the largest packing.

1. The densest packing, largest independent set and minimal vertex cover

1.1. Replica symmetric result

We define the cavity message $\pi_{i\to j}$ to be the probability that a node i is occupied in the cavity graph where the edge (ij) has been deleted. In this modified graph, we then assume that other neighbors of node i are independent and write the belief propagation equations

$$\pi_{i\to j} = \frac{e^{\mu}\prod_{k\in\partial i\backslash j}(1-\pi_{k\to i})}{1+e^{\mu}\prod_{k\in\partial i\backslash j}(1-\pi_{k\to i})}\,, \tag{2}$$

where $\partial i\backslash j$ are all the neighbors of node i excluding j. On a graph $\mathcal{G}$ the Bethe free energy is given using the fixed point of eq. (2) as

$$f_{RS}^{\mathcal{G}} = -\sum_{i\in V}\log\left[1+e^{\mu}\prod_{j\in\partial i}(1-\pi_{j\to i})\right] + \sum_{(ij)\in E}\log\left(1-\pi_{j\to i}\pi_{i\to j}\right). \tag{3}$$

On random K-regular graphs the above equation is satisfied by $\pi_{i\to j} = \bar{\pi}$ that solves the following equation

$$\bar{\pi} = \frac{e^{\mu}(1-\bar{\pi})^{K-1}}{1+e^{\mu}(1-\bar{\pi})^{K-1}} = e^{\mu}(1-\bar{\pi})^{K}\,. \tag{4}$$

The corresponding densities of free energy, density, and entropy reads (at the fixed point):

$$f_{RS}(\mu) = -\log\left(1+\bar{\pi}\right) + \frac{K}{2}\log\left(1-\bar{\pi}^2\right), \tag{5}$$

$$\rho_{RS}(\mu) = -\frac{\partial f_{RS}^{\mathcal{G}}}{\partial\mu}\bigg|_{\bar{\pi}} = \frac{\bar{\pi}}{1+\bar{\pi}}\,, \tag{6}$$

$$s_{RS}(\rho) = -f_{RS} - \mu\rho_{RS}\,. \tag{7}$$

where again ρ is the particle density. Solving $s_{RS}(\rho_s^{\mathrm{RS}}) = 0$, we get the replica symmetric (RS) result for the densest packing evaluated in Table 1. Due to a general property that in the absence of quenched disorder the replica symmetric solution on random regular graphs is equivalent to the first moment (annealed) calculations [27] this replica symmetric result agrees with the rigorously known upper bounds.

However, as in many other optimization problems the replica symmetric result is not exact. A necessary condition for its correctness is referred to as the local stability. The condition that needs to be satisfied for the RS approach to be stable is [12, 16]

$$\lambda = (K-1)\left(\frac{\partial\pi_{i\to j}}{\partial\pi_{k\to i}}\right)^2_{\bar{\pi}} = (K-1)\bar{\pi}^2 < 1\,. \tag{8}$$

Using (7) we get the local stability threshold of the RS solution on random K-regular graphs

$$\mu_l = -\frac{1}{2}\log(K-1) - K\log\left(1-\frac{1}{\sqrt{K-1}}\right),\quad \rho_l = \frac{1}{1+\sqrt{K-1}}\,. \tag{9}$$

We hence conclude that close to the densest packings, the replica symmetric results for the densest packing are never stable and hence not exact. They only provide an upper bound. We will explain further the physical meaning of the density ρ_l in Section 2.1.

K	$\rho_l^{\rm RS}$	$\mu_l^{\rm RS}$	$\rho_s^{\rm RS}$	$\rho_s^{\rm 1RSB}$
3	0.4142	3.3373	0.4591	0.4509
4	0.3660	2.8955	0.4206	0.4112
5	0.3333	2.7726	0.3887	0.3793
6	0.3090	2.7520	0.3620	0.3530
7	0.2899	2.7768	0.3394	0.3309
8	0.2743	2.8251	0.3200	0.3120
9	0.2612	2.8867	0.3031	0.2956
10	0.2500	2.9560	0.2882	0.2811
15	0.2109	3.3450	0.2338	0.2286

Table 1. The replica symmetric clustering transition given by (9), the replica symmetric densest packing $\rho_s^{\rm RS}$ and the 1RSB densest packing $\rho_s^{\rm 1RSB}$ which are obtained from (7) and (12) respectively.

1.2. 1RSB: exact result for $K \geq 20$

The technique that provides next candidate for the densest packing is derived from the survey propagation (SP) equations (also called energetic 1RSB cavity equations) [28]. For vertex cover (equivalent to the hard-core model up to a dual transformation) on random graph the survey propagation equations were first derived in [12, 9], and read

$$q_{i\to j}\left(\{q_{k\to i}\}\right) = \frac{e^y \prod_{k\in\partial i\setminus j}(1-q_{k\to i})}{1+(e^y-1)\prod_{k\in\partial i\setminus j}(1-q_{k\to i})}\,, \tag{10}$$

where $q_{i\to j}$ is the (cavity) fraction of vertices that are frozen to a value 1 (or "occupied") within clusters. For random K-regular graphs the survey propagation equations admit a factorized solution with a fixed point

$$\bar q = \frac{e^y(1-\bar q)^{K-1}}{1+(e^y-1)(1-\bar q)^{K-1}}\,. \tag{11}$$

This leads to the following complexity (i.e. logarithm of number of clusters of solution):

$$\Sigma(y) \;= \tfrac{ye^{-y}}{Z^f}\left[1-(1-\bar q)^K\right] - \tfrac{Kye^{-y}\bar q^2}{2Z^e} + \log(Z^f) - \tfrac{K}{2}\log(Z^e)\,, \tag{12}$$

$$\text{with}\quad Z^f = (1-\bar q)^K + e^{-y}[1-(1-\bar q)^K],\; Z^e = 1+\bar q^2(e^{-y}-1)\,. \tag{13}$$

The 1RSB result for the densest packing (largest independent set) is then given as

$$\rho_s^{\rm 1RSB} = \frac{e^{-y_0}[1-(1-\bar q)^K]}{Z^f} - \frac{Ke^{-y_0}\bar q^2}{2Z^e}\,, \tag{14}$$

where y_0 is such that the complexity $\Sigma(y_0) = 0$. The values obtained by evaluating these expressions for $K \leq 100$ are summarized in Tables 1 and 2. In general one needs to take into account more levels of replica symmetry breaking in order to obtain exact results for the densest packing. The 2RSB calculation is, however, somewhat involved and was done in [12] in the $K=3$ case, in which case one obtains $\rho_s^{\rm RS} = 0.45906$, $\rho_s^{\rm 1RSB} = 0.45086$, $\rho_s^{\rm 2RSB} = 0.45076(7)$.

Note that the Franz-Leone-Panchenko-Talagrand theorem implies that this 1RSB result is an upper bound [29, 30]. We shall now argue when the 1RSB solution is stable against perturbation towards a 2RSB solution for large enought K. While this does not prove the absence of a higher RSB solution, it is a very good hint that the 1RSB solution is the exact one, and we thus conjecture that it provides *exact* values for the densest packing for $K \geq 20$. These two stability

K	ρ_d	ρ_c	ρ_l	ρ_s^{RS}	ρ_s^{1RSB}	μ_d	μ_c	μ_l
16	0.2051	0.2051	0.2052	0.2257	0.2207	3.421	3.421	3.425
17	0.1989	0.1989	0.2000	0.2182	0.2135	3.459	3.459	3.504
18	0.1933	0.1933	0.1952	0.2113	0.2068	3.502	3.502	3.583
19	0.1880	0.1880	0.1907	0.2048	0.2006	3.539	3.539	3.662
20	0.1830	0.1833	0.1866	0.1989	0.1948	3.576	3.591	3.740
30	0.1453	0.1465	0.1566	0.1555	0.1529	3.818	3.886	4.479
40	0.1214	0.1231	0.1380	0.1292	0.1273	3.966	4.086	5.148
50	0.1047	0.1068	0.1250	0.1113	0.1098	4.072	4.241	5.762
60	0.0924	0.0947	0.1152	0.0981	0.0970	4.155	4.369	6.330
70	0.0828	0.0853	0.1075	0.0881	0.0871	4.221	4.479	6.862
80	0.0752	0.0778	0.1011	0.0800	0.0792	4.279	4.575	7.364
90	0.0690	0.0716	0.0958	0.0735	0.0728	4.329	4.658	7.840
100	0.0638	0.0664	0.0913	0.0680	0.0674	4.372	4.736	8.294

Table 2. Values for the dynamical ρ_d, condensation ρ_c, RS stability ρ_l transitions (and the corresponding values of μ), and for the RS and the 1RSB densest packings, the last is exact for $K \geq 20$. There is an estimated numerical error ± 1 on the last digit for the values of ρ_d, μ_d, ρ_c, μ_c since those were evaluated by by population dynamics with population size $N = 10^5$, equilibration time $T_{eq} = 10^3 \times N$ and averaging time $T = 10^4 \times N$. Given this numerical solution, we are not entirely certain that the transition at $K = 16$ is not actually still a second order one. Data for $K \geq 20$ are separated by a line because they are in the regime where the 1RSB solution is expected to be exact (see section 1.2).

checks have been introduced and used extensively by [31, 14, 12, 32, 33, 26] and we refer to these work for more details. In the present case, they have been considered in [12, 16] but evaluated only for small values of K.

1.2.1. SP type-I stability The first type of instability to check is if the fixed point of eq. (10) is stable against small perturbations on a given single instance. This is the case as long as

$$(K-1)\left[\frac{\partial q_{i\to j}\left(\{q_{k\to i}\}\right)}{\partial q_{k_1\to i}}\bigg|_{\bar{q}}\right]^2 = (K-1)\left[\frac{\bar{q}}{1-(1-\bar{q})^{d-1}}\right]^2 < 1 . \quad (15)$$

Performing the numerical computation at the value of y_0 such that the complexity vanishes, we find that the SP solution is stable as long as $K > 3$ (and is unstable for $K = 3$).

1.2.2. SP type-II stability A more involved stability computation is the Type-II stability, also called bug proliferation, that has been considered first by Rivoire for the present problem [12]. Suppose that a message of type $\sigma = 0/1$ (empty/occupied) is turned into another message $\tilde{\sigma} = 1/0$ with small probability $\pi^{\sigma\to\tilde{\sigma}} \ll 1$. Then, some output messages may change, and the system is said unstable with respect to Type-II perturbations if such a bug propagates through the whole system. In the linear response regime, for a generic change of the first input message from the value τ to $\tilde{\tau}$ in eq. (10) happening with probability $\pi_1^{\sigma\to\tilde{\sigma}}$ we can calculate the probability of changing the output message $\pi_0^{\tau\to\tilde{\tau}}$ from the value τ to $\tilde{\tau}$ as

$$\pi_0^{\tau\to\tilde{\tau}} = \frac{1}{\mathcal{C}} \sum_{\substack{(\sigma,\sigma_2,\dots,\sigma_d)\to\tau \\ (\tilde{\sigma},\sigma_2,\dots,\sigma_d)\to\tilde{\tau}}} \pi_1^{\sigma\to\tilde{\sigma}} \eta_2^{\sigma_2} \cdots \eta_d^{\sigma_d} \; e^{y\tilde{\tau}} , \quad (16)$$

where $\mathcal{C}$ is the denominator in eq. (10), $e^{y\bar{\tau}}$ is the "reweighting" connected to an energy change that appears in the survey propagation equations, and the η are edge-variables related with the SP ones such that $\eta^1 = q$ and $\eta^0 = 1 - q$. This defines a matrix V with entries

$$V_{\tau\to\bar{\tau},\sigma\to\bar{\sigma}} = \frac{\partial \pi_0^{\tau\to\bar{\tau}}}{\partial \pi_1^{\sigma\to\bar{\sigma}}}\,, \tag{17}$$

evaluated at the 1RSB solution. Noting λ the largest eigenvalues of this matrix, the system is then stable if $(K-1)\lambda < 1$. In the present case, the matrix reads

$$\begin{bmatrix} V_{0\to1,0\to1} & V_{0\to1,1\to0} \\ V_{1\to0,0\to1} & V_{1\to0,1\to0} \end{bmatrix} = \begin{bmatrix} 0 & \bar{q}/(1-\bar{q}) \\ e^{-y/2}\bar{q}/(1-\bar{q}) & 0 \end{bmatrix}. \tag{18}$$

SP is thus type-II stable as long as [12]:

$$(K-1)^2 \left(\frac{\bar{q}}{1-\bar{q}}\right)^2 e^{-y} < 1\,, \tag{19}$$

which, for the values of y_0 corresponding to the packing transition, is verified for $K \geq 20$.

From these two checks, we thus conclude that the 1RSB value is consistent as long as $K \geq 20$, and conjecture that it gives the exact values for densest packings on random regular graphs. In fact, from what has been observed in other models (see for instance [34] for the coloring problem) we expect the 1RSB approach to be exact for the whole range of the chemical potential μ in this case. In fact, this seems to be consistent with the numerical results of [16]. Proving this conjecture for finite K remains a challenging problem.

2. The clustering and condensation phase transitions

The structure and properties of dense packings is interesting even for densities well below the largest one, as proven for K large enough in [23, 25]. Statistical physics of models such as the present one describes two important phase transitions - clustering and condensation [35]. In the physics of glassy materials the clustering transition corresponds to the so-called dynamical or mode coupling transition and the condensation transition to the Kauzmann transition. In some models these two transitions coincide, and many authors claimed that the hard-core model is in this class [12, 13, 14, 15, 16]. However, the rigorous results of [23, 25] clearly suggest that for large K the clustering and condensation phase transitions are distincts and far away from each other ($\log K/K$ versus $2\log K/K$).

2.1. Continuous clustering transition

When the clustering and condensation transitions coincide then they also coincide with the local stability of the replica symmetric solution eq. (9). In such a case the density ρ_l marks a point below which the replica symmetric solution is exact. In particular below which the annealed (first moment) calculation counts correctly the number of configurations $\mathcal{N}(\rho)$ at a given density ρ in the sense that

$$s_{\rm RS}(\rho) = \lim_{N\to\infty} \log \langle \mathcal{N}(\rho)\rangle/N = \lim_{N\to\infty} \langle \log \mathcal{N}(\rho)\rangle/N = s(\rho)\,. \tag{20}$$

Above ρ_l the number of configurations of a given density is strictly smaller that the replica symmetric result. There is a non-analyticity in the function $s(\rho)$ at ρ_l. Note that the local stability of the replica symmetric solution is equivalent to the Kesten-Stigum bound for the reconstruction on graphs [36].

2.2. Discontinuous clustering transition and condensation transition

In order to investigate whether the clustering transition is distinct from the condensation one and from the ρ_l transition, we need to investigate the 1RSB solution at the value of Parisi parameter $m = 1$. This is again a rather involved heuristic method, that is well known and established in statistical physics [37]. For vertex cover the corresponding equations were studied in [16]. The 1RSB approach at $m = 1$ on random regular graphs in the absence of quenched disorder simplifies significantly due to its relation to the properties of the planted ensemble [38].

To locate the dynamical transition ρ_d we need to do the following: We plant a configuration of density corresponding to the replica symmetric result. We initialize belief propagation in this configuration and iterate till convergence. The density above which the resulting BP fixed point starts to be correlated with the initial condition marks the clustering transition. To measure the level of correlation we define a so-called overlap between the fixed point and the initial configuration. When $\rho_d \neq \rho_l$ the overlap changes discontinuously and we speak about a 1st order phase transition, whereas when $\rho_d = \rho_l$ it changes continuously and we speak about a second order phase transition. The density above which the Bethe free energy of the resulting BP fixed point becomes smaller than the replica symmetric free energy marks the condensation transition ρ_c. Indeed, the difference between these two values is simply the "complexity" of the equilibrium states, that is, the logarithm of the number of equilibrium states Σ (per spins). When $\Sigma(s) = 0$, the condensation transition appears and a non-extensive number of clusters of solutions starts to dominate the Gibbs measure. The local stability of the replica symmetric solution at ρ_l marks the density above which it become algorithmically easy to find the planted configuration.

In [16] the authors realized that for $K < 16$ the three transitions coincide, $\rho_d = \rho_c = \rho_l$, whereas for $K > 16$ one has $\rho_l > \rho_d$. They, however, did not observe $\rho_d \neq \rho_c$. Note that if $\rho_l > \rho_d$ then the transition must be of 1st order with $\rho_c > \rho_d$ even if the difference might be numerically indistinguishable. Indeed, if the transition was continuous then we could expand around the replica symmetric solution to describe the behavior around the transition, but such an expansion is exactly what leads to the computation of ρ_l, hence either we have $\rho_d = \rho_c = \rho_l$ or $\rho_d < \rho_c < \rho_l$.

In the present work we hence evaluated the values of ρ_d and ρ_c more carefully and found that indeed for $K > 16$ the transition is discontinuous (and our data indicates that it is likely to be the case as well for $K = 16$ although this is difficult to ascertain), and for $K \geq 20$ we were able to see numerically a clear difference between ρ_d and ρ_c, the values are summarized in Table 2.

From a large K expansion of the corresponding equations we found that $\rho_c \approx 2(\log K - \log\log K + 1 - \log 2)/K + o(1/K)$ which agrees in these 3 orders with the expansion of the 1RSB result for the densest packing. This is no surprise, because these 3 orders can be obtained even from the rigorously known lower and upper bound of [20] and it is well known that the second moment technique that [20] used cannot surpass the condensation transition. For the large K scaling of the dynamical transition we obtained $\rho_d \approx \log K/K + o(\log K/K)$ which also agrees with the rigorous results in the clustering [23] and with the results of [25] on reconstruction on trees.

3. Conclusion

In this paper, we matched the two extreme regimes of the hard-core model on random regular graphs to give a picture of the behavior of the solution space for the full range of average degrees. For small degrees, the RS solution is valid at low enough chemical potential and becomes continuously unstable towards a full-RSB solution as μ increases. When $K \geq 16$, however, the picture changes: A stable 1RSB solution appears above a sharp dynamical transition and then, as μ continues to increase, a condensation over a finite set of clusters of the probability measure happens. A full-RSB solution is still found for large enough μ until, for $K \geq 20$, even the densest

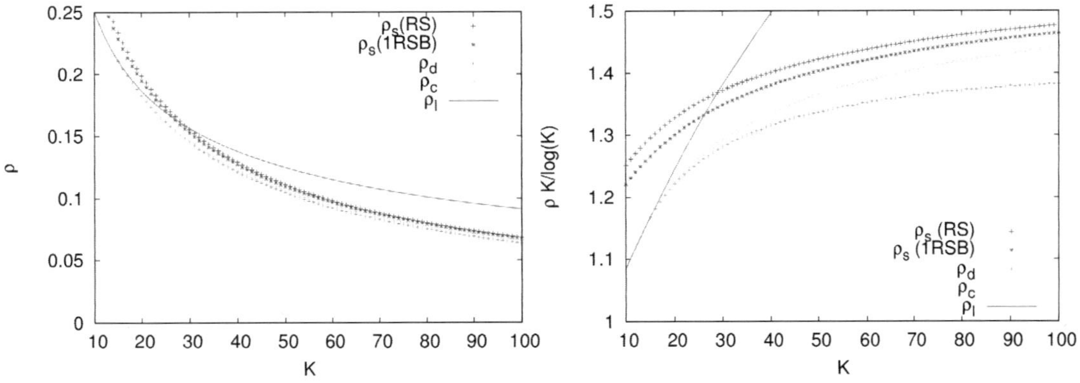

Figure 1. Phase diagram of the hard-core model on K-regular random graphs. The critical densities for clustering ρ_d, condensation ρ_c, replica symmetric stability ρ_l, replica symmetric densest packing ρ_s^{RS} and 1RSB densest packing ρ_s^{1RSB}. The main result of this paper is that the dynamical and condensation transitions are distinct for $K > 16$ and the 1RSB results are exact for $K \geq 20$. The right part of the figure is the rescaled density $\rho K/\log K$ plotted as a function of K. We can see that even for $K = 100$ we are still extremely far from the asymptotic values $\lim_{K\to\infty} \rho_d K/\log K \to 1$, and $\lim_{K\to\infty} \rho_c K/\log K \to 2$.

packing is 1RSB stable. We expect this picture to be qualitatively similar for random graphs with other degree distributions (with finite second moment), although the numbers will change slightly. We hope our paper will help to clarify the different pictures and results in the present literature, and stimulate rigorous work towards proving these conjectures.

Another interesting remark is that since the phase transition is, for large enough K, of the random first order type type, it has the glassy phenomenology associated with hard spheres in physics [4]. In fact, by associating the graph degree with the kissing number via a very rough estimate $K \approx 2^d$, the hard-core dynamical transition arises at $\rho_d \propto d/2^d$ which coincide with the scaling for the dynamical transition in hard spheres in large dimension [4]. The analogy extends less nicely to the scaling of the condensation and densest packings where for hard spheres one expects $\rho \propto d \log d/2^d$ [4], but these are entropy driven quantities and hence differences between continuous and discrete models are to be expected (there are no small vibration in the discrete description). Still this rough agreement is remarkable and the hard-core model should therefore be seen as the simplest mean field lattice model for the glass and jamming transitions.

Acknowledgments
This work has been supported in part by the ERC under the European Union's 7th Framework Programme Grant Agreement 307087-SPARCS, by the Grant DySpaN of "Triangle de la Physique" and by the French Ministère de la defense/DGA.

References

[1] Ambuhl C and Mastrolilli M 2009 *Algorithmica* **53** 488–503
[2] Auyeung A and Abraham A 2004 *Preprint* cs/0405025
[3] Safar M and Habib S 2007 Hard constrained vertex-cover communication algorithm for wsn *Embedded and Ubiquitous Computing* (*Lecture Notes in Computer Science* vol. 4808) (Berlin: Springer) pp 635–49
[4] Parisi G and Zamponi F 2010 *Rev. Mod. Phys.* **82** 789
[5] Weigt M and Hartmann A K 2001 *Phys. Rev.* E **63** 056127
[6] Weigt M and Hartmann A 2000 *Phys. Rev. Lett.* **84** 6118–21
[7] Hartmann A K and Weigt M 2003 *J. Phys. A: Math. Gen.* **36** 11069
[8] Hartmann A K and Weigt M 2001 *Theoretical Computer Science* **265** 6316
[9] Hartmann A and Weigt M 2006 *Phase Transitions in Combinatorial Optimization Problems* (Weinheim: Wiley-VCH)
[10] Zhou H 2003 *Eur. Phys. J.* B **32** 265–70
[11] Weigt M and Zhou H 2006 *Phys. Rev.* E **74** 046110
[12] Rivoire O 2005 *Phases vitreuses, optimisation et grandes déviations* Ph.D. thesis Université Paris-Sud http://tel.archives-ouvertes.fr/tel-00009956/en/
[13] Biroli G and Mézard M 2001 *Phys. Rev. Lett.* **88** 025501
[14] Rivoire O, Biroli G, Martin Olivier C and Mézard M 2004 *Eur. Phys. J.* B **37** 55–78
[15] Mézard M and Tarzia M 2007 *Phys. Rev.* E **76** 041124
[16] Zhang P, Zeng Y and Zhou H 2009 *Phys. Rev.* E **80**(2) 021122
[17] Berthier L and Biroli G 2011 *Rev. Mod. Phys.* **83** 587
[18] Bollobás B and Erdös P 1976 Cliques in random graphs *Mathematical Proceedings of the Cambridge Philosophical Society* vol 80 (Cambridge: Cambridge University Press) pp 419–27
[19] Frieze A M 1990 *Discrete Mathematics* **81** 171–75
[20] Dani V and Moore C 2011 Independent sets in random graphs from the weighted second moment method *Approximation, Randomization, and Combinatorial Optimization. Algorithms and Techniques* (Berlin: Springer) pp 472–82
[21] Grimmett G R and McDiarmid C J 1975 On colouring random graphs *Mathematical Proceedings of the Cambridge Philosophical Society* vol 77 (Cambridge: Cambridge University Press) pp 313–24
[22] Frieze A and McDiarmid C 1997 *Random Structures & Algorithms* **10** 5–42
[23] Coja-Oghlan A and Efthymiou C 2011 On independent sets in random graphs *Proceedings of the Twenty-Second Annual ACM-SIAM Symposium on Discrete Algorithms* SODA '11 (SIAM) pp 136–44
[24] Gamarni D and Sudan M 2013 *Preprint* arXiv:1304.1831
[25] Bhatnagar N, Sly A and Tetali P 2010 Reconstruction threshold for the hardcore model *Approximation, Randomization, and Combinatorial Optimization. Algorithms and Techniques* (Berlin: Springer) pp 434–47
[26] Zdeborová L 2009 *Acta Physica Slovaca* **59** 169–303
[27] Mora T 2007 *Géométrie et inférence dans l'optimisation et en théorie de l'information* Ph.D. thesis Université Paris-Sud http://tel.archives-ouvertes.fr/tel-00175221/
[28] Mézard M and Zecchina R 2002 *Phys. Rev.* E **66** 056126
[29] Franz S, Leone M and Toninelli F L 2003 *J. Phys. A: Math. Gen.* **36** 10967–85
[30] Panchenko D and Talagrand M 2004 *Probability Theory and Related Fields* **130** 319–36
[31] Müller M, Mézard M and Montanari A 2004 *J. Chem. Phys.* **120** 11233
[32] Mertens S, Mézard M and Zecchina R 2006 *Random Structures & Algorithms* **28** 340–73
[33] Krzakala F, Pagnani A and Weigt M 2004 *Phys. Rev.* E **70** 046705
[34] Krzakala F, Zdeborová L *et al.* 2008 *Europhys. lett.* **81** 57005
[35] Krzakała F, Montanari A, Ricci-Tersenghi F, Semerjian G and Zdeborová L 2007 *Proceedings of the National Academy of Sciences* **104** 10318–23
[36] Kesten H and Stigum B P 1966 *J. Math. Anal. Appl.* **17** 309
[37] Mézard M and Montanari A 2009 *Information, physics, and computation* Oxford graduate texts (Oxford: Oxford University Press) ISBN 978-0-19-857083-7
[38] Krzakala F and Zdeborová L 2009 *Phys. Rev. Lett.* **102** 238701

Performance of simulated annealing in p-spin glasses

Florent Krzakala[†,‡] **and Lenka Zdeborová**[⋆]

† Laboratoire de Physique Statistique, CNRS UMR 8550, Université P. et M. Curie, Ecole Normale Supérieure, Paris, France.
‡ ESPCI and CNRS UMR 7083, 10 rue Vauquelin Paris 75005 France
⋆ Institut de Physique Théorique, IPhT, CEA Saclay and URA 2306, CNRS, 91191 Gif-sur-Yvette, France.

E-mail: `florent.krzakala@ens.fr, lenka.zdeborova@cea.fr`

Abstract. We perform careful numerical simulations of slow Monte-Carlo annealings in the dense 3-body spin glass model and compare with the predictions from different theories: thresholds states, isocomplexity, following state. We conclude that while isocomplexity and following state both provide excellent agreement the numerical data, the influence of threshold states – that is still the most commonly considered theory – can be excluded from our data.

Predicting the dynamical behavior from static calculations is the holy grail of theoreticians in statistical physics. After all, this is precisely what statistical mechanics is about from the very beginning: we prefer to consider the static average over all configurations rather than the complex dynamics of all atoms. A particularly important case, both in classical and quantum thermodynamics, is given by the dynamics after very slow variations in an external parameter so that the system remains at equilibrium. Such changes are said to be "adiabatic". When a macroscopic system is in a given phase and if one tunes a parameter, say the temperature, or the external field, very slowly then all observables, such as the energy or the magnetization in a magnet, will be given by the equilibrium equation of state. Given a system at equilibrium in a well-defined phase, it is always possible to consider the adiabatic evolution: In the low-temperature phases of a ferromagnet, for instance, the evolution of the magnetization is different in the two phases or Gibbs states corresponding to the positive or negative magnetization.

To describe this theoretically, one can force the system to be in the Gibbs state of choice for instance, by adding an external infinitesimal field or fixing the boundary conditions and then study the adiabatic evolution for each of these phases. This simplicity breaks down in glassy systems: they have a complicated rugged, many-valleys energy landscape with an enormous number of different Gibbs states. The statics picture and the statistical features of the landscape are well known in the mean field case. However, with the exception of few models [1, 2, 3, 4], an analytical description of the dynamics, and of the way the Gibbs states are evolving upon adiabatic changes, has been missing. Note that by "adiabatic" we mean here a slow dynamics that takes time slower than any power of the system size, but we want to avoid exponentially slow change, where of course we stay always at equilibrium but that has little relevance experimentally if one is not ready to wait a time longer than the age of the universe.

This is by no means an academic exercice. In fact, it is a fundamental question one needs to ask in both physics and computer science. Consider the latter: simulated annealing [5] is one of the most important contribution from statistical physics to optimization and computer science. Consider an annealing experiment where temperature T is changed slowly in time: what is the limiting energy of such an annealing? If one changes slowly the pressure and compresses a set of hard spheres, using a mixture to avoid crystallization, what is the density of the final packing (this is the jamming problem)? Clearly, it would be nice to answer such questions without having to solve the dynamics itself, which is a very difficult and almost intractable problem in general. In fact, since glassy systems are never in equilibrium, one may in fact argue that getting the answer of such questions is more important, and more relevant, than the static solution corresponding to infinite waiting times.

With this motivation in mind, the present authors have discussed a formalism to describe the adiabatic evolution of these glassy Gibbs states as an external parameter is tuned, and applied it to mean field spin models in [6, 7]. The purpose of the present contribution is to discuss numerically the validity of this approach —and of related ones— for the well known and studied fully connected p-spin glass model [8].

1. Following state adiabatically

Let us start by briefly reminding the principle of the state following approach. It was pioneered by Franz and Parisi when they derived the "potential" [9, 10, 11]. It was then discussed thoroughly in [6, 7] and some suggested by [12] have been recently constructed in [13]. Here we shall consider only the simplest version.

How can one follow adiabatically a given Gibbs state? Consider again the example of the ferromagnet with the two "up" and "down" equilibrium states. We can force the system to be in the Gibbs state of choice by fixing the all negative or the all positive boundary conditions. Even far away from the boundaries, the system will stay in the selected state for all $T < T_c$ (above the Curie point any boundary condition will result in a trivial paramagnetic state). By solving the thermodynamics conditioned to the boundaries, we can thus obtain the adiabatic evolution of each of the two states. What boundary conditions should be applied in glassy systems where the structure of Gibbs states is very complicated? The answer is provided by the following gedanken experiment: consider an equilibrium configuration of the system at temperature T_p. Now freeze the whole system except a large hole in it. This hole is now a subsystem with a boundary condition typical for temperature T_p. If the system is in a well-defined state, then no matter the size of the hole, it will always remain correlated to the boundaries and stay in the same state. One may now change the temperature and study the adiabatic evolution of this state[1].

In mean-field systems, this construction allows for an analytic treatment in the spirit of the cavity and replica method. We refer the reader to [6, 7] description of the method and of the equation, and some comparison with Monte-Carlo simulations for sparse systems. Here, we shall consider the fully-connected p-spin model, where the following state method was first put forward by Franz and Parisi in the spherical case [10], although we will consider the Ising version. Again, the computation were done in [6, 7].

For a temperature annealing in a one-step replica symmetry breaking model such as the p-spin, the idea is the following: At high T, a paramagnetic/liquid state exists and therefore a slow annealing should be able to follow the equilibrium computation. Below the dynamical glass temperature T_d, this state shatters into exponentially many Gibbs states, all well separated by extensive energetic or entropic barriers, leading to a breaking of ergodicity and to the divergence of the equilibration time. The idea of the following state (and of the isocomplexity computation as well,see [16]) is thus to follow the state that appear at T_d as the temperature is lowered; this should mark the limit of any non-exponentially slow annealing.

[1] The experienced reader will have recognized the usual construction for the nucleation argument adapted for glassy systems as in the "mosaic picture" of [14, 15], but generalized with different temperature. This construction is also intimately related to the Franz-Parisi potential [9].

2. The p-spin model: a standard benchmark, and a few tricks

The fully-connected 3-spin glass model reads [8]:

$$\mathcal{H} = \sum_{ijk} J_{ijk} S_i S_j S_k \,, \tag{1}$$

where the sum is over all possible triplets of spins, and where the J_{ijk} are quenched random variables with distribution

$$P(J) = \sqrt{\frac{N^2}{\pi p!}} \exp\left[-\frac{N^2}{p!} J^2\right]. \tag{2}$$

This is the mean field spin glass model at the basis of the mean-field theory of glasses [17].

Our goal here is to perform careful Monte-Carlo simulations of annealings for this densely connected model which as far as we know, and perhaps surprisingly, was not done yet in the literature. One sees that simulating this model is quite computationally demanding since computing the energy takes N^3 operations.

The first trick we will use is to *not* use this model, but rather a diluted version of it where instead of using all the N^3 possible triplets of spin, we will use $N\sqrt{N}$ of them. In fact, this is like a simulation of the XOR-SAT model (which is nothing else than the diluted p-spin, that is, the p-spin on random hyper-graphs) with a connectivity $c = \sqrt{N}$. Both this model and the fully connected one have exactly the same static solution as $N \to \infty$ (up to rescaling). For the sizes we will consider, the difference with the fully connected model is negligible, but this reduced connectivity allows for efficient simulations. All the present data are made with $N = 50000$ (for annealing) or $N = 20000$ (for quenches). These size are sufficiently large such that the results do not change visibly when the size is changed by a multiplicative factor two, hence there is no need for finite size scaling analysis which simplified greatly our analysis.

The second trick will be to start with initially equilibrated configuration. That may seem a hard task in such a glassy model, however, using the planting trick [18, 19, 20, 21, 22, 23, 24], this is something easy to achieve. In a nutshell, the idea of quiet planting is to *first* generate a configuration of spins that we want to be an equilibrium one, and *then* to create the disorder in the Hamiltonian such that this is precisely an equilibrium configuration. Of course, an instance of the problem created in such a way have no reason to be a "typical" one; however, as discussed in the references above, if the problem admit a "annealed solution" (that is, if the annealed solution is the correct one, as it is for the p-spin as long as $T \geq T_K$), then instances created by the planting tricks are typical ones in the large N limit. We refer to the aforementioned publication for details.

3. An historical perspective

Let us remind the different type of predictions for behavior of slow annealing; they are summarized in Fig.1.

For a long time, the dominating idea has been that one simply has to compute the so called "thresholds states". There high energy energy minima in the energy landscape. The idea of such computation is count how many minima they are at a given energy, and to compute for which energy they are most numerous. This, according to this idea, allows to locate at which energy there are minima that can, most likely, trap the dynamics. The corresponding energy is depicted as "thresholds states" in Fig. 1. This idea is popular, mostly because it works perfectly in the simple spherical p-spin model [1, 2, 3, 4], however, as discussed for instance in [25], the situations is much more complicated already in the mixed spherical p-spin glass model.

It was soon realized [26, 16] that in fact most of these states are actually the result of a computation that is not self-consistent (unstable towards more steps of repica symmetry breaking), and that the threshold states of "stable" (or at least marginally stable) states is

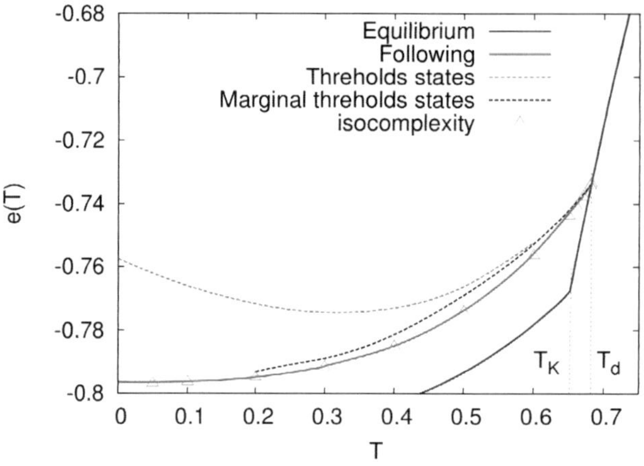

Figure 1. Results of different computations for the annealing in the 3-spin glass problem (see text).

much lower. Computing correctly those turned out to be a complicate task that has been finally achieved by Rizzo in [27]. The result is shown in Fig. 1 as "marginal states". To be precise, we refer as threshold states the (unstable) 1RSB ones, and to the marginal states the (marginal) full RSB ones [27].

However, as discussed in e.g. [10, 11, 26, 12, 16, 6, 7], it may be a bit naive to assume that a Monte-Carlo annealing will just end up in the most numerous states, and that was the motivation for the idea for following states that started at T_d. This is why iso-complexity was introduced: It is proposed to count the number of equilibrium states at T_d, and then to consider the energies at $T < T_d$ for which the number of states is equal to the one at T_d. Iso-complexity leads indeed to a lower bound on adiabatic annealings, because in order to end up at lower energies one would have to be exponentially lucky. Indeed, if one is trap in one of $e^{N\Sigma_1}$ similar states at an energy e_1, and if the total number of states at $e_2 < e_1$ is $e^{N\Sigma_2}$, with $\Sigma_2 < \Sigma_1$, it is exponentially unlikely that any of these correponds to the one at e_1.

In fact, the iso-complexity is only an approximation of the following state formalism (where we explicitly follow states), and explicit differences can be seen (see for instance [6, 7]) however, since the following state computation itself can be rather involved (see [13]) and in some cases (for low enough temperature) we could only do an approximated computation (see [6, 7, 25]). As shown in Fig. 1, however, the results of the two approach is extremely close in the 3-spin model (see also [6, 7] for the diluted case), at least in the case when one starts from $T = T_d$ (actually, there are more devitaion for the the states starting from $T < T_d$, as seen in [6, 7]). The question is now how these rather different predictions compare with simulations.

4. Annealings follow the following state computation

The results of annealings starting from an equilibrated configuration at $T_1 > T_d$ are shown in Fig.2 for different rates. They are clearly approaching the following state predictions as the rate decreases $\Delta T \to 0$. We also show these annealing using only the smallest rate (and demonstrate that the starting temperature does not matter as long as it is larger than T_d). It is striking how easily the Monte-Carlo dynamics is able to go beyond the thresholds state, marginal or not.

These data clearly demonstrate that annealings are not *at all* sensitive to the thresholds states, and that the following states of isocomplexity consideration are the correct ones in this

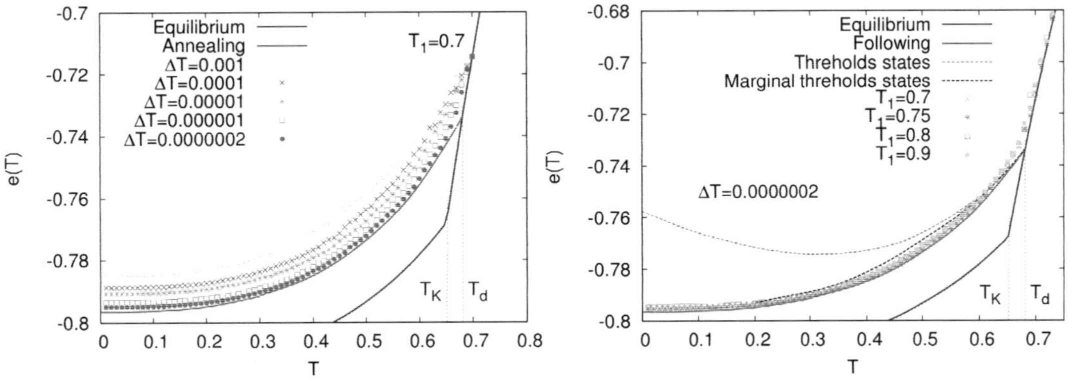

Figure 2. Annealings starting from an equilibrated configuration at high temperature $T_1 > T_d = 0.6825$ in the 3-spin glass problem for $N = 50000$. Left: The energy of annealings from $T_1 = 0.7$ with different cooling rates is shown together with the equilibrium solution of the model (in blue) and the following state prediction (in red). Right: As in the left figure, but with different starting points. For comparison, we also shows the energy of the thresholds states (dashed orange line) and the marginal thresholds states (dashed black line). Annealings have no problems going below these energies that do not seem to affect its behavior.

case. It is also striking how these two are similar (although definitely *not* equal). For models with a discontinuous first order transition, it thus seem that iso-complexity is a very acceptable approximation. It would be interesting to see how generic this is (see [6, 7] for discussion in the XORSAT model, [25] for the spherical model, and the review [28] for hard sphere.).

This can also be clearly seen from annealing from temperature below T_d (see Fig. 3) where the agreement is again perfect with the following state computations. Still, the approximative nature of the computation should be clear from deviation in the overlap (left figure) which display non physical behavior in the following state computation. The point is that this computation is done in a 1RSB formalism, in a region where a more complicated ansatz should be used (see also [25]). In fact, when the state is unstable, one should in principle performs a more involved computation and follows the pseudo-dynamics idea of [12, 13]. This is, however, more complicated, and we shall not attempt to discuss these points here.

5. What about quenches?

Finally, we also performed direct quenches from a high temperature to a small one. The situation here is more complicated as there is no theory equivalent to the following state computation yet (but see [13]).

However, the simulation seems to indicate that the quenches converge to higher energies than the annealing. In fact, they are surprisingly close to the thresholds state (but, again surprisingly, higher from the marginal ones). Although it is difficult to draw firm conclusion from these data, the role of the marginal thresholds in the dynamics is not apparent. Also, it is not clear that the starting temperature is relevant, which also seem contradictory to the idea that quenches are different to annealing in the long runs. The most recent idea, proposed by Rizzo [27], proposed that (up to logarithmic corrections) quenches will tend to a new threshold energy. This is a promising idea that await a numerical confirmation. Surprisingly (given the long lasting interest in spin glasses) we are aware of very few simulations of this problem, and a deep, systematic investigation, would be welcome.

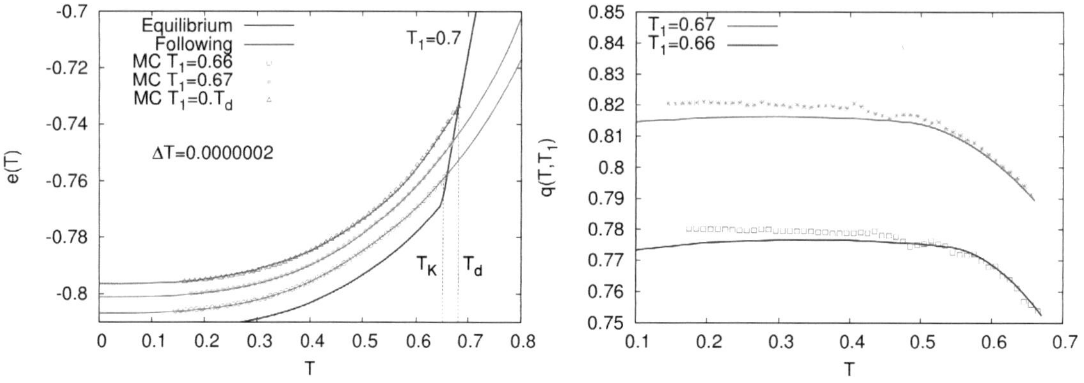

Figure 3. Annealings starting from an equilibrated configuration at low temperature $T_1 \leq T_d$ in the 3-spin glass problem for $N = 50000$. Left: The energy of annealings from three different starting temperatures $T_1 = T_d, 0.67, 0.66$ with slow cooling rate is shown together with the equilibrium solution of the model (in blue) and the following state prediction (in red), demonstrating a perfect agreement. Right: overlap with the starting configuration at $T_1 = 0.67, 0.66$ as a function of the temperature. The approximated nature of the analytical solution (in full line) appears here as it predicts a non monotonous behavior that is not seen in the Monte-Carlo simulation. Indeed, the following state computation is unstable towards more complicated symmetry breaking for $T < 0.55$ ($T_1 = 0.67$) and for $T < 0.49$ ($T_1 = 0.66$).

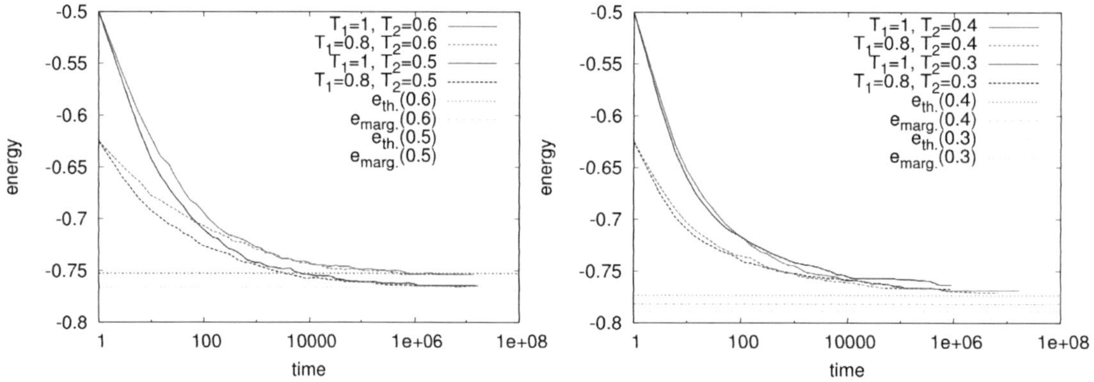

Figure 4. Quenches starting from equilibrated configuration at temperature $T_1 > T_d$ to $T_2 < T_d$ for $N = 20000$. Left: quenches to $T_2 = 0.6$ and $T_2 = 0.5$, Right:quenches to $T_2 = 0.4$ and $T_2 = 0.3$. Although an extrapolation is difficult, it seems —especially in the right figure— that the asymptotic energies are larger than the marginal threshold states prediction. However, perhaps surprisingly, they are close to the standard unstable threshold states, as also noticed in [16].

6. Conclusion and perspective

The conclusion of this short proceeding are two folds: 1) we wanted to illustrate how good the prediction of the following state are (and how good the isocomplexity approximation can be in some case) and 2) nothing seems to happens at energies connected to thresholds or marginal threshold states. We hope that this presentation of the data will convince the community that the bridge between statics and dynamics is not going through thresholds states consideration, but rather through following states considerations.

For quenches, instead, the situations is unclear and more work on the subject is welcome, in particular to test the proposition of [27].

We hope that this short presentation will help to clarify the situation and will help to motivate further works in this direction, perhaps following the idea of pseudo-dynamics [12, 13].

Acknowledgments
This work has been supported in part by the ERC under the European Union's 7th Framework Programme Grant Agreement 307087-SPARCS, by the Grant DySpaN of "Triangle de la Physique", and by the National Science Foundation under Grant No. PHYS-1066293 and the hospitality of the Aspen Center for Physics.

References

[1] Cugliandolo L F and Dean D S 1995 *J. Phys. A* **28** 4213
[2] Cugliandolo L F and Kurchan J 1993 *Phys. Rev. Lett.* **71** 173
[3] Bouchaud J P and Dean D 1995 *J. Phys. I, France* **5** 265–86
[4] Bouchaud J P, Cugliandolo L, Kurchan J and Mézard M 1998 *Spin Glasses and Random Fields* ed Young A P (Singapore: World Scientific)
[5] Kirkpatrick S, Gelatt, Jr C D and Vecchi M P 1983 *Science* **220** 671
[6] Krzakala F and Zdeborová L 2010 *EPL (Europhysics Letters)* **90** 66002
[7] Zdeborová L and Krzakala F 2010 *Phys. Rev.* B **81** 224205
[8] Gross D and Mézard M 1984 *Nucl. Phys.* B **240** 431
[9] Franz S and Parisi G 1997 *Phys. Rev. Lett.* **79** 2486–89
[10] Barrat A, Franz S and Parisi G 1997 *J. Phys. A* **30** 5593–612
[11] Franz S and Parisi G 1998 *Physica A* **261** 317–39
[12] Krzakala F and Kurchan J 2007 *Phys. Rev.* E **76** 021122
[13] Franz S, Parisi G and Urbani P 2012 *Preprint* arXiv:1212.4291
[14] T R Kirkpatrick D T and Wolynes P G 1989 *Phys. Rev.* A **40** 1045
[15] Bouchaud J P and Biroli G 2004 *J. Chem. Phys.* **121** 7347
[16] Montanari A and Ricci-Tersenghi F 2004 *Phys. Rev.* B **70** 134406
[17] Kirkpatrick T R and Thirumalai D 1989 *J. Phys. A* **22** L149
[18] Ozeki Y 1997 *Journal of Physics: Condensed Matter* **9** 11171
[19] Montanari A and Semerjian G 2006 *J. Stat. Phys.* **124** 103–89
[20] Achlioptas D and Coja-Oghlan A 2008 *Foundations of Computer Science, 2008. FOCS '08. IEEE 49th Annual IEEE Symposium on* pp 793–802 ISSN 0272-5428
[21] Krzakala F and Zdeborová L 2009 *Phys. Rev. Lett.* **102** 238701
[22] Krzakala F and Zdeborová L 2011 *J. Chem. Phys.* **134** 034512
[23] Krzakala F and Zdeborová L 2011 *J. Chem. Phys.* **134** 034513
[24] Zdeborová L and Krzakala F 2011 *SIAM J. Discrete Math.* **25** 750–70
[25] Sun Y, Crisanti A, Krzakala F, Leuzzi L and Zdeborová L 2012 *J. Stat. Mech.* **2012** P07002
[26] Montanari A and Ricci-Tersenghi F 2003 *Eur. Phys. J.* B **33** 339–46
[27] Rizzo T 2013 *Preprint* arXiv:1305.2070
[28] Parisi G and Zamponi F 2010 *Rev. Mod. Phys.* **82**(1) 789–845

ELC International Meeting on Inference, Computation, and Spin Glasses (ICSG2013) IOP Publishing
Journal of Physics: Conference Series **473** (2013) 012023 doi:10.1088/1742-6596/473/1/012023

Zeros of the partition function and dynamical singularities in spin-glass systems

K Takahashi[1] **and T Obuchi**[2,3]

[1] Department of Physics, Tokyo Institute of Technology, Tokyo 152-8551, Japan
[2] Cybermedia Center, Osaka University, Osaka 560-0043, Japan
[3] CNRS-Laboratoire de Physique Theórique de l'ENS, 24 rue Lhomond, 75005 Paris, France

Abstract. We study spin-glass systems characterized by continuous occurrence of singularities. The theory of Lee-Yang zeros is used to find the singularities. By using the replica method in mean-field systems, we show that two-dimensional distributions of zeros of the partition function in a complex parameter plane are characteristic feature of random systems. The results of several models indicate that the concept of chaos in the spin-glass state is different from that of the replica symmetry breaking. We discuss that a chaotic phase at imaginary temperature is different from the spin-glass phase and is accessible by quantum dynamics in a quenching protocol.

1. Introduction

Phase transition is one of the most interesting phenomena in statistical physics and has been studied in various ways. The theory of spin glasses has attracted interests for many years as it is important not only for describing real materials but also for realizing the unreasonable effectiveness of the statistical mechanics.

Previous works revealed that the peculiarity of the spin-glass state is attributed to a rugged structure of the free energy landscape [1, 2]. There exist many pure states in the configuration space separated by infinite barriers and the system takes one of their states. By sweeping a parameter such as the temperature, we can find continuous changes of the state, which means that a certain kind of phase transitions occur continuously.

Phase transition is generally expected to involve singularities in thermodynamic functions. However, it is known for example in the analytical result of the Sherrington-Kirkpatrick model [3] that the free energy is a continuous function with respect to the temperature in the spin-glass phase [4, 5]. Although the model exhibits a continuous change of the state known as the temperature chaos [6], such a property cannot be seen in the free energy. This is not a surprising result since a single singularity is hidden in continuous singularities. Thus, from a specific thermodynamic function, one cannot judge whether there exist continuous singularities.

Lee and Yang proposed a theoretical method to find the singularity in the thermodynamic functions in a rather primitive way [7, 8]. They discussed that the singularity is identified by the zeros of the partition function and showed explicitly that the phase transition in the Ising model is described by the zeros. This theory is suitable to our purpose to see the singularity directly. The other advantage of this method is that zeros exist even in finite systems and one can study how the system approaches the thermodynamic limit. We can also argue that the role of zeros is not only in finding the phase transition point but also in calculating any thermodynamic functions.

In this paper, according to our previous works, we develop the theory of zeros in spin-glass systems. We use a reliable systematic method to calculate the zeros of the partition function and discuss the advantage of using zeros for studying spin-glass transitions and the related peculiar behaviours [9, 10]. In order to discuss further possibilities of exploring zeros, we also consider some application on dynamical properties of the systems [11].

2. Zeros of the partition function

2.1. Theory of zeros

In a finite system, the partition function $Z = \mathrm{Tr}\exp(-\beta H)$, where H is the Hamiltonian and β the inverse temperature, is an analytic function with respect to a parameter y such as β and the magnetic field. It can be written in a factorized form

$$Z(y) = \mathrm{e}^{C}\prod_i (y - y(i))^{\eta_i}, \tag{1}$$

where $y(i)$ is ith zeros, η_i is the positive power index and the factor C is an irrelevant constant. Since the partition function is positive by definition, $y(i)$ must be outside the domain where the parameter y is defined. Otherwise, we have an unacceptable result that the partition function becomes zero at $y = y(i)$. Therefore, $y(i)$ is a complex number in general. The free energy density f is extracted from the partition function as $-\beta f = (1/N)\ln Z$ where N is the system size. By using the zeros of the partition function $\{y(i)\}$, we can write

$$-\beta f(y) = \frac{C}{N} + \int \mathrm{d}z_1\mathrm{d}z_2\, \rho(z_1, z_2)\ln(y - z), \tag{2}$$

where z is complex as $z = z_1 + iz_2$ and the density of zeros is defined as

$$\rho(z_1, z_2) = \frac{1}{N}\sum_i \eta_i\delta(z_1 - \mathrm{Re}\, y(i))\delta(z_2 - \mathrm{Im}\, y(i)). \tag{3}$$

The power η_i now plays a weight of the corresponding zero. Thus, we can find the free energy by knowing the density of zeros. All of properties on the thermodynamic state are reflected in zeros. This can also be understood from the fact that the density of zeros is calculated from the partition function with complex y as

$$\rho(y_1, y_2) = \frac{1}{2\pi N}\left(\frac{\partial^2}{\partial y_1^2} + \frac{\partial^2}{\partial y_2^2}\right)\ln|Z(y = y_1 + iy_2)|. \tag{4}$$

Since we can understand all thermodynamic properties by zeros, the phase transition is also found from their distribution. As we mentioned above, in finite systems, the parameter y never coincides with any of $y(i)$ as long as y takes real physical values. However, this is not the case for systems at the thermodynamic limit. If we find a singularity in a thermodynamic function, it must be described by the zeros of the partition function. This implies that the zeros asymptotically approach the physical domain. The location of the phase transition point must be identified by the zeros.

2.2. Replica method

We treat random systems and take an average over random realizations of parameters in the Hamiltonian. In order to calculate the density of zeros, we use the formula in equation (4). We must calculate the random average of the logarithm of the partition function. It is known in spin-glass theory that the replica method is useful for such a calculation. Since we must treat the absolute value of the complex partition function, we use the modified formula

$$\ln |Z| = \lim_{n\to 0} \frac{|Z|^{2n} - 1}{2n}. \tag{5}$$

$2n$-replicated systems are prepared to calculate the density of zeros. This is the main idea in our analysis.

In order to see how the density of zeros is calculated, we sketch the idea by using the Edwards-Anderson model

$$H = -\sum_{\langle ij\rangle} J_{ij} S_i S_j, \tag{6}$$

of Ising spins S_i on a lattice [12]. Here, we take the summation over pairs of spins and J_{ij} is the Gaussian random variable with the variance $[J_{ij}^2] = J^2$. We study zeros in the complex-βJ plane [13]. The average of the $2n$-replicated partition function is calculated as

$$\begin{aligned} [|Z|^{2n}] &= \left[\mathrm{Tr}\exp\left\{\beta\sum_{a=1}^{n}\sum_{\langle ij\rangle} J_{ij}S_i^aS_j^a + \beta^*\sum_{a=1}^{n}\sum_{\langle ij\rangle} J_{ij}S_i^{*a}S_j^{*a}\right\}\right] \\ &= \mathrm{Tr}\exp\left\{\frac{1}{2}\sum_{a,b}^{n}\sum_{\langle ij\rangle}\left(\beta^2J^2Q_i^{ab}Q_j^{ab} + \beta^{*2}J^2Q_i^{*ab}Q_j^{*ab} + 2\beta\beta^*J^2\tilde{Q}_i^{ab}\tilde{Q}_j^{ab}\right)\right\}, \end{aligned} \tag{7}$$

where the trace is taken over Ising spin variables $\{S_i^a\}_{i=1,\cdots,N}^{a=1,\cdots,n}$ and $\{S_i^{*a}\}_{i=1,\cdots,N}^{a=1,\cdots,n}$. We also put

$$Q_i^{ab} = S_i^aS_i^b, \qquad Q_i^{*ab} = S_i^{*a}S_i^{*b}, \qquad \tilde{Q}_i^{ab} = S_i^aS_i^{*b}. \tag{8}$$

These variables are identified with the order parameters of the system. By taking the thermodynamic limit $N \to \infty$, we write the large deviation form $[|Z|^{2n}] = \exp(N\phi_n(\beta,\beta^*))$ and the logarithm of the partition function is calculated as

$$\frac{1}{N}[\ln|Z|] = \lim_{n\to 0}\frac{1}{2n}\phi_n(\beta,\beta^*). \tag{9}$$

The calculation goes along the same way as the standard analysis. Since it is difficult to perform the calculation in finite dimensional systems, we mainly study infinite-range models where the mean-field method is applicable. The additional difficulty in the present case lies in the fact that the function incorporates the complex parameters.

What can we expect for the result? Generally, the result will be written in a form

$$\lim_{n\to 0}\frac{1}{2n}\phi_n(\beta,\beta^*) = \phi^{(1)}(\beta) + \phi^{(1)}(\beta^*) + \tilde{\phi}(\beta,\beta^*). \tag{10}$$

$\tilde{\phi}$ cannot be written as a function of β or β^* and depends on $|\beta|^2 = \beta\beta^*$. In this case, only the last term contributes to the density of zeros:

$$\rho(\beta_1,\beta_2) = \frac{1}{2\pi}\left(\frac{\partial^2}{\partial\beta_1^2} + \frac{\partial^2}{\partial\beta_2^2}\right)\tilde{\phi}(\beta,\beta^*). \tag{11}$$

The density of zeros vanishes if this term is absent. We see from equation (7) that there should exist the correlation $\tilde{Q}_i^{ab}$ to obtain a finite density of zeros. By noting that such a correlation

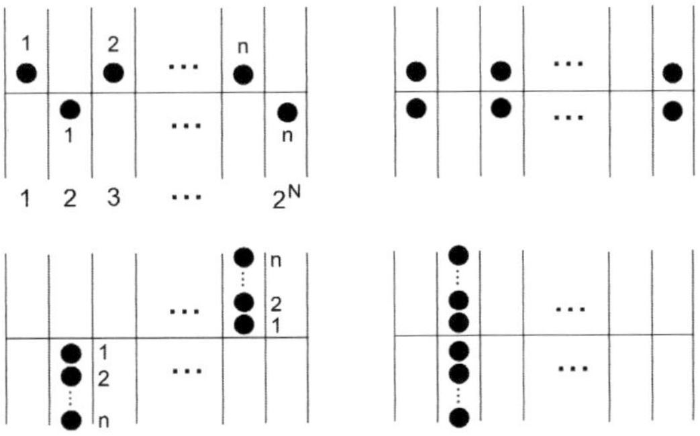

Figure 1. Replica-symmetric configurations of pure states. Each configuration has two kinds of boxes which represent the normal space and the complex-conjugated one respectively. The upper diagrams represent paramagnetic phases and lower spin-glass phases. $\tilde{Q} = 0$ in the left diagrams and $\tilde{Q} \neq 0$ in the right.

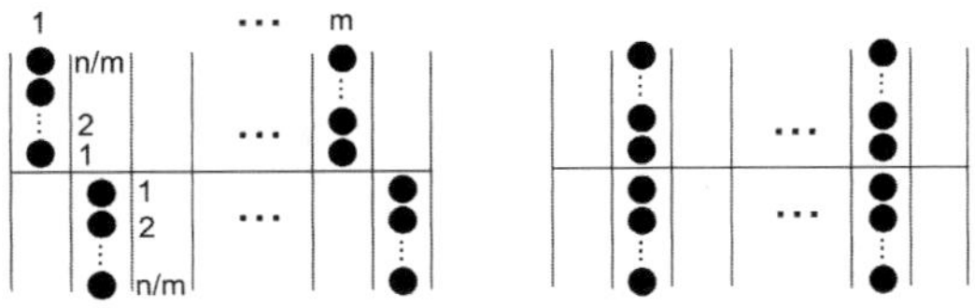

Figure 2. Configurations in one-step replica-symmetry-breaking spin-glass phases. $\tilde{Q} = 0$ in the left diagram and $\tilde{Q} \neq 0$ in the right.

never occurs in pure systems, we see that the two-dimensional distribution of zeros is specific to random systems.

We note the other possibility for obtaining finite density of zeros. When the phase transition occurs, we have a discontinuous behaviour of the thermodynamic functions. Although $\phi_n(\beta, \beta^*)$ is expected to be continuous everywhere, the discontinuity can appear in its derivative. In that case, the zeros exists on phase boundaries in the complex parameter space. An example is known in the Ising model as the circle theorem [8].

2.3. Possible patterns of pure state

A thermodynamic state is described by a typical configuration of the microscopic variables. This is reflected in the structure of the order parameter functions. In random systems, we have correlations between replicas, and we can consider possible patterns of replica symmetry breaking generally. This can be considered schematically in the following way. We represent each replica by a ball and each configuration (pure state) by a box. A typical state is represented by arranging balls in boxes. We can consider various types of configurations, and typical states are summarized in figures 1 and 2. Since we consider complex-conjugated systems, we have two types of boxes, each has 2^N-boxes. They have parameters β and β^* respectively. n-balls are arranged in a one type of boxes and the other n-balls are arranged in the other type of boxes.

In a replica symmetric state, the order parameter is represented by $Q^{ab} = Q$ and $Q^{*ab} = Q^*$ where $a \neq b$. In the upper diagrams of figure 1, a single ball is in a box at most and $Q = Q^* = 0$, which means that the corresponding states are paramagnetic states. In the upper-left diagram of the figure, $\tilde{Q}^{ab} = 0$ which means that there is no correlation between conjugated pairs. We obtain $[|Z|^{2n}] = [Z^n][Z^{*n}]$ and the density of zeros vanishes as a result. On the other hand, for the upper-right type, $\tilde{Q}^{ab} \neq 0$ and we have a finite density of zeros. Thus, we can define two kinds of paramagnetic state in the present analysis. On the other hand, the replica-symmetric spin-glass state is specified by the lower two figures. All balls are inside a single box and the correlations between different replicas are finite, which imply a spin-glass state with $Q \neq 0$ and $Q^* \neq 0$. The replica-symmetric spin-glass state appears in several simple models such as the spherical model.

In a similar way, we can consider spin-glass phases with $Q^{ab} \neq 0$ and $Q^{*ab} \neq 0$ represented by the one-step replica-symmetry-breaking solutions in figure 2. We have two possibilities also in this case with zero- and finite- $\tilde{Q}$. The breaking parameter m is determined by the equation of state and depends on the replica number n and other parameters such as β. These constructions of possible solutions are further generalized to cases of higher-step replica-symmetry-breaking states. In this paper, we only treat systems described by replica-symmetric and one-step replica-symmetry-breaking solutions for simplicity.

The important point to notice here is that states with $\tilde{Q} \neq 0$ can appear in both the paramagnetic and spin-glass phases. Since it is not clear whether such phases appear in models to be considered, we give explicit analysis in the following.

2.4. Zeros and chaos

We examine mean-field models exhibiting spin-glass phase transitions. All of our systems are described by the p-spin Hamiltonian

$$H = -\sum_{i_1<i_2<\cdots<i_p} J_{i_1 i_2\cdots i_p} S_{i_1} S_{i_2}\cdots S_{i_p}, \tag{12}$$

where the random averaging is taken by the Gaussian distribution

$$\mathrm{Prob}(J_{i_1 i_2\cdots i_p}) = \sqrt{\frac{N^{p-1}}{\pi J^2 p!}} \exp\left(-\frac{N^{p-1}}{J^2 p!} J^2_{i_1 i_2\cdots i_p}\right). \tag{13}$$

Using the standard strategy for mean-field systems [1, 2], we can arrive at

$$\begin{aligned}\phi_n(\beta,\beta^*) = & -\frac{p-1}{4}\sum_{a,b}^{n}\left(\beta^2 J^2 (Q_{ab})^p + \beta^{*2} J^2 (Q^*_{ab})^p + 2|\beta|^2 J^2 (\tilde{Q}_{ab})^p\right) \\ & + \ln \mathrm{Tr} \exp\left\{\frac{p}{4}\sum_{a,b}^{n}\left(\beta^2 J^2 (Q_{ab})^{p-1} S^a S^b + \beta^{*2} J^2 (Q^*_{ab})^{p-1} S^{*a} S^{*b} + 2|\beta|^2 J^2 (\tilde{Q}_{ab})^{p-1} S^a S^{*b}\right)\right\},\end{aligned} \tag{14}$$

where the order parameters Q, Q^* and $\tilde{Q}$ are determined by the condition to extremize $\phi_n(\beta,\beta^*)$. We impose several ansatz on those parameters to solve the saddle-point equations.

In order to compare several results, we study two solvable models: the random energy model and spherical model. The random energy model is defined by the Ising spin $S_i^a = \pm 1$ with the many-spin interaction limit $p \to \infty$ [14, 15]. The spin-glass phase is described by the one-step replica-symmetry-breaking solution. In the spherical model, the spin variable is real and the constraint $\sum_{i=1}^{N}(S_i^a)^2 = N$ is imposed. When $p = 2$, the model has the replica-symmetric spin-glass phase and for $p > 2$ the one-step replica-symmetry-breaking one. We show the result of the random energy model and spherical model with $p = 2$ in figure 3 and 4 respectively. The details are studied in [9, 10].

In both the results, zeros appear in a paramagnetic phase (denoted by DC), characterized by the upper right diagram in figure 1, and not in the spin-glass phase. The spin-glass phases

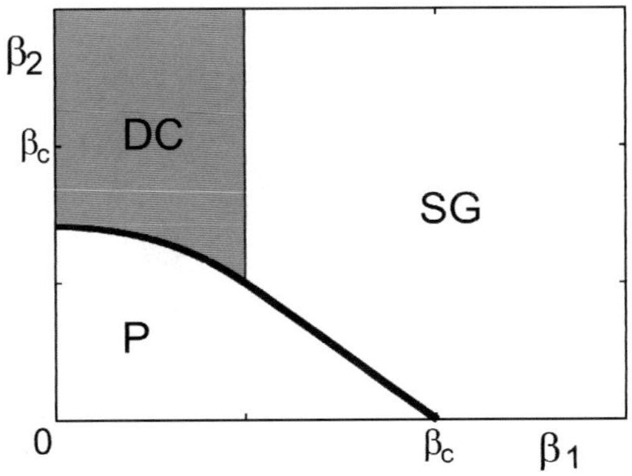

Figure 3. The distribution of zeros in the random energy model. Zeros distribute in the dynamical chaotic (DC) phase uniformly and on the phase boundary of the paramagnetic (P) phase as shown by shaded area and bold line respectively. There are no zeros in the spin-glass (SG) phase. The critical point is given by $\beta_c = 2\sqrt{\ln 2}/J$.

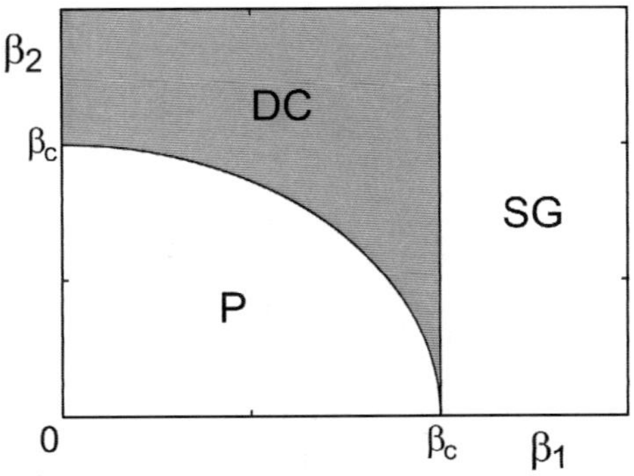

Figure 4. The distribution of zeros in the spherical model at $p = 2$. Zeros distribute uniformly in the DC phase. The critical point is given by $\beta_c = 1/J$.

are described by the left diagrams in figures 1 and 2. Though $\tilde{Q} \neq 0$, we do not see any zeros in the spin-glass phases. The same conclusion is found also in the spherical models with $p > 2$. We do not blame this result on the oversimplified patterns of the replica symmetry breaking. In fact, for the random energy model with complex magnetic field, the zeros appear in the spin-glass phase [18, 19]. The difference lies in the chaotic nature of the spin-glass state [20, 21, 22, 23, 6, 24]. In the spin-glass phase of the random energy model, the system falls into the ground state and nothing spectacular occurs even if we lower the temperature. However, if we change the magnetic field, the pure state changes accordingly. In that case, the state is highly sensitive to the parameter variation. We expect that such a chaotic nature is more important than the step number of the replica symmetry breaking.

To confirm this speculation, we examine the spherical model with multiple interactions [24]. We consider the sum of the Hamiltonian with $p = 3$ and $p = 4$. This model is known to show the temperature chaos effect. We numerically solve the saddle-point equations to draw the phase diagram in figure 5. The spin-glass phase is described by the one-step replica-symmetry-breaking, which is the same as the standard spherical model at $p > 2$. However, in this case,

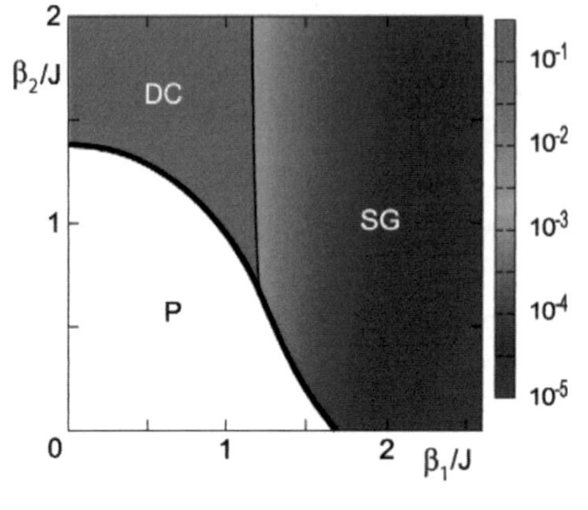

Figure 5. The distribution of zeros in the multiple-interacting spherical model with $p = 3$ and 4. Zeros distribute in the SG and DC phases with values of the density denoted in colour and on the phase boundary of the P phase.

the zeros appear in the spin-glass state which means the relevance of the chaotic effect for the distribution of zeros.

3. Dynamical singularity

3.1. Quantum quench

Our formulation shows that $\tilde{Q} \neq 0$ is necessary to find two-dimensional distributions of zeros. We discussed in the previous section the chaotic (and nonchaotic) nature of the spin-glass phase with $Q \neq 0$ and $Q^* \neq 0$. On the other hand, we also find a different chaotic non-spin-glass phase with $\tilde{Q} \neq 0$ and $Q = Q^* = 0$. The analysis of several systems shows that such a phase appears around the imaginary axis in the complex temperature plane. It does not approach the real axis and cannot describe any thermodynamic phase transitions.

Our idea of exploring this new kind of chaotic phase is to consider real time dynamics. Since the Boltzmann factor with imaginary temperature corresponds to the time evolution operator in quantum mechanics, we expect that the chaotic phase is relevant to quantum dynamics.

In order to introduce the quantum nature, we consider quantum quench dynamics. We consider the quantum Hamiltonian

$$\mathcal{H} = H(\{\sigma_i^z\}) - \Gamma \sum_{i=1}^{N} \sigma_i^x, \tag{15}$$

where $H(\{\sigma_i^z\})$ is the Ising-spin Hamiltonian as we described in the previous section. We first prepare state with the ground state of the Hamiltonian at $\Gamma \to \infty$. Then, we set $\Gamma = 0$ at $t = 0$ and consider the time evolution. The initial state is given by

$$|\psi\rangle = \prod_{i=1}^{N} |+x\rangle_i, \tag{16}$$

where $|+x\rangle_i$ is the eigenstate of σ_i^x with the eigenvalue +1. The state evolution at $t > 0$ is given by $\mathrm{e}^{-iHt}|\psi\rangle$. We focus on the return amplitude

$$G(t) = \langle\psi|\mathrm{e}^{-iHt}|\psi\rangle. \tag{17}$$

$|G(t)|^2$ represents the probability for the state to go back to the initial state. In the σ^z-basis, the initial state is written as

$$|\psi\rangle = \prod_{i=1}^{N} \frac{1}{\sqrt{2}} (|+z\rangle_i + |-z\rangle_i) = \frac{1}{2^{N/2}} \sum_{n=1}^{2^N} |n\rangle, \tag{18}$$

where $|n\rangle$ represents one of eigenstates in the σ^z-basis. Using this representation, we can write the return amplitude as

$$G(t) = \frac{1}{2^N} \sum_{m,n} \langle m|\mathrm{e}^{-iHt}|n\rangle = \frac{1}{2^N} \sum_{n} \langle n|\mathrm{e}^{-iHt}|n\rangle = \frac{1}{2^N} Z(\beta = it). \tag{19}$$

Thus, the return amplitude is written by using the partition function. In random systems, we calculate $[|G(t)|^2] = 2^{-2N}[Z(it)Z(-it)]$, which is obtained by setting $n = 1$ in the formulation of the previous section. In the replica calculation, the state of the system is characterized by the order parameter

$$Q = \begin{pmatrix} 1 & \tilde{q} \\ \tilde{q} & 1 \end{pmatrix}. \tag{20}$$

$\tilde{q}$ represents the correlation between the system at $\beta = it$ and one at $\beta = -it$. We see that this order parameter plays the same role as $\tilde{Q}$ in the previous calculations of zeros.

3.2. Mean-field systems

We examine the random energy model to show that the singularity appears in the return probability. In this case, the calculation is easily done by the energy representation [14, 15]. As the name of the model suggests, the function is written with respect to the sum of random energy levels $G(t) = 2^{-N} \sum_{n=1}^{2^N} \exp(-iE_n t)$. By using the Gaussian random average $[E_n^2] = NJ^2/2$, we find

$$[|G(t)|^2] = \frac{1}{2^N} + \frac{1}{2^{2N}} \sum_{m \neq n} \left[\mathrm{e}^{-i(E_m - E_n)t}\right] \sim \exp(-N \ln 2) + \exp\left(-NJ^2 t^2/2\right). \tag{21}$$

At the thermodynamic limit, the rate function $g(t)$ defined as $[|G(t)|^2] = \exp(-Ng(t))$ is given by

$$g(t) = \begin{cases} \frac{J^2t^2}{2} & \text{for } t \leq t_\mathrm{c} = \frac{\sqrt{2\ln 2}}{J} \\ \ln 2 & \text{for } t > t_\mathrm{c} \end{cases}. \tag{22}$$

The probability freezes at $t = t_\mathrm{c}$, which shows a dynamical singularity. This transition point $t = t_\mathrm{c}$ coincides with that of the P-DC phase transition in the distribution of zeros. This is not an accidental coincidence since the analytical expression is equivalent in the spin representation.

We can also solve the case of the Sherrington-Kirkpatrick model. By using equation (14) with $p = 2$ and $n = 1$, we obtain

$$g(t) = \frac{J^2t^2}{2}(1 + \tilde{q}^2) - \ln\cosh(J^2t^2\tilde{q}), \tag{23}$$

where $\tilde{q}$ is determined by the saddle-point equation

$$\tilde{q} = \tanh(J^2t^2\tilde{q}). \tag{24}$$

This has a solution $\tilde{q} > 0$ at $t > t_\mathrm{c} = 1/J$. Thus, we again have a phase transition and the singularity appears in the rate function. This result implies that the SK model also has a DC phase.

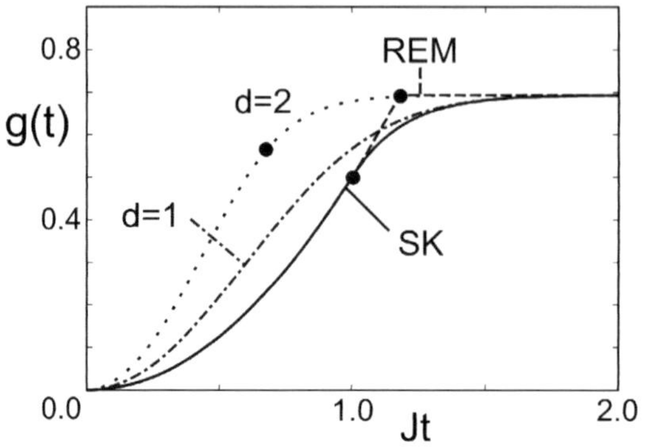

Figure 6. The rate function $g(t)$ in the random energy model (REM), Sherrington-Kirkpatrick model (SK) and Edwards-Anderson model at $d = 1$ and $d = 2$. Dots denote singular points.

3.3. Finite-dimensional systems

We show that the dynamical singularity appears not only in mean field models but also in finite dimensional systems. We use the Edwards-Anderson model defined in equation (6). After the averaging we obtain the expression

$$[|G(t)|^2] = \frac{1}{2^{2N}} e^{-N_B J^2 t^2} \mathrm{Tr} \exp\left(J^2 t^2 \sum_{\langle ij \rangle} S_i S_i^* S_j S_j^* \right) = \frac{1}{2^N} e^{-N_B J^2 t^2} Z_{\mathrm{Ising}}(J^2 t^2), \tag{25}$$

where N_B is the number of the sum $\langle ij \rangle$ and $Z_{\mathrm{Ising}}(\beta)$ is the partition function of the classical pure Ising model. From the result of the Ising model, we can immediately conclude that the dynamical singularity occurs at $d \geq 2$ where d is the dimension of the system. For example, we can write explicit results for the model in one and two dimensional square lattice as

$$g(t) = \begin{cases} J^2 t^2 - \ln\cosh(J^2 t^2) & d = 1 \\ 2J^2 t^2 - \ln\cosh(2J^2 t^2) - \frac{1}{2}\int_0^{2\pi} \frac{d^2 \boldsymbol{k}}{(2\pi)^2} \ln\left[1 - \frac{1}{2}\frac{\sinh(2J^2 t^2)}{\cosh^2(2J^2 t^2)}(\cos k_1 + \cos k_2)\right] & d = 2 \end{cases}. \tag{26}$$

The phase transition at $d = 2$ occurs at the critical point defined by $2J^2 t_c^2 = \ln(1 + \sqrt{2})$. All of our results on the rate function are displayed in figure 6.

For the case of $d = 1$, we do not have any singularity in the rate function. However, this does not mean that there are no zeros on the imaginary axis $\beta = it$. The partition function of the one-dimensional Ising model is written as

$$Z_{\mathrm{Ising}}(\beta) = \prod_{i=1}^{N} (e^{\beta J_i} + e^{-\beta J_i}). \tag{27}$$

This expression shows that the zeros are located on the imaginary axis as $\beta J_i = i\pi/2$. Applying the formula of the density of zeros, we have

$$\rho(\beta_1, \beta_2) = J^2 \int \frac{dz}{\sqrt{2\pi}} e^{-z^2/2} z^2 \delta(\sinh \beta_1 J z) \delta(\cos \beta_2 J z), \tag{28}$$

for the density of zeros. Thus, the zeros distribute continuously on the imaginary axis. However, as we know from the free energy in the spin-glass phase, the singularity does not necessarily appear on the imaginary axis since we are on singularities all the way of the evolution. We expect that the zeros distribute on the imaginary axis also in higher dimensional systems. Therefore, the singularity at $t = t_c$ implies that we enter the DC phase characterized by two-dimensional distributions of zeros.

4. Summary

We developed the theory of zeros in spin-glass systems. We gave evidence that the chaotic effect is relevant to find zeros in the spin-glass phase. The concept of the replica symmetry breaking is not necessarily connected to this effect directly. This is a desirable result since the chaotic effect is expected to occur in non-mean-field systems where the replica-symmetry-breaking ansatz is suspicious to apply.

As an unexpected result, we find zeros in a non-spin-glass phase in almost all models we studied. This phase appears around the imaginary axis in the complex-β plane. Since it is far from the real axis, the state is not relevant to usual spin-glass transition. Instead of that, we expect that this chaotic phase is able to explore by considering dynamics. As an example, we propose a quenching protocol of a quantum system. By relating the quantum amplitude to the partition function at $\beta = it$, we show that the dynamical singularity is described as a phase transitions to the chaotic phase. This occurs not only in mean-field systems but also in finite dimensional systems. To the best of our knowledge, this is the first result that the spin-glass systems in finite dimensions exhibit a dynamical singularity even in systems without the thermodynamic phase transition.

We note that our dynamical phase transition is not directly related to the one discussed in context of glassy dynamics, at this moment. As explained in Introduction, the zeros seem to be closely connected to change of dominating pure states, but we are not sure that this description is useful to discuss the DC phase and the related dynamical behavior. Clarifying this point will unveil some characteristic behavior in quantum quench and possibly lead a connection to other dynamical phenomena in glass systems.

We calculated the return probability to find the dynamical behaviour of the system. From a statistical mechanical point of view, this calculation corresponds to the annealed average. It is also possible to consider the quenched average where the average of the rate function is examined. Although we consider that the annealed calculation is appropriate in the present purpose to calculate the return probability, it is not clear whether the result from the annealed average is generally relevant in characterizing dynamical properties of the system. This is an interesting problem and is left for future studies.

Acknowledgments

T. O. acknowledges the support by the Grant-in-Aid for JSPS Fellows.

References

[1] Mézard M, Parisi G and Virasoro M A 1987 *Spin Glass Theory and Beyond* (Singapore: World Scientific)
[2] Nishimori H 2001 *Statistical Physics of Spin Glasses and Information Processing: An Introduction* (Oxford: Oxford University Press)
[3] Sherrington D and Kirkpatrick S 1975 *Phys. Rev. Lett.* **35** 1792
[4] Parisi G 1980 *J. Phys. A: Math. Gen.* **13** L115
[5] Parisi G 1980 *J. Phys. A: Math. Gen.* **13** 1101
[6] Rizzo T and Crisanti A 2003 *Phys. Rev. Lett.* **90** 137201
[7] Yang C N and Lee T D 1952 *Phys. Rev.* **87** 404
[8] Lee T D and Yang C N 1952 *Phys. Rev.* **87** 410
[9] Takahashi K 2011 *J. Phys. A: Math. Theor.* **44** 235001
[10] Obuchi T and Takahashi K 2012 *J. Phys. A: Math. Theor.* **45** 125003
[11] Obuchi T and Takahashi K 2012 *Phys. Rev.* E **86** 051125
[12] Edwards S F and Anderson P W 1975 *J. Phys. F* **5** 965

[13] Fisher M E 1965 *The Nature of Critical Points (Lectures in Theoretical Physics vol 7C* (the University of Colorado Press, Boulder)
[14] Derrida B 1980 *Phys. Rev. Lett.* **45** 79
[15] Derrida B 1981 *Phys. Rev.* B **24** 2613
[16] Derrida B 1991 *Physica A* **177** 31
[17] Moukarzel C and Parga N 1991 *Physica A* **177** 24
[18] Moukarzel C and Parga N 1992 *J. Phys. I France* **2** 251
[19] Moukarzel C and Parga N 1992 *Physica A* **185** 305
[20] Franz S, Parisi G and Virasoro M A 1992 *J. Phys. I France* **2** 1869
[21] Franz S and Ney-Nifle M 1995 *J. Phys. A: Math. Gen.* **28** 2499
[22] Rizzo T 2001 *J. Phys. A: Math. Gen.* **34** 5531
[23] Rizzo T 2002 *Eur. Phys. J.* B **29** 425
[24] Rizzo T and Yoshino H 2006 *Phys. Rev.* B **73** 064416

AUTHOR INDEX